Genetics and Genomics of Soybean

Plant Genetics and Genomics: Crops and Models

Series Editor: Richard A. Jorgensen

Forthcoming and planned volumes

Vol. 1 Genomics of Tropical Crop Plants (eds: Paul Moore/Ray Ming)
Vol. 2 Genetics and Genomics of Soybean (ed: Gary Stacey)
Vol. 3 Genetics and Genomics of Cotton (ed: Andy Paterson)
Vol. 4 Plant Cytogenetics: Genome Structure and Chromosome Function (eds: Hank Bass/Jim Birchler)
Vol. 5 Plant Cytogenetics: Methods and Instruction (eds: Hank Bass/Jim Birchler)
Vol. 6 Genetics and Genomics of the Rosaceae (eds: Kevin Folta/Sue Gardiner)
Vol. 7 Genetics and Genomics of the Triticeae (ed: Catherine Feuillet/Gary Muehlbauer)
Vol. 8 Genomics of Poplar (ed: Stefan Janssen et al.)

Gary Stacey
Editor

Genetics and Genomics of Soybean

Foreword by Bob Goldberg

Springer

Editor
Gary Stacey
University of Missouri
Columbia, MO, USA
staceyg@missouri.edu

Series Editor
Richard A. Jorgensen
University of Arizona
Tucson, AZ, USA
raj@ag.arizona.edu

ISBN: 978-0-387-72298-6 e-ISBN: 978-0-387-72299-3

Library of Congress Control Number: 2008925775

Printed on acid-free paper

9 8 7 6 5 4 3 2 1

springer.com

Foreword

Genetics and Genomics of Soybean, Edited by Professor Gary Stacey, is a remarkable collection of articles by internationally-recognized experts in the field of soybean genomics – many of whom helped to develop the tools and resources necessary to establish soybean as a powerful crop to investigate important basic and applied questions of plant biology. This collection of articles provides a comprehensive up-to-date review of the field of soybean genomics, and documents how far this field has advanced in the last few years. From the vantage point of someone like myself who first began investigating the organization and expression of the soybean genome thirty years ago, the insights provided by the authors in this book indicate that soybean has indeed "come of age," and that decades-old mysteries of the soybean genome are now being illuminated. *Genetics and Genomics of Soybean* is divided into four sections: (1) soybean genome natural history and diversity – which includes chapters on the genetic variation of the soybean genome and its relationship to other legume genomes; (2) tools, resources, and approaches – which includes reviews of technological advances that are being used to study the soybean genome – including the first glimpse of how the soybean genome is being sequenced and assembled; (3) investigations of soybean biology – which contains chapters that review how genomics tools have been used to study important questions – such as seed development, host-pathogen interactions, abiotic stress, and metabolic pathways; and (4) how Roundup Ready soybeans, generated by genetic engineering, have made an impact on global soybean agriculture. The chapters in this book are essential reading for students and investigators interested in basic and applied aspects of soybean biology. They provide a timely, comprehensive review of the field of soybean genomics, document the status of where the field is today, and, most importantly, raise many exciting questions about soybean evolution and biology that can now be answered using the genomics tools and resources outlined in this important book.

Los Angeles, CA 90095-1606 Bob Goldberg

Preface

Plant genomics is revolutionizing our understanding of basic plant biology and, yet, the impact on major crop plant species is still limited. Until recently, emphasis has been placed on 'model' plant species (e.g., Arabidopsis, and for legumes, *Lotus japonicus* or *Medicago truncatula*, see Chapters 3 and 4). However, if these are models, then what are they models of? Where will we apply the knowledge obtained from the 'models'? Clearly, the targets must be crop plants, which ultimately provide the benefit to mankind. However, why work with models and then test these discoveries in crop plants, when the resources are available to make the original discoveries in the crop? In this scenario, application is direct and immediate.

The Fabaceae (leguminosae) comprise the second largest family of flowering plants with 650 genera and 18000 species. The soybean is a member of the tribe Phaseoleae, the most economically important of the legume tribes (Chapter 2). The soybean, *Glycine max* (L.) Merr. is the major source of vegetable oil and protein on earth (see Chapter 1). As described in detail in this volume, knowledge of soybean genomics and genetics has advanced rapidly to the point that many of the resources previously only available for 'model' species are now ready for exploitation in this crop. Soybean has a very detailed genetic map (Chapter 5), a recently completed physical map (Chapter 6) and developing resources for reverse genetics to study gene function (Chapter 9). As this volume goes to press, it is anticipated that the full sequence of the soybean genome is nearing public release through the efforts of the US Department of Energy-Joint Genome Institute (see Chapter 7 for a preview). This represents a major milestone in ***Genetics and Genomics of Soybean*** and will enable practical applications for soybean improvement.

Knowledge of the soybean genome is already enhancing soybean breeding through the application of molecular assisted selection (Chapter 8). In addition, this information is being applied to both basic and applied research in priority areas. For example, the soybean seed is the major product of the plant and detailed studies, using a full repertoire of functional genomic methods, are well underway (Chapter 11). These studies include the analysis of biochemical pathways involved in both oil and protein synthesis (Chapter 12). The recent resurgence of interest in soybean as a biodiesel source makes these studies particular relevant. Soybean is also a 'heart health food', as designated by the US Food and Drug Association. This is in large part due to the production of a wide variety of bioactive secondary products (Chapter 13). Genetics and genomic information also have an important role to play in improving soybean production. For example, efforts are well underway to apply this information to improve stress (both biotic and abiotic) resistance (Chapters 14, 15, 16, and 17).

The world's expanding population, coupled with growing concerns about the environment and climate change, present tremendous challenges for agriculture (Chapter 1). How will we feed the future expanded population of our planet, with decreasing land in the face of rising environmental challenges? Clearly, legumes, especially soybean, can make significant contributions due to the benefits of crop rotation and influences on soil fertility. It is also clear that biotechnology (for example, in the form of transgenic crop plants) will play an ever increasing role in agriculture. However, this remains a controversial area in many parts of the world. The experience of herbicide resistant soybeans, one of the first transgenic crops to be grown on a large scale, may provide insight into the benefits and future use of biotechnology in agriculture (Chapter 19).

This volume represents a compilation of timely topics pertinent to modern genetics and genomics of soybean. The chapters are written by recognized experts and provide an excellent primer for the no-doubt astounding developments that will come in the future from the full knowledge of the soybean genome sequence. I thank all of the authors for their wonderful and timely contributions. I also thank Jinnie Kim, Senior Editor, Springer Science and Business Media, for originally suggesting this idea and aiding in its development. Finally, special thanks to Jillian Slaight, Editorial Assistant, for moving the volume into production.

Columbia, MO, USA Gary Stacey

Contents

Contributors

John All
Department of Entomology, University of Georgia, Athens, GA 30602-6810, USA, e-mail: jall@uga.edu

Thomas J. Baum
Department of Plant Pathology, Iowa State University, Ames, IA 50011, USA, e-mail: tbaum@iastate.edu

Kristin D. Bilyeu
USDA-ARS, Plant Genetics Research Unit, University of Missouri-Columbia, MO 65211, USA, e-mail: bilyeuk@missouri.edu

H. Roger Boerma
Department of Crop and Soil Sciences, University of Georgia, Athens, GA 30602-6810, USA, e-mail: rboerma@uga.edu

Steven Cannon
USDA-ARS, Department of Agronomy, Iowa State University, Ames, IA 50011, USA, e-mail: scannon@iastate.edu

Jarrod Chapman
Department of Energy Joint Genome Institute, 2800 Mitchell Drive, Walnut Creek, CA 94598, USA e-mail: JChapman@lbl.gov

Perry B. Cregan
Soybean Genomics and Improvement Laboratory, U.S. Department of Agriculture, Agricultural Research Service, BARC-West, Beltsville, Maryland, MD 20705, USA, e-mail: creganp@ba.ars.usda.gov

Paul Gepts
Department of Agronomy and Range Science, University of California, Davis, Davis, CA 95616, USA, e-mail: plgepts@ucdavis.edu

Delkin Orlando Gonzalez
Department of Crop Sciences, University of Illinois, Urbana, Illinois, IL 61801, USA, e-mail: dogonzal@uiuc.edu

Madge Graham
Department of Plant Pathology, Ohio State University, Columbus, OH 43210, USA, e-mail: graham.19@osu.edu

Michelle A. Graham
USDA-ARS CICGRU, Department of Agronomy, Iowa State University, Ames, IA 50011, USA, e-mail: magraham@iastate.edu

Terry Graham
Department of Plant Pathology, Ohio State University, Columbus, OH 43210, USA, e-mail: graham.1@osu.edu

David Grant
USDA-ARS CICGRU, Department of Agronomy, Iowa State University, Ames, IA 50011, USA, e-mail: dgrant@iastate.edu

Tomoko Hatanaka
Kobe University Kobe, 657–8501, Japan, e-mail: thata@kobe-u.ac.jb

Uffe Hellsten
Department of Energy Joint Genome Institute, 2800 Mitchell Drive, Walnut Creek, CA 94598, USA, e-mail: UHellsten@lbl.gov

David F. Hildebrand
University of Kentucky, Lexington, KY 40546-0312, USA, e-mail: dhild@uky.edu

Scott A. Jackson
Department of Agronomy, Purdue University, 915 W. State St., West Lafayette, IN 47907, USA, e-mail: sjackson@purdue.edu

Sarah Jones
Department of Crop Sciences, University of Illinois, Urbana, Illinois, IL 61801, USA, e-mail: sijones@uiuc.edu

Nicholas Kalaitzandonakes
The Economics and Management of Agrobiotechnology Center, Department of Agricultural Economics, University of Missouri-Columbia, MO 65211-6200, USA, e-mail: KalaitzandonakesN@missouri.edu

Srinivasa Konduru
The Economics and Management of Agrobiotechnology Center, Department of Agricultural Economics, University of Missouri-Columbia, MO 65211-6200, USA, e-mail: spkdkc@mizzou.edu

John Kruse
The Economics and Management of Agrobiotechnology Center, Department of Agricultural Economics, University of Missouri-Columbia, MO 65211-6200, USA, e-mail: krusej@missouri.edu

Matt Lavin
Department of Plant Sciences, Montana State University, Bozeman, MT 59717, USA, e-mail: mlavin@montana.edu

Runzhi Li
University of Kentucky, Lexington, KY 40546-0312, USA, e-mail: Runzhi.Li@uky.edu

M. A. Saghai Maroof
Department of Crop and Soil Environmental Sciences, Virginia Polytechnic Institute and State University, Blacksburg, VA 24061, USA, e-mail: smaroof@vt.edu

Phillip E. McClean
Department of Plant Sciences, North Dakota State University, Fargo, ND 58105, USA, e-mail: phhillip.mcclean@ndsu.edu

Melissa G. Mitchum
Christopher S. Bond Life Sciences Center, Division of Plant Sciences, University of Missouri-Columbia, MO 65211, USA, e-mail: goellnerm@missouri.edu

Rex T. Nelson
USDA-ARS CICGRU, Department of Agronomy, Iowa State University, Ames, IA 50011, USA, e-mail: nelsonrt@iastate.edu

Henry T. Nguyen
National Center for Soybean Biotechnology, Division of Plant Sciences, 1-31 Agriculture Building, University of Missouri-Columbia, MO 65211, USA, e-mail: nguyenhenry@missouri.edu

Wayne Parrott
Department of Crop and Soil Sciences, University of Georgia, Athens, GA 30602-6810, USA, e-mail: wparrott@uga.edu

M. S. Pathan
National Center for Soybean Biotechnology, Division of Plant Sciences, University of Missouri-Columbia, MO 65211, USA, e-mail: pathanm@missouri.edu

Daniel Rokhsar
Department of Energy Joint Genome Institute, 2800 Mitchell Drive, Walnut Creek, CA 94598, USA; Department of Molecular and Cell Biology, Center for Integrative

Genomics, University of California at Berkeley, Berkeley, CA 94720, USA, e-mail: DSRokhsar@lbl.gov

Jessica A. Schlueter
Department of Agronomy, Purdue University, 915 W. State St., West Lafayette, IN 47907, USA, e-mail: acissej@purdue.edu

Jeremy Schmutz
Department of Energy Joint Genome Institute, Stanford Human Genome Center, 975 California Avenue, Palo Alto, CA 94304, USA, e-mail: jeremy@shgc.stanford.edu

Randy C. Shoemaker
USDA-ARS CICGRU, Department of Agronomy, Iowa State University, Ames, IA 50011, USA, e-mail: rcsshoe@iastate.edu

David A. Sleper
National Center for Soybean Biotechnology, Division of Plant Sciences, University of Missouri-Columbia, MO 65211, USA, e-mail: sleperd@missouri.edu

Françoise Thibaud-Nissen
Department of Crop Sciences, University of Illinois, Urbana, Illinois, IL 61801, USA, e-mail: fthibaud@tigr.org

Sue A. Tolin
Department of Plant Pathology, Physiology and Weed Science, Virginia Polytechnic Institute and State University, Blacksburg, VA 24061, USA, e-mail: stolin@vt.edu

Dominic M. Tucker
Department of Crop and Soil Environmental Sciences, Virginia Polytechnic Institute and State University, Blacksburg, VA 24061, USA, e-mail: dotucke2@vt.edu

Jigyasa Tuteja
Department of Crop Sciences, University of Illinois, Urbana, Illinois, IL 61801, USA, e-mail: tuteja@uiuc.edu

Brett M. Tyler
Virginia Bioinformatics Institute, Virginia Polytechnic Institute and State University, Blacksburg, VA 24061-0477, USA, e-mail: bmtyler@vt.edu

Babu Valliyodan
National Center for Soybean Biotechnology, Division of Plant Sciences, 1-31 Agriculture Building, University of Missouri-Columbia, MO 65211, USA, e-mail: valliyodanb@missouri.edu

Lila Vodkin
Department of Crop Sciences, University of Illinois, Urbana, Illinois, IL 61801, USA, e-mail: l-vodkin@uiuc.edu

David Walker
Department of Crop and Soil Sciences, University of Georgia, Athens, GA 30602-6810, USA; Current address: USDA-ARS & University of Illinois, National Soybean Research Center, USA, e-mail: walkerdr@uiuc.edu

Richard F. Wilson
National Program Leader, Oilseeds & Biosciences LLC, United States Department of Agriculture, Agricultural Research Service, National Program Staff, Beltsville, MD 20705-5139, USA, e-mail: richard.wilson@ars.usda.gov, rfwilson@mindspring.com

Oliver Yu
Donald Danforth Plant Sciences Center, St. Louis, MO 63132, USA, e-mail: OYu@danforthcenter.org

Gracia Zabala
Department of Crop Sciences, University of Illinois, Urbana, Illinois, IL 61801, USA, e-mail: g-zabala@uiuc.edu

Shuquan Zhu
Department of Crop and Soil Sciences, Center for Applied Genetic Technologies, University of Georgia, Athens, GA 30602-6810, USA, e-mail: szhu@uga.edu

Part I
Natural History and Genetic Diversity

Chapter 1
Soybean: Market Driven Research Needs

Richard F. Wilson

Introduction

Soybean (*Glycine max* L. Merrill) is the dominant oil-seed in world trade, accounting for about 56% of global oilseed production. The contribution this crop makes to the current global economy is estimated conservatively at $48.6 billion or about $18.7 billion in the U.S. alone. Demand for soybean remains strong and continues to grow because it is used as an ingredient in the formulation of a multitude of food, feed and industrial products. These applications include a wide range of soyfoods, shortening to biodiesel applications for soybean oil, and feed to vegetable protein substitutes for meat and dairy products for soybean meal/protein. In addition, soybean is a primary source of high-value secondary co-products such as lecithin, vitamins, nutraceuticals and anti-oxidants. The U.S., Brazil and Argentina are the predominant soybean producing countries, but global soybean production area has reached an apparent plateau. If this trend continues or worsens, extreme pressure will be placed on 'genetic-gain' in soybean yielding ability to ensure adequate supply to meet the escalating demand for soybean and soybean products. Failing to provide an ample supply of soybeans would be felt throughout the world, beginning with a decline in soybean exports. Currently, the U.S. and Brazil crush only about half of their annual production; whereas countries like Argentina, the People's Republic of China and the European Union-25 essentially crush their entire annual supply. Thus, the U.S. and Brazil are the only countries with the flexibility to export whole soybeans to major customers such as the People's Republic of China and the European Union. However, future levels of soybean exports likely will be eroded by the need to service greater domestic use. Already there are signs of a transition toward greater production and trade of refined vegetable oil and meal among soybean producing countries. Currently, the U.S. consumes 95% of its domestic production of soybean oil. Because of the emerging market for biodiesel,

R.F. Wilson
United States Department of Agriculture, Agricultural Research Service, National Program Staff, Oilseeds & Biosciences LLC
e-mails: richard.wilson@ars.usda.gov; rfwilson@mindspring.com

G. Stacey (ed.), *Genetics and Genomics of Soybean*,

a deficit in U.S. soybean oil production is projected by 2020. By the same token, U.S. end-stocks typically support only a 4–6 month supply of soybean meal. With anticipated growth in the livestock and aquaculture industries, a deficit in U.S. soybean meal production is predicted by 2020. If genetic-gains in the improvement of U.S. soybean production are not sufficient to ensure an adequate domestic supply of soybean meal, the U.S. may be in jeopardy of losing domestic livestock production to off-shore locations. The prospect for future deficits in U.S. production of soybeans, soybean oil and soybean meal gains credibility from the national emphasis on reduction of U.S. dependence on petroleum and fossil-fuels. It is estimated that about 30% of the U.S. corn crop may be converted to ethanol production by 2010, and the projected goal of 700 million gallons of biodiesel would consume 23–25% of U.S. annual production of soybean oil. Strong demand for ethanol production already has resulted in higher corn prices, which favors future increases in corn acreage at the expense of soybean. Thus, the impact of bioenergy alternatives on the ability to provide an adequate supply of soybean and soybean products is a serious challenge that must be addressed by the U.S. and global soybean industrial and research communities. It should be noted that significant progress is being made to enhance soybean yielding ability, largely through random exploitation of the wealth of genetic diversity that is harbored among accessions of soybean germplasm collections. However, to maintain the current rate of growth in U.S. soybean supply, assuming no change in U.S. soybean production area, it appears that U.S. soybean yields would have to increase to an average 4085 kg/ha (60.8 Bu/acre) by 2020. Optimistically, yielding ability can be enhanced, but the question that now faces the soybean genetics community is whether or not continued genetic-gains of the required magnitude may be attained through a traditional breeding approach alone. Better understanding of the genetic regulation of seed constituent composition also is needed to help ensure an adequate supply of high-quality protein and oil. In addition, effective strategies for protection against crop losses to diseases such as Asian Soybean Rust, pests such as soybean cyst nematode, and environmental stresses will require detailed analysis of the soybean genome. 'Mining' the soybean genome for this information will facilitate the development of useful DNA markers for genes of interest. Integration of those 'genomic tools' with modern breeding programs will lead to more effective utilization of the genetic diversity in *Glycine max*. The 'tools' and knowledge gained from soybean genomics will enable the 'next generation' advances in soybean breeding that are needed now to meet the needs of U.S. and global agriculture.

Origin and Development of Soybean as a Crop

Domestication of Soybean

Cultivated soybean [*Glycine max* (L.) Merr.] appears to draw its origin from a domestication event in the wild soybean (*Glycine soja* Seib. et Zucc.) that may have occurred in ancient central or southern China nearly 5000 years ago (Gai 1997; Gai and Guo 2001). This estimate is derived, in part, from references to soybean which appeared in Chinese literature during the Shang dynasty from 1700 to 1100 BC (Qiu et al. 1999). However, anecdotal evidence and oral traditions recorded during that time also suggest a much older association of soybean in the Chinese culture (Guo 1993).

The versatility of soybean in preparing various soyfoods is perhaps the major factor that favored its cultivation as an agricultural crop. Soyfoods, like *tofu* (thought to be invented during the Han Dynasty), *douchi* (a fermented salty garnish made from whole soybean) and *doujiang* (a thick sauce made from fermented soybean) were then, and remain today, staples of the Chinese diet. In addition, the beginnings of the soybean oil industry may be traced to China, at least 1000 years ago, when historical records report the common practice of frying *tofu* with soy oil (Gai and Guo 2001).

However, the cultivation of domesticated soybean beyond ancient China did not spread rapidly. For example, soybean may have been introduced to Japan from China or Korea only about 2000 years ago (Li and Nelson 2001). Documented reference to soybean cultivation in Japan does not appear until the early Yayoi culture (Kihara 1969; Sugiyama 1992). In any case, soybean has long been important in the Japanese diet, leading to the development of a unique food culture. Japanese innovations in soyfoods include: vegetable soybean (*edamame*), soybean sprouts (*moyashi*), soymilk (*tonyu*), frozen and baked soybean curd (*kori-dofu*, *yaki-dofu*). Small-seeded soybean may be used for fermented soybean (*natto*), and boiled or fermented medium-sized seeds are used for the production of soybean paste (*miso*). Yellow or green soybean meal (*kinako*) is used in confectionery products or can be fermented to produce soy sauce (*shoyu*) (Wilson 1995).

Soybean was first introduced into North America by Samuel Bowen in 1765, principally to manufacture soy sauce. In 1770, Benjamin Franklin also experimented with soybean in the U.S.; however, his interests were limited to its utility as a forage and ground cover (Hymowitz and Harlan 1983). It was not until early in the 20th century, when the impetus for modern U.S. soybean production was discovered. This occurred in 1915, when soybeans were first crushed for oil in Elizabeth City, North Carolina (Wilson 1987).

The discovery of soybean as an important source of vegetable oil permanently changed the perception of soybean from forage to a seed crop. This transition brought the need for more productive or agronomic types of soybean. By the early 1930s, the United States Department of Agriculture (USDA) and the State Agricultural Experiment Stations at land grant universities established soybean breeding programs in the northern and southern states (Bernard et al. 1988). These efforts were enabled and strengthened by the acquisition and identity preservation of over 4000 soybean landraces from China by the USDA. Today, the USDA soybean germplasm collection contains over 18,000 types of *Glycine max* and is actively used to ensure access to a broad range of genetic diversity for cultivated soybean (USDA, ARS 2007).

By 1950, nearly 100% of the U.S. soybean crop was grown for seed, and the U.S. became the world leader in soybean production. Again, it was the functional utility of soybean seed constituents in a wide array of products that provided the basis for development of the U.S. soybean industry. Even today, as a conservative estimate, food manufacturers in the U.S. routinely create over 400 new food products with soy as an ingredient each year (Liu 1997). Products from soybean oil include: margarine, shortenings, baking and frying fats. Soybean oil also is used in industrial products including soap, cosmetics, resins, plastics, inks, crayons, solvents, clothing, and biodiesel. Soybean meal provides the high-protein feed ingredient that sparked an American revolution in poultry and swine production, and more recently the aquaculture industry. Dietary uses for soy flour in the form of soy concentrate and soy protein isolate include formula for lactose-intolerant infants, and vegetable protein substitutes for meats and dairy products. Industrial uses for soy-protein include coatings, adhesives and building materials. In addition, soybean is the primary source of high-value co-products such as lecithin, vitamins, nutraceuticals and anti-oxidants.

World Soybean Production

Within the past 60 years, an infinitesimal period during its domestication, soybean emerged as the dominant oilseed in world trade. In 2005, the USDA (USDA, FAS 2007) estimated world soybean production at 218 million metric tons (MMT); about 56% of total global oilseed production which includes copra, cottonseed, palm kernel, peanut, rapeseed (canola) and sunflower-seed (Table 1.1). Soybean also is distinguished among these oilseed crops as the primary high-energy, high-protein ingredient for livestock feed. The trading standard set by the National Oilseed Processors Association (NOPA) for high-protein soybean meal is 48% crude protein. No other oilseed meal matches that level of protein or possesses a more desirable dietary complement of essential amino acids. Thus, soybean meal commands a dominant position with a 69% share of the world vegetable protein market. However,

Table 1.1 World Production of Major Oilseeds and Oilseed Product, 2005/06

Commodity	Oilseeds		Meal		Oil	
	MMT	%	MMT	%	MMT	%
Copra	5.8	1.5	1.8	0.9	NA	0.0
Coconut	NA	0.0	NA	0.0	3.5	3.0
Cottonseed	42.5	10.9	14.3	6.8	4.6	3.9
Olive	NA	0.0	NA	0.0	2.3	1.9
Palm	NA	0.0	NA	0.0	36.0	30.5
Palm Kernel	10.0	2.6	5.2	2.5	4.4	3.7
Peanut	33.7	8.7	6.0	2.9	5.2	4.4
Rapeseed	48.6	12.5	26.3	12.5	17.2	14.6
Soybean	218.0	56.1	144.7	69.0	34.3	29.1
Sunflowerseed	29.8	7.7	11.2	5.4	10.4	8.8
Total	388.4	100.0	209.6	100.0	117.8	100.0

United States Department of Agriculture, Foreign Agricultural Service, 2007

Table 1.2 World Production of Soybean and Soybean Products, 2005/06

Country of Origin	Seed		Meal		Oil	
	MMT	%	MMT	%	MMT	%
United States	83.4	38.2	37.4	25.9	9.3	27.0
Brazil	55.0	25.2	21.7	15.0	5.4	15.7
Argentina	40.5	18.6	25.0	17.3	6.0	17.5
China, PRC	16.4	7.5	27.3	18.9	6.1	17.9
India	6.3	2.9	4.3	3.0	1.0	2.8
Paraguay	4.0	1.8	na	0.0	na	0.0
Canada	3.2	1.4	na	0.0	na	0.0
Other	9.4	4.3	15.6	10.8	3.5	10.3
EU-25	na	0.0	10.4	7.2	2.4	6.9
Mexico	na	0.0	3.0	2.1	0.7	1.9
Total	218.0	100.0	144.7	100.0	34.3	100.0

United States Department of Agriculture, Foreign Agricultural Service, 2007

there is significantly more competition among sources of vegetable oil. Soybean and palm lead that market with equal shares accounting for about 70% of total vegetable oil production.

The predominant soybean producing countries at this time are the U.S., Brazil, and Argentina. Although the U.S. produces more soybeans than any other single country, the South American countries of Brazil, Argentina, Paraguay plus Mexico collectively have surpassed North American soybean production (Table 1.2). The demographics for world production of soybean meal and oil follow similar trends, where the U.S. leads Brazil, Argentina and the People's Republic of China (PRC). It follows that these four countries account for about 77% of the world's soybean crushing capacity.

The emergence of Brazil and Argentina as major soybean producing countries is attributed to a significant increase in total harvested area between 1996 and 2004

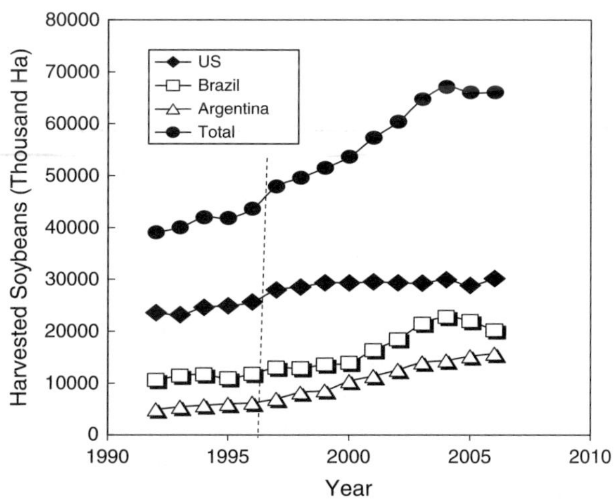

Fig. 1.1 Soybean harvested area in major producing countries

(Fig. 1.1). However, the up-surge in South American soybean production area may be short lived. Since 2004, the world total for harvested soybean area appears to have reached a plateau at about 66.4±0.6 Mha (164±1.5 million acres). This recent trend may be attributed in part to an apparent decline in total area for Brazilian soybean production, coupled with essentially no growth in U.S. area for soybean production. This situation obviously places more pressure on the translation of soybean genetic knowledge into more effective and efficient means to generate the elite yielding varieties that will help to ensure sustained future increases in global soybean production. As an example, global soybean production increased 83.6 MMT between 1996 and 2004. This achievement may be attributed to advances in soybean cultural practices and yielding ability, from a world average 2111.4 kg/ha to 2313.1 kg/ha (31.5 to 34.5 Bu/acre), plus an additional 30.7 Mha (75.8 million acres) in soybean production area. That expansion of world soybean production area exceeded the existing area for U.S. soybean production. If the world soybean production area during that period had not doubled, then it may be deduced that a global average yield of 3446.3 kg/ha (51.3 Bu/acre) would have been required to attain the level of 2004 output. Obviously, such a target is unrealistic given current technology. Hence, constant or declining global acreage is a major constraint.

Without the luxury of expanding production area, unpredictable or uncontrollable events, such as unfavorable weather or epidemics of severe diseases/pests, pose a more severe threat to global soybean production and necessitate significant and timely genetic measures to sustain the ability to keep pace with growing global market demand for soybean and soybean products.

Supply and Demand for Soybean Products

World Trends in Soybean Supply

As a result of the infusion of South American production area plus incremental gains in cultivar yielding ability, world soybean supply (production plus end-stocks) has more than doubled in the past 22 years, to a 2006 total of 282.5 MMT (Fig. 1.2). The rate of increase over that period was 8.5 MMT/yr (R^2, 0.94). At the same time, world use (crush plus exports) of soybean grew at 8.0 MMT/year (R^2, 0.92) from 1984 to 2006 (USDA, FAS 2007).

Thus, these data suggest no eminent limitation in global soybean supply in the foreseeable future, and by the same token, no relent in the growing demand for soybean products. However, there is cause for concern. Closer inspection of these data, 2006 for example, reveals a 37 MMT deficit between 'world soybean use' and 'world soybean production'. Although this difference is covered by the level of end-stocks from 2005 (55.2 MMT), the carryover to 2007 (18.7 MMT, ca. 37 days supply) will be nearly 3-fold less than in 2006.

The economic equilibrium between determinants of 'market price' is maintained by the level of end-stocks, which acts as a necessary buffer to ensure relatively

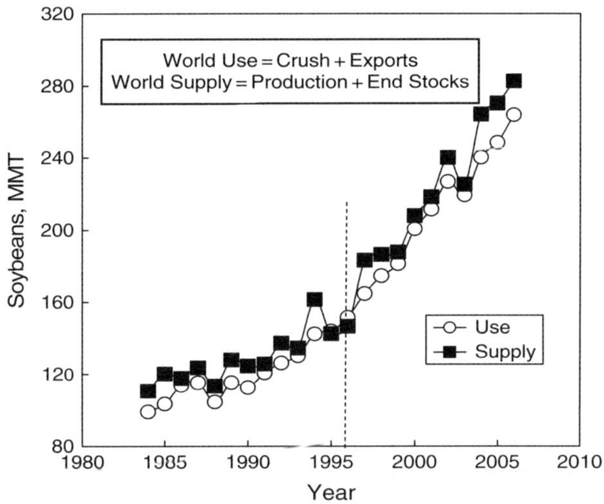

Fig. 1.2 World trends in soybean supply and demand

uninterrupted flow of produce through the marketing system. In fact, there is a very strong negative correlation between soybean end-stocks and the U.S. farm price for soybeans (Fig. 1.3). Hence, a decline in end-stocks relative to U.S. soybean supply typically is accompanied by a rise in the U.S. farm price per bushel. This statistic may be a good predictor of trends in this apparent cycle on a global scale. Although the data for 2006 are incomplete, preliminary estimates of world farm prices have tended higher in 2006–2007.

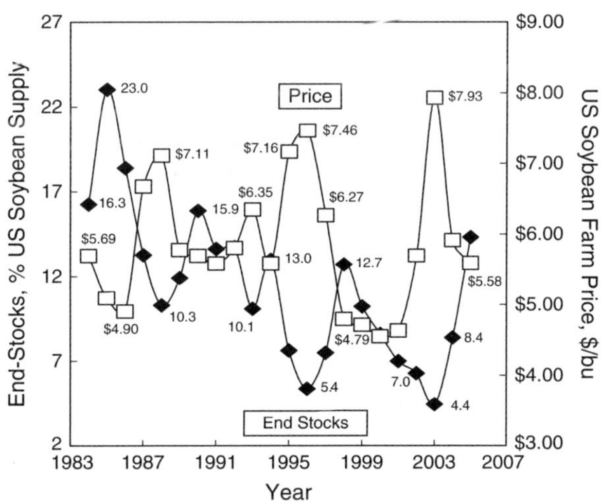

Fig. 1.3 Relation between end-stocks and price

World Trends in Soybean Use

Soybean use (crush plus exports) has increased nearly 3-fold in the past 22 years, to a total of 263.8 MMT. Although world soybean exports account for only about 27% of that total, exports also rose nearly 3-fold from 25.3 MMT in 1984 to 70.7 MMT in 2006. However, there has been a significant shift in demographics within this export market. In the past decade from 1995 to 2005, the U.S. share of the global soybean export market has declined from about 70% to 40% (Fig. 1.4). This change may be attributed to nearly a 5-fold increase (7.5 to 35.8 MMT) in soybean exports from South America, with about 71% coming from Brazil. The level of U.S. soybean exports during that period averaged 26.0 ± 3.0 MMT, which is not significantly different from the mean for the past 22 years. Therefore, Brazilian soybean exports were necessary to maintain the long-term (since 1984) growth rate of global soybean exports at 2.3 MMT/year (R^2, 0.88).

The PRC and the European Union (EU-25) are the recipients of about two-thirds of global soybean exports. Because of escalating demand for protein and oil, world soybean crushing capacity has expanded at a linear rate of 5.74 MMT/year (R^2, 0.96) from 1984 to 2006. The EU-25 crushes 95+% of their soybean imports for protein and oil; the PRC crushes about 78% of their imports plus domestic production. Among soybean exporting countries, Argentina crushes about 79% of their production, while the U.S. and Brazil crush only about half of their annual soybean harvest. Overall, the U.S., PRC, Argentina, Brazil and the EU-25 (in top to bottom order) account for 84% of the world production of soybean oil and meal. However, only the U.S. and Brazil have the apparent flexibility to provide or sustain adequate supply of whole soybeans to the Asian and European processing industries. If the U.S. and/or Brazil deploy greater soybean crushing-capacity in the near future, then the supply of soybeans to the PRC and EU-25 becomes less than certain.

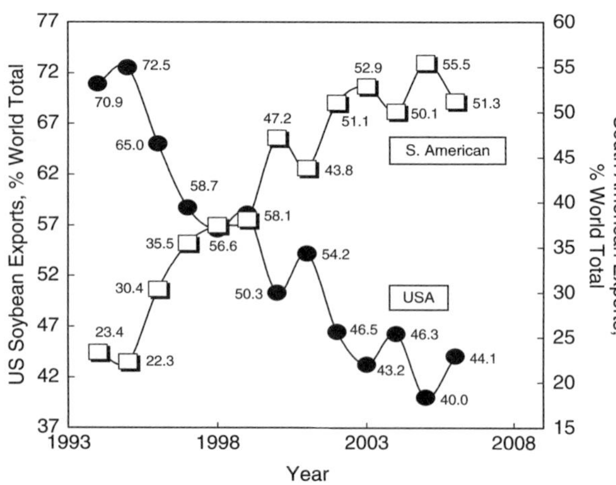

Fig. 1.4 Trends in relative share in the export soybean market

Trends in U.S. Consumption of Soybean Products

Historically, the U.S. crushes about 56% of its annual soybean production. In 2006, this resulted in about 38.5 MMT of meal and 9.2 MMT of oil (USDA, FAS 2007). Domestic consumption of soybean oil has increased at a rate of 0.185 MMT/year (R^2, 0.97) since 1984, to 8.7 MMT (Fig. 1.5), but that rate is expected to accelerate due to use of vegetable oils in the formulation of bio-diesel fuel (Conway et al. 2004). Currently, the U.S. consumes 95.4% of its annual production of soybean oil. Although U.S. end-stocks for soybean oil may exceed 0.5 MMT at this time, the long-term rate of change in domestic soybean oil production is 0.186 MMT/year (R^2, 0.93). Thus by 2020, a deficit of U.S. soybean oil is projected due to increased demand for bio-based alternatives to petroleum (United Soybean Board 2006; Westcott 2007). If genetic-gains in the improvement of U.S. soybean production are not sufficient to ensure an adequate domestic supply of soybean oil, the U.S. may become a substantial customer of Argentina and Brazil, the predominant soybean oil exporting countries.

Domestic consumption of U.S. soybean meal has increased at a rate of 0.67 MMT/year (R^2, 0.97) from 1984 to about 31 MMT in 2006. Approximately 80% of that annual domestic production is used in feeds and vegetable protein products, with the remainder in the international export market. As a result, the U.S. carries an extremely low surplus of soybean meal (Fig. 1.6). Since soybean is the preferred high-protein ingredient for livestock feed, demand for soybean meal in poultry and swine production alone is expected to grow to 29 MMT by 2020 (Westcott 2007). However, new feed markets are emerging. For example, demand for soybean meal/isolate in aquafeed is expected to reach 13 MMT by 2020 (United Soybean Board 2006). These estimates reaffirm forecasts of continued growth in U.S. demand for soybean meal. Yet, at the current rate of increase (0.75 MMT/year; R^2, 0.93), future U.S. soybean meal production, even with projected increases in

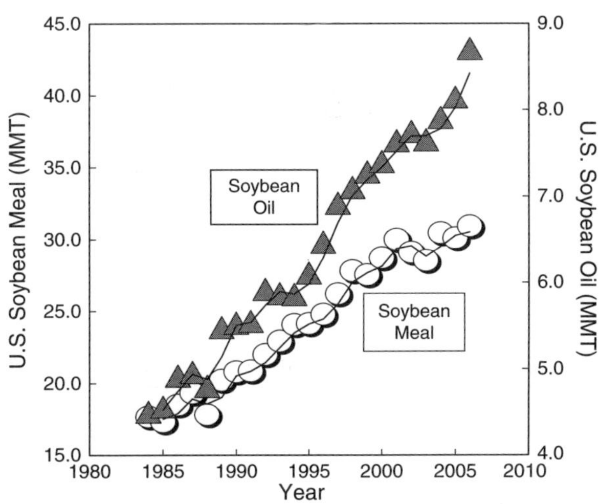

Fig. 1.5 Trend in U.S. consumption of soybean oil and meal

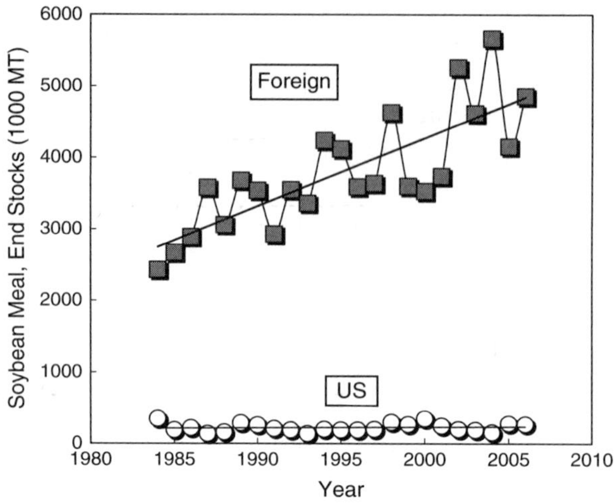

Fig. 1.6 Distribution of soybean meal end-stocks

crush-capacity, probably will not provide an adequate margin. Hence, a deficit of U.S. soybean meal is projected by 2020 due to increased demand for livestock production and aquaculture (United Soybean Board 2006). Once again, if genetic gains in the improvement of U.S. soybean production are not sufficient to ensure an adequate domestic supply of soybean meal, the U.S. may be in jeopardy of losing domestic livestock production to off-shore locations.

Further Constraints to Soybean Production

Soybean Production in an Energy Driven Environment

Although the U.S. produced the largest soybean crop on record in 2006, estimated to be 3.19 billon bushels at an average 2869 kg/ha (42.7 Bu/acre) on 30.2 Mha (74.6 million acres), the prospect for future deficits in U.S. production of soybean, soybean oil and soybean meal gain credibility in view of national energy policy to reduce U.S. dependence on petroleum and fossil-fuels. For example, the Energy Policy Act of 2005 mandates that renewable fuel use in gasoline and diesel reach 7.5 billion gallons by 2012 (Westcott 2007). In practice, higher petroleum costs combined with a variety of tax credits and import tariffs have provided economic incentives for expanded biofuel production capacity that may achieve outputs in excess of the original goal (Ash et al. 2006).

Most of the ongoing and projected biofuel expansion in the U.S. is focused on ethanol. With current technology, one bushel of corn should produce 2.8 gallons of ethanol. It is estimated that about 30% of the U.S. corn crop may be converted to ethanol production by 2010.

Biodiesel production capacity also has increased rapidly in the past five years. About one pound of refined soybean oil is required to formulate one pound of biodiesel. Based on that relation, the projected goal of 700 million gallons of biodiesel would consume 23–25% of U.S. annual production of soybean oil.

Strong demand for ethanol production already has resulted in higher corn prices, which favors future increases in corn acreage. In 2006, the U.S. produced 10.5 billion bushels of corn on 28.6 Mha (70.6 million acres), averaging 10,000 kg/ha (149.1 Bu/acre). With greater potential revenue, corn acreage could reach 36.5 Mha (90 million acres) by 2010. Much of that increase would come by adjusting crop rotations, causing a net decline in soybean acreage and soybean production.

Therefore, the impact of bioenergy alternatives on ability to provide an adequate supply of soybean and soybean products is a serious challenge that must be addressed by the U.S. and global soybean industrial and research communities. It is certain that crushing will continue to be driven by demand for livestock and aquafeeds, and government projections show no slowing of domestic demand for soybean oil, up to 2016, in food and fuel applications. Hence, even with incremental gains in the level of production, U.S. soybean exports may by necessity be significantly eroded in favor of greater crush volume.

Improving the Genetic Efficiency of Soybean Production

Given the significant challenge raised by competition within the U.S. for crop acreage, innovative research must be implemented to ensure there is continued growth in U.S. soybean production to meet the anticipated rise in demand for soybean and soybean products. Enhancement of soybean yielding ability through improved performance and reduced losses to disease and pests are obvious priorities. In that regard, the soybean breeding community already has made significant contributions though development of elite cultivars. Government statistics show fairly steady gains in average U.S. soybean yielding ability, from 2197 to 2896 kg/ha (32.7 to 43.1 Bu/acre), between 1993 and 2006. Foreign soybean production also demonstrated similar advances in yielding ability, although the yields may average 700 kg/ha (10 Bu/acre) less than in the U.S. However, to maintain the current rate of growth in U.S. soybean supply, assuming no change in U.S. soybean production area, it appears that soybean yielding ability in the U.S. would have to increase to an average 4085 kg/ha (60.8 Bu/acre) by 2020. Such a level in yielding ability may be achieved, but the question that now faces the soybean genetics community is whether or not continued genetic-gains of the required magnitude may be attained through a traditional breeding approach alone.

The great reservoir of genetic diversity that is harbored among the accessions of the world's soybean germplasm collections provides a foundation for future advances in genetic technology that are needed to provide elite soybean cultivars with adequate protection from pests and diseases, improved product quality and greater yielding ability. However, these putative genes reside in more than 156,849 accessions of *Glycine max* in about 40 different collections in 20 countries around the world (Carter et al. 2004). The PRC, Taiwan, U.S. and Japan account for about

74% of the world's repository of soybean germplasm (about half of that total is held in the PRC). Many of these accessions are not publicly available to the research community, but even so there has been little effort to characterize the material to improve its utility. Association of phenotypic traits with genotypic markers would be an extremely desirable step that is needed to help distinguish unique accessions, which in turn will facilitate the timely use of valuable genes in variety development.

Soybean genomics research, through analysis and comparison of genomic differences among unadapted and selected populations, will enable a better understanding of the genetic regulation of seed constituent composition. Such knowledge will augment efforts to ensure there is adequate supply of high-quality protein and oil. In addition, effective strategies for protection against crop losses to diseases such as Asian Soybean Rust, pests such as soybean cyst nematode, and environmental stresses will benefit from detailed analysis of the soybean genome. 'Mining' the soybean genome for this information will facilitate the development of useful DNA markers for genes of interest. Integration of those 'genomic tools' with modern breeding programs will lead to more effective utilization of the genetic diversity in *Glycine max*. The 'tools' and knowledge gained from soybean genomics will enable the 'next generation' advances in soybean breeding that are needed now to meet the needs of U.S. and global agriculture.

References

Ash, M., Livezey, J., and Dohlman, E. (2006) *Soybean Backgrounder*. U.S. Department of Agriculture, Economic Research Service, OCS-2006-01, Washington DC.

Bernard, R.L., Juvik, G.A., Hartwig, E.E., and Edwards, C.J., Jr. (1988) *Origins and Pedigrees of Public Soybean Varieties in the United States and Canada*. U.S. Department of Agriculture Technical Bulletin 1746. U.S. Gov. Print. Office, Washington, DC.

Carter, T.E., Jr., Nelson, R.L., Sneller, C.H., and Cui, Z. (2004) Genetic diversity in soybean. In H.R. Boerma and J.E. Specht (Eds.) *Soybeans: Improvement, Production and Uses*. 3rd Edition. American Society of Agronomy, Inc., Madison, Wisconsin, pp.303–416.

Conway, R.K., Erbach, D., Duncan, M., Gallagher, P., Shane, P.L., Smith, P.F., Tumbleson, M.E. (2004). *Bioenergy: Pointing to the Future*, 27A, Council for Agricultural Science and Technology, Ames, IA.

Gai, J. (1997) Soybean Breeding. In: J. Gai (Ed.), *Plant Breeding: Crop Species*. China Agriculture Press, Beijing, pp. 207–251.

Gai, J., and Guo, W. (2001) History of maodou production in China. In: T.A. Lumpkin and S. Shanmugasundaram (Eds.), *Proceedings Second International Vegetable Soybean Conference*, Washington State University, Pullman, WA., pp. 41–47.

Guo, W. (1993) *The History of Soybean Cultivation in China*. Hehai University Press, Nanjing, Jiangsu, China.

Hymowitz, T., and Harlan, J. R. (1983) Introduction of soybean to North America by Samuel Bowen in 1765. Economic Botany. 37, 371–379.

Kihara, H. (1969). History of biology and other sciences in Japan in retrospect. In: C. Oshima (Ed.), *Proceedings XII International Congress of Genetics*, The Science Council of Japan, pp. 49–70.

Li, Z., and Nelson, R. L. (2001) Genetic diversity among soybean accessions from three countries measured by RAPDs, Crop Sci. 41, 1337–1347.

Liu, K. (1997). *Soybeans: Chemistry, Technology, and Utilization*, Springer, New York.

Qiu, L., Chang, R., Sun, J., Li, X., Cui, Z., and Li, Z. (1999) The history and use of primitive varieties in chinese soybean breeding. In: H.E. Kauffman (Ed.), *Proceedings World Soybean Research Conference VI*, Superior Printing, Champaign, IL, pp.165–172.

Sugiyama, S. (1992) On the origin of soybean. *Glycine max* Merrill. J. Brewing Society of Japan 87, 890–899.

United Soybean Board. (2006) Soybean Meal Evaluation to 2020, St. Louis, MO.

U.S. Department of Agriculture, Agricultural Research Service, National Genetic Resources Program. Germplasm Resources Information Network-(GRIN). [Online Database] National Germplasm Resources Laboratory, Beltsville, MD. Available: http://www.ars-grin.gov/var/apache/cgi-bin/npgs/html/ (01 February 2007).

U.S. Department of Agriculture, Foreign Agricultural Service, (2007) Oilseeds: World Markets and Trade, Washington, D.C.

Westcott, P. (2007) Agricultural Projections to 2016, OCE 2007-01. United States Department of Agriculture, Office Chief Economist, Washington, DC.

Wilson, L.A. (1995) Soy Foods. In: D.R. Erickson (Ed.), *Practical Handbook of Soybean Processing and Utilization*, AOCS Press, Champaign, IL, pp. 428–459.

Wilson, R.F. Seed metabolism. p.643–686. In: Wilcox, J.R. (ed.). Soybeans: Improvement, Production and Uses. Second Edition. Am. Soc. Agron., Madison, WI, 888 pp. 1987.

Chapter 2
Soybean Molecular Genetic Diversity

Perry B. Cregan

Introduction

The cultivated soybean [*Glycine max* (L.) Merr.] and the wild soybean (*Glycine soja* Seib. et Zucc.) are annuals and the two members of the *Glycine* subgenus. *G. soja* grows wild in China, Japan, Korea, Russia and Taiwan (Hymowitz 2004). It is generally accepted that cultivated soybean was domesticated 3000–5000 years ago on the Chinese mainland from the wild soybean (Hymowitz and Newell 1981). Cultivated soybean exhibits wide phenotypic variability in terms of seed shape, size, color, and chemical composition; plant morphology and maturity, as well as resistance to a broad range of biotic and abiotic stresses. This genetic diversity and the underlying genetic control of numerous specific traits were described in works such as the recent Third Edition of Soybeans: Improvement, Production and Uses (Boerma and Specht 2004). In particular, Carter et al. (2004) thoroughly documented genetic diversity in terms of the formation, collection, evaluation and utilization of diversity by soybean geneticists and breeders in North American and Asia over 70 years and the impacts of their work on genetic diversity. It is the intent of this review to specifically focus on molecular genetic diversity of the nuclear genome and the multitude of research that was directed at the assessment of molecular diversity of cultivated and wild soybean. This research employed a number of different molecular genetic tools beginning with the analysis of isozyme variation followed by a range of DNA marker types and ultimately variation in DNA sequence. The literature relating to the assessment of isozyme variability in *G. max* and *G. soja* recently received a thorough review by Palmer et al. (2004) and will not be considered here.

The first reports of the assessment of genome-wide molecular genetic diversity of the soybean nuclear genome began in the 1980s with the application of restriction fragment length polymorphism technology (RFLP) (Roth and Lark 1984; Apuya et al. 1988). Subsequent analyses employed RFLP, random amplified polymorphic

P.B. Cregan
Soybean Genomics and Improvement Laboratory, U.S. Department of Agriculture, Agricultural Research Service, Beltsville, Maryland
e-mail: creganp@ba.ars.usda.gov

G. Stacey (ed.), *Genetics and Genomics of Soybean*,

DNA (RAPD) or arbitrary primer PCR, amplified fragment length polymorphism (AFLP), microsatellite or simple sequence repeat (SSR), and DNA sequence analysis for the quantification of genetic diversity in both cultivated and wild soybean. This research had a number of different objectives including (1) the assessment of particular DNA marker systems for appropriately distinguishing and grouping cultivated and wild genotypes, (2) the quantification and comparison of diversity within and among various groups of cultivated and/or wild soybean genotypes (3) the use of genetic diversity estimates as tools in soybean breeding for increasing useful genetic variation, (4) the development of unique DNA fingerprints for genotype and cultivar identification and (5) the assessment of linkage disequilibrium.

Applicability of DNA Marker Types in Soybean

Restriction Fragment Length Polymorphism (RFLP)

Apuya et al. (1988) analyzed 300 RFLP probes selected as low-copy clones in Southern hybridizations to genomic DNA of the genetically distinct soybean cultivars Minsoy and Noir 1. Genomic DNAs were digested with a number of different restriction endonucleases in order to detect RFLP. Of the 300 probes examined only one in five was polymorphic. Despite the low level of polymorphism, 27 loci were analyzed in a population of F_2 plants derived from Minsoy × Noir 1. All loci segregated in a Mendelian fashion and 11 of the 27 loci were contained in four linkage groups. Keim et al. (1989) conducted a survey of RFLP via the analysis of 48 cultivated, eight wild and two *G. gracilis* genotypes using 17 probes to assess the allelic structure of RFLP markers and to identify diverse genotypes that would maximize variability in a resulting mapping population. The *G. gracilis* genotypes were previously joined with *G. max* (Hermann 1962) but were included to maximize morphological diversity in the sampling of genotypes. Extremely low levels of RFLP were recorded despite the diversity of the germplasm analyzed. Two of the 17 probes detected three alleles per locus while the remaining 15 detected only two. The *G. max* genotype A81-356022 and *G. soja* PI 468916 were identified as being particularly diverse with a high level of RFLP that was approximately two-fold higher than that of the Minsoy × Noir 1 cross identified by Apuya et al. (1988). Based upon these data, as well as previous analysis of these two genotypes, a mapping population was created from the interspecific cross of A81-356022 × PI 468916. In a subsequent report, Keim et al. (1992) analyzed 132 RFLP probes in 18 ancestors of U.S. cultivars (ancestral cultivars) as well as 20 adapted cultivars. One objective was to estimate the usefulness of the probes in revealing variation in adapted germplasm. Only one in five markers were informative in any pair of adapted soybean genotypes, again suggesting the relatively low level of RFLP particularly among adapted soybean genotypes.

Skorupska et al. (1993) assessed the feasibility of using the markers from the A81-356022 × PI 468916 RFLP map in the distinct subpopulation of soybean genotypes with maturities adapted to the Southern U.S. A total of 108 genotypes, including older as well as elite cultivars and breeding lines, were analyzed with 83 RFLP probes. Fifty-four percent of the probes were non-informative while 35% had gene diversity values (the probability of detecting polymorphism between any two randomly selected genotypes) of ≥ 0.3. Despite the low levels of molecular diversity in the Southern germplasm pool, the authors indicated that polymorphic probes would serve as a core set for the genetic mapping of agronomic traits in Southern U.S. soybean germplasm. Lorenzen et al. (1995) analyzed 64 soybean ancestral and "milestone" cultivars at 217 RFLP loci to identify a core set of markers that would be useful for pedigree-based analyses of elite soybean cultivars. A set of 97 polymorphic loci were defined that could be used to trace genomic regions contributed by parents to their progeny. Of the 97 loci, 67 had gene diversity scores ≥ 0.30 in the set of 64 cultivars.

Simple Sequence Repeat (SSR) or Microsatellite Markers

The high level of variability and Mendelian inheritance of SSR DNA markers in plants was first reported in soybean by Akkaya et al. (1992) and Morgante and Olivieri (1993). Akkaya et al. (1992) assessed SSR allelic variation at two $(AT)_n$ and one $(ATT)_n$ SSR loci and reported from six to eight alleles among a group of 38 diverse *G. max* and five *G. soja* genotypes and concluded that SSRs would serve as an abundant source of highly polymorphic PCR-based genetic markers in soybean. Morgante and Olivieri (1993) reached similar conclusions regarding the utility and abundance of SSR loci in soybean. Subsequent reports by Rongwen et al. (1995) and Maughan et al. (1995) provided further demonstrations of the usefulness of SSRs for the assessment of genetic diversity in soybean. Rongwen et al. (1995) determined allelic variation at seven SSR loci in a diverse set of 96 soybean genotypes that included N. American cultivars, N. American ancestral cultivars, landraces from the USDA Soybean Germplasm Collection and from China as well as five *G. soja* accessions. From 11 to 26 alleles were found at the seven loci. Gene diversities ranged from 0.71 to 0.95 for the complete set of 96 genotypes and from 0.52 to 0.88 in the set of 28 N. American cultivars. It was concluded that SSRs would be an excellent complement to RFLP loci that were being used by soybean molecular geneticists at the time. Maughan et al. (1995) analyzed a similar set of genotypes including 62 *G. max* lines (landraces, ancestral cultivars, and adapted cultivars) and 32 wild soybeans from diverse Asian origins. From 5 to 21 alleles were detected at five SSR loci with gene diversities ranging from 0.55 to 0.81 in the complete set of 94 genotypes and from 0.29 to 0.62 among the 62 cultivated soybean lines. They suggested that SSRs were the "marker of choice" for a species such as soybean in which molecular genetic diversity is relatively limited. Similar conclusions regarding the usefulness of SSR markers to quantify molecular genetic variation were reached by Song et al. (1998) who analyzed 59 Korean landraces with eight SSR loci.

Amplified Fragment Length Polymorphism (AFLP) Markers

AFLP markers (Vos et al. 1995) are PCR based and permit the multiplex amplification of as many as 50 loci without prior knowledge of DNA sequence. The relatively low level of sequence variation in soybean would make AFLP an attractive alternative to RFLP. Maughan et al. (1996) assessed the use of AFLP in soybean via the analysis of 12 *G. max* and 11 *G. soja* genotypes with 15 AFLP primer pairs. A total of 759 fragments were amplified and 274 (36%) were polymorphic. The number of polymorphic fragments per primer pair varied from 9 to 27 with an average of 18.3. It was concluded that the capacity to rapidly detect thousands of genetic loci at relatively low cost made AFLP an ideal marker for a wide array of genetic investigations in soybean.

Random Amplified Polymorphic DNA or Arbitrary Primer PCR

RAPD (Williams et al. 1990) or AP-PCR markers (Welsh and McClelland 1990) are PCR based, require no prior knowledge of DNA sequence and are analyzed simply as the presence or absence of an amplicon via agarose gel electrophoresis. In order to identify a particularly informative set of RAPD primers, Thompson and Nelson (1998b) analyzed 125 random 10-base primers in 35 soybean genotypes that included 18 ancestral cultivars and 17 maturity group (MG) I-III landraces from the USDA Soybean Germplasm Collection. A total of 281 polymorphic RAPD fragments were identified of which 120 fragments from 64 primers were highly reproducible. A principal-components analysis was used to identify a core set of 35 primers that were critical to the analysis of the 35 genotypes. Thompson and Nelson (1998b) indicated that the correlation of pairwise distances between the 35 genotypes analyzed with the 35 selected primers was highly correlated with those based upon the complete set of RAPD fragment data. This set of 35 RAPD primers was subsequently used in a number of studies to assess molecular genetic variation in cultivated and wild soybean.

Variation in DNA Sequence

The ultimate measure of molecular genetic diversity is the direct comparison of DNA sequence. An important advantage of diversity estimates based upon variation in DNA sequence is the ability to compare across species. Initial estimates of DNA sequence variation were confined to single genes or DNA fragments with the goal of defining gene structure, function, or evolutionary relationships. Scallon et al. (1987) discovered three single nucleotide polymorphisms (SNPs) via the comparison of the 3543 bp sequence of the Gy_4 glycinin locus in the two cultivars Dare and Raiden. Two SNPs were discovered by Zakharova et al. (1989) in the 789 bp of cDNA sequence encoding the A_3B_4 glycinin subunit in the soybean cultivars Mandarin, Mukden and Rannaya-10. Zhu et al. (1995) sequenced a 400 bp fragment of RFLP probe A-199a in three diverse soybean genotypes and found a total of nine SNPs. To compare SNP frequency among DNA fragments of varying length and between populations that vary in size, measures of nucleotide diversity including π (Tajima 1983) and θ (Watterson 1975) were devised, which are normalized for length and adjusted for sample size. Nucleotide diversity from the three aforementioned studies of soybean ranged from $\theta = 0.00085$ (Scallon et al. 1987) to $\theta = 0.015$ (Zhu et al. 1995). This translates into an average of 0.85–10.5 SNPs per kilobase of sequence. In order to provide an estimate of sequence variation in the soybean genome based upon a more extensive analysis of DNA sequence in a large sampling of genotypes, Zhu et al. (2003) analyzed more than 76 kbp of sequence in each of 25 diverse soybean genotypes. The 76 kbp included approximately 28.7 kbp of coding sequence, 37.9 kbp of non-coding perigenic DNA (introns, UTRs and associated genomic DNA), and 9.7 kbp of random non-coding genomic DNA. The mean nucleotide diversity expressed as θ was 0.00097 (an average of slightly less than one SNP per kilobase between any two genotypes in the set of 25 genotypes). Nucleotide diversity was 0.00053, 0.00114, and 0.00179 in coding, non-coding perigenic DNA, and random genomic DNA, respectively. Recent work by Choi et al. (2007) reported the discovery of SNPs in 4240 sequence tagged sites derived from amplicons produced with primers designed to soybean unigenes. In a total of 2.44 mbp of aligned sequence of six diverse genotypes, 4712 single base changes and 839 insertion–deletions (indels) were discovered. This translates to a nucleotide diversity of $\theta = 0.000997$.

The aforementioned reports of nucleotide diversity indicate that as compared to other species, diversity in soybean is relatively low. For example, in rice, Feltus et al. (2004) analyzed 358 mbp of draft sequences of the rice subspecies *Oryza sativa* ssp. *indica* and *japonica* and reported 1.7 single base changes plus 0.11 indels per kbp, which is the equivalent of a nucleotide diversity of $\theta = 0.00181$. Likewise, a calculation of nucleotide diversity in 21.3 kbp of sequence analyzed in five diverse barley cultivars by Kanazin et al. (2002) indicated $\theta = 0.0025$. In sorghum (*Sorghum bicolor*), Hamblin et al. (2004) reported nucleotide diversity of $\theta = 0.0023$, which is more than twice that of soybean. Likewise, Wright et al. (2005) reported nucleotide diversity of $\theta = 0.00627$ in modern maize (*Zea mays* L.) inbreds, while in sugarbeet (*Beta vulgaris* L.) a similarly high nucleotide diversity of $\theta = 0.0077$ was reported in a comparison of two genotypes by Schneider et al. (2001). While the level of DNA sequence variation in soybean is relatively low, it can nonetheless provide an excellent means to compare molecular genetic variability as suggested by the recent report of Hyten et al. (2006) in which nucleotide diversity in *G. soja* was compared with three distinct *G. max* populations.

Molecular Genetic Diversity Within and Among Various Groups of Soybean Genotypes

Numerous studies using a variety of DNA marker types reported on the levels of molecular genetic diversity within and between populations of both cultivated and wild soybean germplasm. These reports included comparisons of (1) adapted cultivars, ancestral cultivars, and landraces, (2) cultivars and landraces from various Asian origins and (3) cultivated versus wild soybean genotypes. A number of studies in each of these categories are briefly summarized.

North American Adapted Cultivars, Ancestral Cultivars and Landraces

Based upon the analysis of 17 RFLP loci in 48 cultivated soybean genotypes including cultivars, ancestral cultivars and landraces, Keim et al. (1989) calculated Euclidean distances as a measure of diversity between individuals in the three groups. The average diversity among the landraces (0.37) was greater than that among the ancestral cultivars (0.26) which were in turn greater than that among the cultivars (0.16). Kisha et al. (1998) analyzed "gene pools" of cultivated soybean genotypes including 53 northern elite, 50 southern elite, 20 N. American ancestral cultivars, as well as 28 southern landraces (MG V-VIII) and 14 northern landraces (MG 0-IV) with 53 RFLP probes. A cluster analysis of these data generally grouped genotypes based upon their gene pool of origin although the ancestral cultivars were dispersed among the clusters. Based upon the average percent heterozygosity across all loci for each pool, the ancestral cultivar pool was determined to be the most diverse while the southern elite cultivars were the least diverse. It was concluded that more diversity was present between the northern elite, southern elite, northern landrace and southern landrace pools than within them. In the AFLP analysis by Maughan et al. (1996) a diverse set of 16 ancestral and adapted cultivars were analyzed along with 11 wild soybean genotypes. In this analysis, the ancestral and adapted cultivars clustered tightly together and separately from the wild soybean accessions. In another study involving ancestral cultivars of the N. American as well as the Chinese soybean germplasm pools, Li et al. (2001) compared the genetic diversity of 18 N. American and 32 Chinese ancestral cultivars using RAPD markers to establish the genetic relationships between the two groups of ancestors. Based upon mean genetic distance among cultivars within the N. American and Chinese ancestral groups, the N. American ancestors were determined to have a slightly lower level of genetic diversity. Cluster analyses generally separated the two gene pools. In particular, large differences were detected between the ancestors of northern U.S. and Canadian soybeans and the Chinese ancestors.

Diwan and Cregan (1997) assayed allelic variation in 35 N. American ancestral cultivars that represented 95% of the allelic variation present in North American cultivated soybean germplasm as determined via pedigree analysis (Gizlice et al. 1994). Twenty SSR loci were analyzed and an average of 10.1 alleles was detected per locus (range 5–17) with a mean gene diversity of 0.80 (range: 0.50–0.87). In an extensive study of SSR allelic diversity, Narvel et al. (2000) analyzed 39 adapted cultivars and 40 MG I-IV landraces which were selected for their yield potential in a replicated field trial. Each genotype was analyzed with 74 SSR loci distributed across the 20 consensus linkage groups and a total of 397 alleles were detected. There were 138 alleles specific to the landraces and only 32 alleles specific to the cultivars. Average gene diversity among the landraces was 0.56 and ranged from 0.0 to 0.84 while gene diversity among the adapted lines was 0.50 and ranged from 0.0 to 0.79. As would be anticipated, genetic similarity estimates based on simple matching coefficients revealed more genetic diversity among the landraces than among the cultivars.

Cultivated Soybean with Different Asian Origins

Li and Nelson (2001) used RAPD analysis with the objective of comparing genetic variation within and among 120 cultivated soybean accessions from eight Chinese and three South Korean provinces and three Japanese districts in an attempt to relate patterns of diversity to geographic origin. Of 115 polymorphic RAPD fragments, all were present in the Chinese accessions while only eight were not present in the S. Korean or Japanese accessions. Thus, divergence among the three national genepools was mainly a function of fragment frequencies. Genetic distances among genotypes ranged from 0.14 to 0.55 with a mean of 0.42. The highest genetic distances were between accessions from China versus those from Japan and S. Korea. The accessions from Japan and S. Korea had similar but much lower genetic distances than those from China. Cluster analyses generally put the Korean and Japanese genotypes together and separate from the Chinese accessions. It was concluded from these data that the S. Korean and Japanese gene pools were probably derived from a relatively few introductions from China.

In an extensive analysis of 131 *G. max* landraces and/or pureline selections from 14 Asian countries, Abe et al. (2003) assayed allelic variation at 20 SSR loci, one each from the 20 consensus linkage groups defined by Cregan et al. (1999). The landraces were primarily from China and Japan and were selected to represent the diversity of geographic regions in these two nations. Germplasm from southeast and south central Asia was also included. An extremely high level of allelic diversity was detected with an average of 11.9 alleles per locus and a mean gene diversity of 0.782. A cluster analysis separated the Japanese from the Chinese accessions and suggested their origins from different germplasm pools. Korean accessions clustered in both the Chinese and Japanese groups while the southeast and south central accessions clustered with the Chinese lines. It was concluded that the soybeans from southeast and south central Asia were derivatives of the diverse Chinese germplasm pool.

In an analysis of cultivated soybean of the seven primary ecotypes from the three Chinese production regions (Northern, Yellow River and Southern), Wang et al. (2006) analyzed 122 landraces and seven cultivars selected to represent the range of phenotypic diversity for 14 agronomic and morphological traits. Allelic diversity was determined at 60 SSR loci that were uniformly distributed across the 20 soybean linkage groups. An average of 12.2 alleles per locus was detected and gene diversity ranged from 0.5 to 0.92 with a mean of 0.78. A cluster analysis yielded five major groups, two that contained primarily Northern ecotypes, one

Yellow River ecotypes, one Southern ecotypes and one that contained both Northern and Yellow River ecotypes. The Yellow River ecotypes had the greatest allelic diversity and were present in each of the five clusters supporting the suggestion that the Yellow River is the center of diversity of Chinese soybean.

Cultivated Versus Wild Soybean

The loss of genetic diversity due to domestication of a cultivated species from its wild progenitor is a subject of interest in many crop species. The so-called "genetic bottleneck" of domestication can drastically reduce variability as a result of selection for traits such as non-shattering of seeds, loss of germination inhibition, erect growth habit, seed size and seed composition. In maize, Tenaillon et al. (2004) estimated a 38% loss of genetic variability through the domestication bottleneck from the maize progenitor Teosinte (*Zea mays* ssp. *Parviglumis*) to cultivated maize. Keim et al. (1989) provided one of the first molecular genetic comparisons of cultivated versus wild soybean using 17 RFLP markers. Based upon a Euclidean genetic distance measure they indicated that molecular diversity was least among the 18 cultivars examined and was greatest among a group of 10 *G. max* landraces. Surprisingly, diversity among eight *G. soja* lines was intermediate to the cultivars and landraces.

Powell et al. (1996) used 11 SSR loci to analyze 22 cultivated and 25 wild soybean genotypes. The cultivated genotypes included 10 genotypes, nine of which were ancestral cultivars. These were selected based on a high level of RFLP diversity. An additional 12 accessions were selected from throughout the geographical range of cultivated soybean in Asia. The *G. soja* accessions were similarly selected to represent the geographical range of the wild soybean in Asia. The diversity index, ($\hat{H}$) (Weir 1990) was used to measure variability at each locus within each set of genotypes. $\hat{H}$ was significantly greater among the *G. soja* genotypes for eight of the 11 SSR loci. The mean diversity over loci was $\hat{H} = 0.539$ and $\hat{H} = 0.830$ in *G. max* and *G. soja*, respectively [calculated from (Powell et al. 1996)]. It was concluded that the domestication of *G. max* from *G. soja* was a key factor influencing the lower level of genetic variability in cultivated soybean. Li and Nelson (2002) selected 10 wild soybean genotypes and 10 cultivars from each of four Chinese provinces. With the exception of two of the *G. max* accessions from one province, all of the cultivated soybeans were landraces or primitive varieties. DNA of each of the 80 genotypes was analyzed with a selected set of 35 previously identified RAPD primers (Thompson and Nelson 1998b) along with two additional primers which produced a total of 269 fragments, 172 of which were polymorphic. Euclidean distances (D_{ij}) were calculated between all pairs of lines. The mean distance among the *G soja* lines ($D_{ij} = 0.46$) was significantly greater than that among the *G. max* genotypes ($D_{ij} = 0.40$) indicating greater genetic diversity. RAPD analysis was also used by Xu and Gai (2003) to measure diversity in a set of 27 *G. max* landraces and 21 wild soybean genotypes from China. The cultivated accessions were selected based upon geographical distribution and seasonal type and the wild soybeans based upon geographical origin. Data were collected from 20 RAPD primers producing 177 bands, of which 66 were polymorphic and none was species specific. The mean gene diversity was 0.188 in the cultivated and 0.285 in the wild soybean indicating significantly greater genetic diversity in *G. soja*.

Kuroda et al. (2006) used one randomly selected SSR locus from each of the 20 consensus soybean linkage groups to analyze 77 *G. soja* accessions collected from across Japan as well as 53 currently grown soybean cultivars. A total of 405 alleles were detected in the wild, and 109 in the cultivated genotypes. Mean gene diversity across the 20 loci was 0.870 in the wild accessions and 0.496 in the cultivated genotypes, which was indicative of significantly less genetic variability in cultivated versus wild soybean. Another estimate of genetic variability in cultivated versus wild soybean based on DNA sequence variation in 102 randomly chosen gene fragments was reported by Hyten et al. (2006). More than 55 kbp of sequence from each of 26 wild soybean genotypes was compared with that of 52 Asian landraces. Both sets of genotypes were selected to maximize diversity based upon geographic origin. Hyten et al. (2006) reported nucleotide diversity values of $\theta = 0.00115$ and $\theta = 0.00235$ in cultivated and wild soybean, respectively.

The molecular diversity values from Powell et al. (1996), Li and Nelson (2002), Xu and Gai (2003), Kuroda et al. (2006) and Hyten et al. (2006) each provide estimates of the loss of diversity through the domestication bottleneck (Table 2.1). Estimates of the proportion of diversity retained after domestication range from 0.49 (Hyten et al. 2006) to 0.87 (Li and Nelson 2002) with a mean of 0.65. However, the 37 RAPD loci used by Li and Nelson (2002) included 35 loci that were carefully selected for high levels of molecular diversity and thus probably do not provide unbiased estimates of diversity. In the remaining studies, the loci appeared to be selected at random, although this is not completely clear in the case of Xu and Gai (2003). If the Li and Nelson (2002) estimate is excluded, the proportion of diversity retained after domestication is 0.59 which is very close to the estimates in maize.

Table 2.1 Molecular genetic diversity values reported in studies comparing cultivated and wild soybean genotypes and the ratio of genetic diversity values in *G. max* versus *G. soja* as an estimate of diversity retained through the genetic bottleneck of domestication

	Genetic diversity estimate		Proportion of diversity retained
Data source	*G. max*	*G. soja*	*G. max/G.soja*
Powell et al. (1996)	0.538	0.830	0.65
Li and Nelson (2002)	0.40	0.46	0.87
Xu and Gai (2003)	0.188	0.285	0.66
Hyten et al. (2006)	0.00115	0.00235	0.49
Kuroda et al. (2006)	0.496	0.870	0.57
Mean			0.65

The Search for Increased Genetic Diversity in Soybean Breeding

Plant breeders are continuously searching for new sources of useful genetic variation that will positively impact their breeding programs. Concerns of the lack of genetic variation in soybean resulting from the narrow genetic base of Asian introductions and many years of intense selection (National Research Council 1972) led to a number of assessments of molecular genetic variability aimed at the discovery of unique variation that could facilitate continued gains in soybean yields. Sneller et al. (1997) evaluated molecular diversity of landraces with maturities adapted to the southern U.S., elite southern cultivars and the elite northern parents of north × south elite cultivar crosses at 60 RFLP loci. Agronomic evaluations were also conducted of the landraces and the progeny from the north × south crosses. The RFLP analysis indicated that the landraces and the northern elite cultivars were genetically divergent from the southern elite cultivars and from each other. While the agronomic characteristics of many of the landraces were inferior, some genetically diverse lines (based upon genetic distances calculated from the RFLP data) with better agronomic potential were identified that might serve as sources of useful genetic variability in breeding. Likewise, the genetically divergent northern cultivars were suggested as another source of genetic variation for use in southern U.S. breeding. The similar report by Kisha et al. (1998), described earlier, used 53 RFLP loci to study genetic relationships among northern elite cultivars, southern elite cultivars, ancestral cultivars and landraces with maturities adapted to both the northern and southern U.S. Much like Sneller et al. (1997), it was concluded that northern elite cultivars would be particularly useful for providing increased diversity to the southern U.S. germplasm pool. The southern landraces were also suggested as sources of diversity that would be useful in both the northern and southern germplasm pools.

Thompson et al. (1998) used data derived from 125 RAPD primers to compare 18 ancestral cultivars with 17 maturity group I–III Asian landraces that had produced high yielding progeny in crosses with adapted N. American cultivars. Genetic distances were calculated based upon the RAPD data and the pairwise distances ranged from 0.26 to 0.67 with an average of 0.56. The averages and ranges for the two groups were similar, indicating approximately the same diversity in the two groups of genotypes. However, in cluster analyses the landraces clustered apart from the ancestral cultivars suggesting that they may be a useful source of genetic diversity to be exploited in soybean breeding. Thompson and Nelson (1998a) reported that lines derived from crosses of seven of the diverse landraces identified by Thompson et al. (1998) with adapted cultivars produced progeny that out-yielded their adapted parent. This result indicated that exotic germplasm could contribute genes to enhance yield. Brown-Guedira et al. (2000) compared patterns of genetic diversity in the same set of 18 ancestral cultivars used by Thompson et al. (1998) with 87 lines that included MG 00-IV landraces that had produced progeny with high yields. A few U.S. cultivars with uncertain parentage were also included in this group of 87 genotypes. A total of 46 RAPD markers and 3 SSRs from different linkage groups were used to characterize molecular genetic variation. Genetic distances ranged from 0.08 to 0.76 with a mean of 0.52. Cluster analyses of the distance matrix identified 11 clusters, three of which were composed almost exclusively of landraces that were distinct from the ancestral base of U.S. soybean cultivars. These accessions were considered to be of particular interest as sources of useful genetic variation for yield improvement of northern U.S. cultivars.

An AFLP analysis was used to compare the level of genetic diversity within and between 59 modern cultivars from China, 30 from Japan, 66 from N. America along with and 35 N. American ancestral cultivars (Ude et al. 2004). Genetic distance (GD) between pairs of genotypes was calculated on the basis of the similarity indices determined by the 332 AFLP fragments, 90 of which were polymorphic. Within each of the cultivar groups, the average GD between pairs of genotypes was 6.3% among the 30 Japanese cultivars, 7.1% among the 66 N. American cultivars, 7.3% among the 35 N. American ancestral cultivars and 7.5% among the 59 Chinese cultivars. The average GD between the N. American cultivars and the Chinese and Japanese cultivars was 8.5% and 8.9%, respectively. None of these distances was significantly different; however, the greater genetic distances between the N. American cultivars and those from China and Japan versus the distances among the N. American cultivars indicated that the Asian cultivars may be a useful source of genetic variation for cultivar improvement in N. America. A cluster analysis indicated that the Japanese cultivars were more removed from the N. American cultivars than were the Chinese cultivars and would probably be the better of the two sets of Asian cultivars as a source of genetic diversity for yield improvement in N. American breeding.

Molecular Genetic Diversity for Unambiguous DNA Fingerprinting

The first report of the use of DNA markers to distinguish soybean cultivars used RFLP loci. In their evaluation of RFLP loci, Apuya et al. (1988) discovered multiple locus RFLP probes that distinguished a set of five cultivars. Lorenzen and Shoemaker (1996) used between 37 and 50 RFLP loci to distinguish members of 17 "cultivar groups", where groups were defined as all accessions with a similar common name and the selections made from these cultivars. This analysis successfully identified cultivar groups that were originally a heterogeneous seed mixture such as A.K. (*A*ll *K*inds) and Manchu. In cases where no phenotypic diversity was reported between two members of a group, molecular genetic diversity was detected 40 of 44 times. These results clearly suggested the power of molecular marker technology to detect genetic differences for purposes of distinguishing phenotypically similar genotypes.

SSR Markers for DNA Fingerprinting

Diwan and Cregan (1997) examined the use of SSR loci to distinguish sets of cultivars that were phenotypically indistinguishable. A total of 10 MG I, seven MG II, 10 MG IV and 9 MG VI cultivars were identified such that cultivars within each group were indistinguishable based upon eight morphological and pigmentation traits. The cultivars within the four groups were readily distinguishable using the 20 SSR loci. Keim et al. (1989) previously identified seven cultivars that could not be distinguished using 17 RFLP probes. The seven cultivars were readily distinguished using the 20 SSR loci. Cregan and Diwan (1997) found a mean of 2.95 alleles among the seven cultivars at the 20 loci. Subsequent research by Song et al. (1999) analyzed 48 SSR loci on the set of 35 N. American ancestral cultivars used by Diwan and Cregan (1997) along with a diverse set of 66 elite N. American cultivars. Only loci in which adjacent alleles differed by at least three basepairs were maintained for further statistical analysis via a clustering procedure. A final set of 13 loci from 12 different linkage groups was identified which easily produced unique SSR allele size profiles for each of the 66 elite cultivars. The 13 loci also readily distinguished the members of the four sets of cultivars examined by Diwan and Cregan (1997) that were phenotypically identical. The set of 13 loci was proposed by Song et al. (1999) as a standard set for use in DNA profiling of soybean cultivars for purposes of Plant Variety Protection.

SNP Markers for DNA Fingerprinting

The availability of genetically mapped SNP markers in soybean (Choi et al. 2007) provides an alternative source of DNA markers for genetic fingerprinting. Yoon et al. (2007) determined the allele present at each of 58 mapped SNP loci selected from across the 20 soybean consensus linkage groups. Each was analyzed in a set of cultivars that included 16 N. American ancestral cultivars, 59 elite N. American cultivars, 21 elite Korean cultivars, as well as the same four sets of MG I, II, IV and VI cultivars examined by Diwan and Cregan (1997) that were phenotypically indistinguishable. Based upon a clustering procedure, a set of 23 informative SNPs loci was identified. The 23 loci were spread across 19 of the 20 soybean consensus linkage groups. The 23 loci very efficiently distinguished the N. American ancestral, Korean and N. American cultivars as well as the cultivars within the four sets of phenotypically identical cultivars. This set of SNP markers provide an alternative to SSR markers for the DNA fingerprinting of cultivars or other germplasm.

Linkage Disequilibrium

Linkage disequilibrium (LD) is the non-random association of alleles at two or more loci and is affected by a number of factors. These include the (1) rate of recombination with higher recombination lowering LD, (2) population subdivision or admixture which increase LD, (3) selection which increases LD in the vicinity of selected loci, and (4) mutation rate with high mutation rate decreasing overall LD but increasing LD in proximity to newly mutated loci (Rafalski and Morgante 2004). Mating system is also an important determinant of LD. In selfing species such as soybean, with high homozygosity, recombination occurs between identical haplotypes and thus does not reduce LD. LD is the basis of genetic association analysis for the discovery and fine mapping of genes or quantitative trait loci (QTL) in natural populations (Risch and Merikangas 1996). Genetic association analysis measures correlations between allelic variants and phenotypic differences in naturally occurring populations and depends on historical LD for the detection of significant associations (Flint-Garcia et al. 2003).

There are only two published estimates of LD in soybean. Zhu et al. (2003) indicated that LD significantly decayed at distances of 2.0–2.5 centiMorgans (cM) (roughly equivalent to 1.0–1.25 mbp) across a 12.5 cM region of linkage group G. In a more extensive examination of LD in four sets of soybean germplasm accessions including 26 *G. soja*, 52 landraces, 17 N. American ancestral cultivars and 25 modern cultivars, Hyten et al. (2007a) analyzed LD decay across three genome regions ranging in length from 336 to 574 kbp. In *G. soja*, LD was the least extensive and did not extend past 100 kbp; however, in the three cultivated soybean populations LD extended from 90 kb to 574 kbp. The extent of LD in the three *G. max* populations varied greatly between the three genome regions. The structure of LD was described using haplotype blocks that are consecutive loci in high LD flanked by blocks demonstrating historical recombination (Altshuler et al. 2005; Daly et al. 2001; Gabriel et al. 2002). Using common methods to define haplotype blocks (Barrett et al. 2005; Gabriel et al. 2002; Wang et al. 2002), Hyten et al. (2007a) determined that *G. soja* had haplotype blocks with an average block length of 4.8 kb/block. The largest haplotype block spanned 25 kb with the majority of blocks spanning <1 kb. The 52 landraces and the ancestral cultivar populations had similar size haplotype blocks, which were on average much larger than those of *G. soja.* The average block size in the modern cultivar population was more than twice that of any of the other populations as estimated by each of the three methods of haplotype block determination and there were only a few blocks that were <1 kb in length. Tag SNPs are defined as a subset of SNPs that capture a large fraction of the allelic variation of all SNP loci (Altshuler et al. 2005). Therefore, based upon allelic variation in the 52 landraces Hyten et al. (2007a) calculated the number of tag SNPs needed to capture 100% of alleles in the three genome regions. The estimate ranged from a SNP every 9 kb to a SNP every 51 kb. (Hymowitz 2004) estimated that the euchromatic DNA represented about 64% of the genome or approximately 705 Mb. Thus, to fully capture allelic variation in the euchromatic portion of the genome would require from 13,800 to 78,300 SNPs depending upon which of the three genomic regions is the most representative of the soybean genome.

While sequenced tagged sites containing only a few thousand SNPs have been mapped in soybean to date (Choi et al. 2007), the imminent availability of the Department of Energy, Joint Genome Institute whole genome sequence, as well as the sequence of alternative genotypes will greatly accelerate SNP discovery. In addition, the availability of high throughput SNP detection assays will expedite genotyping. A report at the Plant Animal Genome XV meeting in San Diego, CA (Hyten et al. 2007b) indicated that the Illumina Inc. GoldenGate SNP detection assay (http://www.illumina.com/pages.ilmn?ID=11) functions extremely well in soybean despite its highly duplicated genome. It is likely that the Illumina Infinium assay (http://www.illumina.com/downloads/INFINWKFLOW.pdf) which is capable of the analysis of more than 100,000 SNPs in parallel will also function in soybean. Such analysis platforms have the potential for the rapid characterization of the entire genomes of thousands of diverse soybean genotypes.

Conclusions

The characterization of the molecular genetic diversity of various sets of wild and cultivated soybean genotypes began more than 20 years ago. A number of different DNA marker systems were used in these analyses. The technical advances from Southern hybridization-based analysis to various types of PCR-based markers have increased the speed and reduced the cost of data acquisition. These analyses were undertaken to meet wide ranging objectives from simply testing the usefulness of a particular marker system to identifying exotic germplasm accessions to expand the genetic diversity of the elite germplasm pool in order to permit genetic improvement for increased soybean yield. Recent advances in high throughput DNA sequencing technology for inexpensive SNP discovery and tools for the detection of tens of thousands of SNP DNA markers in parallel suggest that in the near future there will be few limits to our ability to characterize genetic diversity in very fine detail. In the next few years it is likely that the entire USDA Soybean Germplasm Collection of more than 17,000 accessions will be characterized at each of 100,000 or more loci and that this level of characterization will define the entire haplotype variation of each accession. At that point, the major question facing soybean genomicists will be how to most effectively mine this large dataset for the genetic improvement of soybean. Furthermore, the availability of such a dataset increases the need for rapid and accurate phenotypic analysis. The analysis of molecular diversity is, and will remain important, but until extensive and accurate phenotypic data are available with which to associate genotypic diversity we will not realize the genetic progress that appears to be possible.

References

Abe, J., Xu, D.H., Suzuki, Y., Kanazawa, A., and Shimamoto, Y. (2003) Soybean germplasm pools in Asia revealed by nuclear SSRs. Theor. Appl. Genet. 106, 445–53.

Akkaya, M.S., Bhagwat, A.A., and Cregan, P.B. (1992) Length polymorphisms of simple sequence repeat DNA in soybean. Genetics 132, 1131–9.

Altshuler, D., Brooks, L.D., Chakravarti, A., Collins, F.S., Daly, M.J., and Donnelly, P. (2005) A haplotype map of the human genome. Nature 437, 1299–320.

Apuya, N.R., Frazier, B.L., Keim, P., Roth, E.J., and Lark, K.G. (1988) Restriction fragment length polymorphisms as genetic markers in soybean, *Glycine max* (L.) Merrill. Theoretical and Applied Genetics. 75, 889–901.

Barrett, J.C., Fry, B., Maller, J., and Daly, M.J. (2005) Haploview: analysis and visualization of LD and haplotype maps. Bioinformatics 21, 263–5.

Boerma, H.R., and Specht, J.E. (2004) *Soybeans: Improvement, Production, and Uses*. 3rd ed. American Society of Agronomy, Crop Science Society of America, Soil Science Society of America, Madison, WI.

Brown-Guedira, G.L., Thompson, J.A., Nelson, R.L., and Warburton, M.L. (2000) Evaluation of genetic diversity of soybean introductions and North American ancestors using RAPD and SSR markers. Crop Sci. 40, 815–823.

Carter, T.E., Nelson, R., Sneller, C.H., and Cui, Z. (2004) Genetic diversity in soybean:In: H. R. Boerma and J. E. Specht (Eds), *Soybeans: Improvement, Production, and Uses*. American Society of Agronomy, Crop Science Society of America, Soil Science Society of America, Madison, WI, pp. 303–416.

Choi, I.-Y., Hyten, D.L., Matukumalli, L.K., Song, Q.-J., Chaky, J.M., Quigley, C.V., Chase, K., Lark, K.G., Reiter, R.S., Yoon, M.-S., Hwang, E.-Y., Yi, S.-I., Young, N.D., Shoemaker, R.C., van Tassell, C.P., Specht, J.E., and Cregan, P.B. (2007) A soybean transcript map: gene distribution, haplotype and SNP analysis. Genetics 176, 685–696.

Cregan, P.B., Jarvik, T., Bush, A.L., Shoemaker, R.C., Lark, K.G., Kahler, A.L., Kaya, N., VanToai, T.T., Lohnes, D.G., and Chung, J. (1999) An integrated genetic linkage map of the soybean genome. Crop Sci. 39, 1464–1490.

Daly, M.J., Rioux, J.D., Schaffner, S.F., Hudson, T.J., and Lander, E.S. (2001) High-resolution haplotype structure in the human genome. Nat Genet 29, 229–32.

Diwan, N., and Cregan, P.B. (1997) Automated sizing of fluorescent-labeled simple sequence repeat (SSR) markers to assay genetic variation in soybean. Theor. Appl. Genet. 95, 723–733.

Feltus, F.A., Wan, J., Schulze, S.R., Estill, J.C., Jiang, N., and Paterson, A.H. (2004) An SNP resource for rice genetics and breeding based on subspecies indica and japonica genome alignments. Genome Res. 14, 1812–9.

Flint-Garcia, S.A., Thornsberry, J.M., and Buckler, E.S., IV. (2003) Structure of linkage disequilibrium in plants. Annu. Rev. Plant Biol. 54, 357–74.

Gabriel, S.B., Schaffner, S.F., Nguyen, H., Moore, J.M., Roy, J., Blumenstiel, B., Higgins, J., DeFelice, M., Lochner, A., Faggart, M., Liu-Cordero, S.N., Rotimi, C., Adeyemo, A., Cooper, R., Ward, R., Lander, E.S., Daly, M.J., and Altshuler, D. (2002) The structure of haplotype blocks in the human genome. Science 296, 2225–9.

Gizlice, Z., Carter, T.E., Jr., and Burton, J.W. (1994) Genetic base for North American public soybean cultivars released between 1947 and 1988. Crop Sci. 34, 1143–1151.

Hamblin, M.T., Mitchell, S.E., White, G.M., Gallego, J., Kukatla, R., Wing, R.A., Paterson, A.H., and Kresovich, S. (2004) Comparative population genetics of the panicoid grasses: sequence polymorphism, linkage disequilibrium and selection in a diverse sample of sorghum bicolor. Genetics 167, 471–83.

Hermann, A. 1962. A revision of genus *Glycine* and its immediate allies. USDA Tech. Bull. 1268, 1–79.

Hymowitz, T. (2004) Speciation and cytogenetics:In: H. R. Boerma and J. E. Specht (Eds), *Soybeans: Improvement, Production, and Uses*. American Society of Agronomy, Crop Science Society of America, Soil Science Society of America, Madison, Wis., pp. 97–136.

Hymowitz, T., and Newell, C.A. (1981) Taxonomy of the genus *Glycine*, domestication and uses of soybeans. Economic botany. 35, 272–288.

Hyten, D.L., Choi, I.-Y., Song, Q., Shoemaker, R.C., Nelson, R.L., Costa, J.M., Specht, J.E., and Cregan, P.B. (2007a) Highly variable patterns of linkage disequilibrium in multiple soybean populations. Genetics 175, 1937–1944.

Hyten, D.L., Song, Q., Zhu, Y., Choi, I.Y., Nelson, R.L., Costa, J.M., Specht, J.E., Shoemaker, R.C., and Cregan, P.B. (2006) Impacts of genetic bottlenecks on soybean genome diversity. Proc. Natl. Acad. Sci. U.S.A. 103, 16666–71.

Hyten, D.L., Choi, I.-Y., Yoon, M.-S., Song, Q.-J., Specht, J.E., Nelson, R.L., Chase, K., Young, N.D., Lark, K.G., Shoemaker, R.C., and Cregan, P.B. 2007b. *An Assessment of Genome-wide Linkage Disequilibrium in Soybean*. Plant & Animal Genome XV, San Diego, CA.

Kanazin, V., Talbert, H., See, D., DeCamp, P., Nevo, E., and Blake, T. (2002) Discovery and assay of single-nucleotide polymorphisms in barley (*Hordeum vulgare*). Plant Mol Biol 48, 529–37.

Keim, P., Shoemaker, R.C., and Palmer, R.G. (1989) Restriction fragment length polymorphism diversity in soybean. Theor. Appl. Genet. 77, 786–792.

Keim, P., Beavis, W., Schupp, J., and Freestone, R. (1992) Evaluation of soybean RFLP marker diversity in adapted germ plasm. Theor. Appl. Genet. 85, 205–212.

Kisha, T.J., Diers, B.W., Hoyt, J.M., and Sneller, C.H. (1998) Genetic diversity among soybean plant introductions and North American germplasm. Crop Sci. 38, 1669–1680.

Kuroda, Y., Kaga, A., Tomooka, N., and Vaughan, D.A. (2006) Population genetic structure of Japanese wild soybean (*Glycine soja*) based on microsatellite variation. Mol. Ecol. 15, 959–74.

Li, Z., and Nelson, R.L. (2001) Genetic diversity among soybean accessions from three countries measured by RAPDs. Crop Sci. 41, 1337–1347.

Li, Z., and Nelson, R.L. (2002) RAPD marker diversity among cultivated and wild soybean accessions from four Chinese provinces. Crop Sci. 42, 1737–1744.

Li, Z., Qiu, L., Thompson, J.A., Welsh, M.M., and Nelson, R.L. (2001) Molecular genetic analysis of U.S. and Chinese soybean ancestral lines. Crop Sci. 41, 1330–1336.

Lorenzen, L.L., and Shoemaker, R.C. (1996) Genetic relationships within old U.S. soybean cultivar groups. Crop Sci. 36, 743–752.

Lorenzen, L.L., Boutin, S., Young, N., Specht, J.E., and Shoemaker, R.C. (1995) Soybean pedigree analysis using map-based molecular markers. I. Tracking RFLP markers in cultivars. Crop Science. 35, 1326–1336.

Maughan, P.J., Saghai Maroof, M.A., and Buss, G.R. (1995) Microsatellite and amplified sequence length polymorphisms in cultivated and wild soybean. Genome 38, 715–723.

Maughan, P.J., Saghai-Maroof, M.A., Buss, G.R., and Huestis, G.M. (1996) Amplified fragment length polymorphism (AFLP) in soybean: species diversity, inheritance, and near-isogenic line analysis. Theor. Appl. Genet. 93, 392–401.

Morgante, M., and Olivieri, A.M. (1993) PCR-amplified microsatellites as markers in plant genetics. Plant J. 3, 175–82.

Narvel, J.M., Fehr, W.R., Chu, W.C., Grant, D., and Shoemaker, R.C. (2000) Simple sequence repeat diversity among soybean plant introductions and elite genotypes. Crop Sci. 40, 1452–1458.

National Research Council. Committee on Genctic Vulnerability of Major Crops. (1972) *Genetic Vulnerability of Major Crops* National Academy of Sciences, Washington.

Palmer, R.G., Pfeiffer, T.W., Buss, G.R., and Kilen, T.C. (2004) Qualitative genetics:In: H. R. Boerma and J. E. Specht (Eds), *Soybeans: Improvement, Production, and Uses*. American Society of Agronomy, Crop Science Society of America, Soil Science Society of America, Madison, WI, pp. 137–233.

Powell, W., Morgante, M., Doyle, J.J., McNicol, J.W., Tingey, S.V., and Rafalski, A.J. (1996) Genepool variation in genus *Glycine* subgenus *Soja* revealed by polymorphic nuclear and chloroplast microsatellites. Genetics 144, 793–803.

Rafalski, A., and Morgante, M. (2004) Corn and humans: recombination and linkage disequilibrium in two genomes of similar size. Trends Genet. 20, 103–11.

Risch, N., and Merikangas, K. (1996) The future of genetic studies of complex human diseases. Science 273, 1516–7.

Rongwen, J., Akkaya, M.S., Bhagwat, A.A., Lavi, U., and Cregan, P.B. (1995) The use of microsatellite DNA markers for soybean genotype identification. Theor. Appl. Genet. 90, 43–48.

Roth, E.J., and Lark, K.G. (1984) Isopropyl-N(3-chlorophenyl) carbamate (CIPC) induced chromosomal loss in soybean: a new tool for plant somatic cell genetics. Theoretical and Applied Genetics. 68, 421–431.

Scallon, B.J., Dickinson, C.D., and Nielsen, N.C. (1987) Characterization of a null-allele for the Gy_4 glycinin gene from soybean. Mol. Gen. Genet. 208, 107–113.

Schneider, K., Weisshaar, B., Borchardt, D.C., and Salamini, F. (2001) SNP frequency and allelic haplotype structure of Beta vulgaris expressed genes. Molecular Breeding: New Strategies in Plant Improvement. 8, 63–74.

Skorupska, H.T., Shoemaker, R.C., Warner, A., Shipe, E.R., and Bridges, W.C. (1993) Restriction fragment length polymorphism in soybean germplasm of the southern USA. Crop Sci. 33, 1169–1176.

Sneller, C.H., Miles, J.W., and Hoyt, J.M. (1997) Agronomic performance of soybean plant introductions and their genetic similarity to elite lines. Crop Sci. 37, 1595–1600.

Song, Q.-J., Choi, I.-Y., Heo, N.-K., and Kim, N.-S. (1998) Genotype fingerprinting, differentiation and association between morphological traits and SSR loci of soybean landraces. Plant Res. 1, 81–91.

Song, Q.J., Quigley, C.V., Nelson, R.L., Carter, T.E., Boerma, H.R., Strachan, J.L., and Cregan, P.B. (1999) A selected set of trinucleotide simple sequence repeat markers for soybean cultivar identification. Plant Varieties and Seeds 12, 207–220.

Tajima, F. (1983) Evolutionary relationship of DNA sequences in finite populations. Genetics 105, 437–60.

Tenaillon, M.I., U'Ren, J., Tenaillon, O., and Gaut, B.S. (2004) Selection versus demography: a multilocus investigation of the domestication process in maize. Mol. Biol. Evol. 21, 1214–25.

Thompson, J.A., and Nelson, R.L. (1998a) Utilization of diverse germplasm for soybean yield improvement. Crop Sci. 38, 1362–1368.

Thompson, J.A., and Nelson, R.L. (1998b) Core set of primers to evaluate genetic diversity in soybean. Crop Sci. 38, 1356–1362.

Thompson, J.A., Nelson, R.L., and Vodkin, L.O. (1998) Identification of diverse soybean germplasm using RAPD markers. Crop Sci. 38, 1348–1355.

Ude, G.N., Kenworthy, W.J., Costa, J.M., Cregan, P.B., and Alvernaz, J. (2004) Genetic diversity of soybean cultivars from China, Japan, North America, and North American ancestral lines determined by amplified fragment length polymorphism. Crop Sci. 43, 1858–1867.

Vos, P., Hogers, R., Bleeker, M., Reijans, M., van de Lee, T., Hornes, M., Frijters, A., Pot, J., Peleman, J., Kuiper, M., and et al. (1995) AFLP: a new technique for DNA fingerprinting. Nucleic Acids Res 23, 4407–14.

Wang, L., Guan, R., Zhangxiong, L., Chang, R., and Qui, L. (2006) Genetic diversity of Chinese cultivated soybean revealed by SSR markers. Crop Sci. 46, 1032–1038.

Wang, N., Akey, J.M., Zhang, K., Chakraborty, R., and Jin, L. (2002) Distribution of recombination crossovers and the origin of haplotype blocks: the interplay of population history, recombination, and mutation. Am J Hum Genet 71, 1227–34.

Watterson, G.A. (1975) On the number of segregating sites in genetical models without recombination. Theor. Popul. Biol. 7, 256–76.

Weir, B.S. (1990) *Genetic Data Analysis.* Sinauer Associates, Sunderland, MA.

Welsh, J., and McClelland, M. (1990) Fingerprinting genomes using PCR with arbitrary primers. Nucleic Acids Res. 18, 7213–8.

Williams, J.G., Kubelik, A.R., Livak, K.J., Rafalski, J.A., and Tingey, S.V. (1990) DNA polymorphisms amplified by arbitrary primers are useful as genetic markers. Nucleic Acids Res. 18, 6531–5.

Wright, S.I., Bi, I.V., Schroeder, S.G., Yamasaki, M., Doebley, J.F., McMullen, M.D., and Gaut, B.S. (2005) The effects of artificial selection on the maize genome. Science 308, 1310–4.

Xu, D.H., and Gai, J.Y. (2003) Genetic diversity of wild and cultivated soybeans growing in China revealed by RAPD analysis. Plant Breeding 122, 503–506.

Yoon, M.S., Song, Q.J., Choi, I.Y., Specht, J.E., Hyten, D.L., and Cregan, P.B. (2007) BARC-SoySNP23: a panel of 23 selected SNPs for soybean cultivar identification. Theor. Appl. Genet. 114, 885–99.

Zakharova, E.S., Epishin, S.M., and Vinetski, Y.P. (1989) An attempt to elucidate the origin of cultivated soybean via comparison of nucleotide sequences encoding glycinin B4 polypeptide of cultivated soybean, *Glycine max*, and its presumed wild progenitor, *Glycine soja*. Theoretical and Applied Genetics. 78 (6), 852–856.

Zhu, T., Shi, L., Doyle, J.J., and Keim, P. (1995) A single nuclear locus phylogeny of soybean based on DNA sequence. Theor. Appl. Genet. 90, 991–999.

Zhu, Y.L., Song, Q.J., Hyten, D.L., Van Tassell, C.P., Matukumalli, L.K., Grimm, D.R., Hyatt, S.M., Fickus, E.W., Young, N.D., and Cregan, P.B. (2003) Single-nucleotide polymorphisms in soybean. Genetics 163, 1123–34.

Chapter 3
Legume Comparative Genomics

Steven Cannon

Introduction

The ability to make comparisons between genome sequences will be crucial for leveraging and exchanging knowledge learned in these model systems, and applying that knowledge to a wide range of agronomically important species. Sequence comparisons are also a key tool for the evolutionary trajectories giving rise to new plant functions, structures, chemistries, and physiologies.

At the time of writing, four plant genome sequences are almost completely determined: *Arabidopsis thaliana* (At), two rice cultivars (*Oryza sativa*; Os), and *Populus trichocarpa* (Pt; black cottonwood or western balsam poplar). Genome sequencing is well underway for three legumes genomes: the model forage legumes *Medicago truncatula* (Mt) and *Lotus japonicus* (Lj), and *Glycine max* (Gm; soybean). Numerous other plant genome sequencing projects are also underway or planned, including tomato, corn, *Mimulus guttatus* (monkey flower), *Physcomitrella patens* (a moss), *Miscanthus* (switchgrass), *Citrus sinensis* (orange), Sorghum, cotton, cassava, *Brachypodium distachyon* (a model grass), and *Aquilegia formosa* (columbine).

The extent to which knowledge can be extrapolated between genomes depends in large part on this fundamental question: how do genomes change, and do they all change the same way and at roughly similar rates? This very broad question can be divided and made more specific: what are (1) the organization of genes and non-genes; (2) the mechanisms of large-scale genome change; (3) the pace of synteny loss?

Restating and elaborating these questions, (1) How similar are various genomes in organization of their component small parts: genes, regulatory regions, repetitive DNA, low-copy intergenic material, centromeric repeats, pericentromeric and telomeric sequences, etc? Do these elements behave essentially the same in all plant genomes? (2) Do all genomes change via similar mechanisms – acting in

S. Cannon
USDA-ARS, Department of Agronomy, Iowa State University, Ames, IA 50011, USA
e-mail: scannon@iastate.edu

G. Stacey (ed.), *Genetics and Genomics of Soybean*,

similar proportion? That is, do genomes in all lineages change primarily by the same mechanisms, such as polyploidy, breakages, fusions, inversions, translocations, and transposon insertions? (3) How similar are various genomes in degree of gene-order conservation (synteny) across taxa? That is, at what pace is synteny lost?

These questions are only partly orthogonal. Gene order might be generally retained between two genomes (3), with gene density differing greatly between genomes, or even in different parts of a single genome (1). The degree to which synteny is retained across various lineages (3) depends on the rates and mechanisms of genome change in these lineages (2) – though, conceivably, different mechanisms could produce similar changes in gene distribution of synteny.

This chapter will briefly review some of the key literature on these fundamental questions, with organization generally consisting of brief summaries of comparative genomic findings from other plant families, followed by a summary of results of comparisons among the legumes. As much of the early comparative genomic work was first carried out in the grasses, these will be a featured "comparison" for this paper on legume comparative genomics. The remainder of the Introduction briefly describes background information on legume sequencing projects and legume systematics.

One certainty in plant comparative genomics is that there will be exceptions and surprises as we examine more genomes. Nevertheless, a theme that emerges from comparisons so far is surprising commonality and similarity among plant genomes, even at great evolutionary distances.

Legume Genome Sequencing Strategies and Status

The international *Medicago truncatuala* (Mt) genome sequencing consortium, initiated by early funding from the Samuel Roberts Noble Foundation, and now funded by the National Science Foundation and the European Union, is scheduled to complete the euchromatic genome regions (16 chromosome arms) by the end of 2008. This project is using a clone-by-clone approach, in which bacterial artifical chromosomes (BACs), with average insert size of approximately 120 kb, are sequenced and used to extend BAC-contig tiling paths to produce increasingly large sequence contigs. Contigs are anchored and oriented using genetic markers developed from a large proportion of the BAC sequences. As of early 2007, BAC contigs and sequences covered approximately 60% of the major euchromatic regions of the Mt genome (Cannon et al. 2006 and unpublished data).

The *Lotus japonicus* (Lj) genome sequencing project is being carried out by the Kazusa DNA Research Institute in Japan. This project is also primarily using a clone-by-clone approach, sequencing transformation-competent artificial bacterial chromosomes (TACs), with average insert size of approximately 100 kb. The clone-by-clone sequence is also being augmented by a combination of whole genome shotgun (WGS) and low-coverage TAC sequencing. As of last published reports in 2006, Lj sequence coverage was also approximately 60% of the euchromatic regions of the Lj genome (Young et al. 2005; Cannon et al. 2006).

The *Glycine max* (Gm) genome is being sequenced primarily with a WGS approach, with sequence coming from a combination of random reads, paired fosmid ends, and paired BAC end sequences. Additionally, approximately 500 BACs will be sequenced to high coverage; these will be a mix of BACs selected for biological interest by members of the soybean research community, and to span gaps where necessary.

Legume Systematics and Consequences for Comparative Studies

The legume family is extremely diverse, with around 20,000 species and 700 genera, found in every terrestrial and some aquatic environments (Doyle and Luckow 2003). The majority of species are in the papilionoid subfamily, with 476 genera and about 14,000 species (Lewis et al. 2003). The Mimosoideae subfamily contains 77 genera and around 3,000 species. The remainder fall in the casalpinoideae subfamily – something of a grab-bag, of 162 genera and around 3,000 species, including diverse early-diverging legume taxa (Fig. 3.1).

This papilionoid subfamily includes the crop legumes and the major model legume species and, thus, is the taxonomic space across which much of legume comparative genomics and "translational genomics" will take place. Most papilionoid species of agronomic interest fall within one of two large clades: first, the Hologalegina clade, containing most of the temperate herbaceous legumes (thus, the colloquial shorthand "temperate herbaceous legumes"), including clovers, vetches, pea, lentil, *Medicago*, and *Lotus*; and second, the Millettioid clade, mostly consisting of tropical and subtropical species, and including common beans, soybean, and cowpea (Maddison and Schulz 1996–2006; Doyle and Luckow 2003; Doyle et al. 1997; Hu et al. 2000). Some commonly encountered genera in Hologalegina are *Vicia, Medicago, Pisum, Trifolium, Cicer, Lens, Astragalus, Wisteria, Lotus, Robinia*, and *Sesbania*. Some commonly encountered genera in the Millettioid clade are *Glycine, Phaseolus, Vigna, Erythrina* (coral bean), and *Apios americana* (groundnut), as well as some earlier-diverging clades, one with the eponymous genus *Milletia*, and the other with *Indigofera* (containing the shrub that was used to produce indigo dye). Beyond of these large clades, basal genera in the papilionoid subfamily include the "dalbergioid" clade, including numerous tropical trees (e.g. rosewood) and *Arachis* (peanut), and the "genistoid" clade, including *Lupinus* (lupine).

Fossil and molecular *dating* methods indicate that most morphological and species diversity in the legumes originated during a burst of speciation early in the Tertiary, ~ 60–50 mya (Lavin et al. 2005; Cronk et al. 2006). This is after the Cretaceous and the major extinction event that ended the "age of the dinosaurs." This early radiation means that, perhaps surprisingly, many early-diverging genera – including those in the caesalpnoideae and mimosoideae – did not originate a great deal earlier than early-diverging lineages in the papilionoidae. Lavin et al. (2005) date the genistoid clade (*Lupinus*) at ~56 mya; the dalbergioid clade (*Arachis*) at ~55 mya; the millettoid clade (*Glycine*) at ~45 mya; and the Hologalegina clade (*Medicago,*

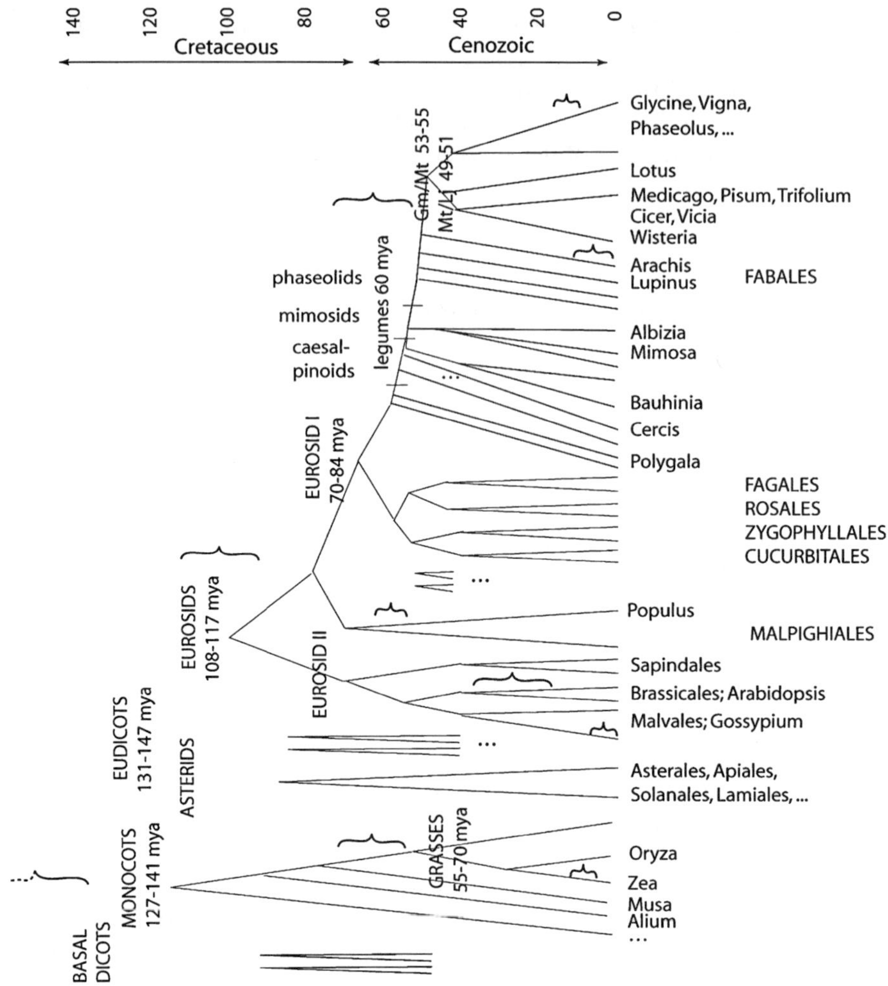

Fig. 3.1 Legume phylogeny, outgroups, and estimated WGD timeframes

Lotus) at ~51 mya. The *Glycine-Medicago* split occurred ~54 mya. And *Medicago* and *Lotus* separated early in Hologalegina, so they diverged at ~51 mya.

These dates and the likely early burst of legume speciation have important implications for comparisons between the model legumes (*Glycine, Medicago, Lotus, Phaseolus*, pea) and other agronomic species. Comparisons between soybean and *Medicago*, or between *Medicago* and *Lotus*, actually require traversing substantial evolutionary time (~50–55 million years to common ancestors). Additionally, evolutionary events that may have occurred "early" in the legumes (most prominently, nodulation or polyploidy) may actually have occurred within a relatively short evolutionary timeframe – of, say, ~10 million years in the early Cenozoic.

Polyploidy and Consequences for Genome Comparisons

Definitions and History

The terms "polyploidy" "paleopolyploidy," and "whole genome duplication" (WGD) all point to the same process of doubling of chromosomal number, and are essentially interchangeable. Over time, with rearrangements and loss of genes and chromosomal segments, the genome "diploidizes," losing most evidence of the original duplication. The terms paleopolyploidy or WGD are perhaps used more frequently to describe ancient events, describing genomes that are a long way towards a diploid state. For particularly old events, duplication remnants may be difficult to distinguish from aneuploid (partial) duplications or other causes of duplication of multiple genomic segments. In these cases, WGD is an inferred, hypothetical event.

Plant genome comparisons established that polyploidy occurred early in angiosperm evolution, and occurred numerous times independently in subsequent plant lineages. Thus, most if not all angiosperms retain remnants of several rounds of WGD. (Masterson 1994; Bowers et al. 2003; De Bodt et al. 2005; Cui et al. 2006).

Ranges for genome duplications (WGD) are shown with braces. References for WGD and clade timings are given in Table 3.1 (following page). Note early radiation in the legumes, indicated by long branch terminal lengths for many lineages (from

Table 3.1 References for estimated dates of clades (A) and genome duplications (B) References and notes in right-hand column are: (1) Tuskan et al. 2006; (2) Sanderson et al. 2004; (3) Wikstrom et al. 2001; (4) Wikstrom et al. 2003; (5) Davies et al. 2004; (6) Lavin et al. 2005; (7) Kellogg 2001; (8) Blanc et al. 2003; (9) Bowers et al. 2003; (10) Paterson et al. 2004; (11) Schlueter et al. 2004; (12) Rauscher et al. 2004; (13) extrapolated from Lavin et al. between papilionid crown and poplar/legume split; also 44–58 mya in Schlueter et al. (2004)

A. Speciations	Example	Date (mya)	Ref.
Rosid I/Rosid II	soybean/Arabidopsis	100–120	1,2
Fabaceae/Salicaceae	soybean/poplar	70–84	3,4,5
Hologalegina/Millettoid	*Medicago*/soybean	54.3 ± 0.6	6
Medicago/Lotus	*Medicago/Lotus*	50.6 ± 0.8	6
dicots/monocots	soybean/rice	131–147	2,3,4
monocot crown age	*Joinvillea* (outgroup)	127–141	3,4,5,7
eudicot crown age	*Ranunculus* (buttercup)	125–147	3,4,5
eurosid I crown age	cucumber	70–84	3,4,5
rosid crown age	*Cercis* (redbud)	108–117	3,4,5
B. Duplications		**Date (mya)**	
Brassicaceae		24–40	8,9
Salicaceae		60–65	1
Fabaceae		55–80	13
grasses		~70	10
corn		11	10
Glycine max		14.5	11
Glycine tomentella		< 50 kya	12

Lavin et al. 2005). A small number of other lineages are included for reference, including Malpighiales (with poplar), Brassicales (with *Arabidopsis*).

Polyploidy expands allelic variation and phenotypic diversity, it opens the door to functional divergence and innovation in entire metabolic and developmental pathways, and it creates a reproductive barrier and evolutionary bottleneck. Polyploidy also has effects similar to heterosis, with transient increases in measures such as stature, total dry weight, and seed size (e.g. Guo et al. 1996; Bretagnolle and Thompson 2001; Birchler et al. 2003). This effect may be due in part to gene dosage effects: every gene is immediately present in at least two copies. Polyploidy is also of interest because it complicates gene positional comparisons between related species, whereas species with a shared polyploidy history are more likely to have simple chromosomal relationships. All of these characteristics make it important to determine the history of polyploidy events in the legumes.

Prominent Angiosperm Genome Duplications

As it will frequently be helpful to make comparisons between legume and non-legume model genomes, it is important to identify plant lineages in which polyploidy has occurred. Some important WGD events are shown in Fig. 3.1, with timing estimates indicated with blue brackets. Approximate timings of speciation and WGD events are inferred from literature shown in Table 3.1.

In the monocots, all members of the grasses have undergone at least one round of polyploidy, due to an event that affected a progenitor of the grasses at ~70 mya (Paterson et al. 2004; Yu et al. 2005), before radiation of the grasses at ~55–70 mya (Kellogg 2001). Maize has undergone an additional round of polyploidy, as have some other grass lineages such as wheat. Timing of WGD in maize depends on the silcnt-site/time rate constants used, but the range is ~4.8–11 mya (Swigonova et al. 2004; Paterson et al. 2004).

In the Arabidopsis genome, remnants of three probable WGD are visible: the most recent around 24–40 mya (Blanc et al. 2003; Bowers et al. 2003).

In the poplar genome, WGD occurred near the origin of the Salicaceae, at around 60–65 mya (Tuskan et al. 2006). Interestingly, this genome is changing much more slowly than many plant lineages, likely because gametes from millenia-old clonal populations are regularly pumped into the gene pool in this genus of wind-pollinated trees (Tuskan et al. 2006). As a consequence, many internal duplications ("synteny blocks") are larger and clearer than those in the Arabidopsis genome, even though the WGD episode is estimated to have occurred significantly earlier than the WGD in Arabidopsis.

Remnants of two additional earlier WGD appear to be evident in the poplar and Arabidopsis genomes. The timings of these events is unclear, but recent studies place the middle duplication near the origin of the eurosids, around the split between Eurosid I (with legumes and poplar) and Eurosid II (with Arabidopsis and cotton) (Bowers et al. 2003; Chapman et al. 2006; Sanderson et al. 2004; De Bodt et al.

2005; Tuskan et al. 2006). If the event didn't predate this speciation, then independent events after the speciation are required.

Timing of oldest angiosperm WGD event(s) is still obscure, but likely predates the monocot-dicot split (Bowers et al. 2003; Simillion et al. 2002; Blanc et al. 2003).

Genome Duplications in the Legumes

In the legumes, most evidence points to one round of WGD very early in or shortly preceding the origin of the family. Studies in the 1990s of chromosomal correspondences within the soybean genome, using genetic marker comparisons, suggested that the soybean genome contained at least some regions present in more than two copies (Shoemaker et al. 1996; Lee et al. 2001; Yan et al. 2003).

Self-comparisons of large ESTs data sets from soybean or *Medicago* show a clear recent duplication in soybean, and somewhat weaker evidence for older duplications in soybean and *Medicago* (Schlueter et al. 2004; Blanc and Wolfe 2004). The basis for these studies is that silent-site mutations in homologous gene pairs give a distribution of changes per silent site (often called a "Ks" measurement). The older Ks peaks in soybean and *Medicago* were dated to ~44–64 mya. Schlueter et al. (2004) places a duplication event in *Medicago* at ~58 mya, consistent with ~54 mya estimated by Lavin et al. (2005). Interestingly, Schlueter et al. (2004) also place the early duplication in soybean at ~44 mya. This is significantly earlier than the (likely same) event in *Medicago*, but this might be explained by variance in the Ks peaks and/or different rates of silent-site change in the two lineages. The rate used for these calculations (and in the next paragraph) is 6.1×10^{-9} substitutions per synonymous site per year (Lynch and Conery 2000).

Using similar dating of gene pairs, but taking gene pairs from internal synteny blocks, Mudge et al. (2005) estimated the duplication in *Medicago* occurred at ~64 mya (0.79 synonymous substitutions per site), compared with and *Glycine/Medicago* speciation at 48–50 mya. Similarly, Cannon et al. (2006) found a peak at 0.80 synonymous substitutions per site, corresponding to ~65 mya, using a whole-genome comparison of *Medicago* to itself. This is significantly before the split with *Lotus*, estimated in the same study at ~51 mya (0.64 synonymous substitutions per site) – consistent with the range 50.6 ± 0.9 mya in Lavin et al. (2005). A duplication date of ~65 mya would place the duplication before the ~60 mya origin of the legumes proposed by Lavin et al. (2005) – though all of these molecular rate conversions need to be treated cautiously, as silent-site variation may not always be entirely silent, and rates should not be assumed to be the same in different lineages.

Analyses of the relative phylogenetic positions of genes from several species from a gene family can also be used to determine the relative timings of speciations and duplications. Pfeil et al. (2005) used this approach to establish that an early legume duplication predated the *Medicago/Glycine* split. Using this approach and a comparison of synteny in and between *Medicago* and *Lotus*, Cannon et al. (2006)

confirm that the early legume duplication occurred after the split with poplar and well before the split between *Medicago* and *Lotus*. This brackets the early legume duplication between about 55 and 84 mya. The upper limit (55 mya) is bounded by the estimated divergence time between soybean and *Medicago* (Lavin et al. 2005), and the lower limit (84 mya) is bounded by the split between Fabales (legumes) and Malpighiales (including poplar) (Sanderson et al. 2004).

It should be said that some aspects of the nature and timing of the "early legume duplication" remain unclear. Timing of the Ks frequency peaks is uncertain because the peaks are broad, and we lack molecular clock calibrations for most lineages. In the synteny comparisons of Mt and Lj, the genome self-comparisons (Mt × Mt or Lj × Lj) show extensive fragmentation and loss (absence of synteny) (Cannon et al. 2006 and unpublished results). Only one region of synteny in either of the self-comparisons extends beyond about a megabase; this is between Mt chromosomes 5 and 8, and between corresponding Lj chromosomes 4 and 2. The remaining synteny blocks are scattered and small, occurring on 31 of 36 possible Mt × Mt chromosome pairings (Cannon et al. 2006 and unpublished results). Some of the absence of synteny is undoubtedly explained by the incomplete state of both genome sequences, but the synteny fragmentation is more extensive than is seen in genome self-comparisons of poplar, Arabidopsis, or rice in simulations with similar amounts of loss from these genomes (Wang and Young, unpublished results).

Much of the fragmentation evident in genome self-comparisons may also be due to the large amount of time that has elapsed since this event. If the duplication did occur ~65 mya, as implied by Mudge et al. (2005) and Cannon et al. (2006), this would place it before the origin of the legumes, making it possibly twice as ancient as the most recent duplication in Arabidopsis (at 24–40 mya; see references above).

Taken together, the Ks, phylogenetic, and synteny evaluations do appear to point to an event that affected the majority of the genome, early in or preceding the origin of the legume family. However, a clearer picture of the nature and timing of the early duplication and its aftermath will require completion of the genome sequences, and probably additional partial sequencing from earlier-diverging taxa to better pinpoint the timing of the event.

In addition to the early legume genomic duplication, polyploidy has occurred in several legume lineages. Polyploidy has occurred several times in *Arachis* (peanut, possibly in the course of domestication by early agriculturists (Kochert et al. 1996; Moretzsohn et al. 2004). Polyploidy has also occurred early in the *Glycine* genus (Schlueter et al. 2004; Shoemaker et al. 1996; Lee et al. 1999, 2001; Yan et al. 2003). Additionally, *Glycine tomentella*, a perennial Australian complex of several diploid and allotetraploid "races", has repeatedly undergone polyploidy, ongoing in historical time (Rauscher et al. 2004; Doyle et al. 2004).

As with the grass genomes, polyploidy adds a complication to comparative studies. In *Glycine*, after two rounds of duplication, as many as four homoeologous genomic segments should be expected relative to a corresponding genomic segment from outside the legumes. For comparisons to other plants that have undergone their own independent duplications, such as poplar (Tuskan et al. 2006) or

Arabidopsis (The The Arabidopsis Genome Initiative 2000), the relationships are further complicated: four soybean homoeologs could correspond equally to two poplar homoeologs (and to two more distant poplar homoeologs originating from an earlier WGD). Similarly, four soybean homoeologs could correspond equally to at least four Arabidopsis homoeologs, since Arabidopsis has undergone two rounds of WGD since separation from the Rosid I clade (the clade containing the legumes and poplar) (Tuskan et al. 2006 and unpublished data).

The kinds of many-to-many relationships predicted by two rounds of duplications in the legumes and independent duplications in ancestors of poplar and Arabidopsis are, in fact, seen in such comparisons. Several of these are described below in the context of microsynteny studies.

Synteny

Definitions and History

The term "synteny" was coined to describe genes on the same chromosome (regardless of whether they show linkage in classical tests for recombination). More frequently now, the term is used to indicate conserved gene order between chromosomal regions (either between species or within a duplicated region of one genome).

The paper by Gale and Devos (1998) describing conservation among nine genomes in the grasses provided a conceptual model of the grasses as a coherent genetic system (also reviewed in Freeling 2001; Devos 2005). The key observation – drawing on the work of several groups over the preceding decade – was that large chromosomal blocks have been conserved across most the diverse grass species in the study. Most of the DNA across these genomes could be described in terms of 25 "rice linkage blocks," or portions of the rice genome that retained homologs in the comparison genomes. Some of these blocks are large, corresponding to whole chromosomes in several of the species. In the paper's "circle diagram," showing chromosomal correspondences across seven species, most of rice chromosome 1, for example, is homologous to millet V, sugar cane II and III, Sorghum LG G, parts of maize 3 and 8, and most of oat LG C.At greater evolutionary distances, numerous small and degraded synteny blocks can still be observed, for example, between tomato and Arabidopsis (Ku et al. 2000).

Macrosynteny in the Legumes

Through the 1990s, it was unclear whether synteny was as extensive within the legumes as in the grasses. With the summary by Choi et al. (2004) of more than a decade worth of synteny and comparative marker studies in the legumes, it became clear that synteny extended across broad swaths of diverse species in at least the

Papilionoid subfamily of the legumes (the subfamily containing the majority of legume species, and nearly all of the agronomically important species).

Comparisons between draft sequence from the Mt and Lj genome sequencing projects shows conservation of regions nearly to the length of entire chromosomes in some cases – for example, between Mt 1 and Lj 5 or Mt 2 and Lj 6 (Cannon et al. 2006). However, interestingly, this study finds very little synteny between Mt 6 and any Lj chromosome. The Mt 6 also has an unusually high transposon density, with several density peaks across the chromosome, perhaps indicating knobs such as seen on Arabidopsis chromosome 4. The Mt chromosome 6 also includes an unusually large cluster of genes from the TIR-NBS-LRR disease resistance family. The cause of these unusual patterns is not yet known – whether, for example, the chromosome is new, or is not removing transposons because of suppressed recombination, or is hypermethylated, as is the knob on Arabidopsis chromosome 4 (Gendrel et al. 2002).

The Mt-Lj comparison also confirms the lack of WGD in either lineage following their split at ~40 mya, but does help confirm an earlier WGD event (described further below).

In both the grasses and the legumes, despite extensive conservation for many regions, there are exceptions to these simple mappings. In particular species, some genomic regions have undergone extensive and rapid rearrangement (e.g. Mt 6 or knob on Arabidopsis chromosome 4); polyploidy appears to have occurred early in the legumes; some taxa experienced at least on additional round of polyploidy; and the genomes of some species expanded many-fold through transposon activity.

Effect of Polyploidy on Synteny Comparisons

One factor significantly complicating synteny studies between genomes that have undergone WGD is that polyploidy generates multiple corresponding regions, and appears to spur rearrangement and segmental chromosomal losses (Song et al. 1995; Pontes et al. 2004; Adams and Wendel 2005; Comai et al. 2003).

This pattern of gene loss from duplicated segments is one that Mike Freeling described in terms of "fractionation and consolidation" (Freeling 2001; Langham et al. 2004). Following duplication, selection is initially relaxed for any given gene in a pair. With stochastic loss of one or another of the duplicated genes or regulatory regions, a part ("fraction") of the original complement of genes remains on one homoeolog, and others remain on the other homoeolog. The approximate original complement can be inferred by "consolidating" the genes from the two homoeologous regions.

The problem presented by fractionation following polyploidy is that over time, it would be possible to lose all homology from two regions that nevertheless together contain all the genes from the ancestral region. This high level of fractionation occurred for the homoeologous regions around the loci for maize liguleless2 (*lg2*) and its genomic duplicate, liguleless related sequence1 (*lrs1*) (Langham et al. 2004). Together, these regions contain 13 genes, with only *lg2* and *lrs1* remaining as duplicated genes. However, 12 of the 13 genes can be found in a corresponding region in rice (Langham et al. 2004). Only by constructing the ancestral gene state is it possible to see the synteny in between the homoeologous regions. A method for inferring such ancestral states is described by Odland et al. (2006). Odland et al. compare rice homoeologous regions (separated ~70 mya), "collapsing" them into simulated ancestral chromosome blocks, and then use these ancestral blocks to make more robust comparisons to collinear maize sequence-based genetic markers.

Synteny studies can provide a means of testing hypothesized genome duplication histories. A study of a 10-cM region from soybean linkage group G homoeologous regions (on linkage group D2 and at least one other linkage group) finds correspondences to six regions in the Arabidopsis genome (Foster-Hartnett et al. 2002). The regions of correspondence probably are short, from several genes up to ~2 Mbp in Arabidopsis (the paper compared sampled sequences from soybean, so has lower resolution than complete sequence will provide). The two longest corresponding regions in Arabidopsis come from the largest contiguous internal Arabidopsis duplication fragment, between Arabidopsis chromosomes 2 and 3. The most likely explanation for this pattern of correspondences is one of at least two independent rounds of polyploidy in the Arabidopsis and soybean lineages, followed by selective gene loss from all resulting regions. A comparison of ~1 Mbp of genomic sequence from two regions in soybean to *Medicago* and Arabidopsis showed similar patterns: the soybean region(s) matching one to two *Medicago* regions and two to four Arabidopsis regions (Mudge et al. 2005). Another study shows correspondences between *Lotus, Medicago*, poplar, and Arabidopsis in regions containing the *Lotus* SYMRK and *Medicago* NORK receptor kinase genes (Kevei et al. 2005). In this region, Kevei et al. find synteny with four regions in Arabidopsis and three in poplar. As with the regions described by Foster-Hartnett et al. (2002) and Mudge et al. (2005), synteny is interrupted between any two regions by interspersed gene losses or local duplications.

Not all classes of genes respond to polyploidy in the same way. In some gene families, such as transcription factors, most genes are retained, whereas other gene families (such as those involved in defense recognition) undergo rapid turnover (Cannon et al. 2004). In a more comprehensive analysis of Arabidopsis genes, Maere et al. (2005) argue that three whole-genome duplications in that genome were directly responsible for >90% of the increase in transcription factors, signal transducers, and developmental genes in the last 350 million years. Chapman et al. (2006) propose that genes retained after polyploidy may buffer critical functions, and further, that gradual loss of this buffering capacity of duplicated genes may contribute to the cyclicality of genome duplication over time.

Microsynteny in the Legumes

Substantial microsynteny is seen in comparisons between *Medicago* and *Lotus* (Choi et al. 2004, 2006; Zhu et al. 2006; Kevei et al. 2005; Cannon et al. 2006), between *Medicago* and *Glycine* (Mudge et al. 2005), and between *Lotus* and *Glycine* (Hwang et al. 2006).

Quantifying synteny is complicated by tandem duplications and by gene-calling parameters and accuracy. For example, inclusion of coding sequences from transposons would decrease apparent synteny, as would counting of differential tandem expansions in one region vs. the other.

Defining "synteny quality" as twice the number of gene matches divided by the total number of genes in both segments (after excluding transposable elements and collapsing tandem duplications), Cannon et al. (2006) report that "synteny quality" for Mt × Lj is 62% for an extended syntenic block (58/94 of genes exhibit corresponding homologs within these regions). This region is shown in Fig. 3.2. This figure also shows homoeologous segments from Mt and Lj self-comparisons. Synteny quality in the Mt × Mt comparison is just 36%, and is 30% in the Lj × Lj region. The synteny in the self-comparisons of either the Mt or Lj is highly degraded, consistent with a history of very early polyploidy. The synteny seen within *G. max* was variable in the regions examined by Schlueter et al. (2007), but at the high ends, was far less degraded than between any duplications within Mt or Lj. Again, this would be consistent with polyploidy in *G. max* much more recent than in the ancestral legume duplication.

Mudge et al. (2005) made a comparison of several corresponding regions in *Medicago*, soybean, and Arabidopsis. The comparison illustrates several important points. In one synteny comparison spanning ~400 kb from each of two homoeologous regions in *Medicago* and a corresponding soybean region, phylogenetic analysis of each gene in the region shows one of the *Medicago* homoeologs is more closely related to the soybean region (in other words, is separated by speciation and so is orthologous); and the other *Medicago* region separated much earlier and is paralogous to both *Medicago* and soybean regions. This clearly fits a model of an early legume polyploidy, significantly predating the soybean-*Medicago* split. Extent of microsynteny is consistent with this model. Synteny quality between the soybean and *Medicago* orthologous regions is 60% but between the soybean and *Medicago* paralogous regions is 27%. And between the *Medicago* homoeologs, the synteny

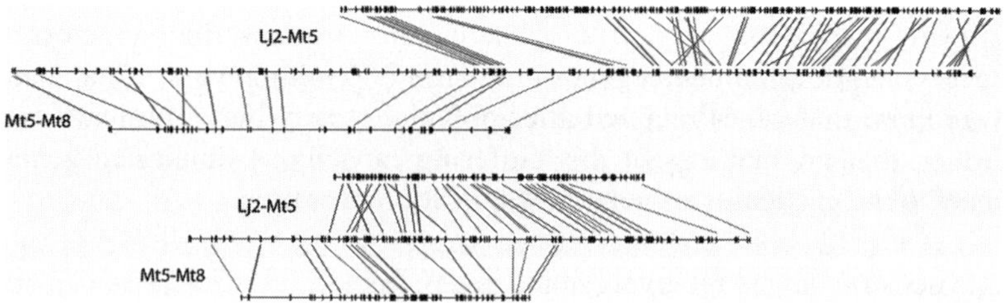

Fig. 3.2 Synteny in selected chromosomal regions between *Medicago* (Mt) and *Lotus* (Lj), and within a duplication in Mt compared with itself. *Top pair*: Lj 2 × Mt 5; *second pair*: Mt 5 × Mt 8; *third pair*: Lj 2 × Mt 5 (another region, within 1 Mbp of first regions); *last pair*: Mt 5 × Mt 8 (also within 1 Mbp of first regions). Note higher densities of collinear genes in the Mt × Lj comparison than in the internal genome duplication. Figure is adapted from Cannon et al. (2006)

quality is only 18% (only nine genes shared of approximately 50 in either *Medicago* homoeolog).

It may be significant in the Mudge et al. study that the synteny quality is lower in the *Medicago* paralogous regions than in the *Medicago-Glycine* homoeolog. Within a single genome, selection pressure should be lowered for duplicated genes, so rate of loss of either gene should be higher than in the stochastic, independent losses between two different genomes. An important conclusion is that internal synteny, remaining after polyploidy, may be more difficult to detect than synteny between two species at separated by a similar amount of time. For example, we might expect much clearer synteny between *Glycine* and *Phaseolus* than between two *Glycine* homoeologs. Relatedly, synteny may turn out to be cleaner between *Medicago* and *Lotus* than between *Medicago* and *Glycine* (which has undergone polyploidy), even though all three species diverged in similar time frames (~40–50 mya).

Genome Sizes, Gene–Space Organization, and Consequences

An important question for genome comparisons – and for genome studies generally – is the nature of the enormous variation in genome sizes and, relatedly, the nature of gene organization within genomes.

It is now generally accepted that most genome size variation (besides the effect of polyploidy) is explained by expansions of transposons (general: Bennetzen et al. 2005; corn: Du et al. 2006; wheat: Devos et al. 2005; Vitte and Bennetzen 2006; pea: Jing et al. 2005; Vicia: Neumann et al. 2006). Both the grasses and the legumes contain species differing more than 40-fold in genome size. Legume genomes range from *Leucaena macrophylla* (299 Mbp) to *Vicia faba* (13,059 Mbp); and grass genomes range from *Oropetium thomaeum* (245 Mbp) to *Triticum aestivum* (~16,979 Mbp) (Kew C-values database, Bennett and Leitch 2004). The nature of gene distribution in large genomes will strongly affect the ability to sequence large genomes, and to exchange information between them. This matters in the legumes, as the genomes of several agronomically important species are large. Soybean is ~1,103 Mbp, pea is ~4,778 Mbp, and *Vicia faba* is 13,059 Mbp (Bennett and Leitch 2004).

Less well understood is how genes, repetitive sequences, and other DNA are organized genome-wide. What is the range of variation in gene organization in euchromatic regions? What is the range of variation in organization of centromeres, pericentromeres, telomeres, and euchromatin?

In terms of organization in euchromatic regions, accumulating evidence suggests that gene density may be relatively homogeneous across very large regions in most plant genomes (or at least in the diverse genomes under active study) – though can vary enormously between genomes. This contrasts with a "gene islands" model described in several influential papers at the end of the 1990s, which suggested that in large genomes, genes might nevertheless be located in a relatively small "gene-space" – for example as "gene islands in [a] great sea of maize repetitive

DNAs" (SanMiguel et al. 1998; Bennetzen et al. 1998). A concise statement of this model is that "complex cereal genomes are largely composed of small gene-rich regions intermixed with 5–200 kb blocks of repetitive DNA" (Yuan et al. 2002).

In maize, the comparison of the largest contiguous sequence available to-date suggests that the "island" model was, in fact, not an apt description. In a comparison of 7.8 Mbp and 6.6 Mbp from corresponding (homoeologous) regions of the corn genome and 4.9 Mbp from the corresponding region of rice, Bruggmann et al. (2006) report relatively homogeneous gene density within each region, but large differences between homoeologs. They report "Analysis of these two large regions does not reveal evidence of large gene islands separated by retrotransposon blocks. As previously reported, most gene islands are small (one to two genes; Bennetzen et al. 2005) and vary between the different homoeologous regions. A picture is emerging in which different chromosomal regions evolve into a mosaic of syntenic blocks with differential expansion caused by the contraction of genic and intergenic space in combination with the addition of different combinations of repeat elements." (Bruggmann et al. 2006). While one of the maize homoeologs is 95% the size of the rice segment, the other maize homoeolog has expanded over most of its length, to 299% the size of the rice segment. The expansion is due to expansion by factors of 1.2–1.4 in genic regions (UTRs, exons, introns), and 2.3–2.7 for the remaining space (nongenic and repetitive sequence) (Bruggmann et al. 2006). In wheat, with a genome six times larger than corn's, early indications also point to genes widely dispersed rather than in gene-rich "islands." In four sequenced wheat BACs, corresponding to gene-rich regions in rice, each wheat BAC contained one-two genes, with a gene density of ~1 gene per 75 kb.

What is the range of variation in organization of centromeres, pericentromeres, telomeres, and euchromatin? In sequenced genomes to-date, there is a gradual transition on nearly all chromosomes from gene-rich euchromatic regions to multi-megabase, gene-poor, transposn-rich "pericentromeric" regions, and finally to tandem arrays of many "satellite repeats" of ~180 bp. However, the sizes of these regions (both pericentromeric and satellite repeats) vary greatly (reviewed in Ma et al. 2007; Hall et al. 2004; Lam et al. 2004).

Sizes of the satellite repeat regions ranges from 0.4 to 1.4 Mbp in Arabidopsis chromosomes, and 60 kb–1.9 Mbp in rice (Ma et al. 2007). Sizes of these regions vary even within ecotypes in a species, as shown in Arabidopsis, maize, rice, and *Medicago* (reviewed in Ma et al. 2007). In Mt, measurable genome size differences in accessions Jemalong A17 and R108-1 are due, at least in part, to much shorter satellite repeat regions in R108-1 (Kulikova et al. 2004). In Mt Jemalong A17 (the variety being sequenced), three types of satellite repeats were identified: MtR1 and MtR2, which are found in pericentromeric regions of all chromosomes, and MtR3 which is suspected to be the functional centromere domain, varying in size from ~450 bp to more than 1 Mbp (Kulikova et al. 2004). Together, these three types of satellite repeats are estimated to comprise ~6.5–8% of the Mt Jemalong A17 genome. Although these core satellite repeat regions are generally expected to contain few or no genes, there may be exceptions, as in the ~45 transcribed genes in the core centromeric region of rice chromosome 8 (Nagaki et al. 2004).

The sizes of pericentromeric regions also vary significantly between species and between chromosomes within a genome. In Arabidopsis, roughly 93% of the genome was characterized as euchromatic (Koornneef et al. 2003). However, pericentromere boundaries are not clear, and are probably better characterized as "pericentromeric gradients," with gene densities declining and transposon densities increasing (particularly class I LTR transposons) over spans of ~2–5 Mbp on approaches to centromeres (The Arabidopsis Genome Initiative 2000). Thus, depending on definition of pericentromeric border, roughly 25 Mbp (20%) of the Arabidopsis genome could be considered pericentromeric. In poplar, approximately 70% of the genome is primarily euchromatic, with most of the remainder being in pericentromeric gradients (Tuskan et al. 2006). Similarly, in rice, regions of ~2–10 Mbp are in pericentromeric gradients (Yu et al. 2005, e.g. Fig. S5).

The nature of the pericentromere will be of critical importance for every plant genome sequencing project and for comparative genomic work. Sequencing and genome assembly are made difficult by the high repeat content in pericentromeric heterochromatin, and gene content is low. Similarly, genome comparisons in these regions are complicated by large numbers of transposon sequences and high rates of turnover in transposons. There are a number of interesting unanswered questions. How frequently do centromere locations change? What genes tolerate location in pericentromeric heterochromatin? What range of gene densities exist in pericentromeric regions? Further sequencing in *Medicago, Lotus*, tomato, soybean and numerous other plant genomes will help answer these and many other unanswered questions.

Conclusions

Comparative genomics will be crucial for translating knowledge between model species, and between models and crop species. The transfer won't just be a matter of research efficiency, but will be important for making best use of the enormously inventive germplasm and phenotypic variation across the legumes. There is a great deal of benefit in considering the legumes as a coherent, broad genetic system. This concept was stated succinctly in a report on the 2004 meeting on the "Legume Crops Genome Initiative" (LCGI): "Cross-legume genomics seeks to advance: (1) knowledge about the legume family as a whole; (2) understanding about the evolutionary origin of legume-characteristic features such as rhizobial symbiosis, flower and fruit development, and its nitrogen economy; and (3) pooling of genomic resources across legume species to address issues of scientific, agronomic, environmental, and societal importance." (Gepts et al. 2005). What is the "model" will depend on the trait being studied. For example, it will make more sense to study oil biosynthesis in soybean, cold-tolerance in alfalfa, and phosphate uptake in lupin. Lessons learned in any of these "models" will be applicable, however, in the other species.

Comparative genomic techniques will not be useful solely as a means of positional cloning or gene-finding in related species. The techniques have the capacity to elucidate how traits have evolved and continue to evolve. For example, it is only by comparing nodulation in diverse species that we will learn how this important trait originated, whether once or several times, using what existing molecular machinery; etc. Similarly, comparisons will show how the diversity – and capacity for change – in defense response mechanisms. The same can be said of a very large number of traits: flavanoid biosynthesis, perenniality, nutrient uptake; etc.

Although we have a great deal to learn about comparative structural genomics, some tentative general conclusions can be drawn about the three questions posed in the introduction. First: what is the organization of genes and non-genes? In genomes sequenced to-date, gene organization has been essentially similar: locally relatively homogeneous gene densities across most parts of most euchromatic arms, and declining gradually on approach to the centromere. However, some regions break this pattern – for example, knobs in Arabidopsis, or transposon-dense chromosome 6 in *Medicago*. Second, what are the mechanisms of large-scale genome change? Undoubtedly, this question will be answered negatively in interesting, particular ways. Broadly, though, it appears that all lineages do change primarily by the same mechanisms, such as polyploidy, breakages, fusions, inversions, translocations, and transposon insertions. However, not all lineages experience these effects in similar dose. Some lineages experience multi-fold expansion due to transposon activation, some undergo more episodes of polyploidy, and some have experienced greater levels of rearrangement. Third, what is the pace of synteny loss? Clearly, rates of rearrangement differ in various lineages (e.g. slower in poplar, more rapidly in Arabidopsis), but detectable macrosynteny remains over the range of 100 million years. However, microsynteny is strongly affected by fractionation following large-scale genomic duplications. Within a duplicated region, many single gene pairs may be degraded.

Comparisons of genomes sequenced to-date show both remarkable conservation and change. It is amazing to realize that through phylogenetic comparisons, or structural-genomic comparisons, it is possible to see the trace of events that occurred in early angiosperm evolultion, or earlier – in effect, "genome archaeology." At the same time, it is clear that genomes are dynamic, creative, rapidly-changing environments in particular ways. By understanding processes such as polyploidy, local gene duplications and losses, rearrangements, and rapid turnover of repetitive elements, we have the opportunity to see plant biology more clearly, both distant-past and present.

References

Adams, K.L., and Wendel, J.F. (2005) Polyploidy and genome evolution in plants: Genome studies and molecular genetics. Curr. Opin. Plant Biol. 8:135–141.

Arabidopsis Genome Initiative (2000) Analysis of the genome sequence of the flowering plant *Arabidopsis thaliana*. Nature 408(6814):796–815.

Bennett, M.D. and Leitch I.J. (2004) Angiosperm DNA C-values database (release 5.0, Dec. 2004) http://www.rbgkew.org.uk/cval/homepage.html

Bennetzen, J.L., SanMiguel P., Chen, M., Tikhonov, A., Francki, M., Avramova, Z. (1998) Grass genomes. Proc. Natl. Acad. Sci. USA 95:1975–1978

Bennetzen, J.L., Ma, J., Devos, K.M. (2005) Mechanisms of recent genome size variation in flowering plants. An. Bot. 95:127–132.

Birchler, J.A., Auger, D.L., and Riddle, N.C. (2003) In search of a molecular basis of heterosis. Plant Cell 15: 2236–2239.

Blanc, G., Hokamp, K.H., and Wolfe, K.H. (2003) A recent polyploidy superimposed on older large-scale duplications in the Arabidopsis genome. Genome Res. 13,137–144.

Blanc, G., Wolfe, K.H. (2004) Widespread paleopolyploidy in model plant species inferred from age distributions of duplicate genes. Plant Cell 16:1667–1678.

Bowers J.E., Chapman B.A., Rong J., and Paterson A.H. (2003) Unravelling angiosperm genome evolution by phylogenetic analysis of chromosomal duplication events. Nature 422: 433–436.

Bretagnolle, F. and Thompson, J.D. (2001) Phenotypic plasticity in sympatric diploid and autotetraploid *Dactylis glomerata*. Int. J. Plant Sci. 162: 309–316.

Bruggmann, R., Bharti, A.K., Gundlach, H., Lai, J., Young, S., Pontaroli, A.C., Wei, F., Haberer, G., Galina, F., Du, C., Raymond, C., Estep, M.C., Liu, R., Bennetzen, J.L., Chan, A.P., Rabinowicz, P.D., Quackenbush, J., Barbazuk, W.B., Wing, R.A., Birren, B., Nusbaum, C., Rounsley, S., Mayere, K.F.X., Messing, J. (2006) Uneven chromosome contraction and expansion in the maize genome. Genome Res. 16:1241–1251.

Cannon, S.B., Sterck, L., Rombauts, S., Sato, S., Cheung, F., Gouzy, J.P., Wang, X., Mudge, J., Vasdewani, J., Scheix, T., Spannagl, M., Nicholson, C., Humphray, S.J., Schoof, H., Mayer, K.F.X., Rogers, J., Quetier, F., Oldroyd, G.E., Debelle, F., Cook, D.R., Retzel, E.F., Roe, B.A., Town, C.D., Tabata, S., Van de Peer, Y., and Young, N.D. (2006) Legume genome evolution viewed through the *Medicago truncatula* and *Lotus japonicus* genomes. Proc. Natl. Acad. Sci. USA 103(40):14959–64.

Cannon, S.B., Mitra, A., Baumgarten, A., Young, N.D., May, G. (2004) The roles of segmental and tandem gene duplication in the evolution of large gene families in Arabidopsis thaliana. BMC Plant Biol. 4:10

Chapman B.A., Bowers J.E., Feltus F.A., and Paterson A.H. (2006) Buffering crucial functions by paleologous duplicated genes may contribute to cyclicality to angiosperm genome duplication. Proc. Natl. Acad. Sci. USA. 103(8):2730–2735.

Choi, H-K., Mun, J-H., Kim, D-J., Zhu, H., Baek, J-M., Mudge, J., Roe, B.A., Ellis, N., Doyle, J., Kiss, G.B., Young, N.D., Cook, D.R. (2004) Estimating genome conserevation between crop and model legume species. Proc. Natl. Acad. Sci. USA. 101(45):15289–15294.

Choi, H-K., Luckow, M.A., Doyle, J.J., and Cook, D.R. (2006) Development of nuclear gene-derived markers linked to legume genetic maps. Mol. Genet. Genom. 276(1):56–70.

Comai, L., Madlung, A., Josefsson, C., and Tyagi, A. (2003) Do the different parental 'heteromes' cause genomic shock in newly formed allopolyploids? Philos. Trans Royal Soc. 358: 1149–55.

Cui, L., Wall, P.K., Leebens-Mack, J.H., Lindsay B.G., Soltis, D.E., Doyle, J.J., Soltis, P.S., Carlson, J.E., Arumuganathan, K., Barakat, A., Albert, V.A., Ma, H., and dePamphilis, C.W. (2006) Widespread genome duplications throughout the history of flowering plants. Genome Res. 16: 738–749.

Cronk Q., Ojeda I., and Pennington R.T. (2006) Legume comparative genomics: progress in phylogenetics and phylogenomics. Curr. Opin. Plant Biol. 9(2):99–103.

Davies, T.J., Barraclough, T.G., Chase, M.W., Soltis, P.S., Soltis, D.E., Savolainen, V. (2004) Darwin's abominable mystery: insights from a supertree of the angiosperms. Proc. Natl. Acad. Sci. USA 101: 1904–1909.

De Bodt S., Maere S., and Van de Peer Y. (2005) Genome duplication and the origin of angiosperms. Trends Ecol. Evol. 20(11):592–597.

Devos, K.M. (2005) Updating the 'crop circle'. Curr. Opin. Plant Biol. Apr;8(2):155–62.

Devos, K.M., Ma J., Pontaroli A.C., Pratt L.H., and Bennetzen J.L. (2005) Analysis and mapping of randomly chosen bacterial artificial chromosome clones from hexaploid bread wheat. Proc. Natl. Acad. Sci. U S A. 2005 102(52):19243–8

Doyle, J.J. and Luckow, M.A. (2003) The rest of the iceberg. Legume diversity and evolution in a phylogenetic context. Plant Physiology 131:900–910.

Doyle, J.J., Doyle, J.L., Rauscher, J.T., and Brown, A.H.D. 2004. Evolution of the perennial soybean polyploid complex (*Glycine* subgenus *Glycine*): A study of contrasts. Biol. Journal Linnean Soc. 82:583–597.

Doyle, J.J., Doyle, J.L., Ballenger, J.A., Dickson, E.E., Kajita, T., and Ohashi, H. (1997) A phylogeny of the chloroplast gene *rbcL* in the Leguminosae: taxonomic correlations and insights into the evolution of nodulation. Am. J. Bot. 84: 541–554.

Du, G., Swigonova, Z., and Messing, J. (2006) Retrotranspositions in orthologous regions of closely related grass species. BMC Evol Biol 6:62

Foster-Hartnett, D., Mudge, J., Larsen, D., Danesh, D., Yan, H., Denny, R., Penuela, S., and Young, N.D. (2002) Comparative genomic analysis of sequences sampled from a small region on soybean (*Glycine* max) molecular linkage group G. Genome. 45(4):634–45.

Freeling, M., (2001) Grasses as a single Genetic System: Reassessment 2001. Plant Physiol. 125:1191–1197.

Gale, M.D., Devos, K.M. (1998) Comparative genetics in the grasses. Proc. Natl. Acad. Sci. USA 95:1971–1974.

Gendrel, A-V., Lippman, Z.Z., Yordan, C.C., Colot, V., and Martienssen, R. (2002) Heterochromatic histone H3 methylation patterns depend on the Arabidopsis gene *DDM1*. Science 297: 1871–1873.

Gepts, P., Beavis, W.D., Brummer, E.C., Shoemaker, R.C., Stalker, H.T., Weeden, N.F., Young, N.D. (2005) Legumes as a model plant family. Genomics for food and feed report of the cross-legume dvances through genomics conference. Plant Physiol. 137:1228–1235.

Guo, M., Davis, D., and Birchler, J.A. (1996) Dosage Effects on Gene Expression in a Maize Ploidy Series. Genetics 142:1349–1355.

Hall, A.E., Keith, K.C., Hall, S.E., Copenhaver, G.P., Preuss, D. (2004) The rapidly evolving field of plant centromeres. Curr Opin Plant Biol 7:108–114.

Hu, J.-M., Lavin, M., Wojciechowski, M. and Sanderson, M.J. (2000) Phylogenetic systematics of the tribe Millettieae (Leguminosae) based on trnK/matK sequences, and its implications for the evolutionary patterns in Papilionoideae. American Journal of Botany 87(3): 418–430.

Hwang, T-Y., Moon, J-K., Yu, S., Yang, K., Mohankumar, S., Yu, Y.H., Lee, Y.H., Kim, H.S., Kim, H.M., Maroof, M.A.S., Jeong, S-C. (2006) Application of comparative genomics in developing molecular markers tightly linked to the virus resistance gene Rsv4 in soybean. Genome 49:380–388.

Jing, R., Knox, M.R., Lee, J.M., Vershinin, A.V., Ambrose, M., Ellis, N.E., and Flavell, A.J. (2005) Insertional polymorphism and antiquity of PDR1 retrotransposon insertions in Pisum species. Genetics 171:741–752.

Kellogg E.A. (2001) Evolutionary history of the grasses. Plant Physiol 125:1198–1205.

Kevei, Z., Seres, A., Kereszt, A., Kalo, P., Kiss, P., Toth, G., Endre, G., Kiss, G.B. (2005) Significant microsynteny with new evolutionary highlights is detected between Arabidopsis and legume model plants despite the lack of macrosynteny. Mol Gen Genomics (2005) 274: 644–657.

Kochert, G., Stalker, H.T., Gimenes, M., Galgaro, L., Lopes, C.R., and Moore, K. (1996) RFLP and cytogenetic evidence on the origin and evolution of allotetraploid domesticated peanut, Arachis hypogaea (Leguminosae). Am J Bot 83:1282–1291.

Koornneef, M., Fransz, P., de Jong, H. (2003) Cytogenetic tools for Arabidopsis thaliana. Chromosome Res. 11(3)183–194.

Ku, H.M., Vision, T., Liu J.P., and Tanksley, S.D. (2000). Comparing sequenced segments of the tomato and Arabidopsis genomes: large-scale duplication followed by selective gene loss creates a network of synteny. Proc. Natl. Acad. Sci. U.S.A. 97: 9121–9126.

Kulikova, O., Geurts, R., Lamine, M., Kim, D-J., Cook, D.R., Leunissen, J., de Jong, H., Roe, B.A., Bisseling, T. (2004) Satellite repeats in the functional centromere and pericentromeric heterochromatin of *Medicago* truncatula. Chromosoma 113:276–283.

Lam, E., Kato, N., Watanabe, K. (2004) Visualizing chromosome structure/organization. Annu. Rev. Plant Biol. 55:537–54.

Langham, R.J., Walsh, J., Dunn, M., Ko, C., Goff, S.A., and Freeling, M. (2004). Genomic duplication, fractionation and the origin of regulatory novelty. Genetics,

Lavin, M., Herendeen, P.S., and Wojciechowski, M.F. (2005) Evolutionary rates analysis of Leguminosae implicates a rapid diversification of lineages during the Tertiary. Systematic Biology 54: 530–549.

Lee, J.M., Bush, A., Specht, J.E., and Shoemaker, R. (1999). Mapping duplicate genes in soybean. Genome, 42: 829–836.

Lee, J.M., Grant, D., Vallejos, C.E., and Shoemaker, R. (2001) Genome organization in dicots. II. Arabidopsis as a 'bridging species' to resolve genome evolution events among legumes. Theor. Appl. Genet. 103: 765–773.

Lewis, G.P., Schrire, B.D., Mackinder, B.A., Lock, J.M., ed. (2003) Legumes of the World. Royal Botanic Gardens, Kew, UK

Lynch M. and Conery J.S. (2000) The evolutionary fate and consequences of duplicate genes. Science 290:1151–1155.

Ma, J., Wing, R.A., Bennetzen, J.L., Jackson, S.A. (2007) Plant centromere organization: a dynamic structure with conserved functions. Trends in Genetics 23(3):134–139.

Maddison, D.R. and K.-S. Schulz (eds.) 1996–2006. The Tree of Life Web Project. Internet address: http://tolweb.org

Maere, S., De Bodt, S., Raes, J., Casneuf, T., Montagu, M.V., Kuiper, M., Van de Peer, Y. (2005) Modeling gene and genome duplications in eukaryotes. Proc Natl Acad Sci USA 102(15):5454–5459.

Masterson, J. (1994) Stomatal size in fossil plants: evidence for polyploidy in majority of angiosperms. Science (Washington, D.C.), 264: 421–424.

Moretzsohn M.C., Hopkins M.S., Mitchell S.E., Kresovich S, Valls J.F.M., and Ferreira M.E. (2004) Genetic diversity of peanut (Arachis hypogaea L.) and its wild relatives based on the analysis of hypervariable regions of the genome. BMC Plant Biology 2004, 4:11.

Mudge J., Cannon, S.B., Kalo, P., Oldroyd, G.E.D., Roe, B.A., Town, C.D., Young, N.D. (2005) Highly syntenic regions in the genomes of soybean, *Medicago* truncatula, and Arabidopsis thaliana. BMC Plant Biology 2005, 5:15.

Nagaki, K., Cheng, Z., Ouyang, S., Talbert, P.B., Kim, M., Jones, K.M., Henikoff, S., Buell, R.C., Jiang, J. (2004) Sequencing of a rice centromere uncoveres active genes. Nature Genetics 36(2):139–145.

Neumann, P., Koblizkova, A., Navratilova, A., Macas, J. (2006) Significant Expansion of Vicia pannonica Genome Size Mediated by Amplification of a Single Type of Giant Retroelement. Genetics 173:1047–1056.

Odland W., Baumgarten A., and Phillips R. (2006) Ancestral rice blocks define multiple related regions in the maize genome. Crop Science 46:41–48.

Paterson, A.H., Bowers, J.E., and Chapman, B.A. (2004) Ancient polyploidization predating divergence of the cereals, and its consequences for comparative genomics. Proc. Natl. Acad. Sci. USA 101:9903–9908.

Pfeil, B.E., Schlueter, J.A. Shoemaker, R.C., and Doyle, J.J. (2005) Placing paleopolyploidy in relation to taxon divergence: a phylogenetic analysis in legumes using 39 gene families. Systematic Biology 54:441–454.

Pontes, O., Neves N., Silva M., Lewis M.S., Madlung A., Comai L., Viegas W., and Pikaard C.S. (2004) Chromosomal locus rearrangements are a rapid response to formation of the allotetraploid Arabidopsis suecica genome. Proc. Natl. Acad. Sci. USA 101: 18240–18245.

Rauscher J.T., Doyle J.J., and Brown A.H. (2004) Multiple origins and nrDNA internal transcribed spacer homeologue evolution in the *Glycine* tomentella (Leguminosae) allopolyploid complex. Genetics 166(2):987–98.

Sanderson M.J., Thorne J.L., Wikstrom N, and Bremer K. (2004) Molecular evidence on plant divergence times. Am. J. Bot. 91(10):1656–1665.

SanMiguel P., Gaut B.S., Tikhonov A., Nakajima Y., and Bennetzen J.L. (1998) The paleontology of intergene retrotransposons of maize. Nat Genet. 20(1):43–5.

Schlueter, J.A., Dixon P., Granger, C., Grant, D., Clark, L., Doyle, J.J. and Shoemaker, R.C. (2004) Mining EST databases to resolve evolutionary events in major crop species. Genome 47: 868–876.

Schlueter, J.A., Lin, J-Y, Schlueter, S.D., Vasylenko-Sanders, I.F., Deshpande, S., Yi, J., O'Bleness, M., Roe, B.A., Nelson, R.T., Scheffler, B.E., Jackson, S.A., Shoemaker, R.C. (2007) Gene duplication and paleopolyploidy in soybean and the implications for whole genome sequencing. BMC Genomics 2007, 8:330.

Shoemaker, R.C., Polzin, K., Labate, J., Specht, J., Brummer, E.C., Olson, T., Young, N.D., Concibido, V., Wilcox, J., Tamulonis, J.P., Kochert, G., and Boerma, H.R. (1996) Genome duplication in soybean (*Glycine* subgenus soja). Genetics, 144: 329–338.

Simillion, C., Vandepoele, K., Van Montagu, M.C.E., Zabeau, M., and Van de Peer, Y. (2002) The hidden duplication past of Arabidopsis thaliana. Proc. Natl Acad. Sci. USA 99, 13627–13632.

Song, K., Lu, P., Tang, K., and Osborn, T.C. 1995. Rapid genome change in synthetic polyploids of Brassica and its implications for polyploid evolution. Proc. Natl. Acad. Sci. USA 92: 7719–7723.

Swigonova, Z., J. Lai, J. Ma, W. Ramakrishna, V. Llaca, J.L. Bennetzen, and J. Messing. (2004) Close split of sorghum and maize genome progenitors. Genome Res. 14:1916–1923.

Tuskan G.A., DiFazio S., Jansson S., Bohlmann J. et al. (2006) The genome of black cottonwood (Populus trichocarpa (Torr. & Gray). Science 313:1596–1604.

Vitte, C. and Bennetzen, J.L. (2006) Analysis of retrotransposon structural diversity uncovers properties and propensities in angiosperm genome evolution. Proc Natl Acad Sci U S A. 103(47):17638–43.

Wikstrom, N., Savolainen, V., and Chase, M.W. (2001) Evolution of the angiosperms: calibrating the family tree. Proceedings of the Royal Society of London, series B 268: 2211–2220.

Wikstrom, N., Savolainen, V., and Chase, M.W. (2003) Angiosperm divergence times: congruence and incongruence between fossils and sequence divergence estimates. In P. C. J. Donoghue and M. P. Smith [eds.], Telling the evolutionary time: molecular clocks and the fossil record, 142–165. Taylor & Francis, London, UK.

Yan H.H., Mudge J., Kim D.J., Larsen D., Shoemaker R.C., Cook D.R., and Young N.D. (2003) Estimates of conserved microsynteny among the genomes of *Glycine* max, *Medicago* truncatula and Arabidopsis thaliana. Theor Appl Genet 106(7):1256–65.

Young N.D., Cannon, S.B., Sato, S., Kim, D.J., Cook, D.R., Town, C.D., Roe, B.A., Tabata, S. (2005) Sequencing the genespaces of *Medicago* truncatula and *Lotus* japonicus. Plant Physiology 137:1174–1181.

Yu J., Wang J., Lin W., Li S., Li H., et al. (2005) The genomes of Oryza sativa: A history of duplications. PLoS Biol 3(2):e38.

Yuan Y., SanMiguel P.J., and Bennetzen J.L. (2002) Methylation-spanning linker libraries link gene-rich regions and identify epigenetic boundaries in Zea mays. Genome Res. 12(9):1345–9.

Zhu, H., Riely, B.K., Burns, N.J., Ane, J-M. (2006) Tracing Nonlegume Orthologs of Legume Genes Required for Nodulation and Arbuscular Mycorrhizal Symbioses. Genetics 172: 2491–2499.

Chapter 4
Phaseolus vulgaris: A Diploid Model for Soybean

Phillip E. McClean, Matt Lavin, Paul Gepts, and Scott A. Jackson

The Importance and Structure of Modern Common Bean Germplasm

Social, Economic, and Agronomic Importance of Common Bean

Along with maize and cassava, common bean is a critical component of diets for many of the developing countries in the world. Beans are an important source of family income and a critical component of the daily diet within African countries where the population is projected to double by 2020. As such, it is the most important edible food legume in the world's diet. It represents 50% of the grain legumes consumed worldwide, and its production is nearly twice that of chickpeas, the second most consumed food legume. Because poverty limits access to animal protein in developing countries, these peoples turn to common bean as a protein source. From a dietary perspective, it accounts for 40%, 31%, and 15% of the daily intake of total protein in some of the least developed countries, such as Burundi, Rwanda, and Uganda, respectively. And even for a major producer like Brazil, 9% of the dietary protein is provided beans.

Common bean is grown in monocultures or as the primary or secondary species in a multicropping system. While the cash value of the crop exceeds $1 billion in the United States, in many ways it is more important elsewhere. For a very poor country such as Myanmar, bean is its most important agricultural export commodity accounting for 10% of their total export income (http://faostat.fao.org/faostat/; verified July 14, 2006). Yield varies significantly, from 638 kg/ha, 671 kg/ha, and 918 kg/ha in Uganda, Rwanda and Burundi, respectively, to 1,944 kg/ha in developed countries. Improving agricultural productivity in Africa and the developing Americas is seen as a means to reverse the trend of increasing poverty and hunger in these regions. Therefore, identifying and minimizing yield limiting factors is an on-going concern for many bean improvement programs. For example, aluminum

P.E. McClean
Department of Plant Sciences, North Dakota State University, Fargo, ND 58105, USA
e-mail: phhillip.mcclean@ndsu.edu

G. Stacey (ed.), *Genetics and Genomics of Soybean*,

toxicity and phosphate deficiency affect yields in both the Americas and Africa and as such improvement programs are focusing on these traits (Beebe et al. 2006; Ochoa et al. 2006). In addition, given the prevalence of bean in the diets of these countries, modifying the nutrient content of common bean to make it a more balanced and nutritious food source is also receiving emphasis. Specifically, zinc content is a focus because of the recognition that this mineral is an important dietary component for those individuals infected with the AIDS virus. Understanding the genetic and genomic aspects of these traits and developing tools for their improvement will accelerate improvement of this important societal crop.

Population Structure of Domesticated Common Bean

The major subdivisions of wild common bean progenitors are known, and the domesticated gene pools have been defined. Based on phaseolin seed storage protein variation (Gepts and Bliss 1986; Gepts 1990), marker diversity (Becerra Velásquez and Gepts 1994; Koenig and Gepts 1989; Tohme et al. 1996; McClean et al. 2004b), and morphology (Gepts and Debouck 1991), two major gene pools of wild common bean were identified. The Middle American gene pool extends from Mexico through Central America and into Colombia and Venezuela, while the Andean gene pool is found in southern Peru, Chile, Bolivia and Argentina. The two gene pools appear to converge in Colombia (Gepts and Bliss 1986). A third, possibly ancestral gene pool based in southern Ecuador and northern Peru, was described (Debouck et al. 1993; Kami et al. 1995). Two major domestication events appear to have resulted in the Middle American and Andean gene pools (Kaplan and Lynch 1999) that mirror the geographic distribution of the wild progenitors (Gepts 1998; Islam et al. 2002; Blair et al. 2006). These domestication events appear to have arisen from wild beans in the region where the domestication occurred. This domestication event is in contrast to species such as maize (Matsuoka et al. 2002) and soybean (Powell et al. 1996; Xu and Gai 2003) where a derived domesticated population emerged from a single subspecies or related species, respectively. Following domestication, gene pool divergence led to the appearance of three races within the Andean gene pool, Nueva Granada, Peru and Chile (Singh et al. 1991), and four Middle American races, Durango, Jalisco, Mesoamerica, and Guatemala (Singh et al. 1991; Beebe et al. 2000).

Phylogenetic Evolution of Legumes and the Relationship Between *Glycine max* and *Phaseolus vulgaris*

Phylogenetic Relationship of Legumes to Other Angiosperms

The legume family is well supported by many recent molecular phylogenetic studies as a member of the Eurosid I clade of dicot flowering plants (Fig. 4.1). Although

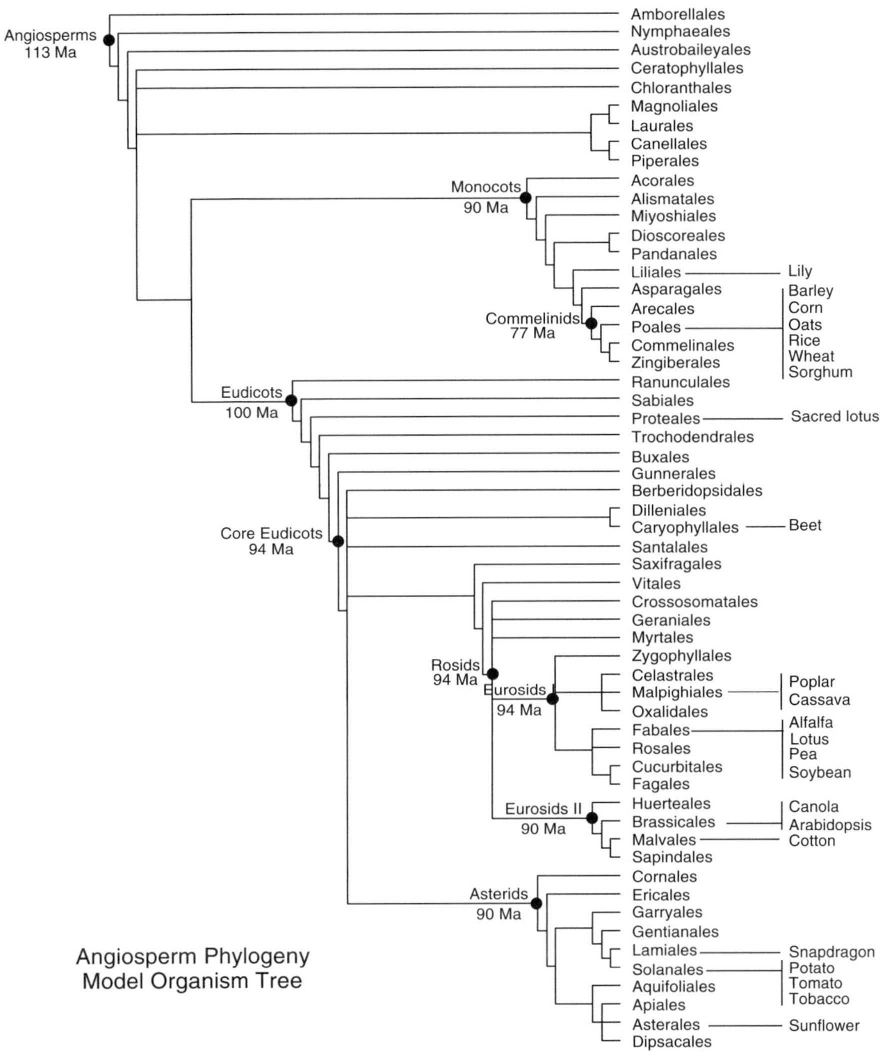

Fig. 4.1 Synopsis of the model organism tree adapted from the Angiosperm Phylogeny web site (Stevens 2001; www.mobot.org/MOBOT/research/APweb). The minimum ages of marked clades in millions of years (Ma) are taken from Crepet et al. (2004), except for the minimum age of the Commelinid clade, which is taken from Magallón and Sanderson (2001). Crepet et al. (2004) take a conservative approach to assigning fossils to extant clades only with unequivocal morphological evidence. Some clades deeper in the phylogeny have equivalent minimum ages to those nested higher in the phylogeny. This is because the fossils used to make these age assignments come from strata with similar minimum ages (*See* also Color Insert)

most of these Eurosid families have an exceptional fossil record, no obvious morphological features characterize them (Crepet et al. 2004). The minimum age of the Eurosid I clade is 94 Ma (Crepet et al. 2004), which is significantly older than the age of the origin of legumes, estimated at about 60 Ma (summarized in Lavin et al. 2005).

Legumes belong to the Fabales, which comprise four families, 754 genera, and 20,055 species (Stevens 2001), or nearly 10% of the Eudicot species diversity (Magallón et al. 1999). Strongly zygomorphic or bilaterally symmetric flowers first appear during the early Tertiary by about 60 Ma in age with the appearance of the Fabales. Wikström et al. (2001) estimated the age of origin of Fabales at 94-89 Ma, and the extant diversification at 79-74 Ma. These age estimates are biased old (Lavin et al. 2005). Regardless, Fabales is an unanticipated group in not being characterized by morphology even though quite strongly supported by molecular phylogenetic analysis (summarized in Stevens 2001). The constituent families of Fabales, as reviewed by Lewis et al. (2005) and Stevens (2001), include the Polygalaceae with 18 genera and 1,045 species having a world wide distribution. Surianiaceae with 5 genera and 8 species are mostly Australian, but *Recchia* is endemic to Mexico and *Suriana maritima* is pantropical. Quillajaceae with one genus (*Quillaja*) and 3 species is endemic to temperate Chile. Fabaceae with 730 genera and 19,400 species (13,855 of which belong to the papilioniod legume clade) is world wide in distribution but with a predilection to deserts, grasslands, savannas, and seasonally dry tropical forests, although certain groups are diverse in tropical wet forests. The relationships among these four families of Fabales are not well supported (Stevens 2001). Polygalaceae and legumes are the only two families of Fabales that produce bilaterally symmetric flowers, but this sister relationship is otherwise tenuously or not at all resolved with molecular phylogenetic analyses. For example, the *rpl22* gene was transferred from the chloroplast to the nucleus in Polygalaceae and Fabaceae, but this condition has not been analyzed for any Quillajaceae and Surianiaceae, as well as many members of Fabaceae. Many papilionoid legumes have lost the *rps16* gene, which is also known to be absent from *Polygala*. Stevens (2001), however, indicates a lack of sampling for these traits from most Fabales.

Phylogenetic Relationships of Legumes to Model Plant Genome Species

Closely related to the Fabaceae within the Eurosid I angiosperm group are many other economically important plant groups, particularly the Rosales (including *Fragaria, Malus, Prunus, Rubus*, etc.), Cucurbitales (*Citrullus, Cucumis, Cucurbita, Luffa*, etc.), and Fagales (*Fagus, Quercus, Juglans*, etc.; Fig. 4.1). More distantly related to legumes, yet within this same Eurosid I clade, is the Malpighiales, which includes cassava (Euphorbiaceae) and poplar (Salicaceae). *Arabidopsis* and canola (Brassicaceae), as well as cotton (Malvaceae), belong to the Eurosid II clade that is, in part, sister to the Eurosid I clade. These collective Rosids have a minimum age of 94 Ma, according to the best fossil evidence (Crepet et al. 2004). As with most molecular phylogenetic analyses, these groups are well supported but not necessarily by obvious morphological characters. Eurosid I constituents, for example, show a predisposition to fix nitrogen via root-dwelling associates (e.g., actinomycetes, rhizobia, etc.), but this is certainly not a uniform attribute of this clade (Stevens 2001). Legumes are much more distantly related to other model plant groups, such as those that belong to the Asterids (e.g., snapdragon, potato, tomato, tabacco, and sunflower), Caryophyllales (e.g. beet), or Monocots (e.g., lily), including the Commelinids (e.g., the grasses; Fig. 4.1).

Phylogenetic Relationships of Glycine and Phaseolus to Model Legumes

Although many economically important species come from the legume family, a small sample of model legume species are illustrated to emphasize the great age that separates the many model and economically important legume species (Fig. 4.2). For example, tracing backwards in time along the genealogical paths that lead to the most recent common ancestor of *Lupinus* or *Arachis* with any of the other domesticated or model legume species (e.g., *G. max* or *Pisum sativum*) reveals that these species evolved along separate lines for at least 56.5 Ma. The early Tertiary age of these lineages is coeval with the Paleocene-Eocene boundary, or about the same time that flowering plants were beginning to dominate terrestrial vegetation. Even seemingly closely related pairs of model legume species, such as *G. max* and *P. vulgaris*, or *P. sativum* and *Vicia faba*, evolved along separate lines for well over 15 Ma. As reference for emphasizing the magnitude of these ages, humans and chimpanzee have been evolving along separate lines for at most 7 Ma (Kumar et al. 2005). In spite of the nearly 20 Ma age that separates *G. max* from *P. vulgaris*, these two species belong to the same group of papilionoid legumes, which is referred to as the core Phaseoleae (Lewis et al. 2005).

Phaseolus Genomic Tools

Cytogenetics of Phaseolus

Common bean is a diploid species with $2n = 22$ chromosomes and a medium-sized genome. Estimates of the size of the haploid genome of *P. vulgaris* range from 588 (Bennett and Leitch 1995, 2005) to 637 Mbp (Arumuganathan and Earle 1991). Most of the species of the genus *Phaseolus* have the same chromosome number with the exception of the *P. leptostachyus* Benth. clade, which has $2n = 20$ chromosomes (Delgado-Salinas et al. 2006). With some exceptions, including soybean, most other members of the Phaseoleae tribe have the same or a similar chromosome number around $2n = 22$. These include, in the genus Vigna, which is closely related to the

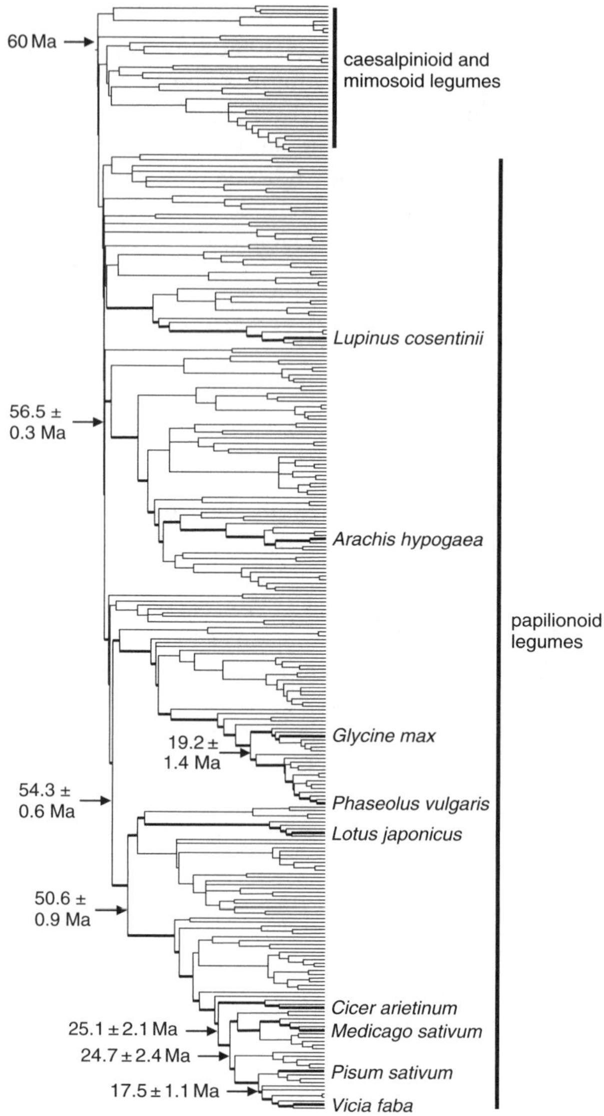

Fig. 4.2 A phylogeny with branch lengths scaled to time (a chronogram) derived from the analysis of Lavin et al. (2005) in which selected crop species are shown and the lineages leading to these highlighted. Tracing two of these highlighted lineages back in time to where they intersect or coalesce reveals the age of their most recent common ancestor (MRCA). For example, the age of the MRCA of *Glycine* and *Phaseolus* is about 19 Ma, whereas that of *Arachis* and any of the other species shown is about 56.5 Ma. This illustrates that many of the cultivated or model legume species are deeply divergent in time from one another. The minimum age of legumes, 60 Ma, is fixed for the root of the legume clade, and the evidence for this is summarized in Lavin et al. (2005)

genus *Phaseolus*, cowpea [*Vigna unguiculata* (L.) Walp], mung bean [*Vigna radiata* (L.) R. Wilczek], and rice bean [*Vigna umbellata* (Thunb.) Ohwi & H. Ohashi]. In pigeon pea [*Cajanus cajan* (L.) Millsp.], the chromosome number is also $2n = 22$, as well as in hyacinth bean [*Lablab purpureus* (L.) Sweet].[1]

Compared to these diploid Phaseoleae species, the genus *Glycine* has undergone genome duplications followed by rearrangements leading to diploidization (XX); accordingly, its chromosome number is $2n = 40$ [e.g., soybean (*G. max*). Among its wild relative $2n = 40$ (*G. soja* Siebold & Zucc.), or $2n = 80$ [*G. tabacina* (Labill.) Benth.], or $2n = 38$, 40, 78, and 80 (*G. tomentella* Hayata). Within the phaseoloid group, the closest generic ally of *Glycine* is the genus *Teramnus* with a chromosome number of $2n = 28$ (Doyle and Luckow 2003). Thus, the tetraploid nature of soybean is unusual compared to most of its relatives in the millettioid/phaseoloid clade. This situation has to be taken into consideration when attempting to establish macro- and microsynteny between the genome of soybean and related species. It is also important to consider ancestral polyploidization and diploidization events as the physical map (Shultz et al. 2006) and the soybean genome sequence are assembled.

The mitotic and meiotic metaphase chromosomes of *Phaseolus* are small and relatively undifferentiated, making identification of individual chromosomes difficult until recently (Maréchal 1971). The presence of polytene chromosomes in the embryonic suspensor cells has not provided a solution to chromosome identification because they remain relatively uncondensed compared to the polytene chromosomes of *Drosophila* (Nagl 1974). Nevertheless, fluorescent in situ hybridization (FISH) was used to locate rRNA and phaseolin loci on these polytene chromosomes (Nenno et al. 1994). More recently, fluorescently-labeled pooled RFLP and individual BAC probes (Kami et al. 2006) were used to perform FISH on mitotic chromosomes of common bean. Idiograms of two European cultivars (Saxa and Tschermak's fadenlose Wachs) were established using a combination of double FISH with 5S and 45S rDNA probes, chromosome morphology, and heterochromatin distribution. Similar idiograms were established for three other *Phaseolus* domesticated species, *P. coccineus, P. acutifolius*, and *P. lunatus* (Moscone et al. 1999). FISH with pooled fluorescently-labeled RFLP probes mapping to the same linkage group (Vallejos et al. 1992) led to a correlation between the genetic and chromosomal maps in two cultivars, Saxa and Diacol Calima (Pedrosa et al. 2003). More recently, Pedrosa-Harand et al. (2006) demonstrated that the two 5S rRNA loci were conserved across the species. In contrast, the number of 45S rRNA loci varied between two and five in the Middle American gene pool and six and nine in the Andean gene pool. From an evolutionary perspective, ancestral wild beans from northern Peru and Ecuador (Kami et al. 1995) resemble the Middle American gene pool in the number of 45S rRNA loci. Thus, these results reveal a major amplification event in the Andean lineage after its separation from the Middle American lineage. Since hybrids between the two gene pools exhibit a full range of the number of 45S rRNA loci, the number

[1] Chromosome numbers: Missouri Botanical Garden Index to Plant Chromosome Numbers – IPCN: http://mobot.mobot.org/W3T/Search/ipcn.html

of 45S rRNA loci will not be a reliable reference loci for synteny mapping between common bean and soybean.

Genetic Maps

The importance of genetic linkage maps in understanding the inheritance of a trait has led bean breeders to develop over 25 linkage maps in different crosses of common bean (Kelly et al. 2003; Miklas et al. 2006). Most are low-density linkage maps with markers on average every 10 cM and incomplete coverage of the genome. In order to maximize polymorphism at the molecular level, the majority of mapping populations were derived from crosses between domesticated parents belonging to the Andean vs. Middle American gene pools. In some cases, maps were developed in crosses involving between parents of the same gene pool or between a domesticated and wild parent.

To correlate the mapping results of these different maps, a core map was established in the recombinant inbred population resulting from the cross BAT93 × Jalo EEP558 (Freyre et al. 1998). BAT93 is a breeding line from the Mesoamerican gene pool, and Jalo EEP558 is an Andean gene pool member resulting from selection in a Brazilian landrace. The two parents show contrasting resistances to pathogens. This population consists of some 75 lines showing a high level of polymorphism (Nodari et al. 1992). Some 600 markers were mapped directly in this population, including 71 RFLPs, 161 AFLPs, 158 RAPDs, 50 ISSRs, and 200 microsatellites (Freyre et al. 1998; Papa and Gepts 2003; González et al. 2005; Blair et al. 2003; Mattos de Grisi et al. unpubl. results). In addition, several of the markers, principally RFLPs and sequence-tagged markers, are shared with other maps, allowing a general correlation among linkage groups of different maps. The linkage group numbering adopted by Freyre et al. (1998) has become the standard across linkage mapping studies in common bean.

A new set of gene-based markers is being implemented for genetic mapping in common bean including TRAP (Targeted Region Amplification Polymorphism) and RGA (Resistance-Gene Analogs)-based markers targeting disease resistance genes (Divkin et al., 1999; López et al. 2003; Mutlu et al. 2005; Miklas et al. 2006). In addition, gene-based SNP markers are being mapped at CIAT, and mapping of EST sequences at North Dakota State University and at the University of Saskatchewan will lead to the development of a transcriptional map for beans that could help in the establishment of correlations between candidate genes and specific QTLs (Gepts et al. 2007). A deliberate effort needs to be developed to identify markers that identify homologous sequences in both common bean and soybean to delineate regions of synteny.

Gene Cloning and Sequencing

As with many species, investigators have studied genes on an individual basis in common bean. One of the earliest examples is the cloning of the phaseolin gene in which the presence of introns was first demonstrated in plants (Sun et al. 1981). The breadth of research for a species can be estimated in a relative manner by comparing the number of complete coding sequences (CDSs) deposited in GenBank. These sequences represent full gene sequence data that was confirmed either genetically or biochemically. Among the *Fabaceae* species, soybean has the most CDSs sequences with 956 (Table 4.1). By comparison, the number for common bean is 195. This number is only half that of *M. truncatula*, an emerging species for legume research. It is important to note that for each of the species in Table 4.1, the number of CDSs represents an almost doubling of those available just three years ago (McClean et al. 2004a).

Disease resistance is one area in which gene cloning is progressing using a variety of methods. The historical pathogenesis-related genes were cloned several years ago (Ryder et al. 1987; Walter et al. 1990; Blyden et al. 1991; Margis-Pinheiro et al. 1994). Recently the region surrounding the *I* gene, a major gene involved in bean common mosaic virus resistance, was sequenced, and genes similar to other disease resistance genes were discovered (Vallejos et al. 2006). Recently, a cDNA-AFLP screen defined a number of genes associated with the induction of resistance including those associated with G-protein and ABA signaling (Cadle-Davidson and Jahn 2006). This led the authors to suggest general rather than specific signaling systems may be recruited for the defense response. Genes potentially associated with anthracnose resistance were also recently described (Melotto et al. 2004).

Research into other important traits for common bean productivity is also identifying genes and their involvement in the expression of the trait. Recent research that focused on drought and other water relations: (1) identified specific ABA 8′-hydroxylases genes involved in the response to and recovery from water stress

Table 4.1 Summary GenBank statistics for the Fabaceae family and specific species (verified December 20, 2006)

	Total[a]	EST[b]	CDS[c]	Complete CDS[d]	Genomic survey[e]
Fabaceae	1, 467, 205	887, 838	9, 519	3, 239	545, 653
Cicer spp.	2, 119	297	451	16	0
Glycine max	645, 347	359, 435	2, 137	956	281, 031
Lotus japonicus	199, 018	150, 631	305	163	46, 569
Medicago truncatula	397, 575	225, 129	481	251	168, 809
Phaseolus vulgaris	27, 222	22, 666	579	158	2, 980
Phaseolus spp.	48, 465	43, 534	715	195	2, 980
Pisum sativum	6, 058	3, 594	1, 114	485	154
Vicia spp.	1, 616	1	418	37	686
Vigna spp.	1, 657	466	406	182	89

[a] PubMed nucleotide query: species [orgn]
[b] PubMed nucleotide query: species [orgn] est NOT cds
[c] PubMed nucleotide query: species [orgn] cds NOT est NOT plastid NOT chloroplast NOT mitochondria
[d] Pubmed nucleotide query: species [orgn] complete cds NOT est NOT plastid NOT chloroplast NOT mitochondria
[e] Pubmed Genome Sequence Survey (GSS) query: species [orgn]

(Yang and Zeevaart 2006); (2) showed that specific bZIP transcription factors are specifically expressed in the root and only during water stress (Rodriguez-Uribe and O'Connell 2006); and (3) revealed the differential expression of aquaporins (Aroca et al. 2006) during the stress. Another example is the cloning of several phototropin genes that encode receptor-type protein kinases and the demonstration that they undergo a change in phosporyaltion state during the phototropism response (Inoue et al. 2005).

Insect resistance is a key to preventing seed deterioration caused by insect predation during storage. Studies have identified members of the lectin gene family as key to the resistance. This family has been extensively studied, and key members with regards to resistance, the arcelins, phytohemmaglutinins, and the amylase inhibitor family members are encoded by a block of genes co-located on linkage group B4 (Freyre et al. 1998; Kami et al. 2006). Sequence comparisons revealed that the evolutionary history of the *Phaseolus* arcelins (Lioi et al. 2003) and phyothemmaglutinins (Hoffman and Donaldson 1985) includes duplications and diversification of function. The recent cloning and sequencing of 156 kb genomic region around the APA family in common bean will further extend our understanding of the roles of these genes in seed predation (Kami et al. 2006). From a comparative genomics perspective, this genomic data can be compared to the emerging genome sequence of soybean to further understand the effects of duplication and diploidization on the structure of the soybean genome.

EST Collections

In recent years, the number of expressed sequence tags (ESTs) in *Phaseolus* has increased markedly (compare Table 4.1 here and Table 4.1 in McClean et al. 2004a). To date some 45,000 ESTs have been developed in *Phaseolus*. These include 25,000 in *P. vulgaris* (Ramirez et al. 2005; Melotto et al. 2005) and 20,000 in the closely related species *P. coccineus*. Three *P. vulgaris* genotypes were used to obtain these ESTs: two of Mesoamerican origin (Negro Jamapa and SEL1308) and one of Andean origin (G19833). Tissues sampled include seedling shoots [with or without *Colletotrichum lindemuthianum* (anthracnose) infection], seedling leaves, nodules elicited by *Rhizobium tropici* strain CIAT899, roots, leaves (three genotypes), and pods. In *P. coccineus*, ESTs were isolated from the suspensor regions in globular-stage embryos six days after pollination (e.g., GenBank: CA916678; http://www.ncbi.nlm.nih.gov/entrez/viewer.fcgi?db=nucleotide& val=27403670). Because of the close relationship between the two species, sequences in *P. vulgaris* can be identified through similarity with *P. coccineus* (Nanni et al. 2005). Gene indices for the *P. vulgaris* ESTs were recently developed (http://biocomp.dfci.harvard.edu/tgi/cgi-bin/tgi/gimain.pl?gudb=p_vulgaris; http://www.plantgdb.org/download/download.php? dir=/Sequence/ESTcontig/Phaseolus_vulgaris).

Additional EST resources in common bean (for example, from different flower, pod, and seed developmental stages, organs, and tissues) would be particularly useful because these are or lead to the harvested organs. A comparison of expression patterns between *P. vulgaris* and *G. max* would also be useful to gauge the effect of polyploidization/diploidization.

BAC Constructions and Their Applications

Eleven BAC libraries have been constructed in the genus *Phaseolus*, ten in *P. vulgaris* and one in *P. lunatus* (Gepts et al. 2007). To increase the concentration of high-molecular weight DNA available for cloning, Kami et al. (2006) developed an extraction procedure based on a novel cell nuclei isolation procedure. The majority of the libraries were developed after restriction digestion with *Hin*dIII, although one library results from *Eco*RI digestion and two from *Bam*HI digestion. Most libraries have a coverage of 5–12x genome, based on their average insert size. The BAT93 library has a coverage of 20x, in part because it has been designated as the standard genotype for *Phaseolus* genomics (Broughton et al. 2003).

The *Phaseolus* BAC libraries are a unique phylogenetically ordered set useful for evolutionary studies (Fig. 4.3) as each represents a key genotype in the intra-specific evolution and domestication of common bean (Gepts et al. 2007). DGD1962 is a wild bean from northern Peru, representing the presumed ancestral

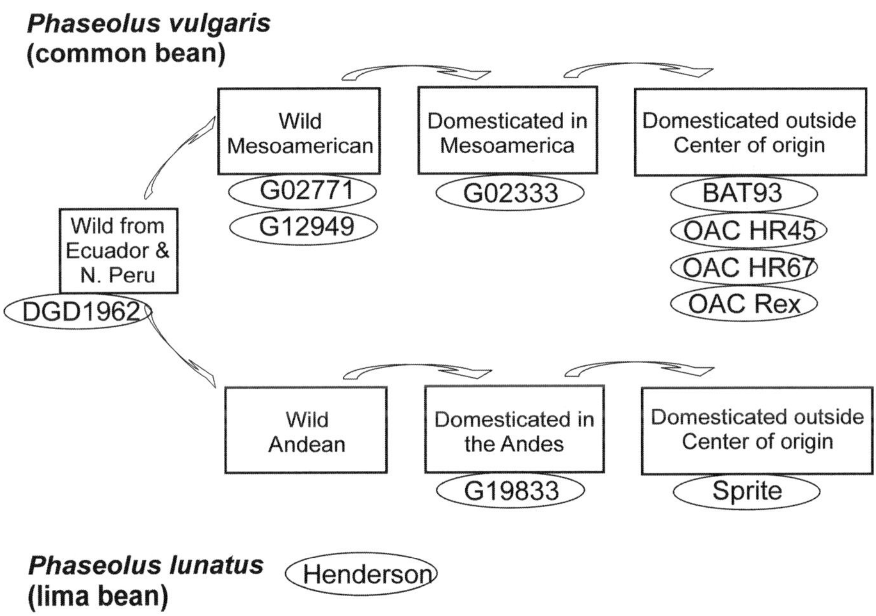

Fig. 4.3 Phylogenetic and genealogical distribution of BAC libraries in *Phaseolus* spp. Boxes represent different segments of the *P. vulgaris* gene pools and the general direction of their evolution. Names surrounded by ellipses are the genotypes in which BAC libraries have been established (from Gepts et al. 2007)

gene pool of the species (Debouck et al. 1993; Kami et al. 1995). The remainder of the libraries is distributed in the two evolutionary lineages that were domesticated. In the Mesoamerican lineage, G02771 and G12946 are wild Mexican beans that contain the three subfamilies of the APA seed proteins, which confer resistance to seed weevils. G2333 is a Mexican landrace highly resistant to anthracnose. BAT93 and OAC-HR45 and OAC-HR67 are breeding lines and OAC-Rex is a cultivar from the Mesoamerican gene pool. In the Andean gene pool, G19833 is a landrace from Peru, whereas Sprite is a bred variety. Thus, using this array of BAC libraries it is possible to study overall structural evolution of the genome in *Phaseolus* both prior and after domestication. It is also possible to analyze phenotypic changes resulting from specific structural modification at the genome level. Single BAC clones were fully sequenced, one around the *Co-4* locus for resistance to anthracnose (Melotto et al. 2004), and the other around the *APA* locus (Kami et al. 2006). Comparative sequence analysis of similar regions in soybean can address the question of the utility of common bean as a diploid model for soybean.

Emerging Procedures: TILLING and Transformation

As defined by McCallum et al. (2000), TILLING (or Targeted Induced Local Lesions IN Genomes) is a tool to identify artificially-induced mutants in specific genes of interest. This reverse genetics approach does not require an effective transformation system. In common bean, a TILLING platform is being developed by W. Broughton, P. Lariguet, T. Porch, and M.W. Blair (unpubl. data and Gepts et al. 2007) in the genotype BAT93, mentioned before. Mutagen (EMS) concentration and the resulting lethality have been determined. Currently, some 1,500 M2 families have been produced, and 900 of these were advanced to the M3 stage. This TILLING platform will benefit from future sequencing efforts in common bean and other species, such as soybean by providing reference sequences of loci of interest. In turn, these sequences will be used to design gene-specific primers.

Common bean is reputed to be a species recalcitrant to transformation. However, recent progress has increased the feasibility of transformation as a tool in *Phaseolus* (Veltcheva et al. 2005). In tepary bean (*Phaseolus acutifolius* A. Gray), Agrobacterium transformation is relatively efficient and was used to test gene function (De Clercq et al. 2002; Zambre et al. 2005). Although transformation of *P. vulgaris* appears to be more difficult, three avenues have been pursued. Biolistics was used to obtain transgenic common bean plants (Aragão et al. 1998, 1999, and 2002), some of which are undergoing field testing in Brazil (F. Aragão personal communication). Liu et al. (2005) developed a protocol based on the combination of sonication and vacuum infiltration to transform common bean with a Late Embryogenesis Abundant gene conferring abiotic stress tolerance. Although the efficiency remains to be improved, this method looks promising. Recently, Estrada-Navarrete et al. (2006) modified a protocol originally developed in soybean for efficient and robust *Agrobacterium rhizogenes* transformation of common bean. The availability of transgenic roots in *Phaseolus* will provide a new method for overexpressing or suppressing endogenous genes, especially those involved in root biology and root-microbe interactions.

Application of Phaseolus Genomic Tools to *Glycine*

Soybean and common bean are valuable legumes in their own regard, and share the feature of being important nitrogen fixing species. They also share an evolutionary lineage that has the potential to foster a research synergy that will benefit both species. *G. max* has a relatively large sized genome of $\sim 1,100$ Mbp organized in 20 chromosome pairs, whereas, *P. vulgaris* has a genome nearly half the size of soybean organized in 11 chromosome pairs. Taxonomically, *P. vulgaris* is closely related to soybean, separated by 19.2 Ma (Lavin et al. 2005), possibly around the time of the last major duplication event in soybean (Schlueter et al. 2004). Genetically, *P. vulgaris* appears to be a true diploid, whereas, soybean is genetically/genomically complicated by multiple rounds of duplication/polyploidization (Blanc and Wolfe 2004; Schlueter et al. 2004). Thus, given the close evolutionary proximity of these two species, it may be possible to exploit the simple genome of *Phaseolus* to understand the organization and evolution of the duplicated soybean genome. To that end, a physical map of *Phaseolus* is being created from fingerprinted BAC clones from the G19833 genotype. This BAC-based physical map, along with a BAC end sequence (BES) database, will be used to help resolve duplication events in soybean as well as to provide a platform for mapping and gene cloning in *Phaseolus*. *Phaseolus*, along with *M. truncatula, Lotus japonicus*, and soybean provide a framework within legumes for highly detailed evolutionary analyses of domestication and polyploidization.

Use of Secondary Species to Assist with Genome Analysis

Soybean has a complicated genome due to multiple rounds of duplication and/or polyploidization (Shoemaker et al. 1996; Blanc and Wolfe 2004; Pagel et al. 2004; Schlueter et al. 2004; Walling et al. 2005). Since *Phaseolus* is diploid, to the extent that we can determine without genome sequencing, and given its phylogenetic proximity to soybean (Doyle and Luckow 2003), it provides an out-group to evolutionarily and structurally help define duplications within the soybean genome that occurred after the divergence of *Phaseolus* and soybean (Fig. 4.4). Shared duplications that occurred prior to the divergence of *Phaseolus* and soybean may be used to compare orthologous regions from either *M. truncatula* or *L. japonicus* to determine the ancestral legume genome structure.

Multiple sequence alignments of orthologous regions from related genomes has proven useful to understand genome dynamics such as gene movement/loss/gain and repeat accumulation/differentiation/loss (Thomas et al. 2003; Lai et al. 2004;

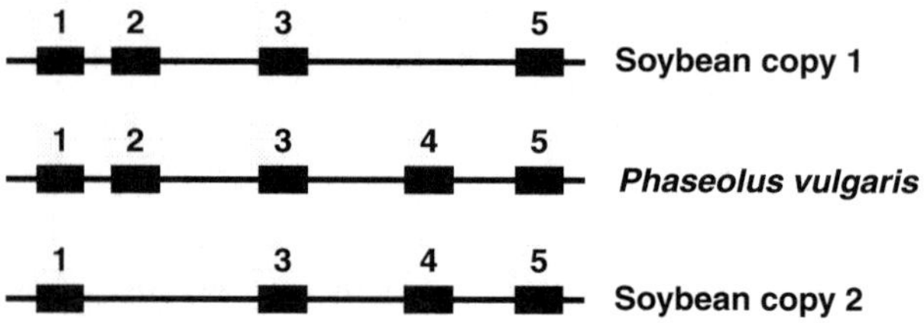

Fig. 4.4 Sequence alignment of a duplicated region in soybean to the orthologous region from *Phaseolus vulgaris*. The putative ancestral state can be determined and gene loss can be seen in both soybean copies

Ma et al. 2005). Comparative sequence alignments were extremely useful to confirm gene structure and predict regulatory elements as seen in comparisons of human to other mammals and among yeast species (Flint et al. 2001; Kellis et al. 2003). These algorithms benefit from having multiple species at varying levels of separation from the genome of interest in order to provide useful annotation. Closely related genomes can be useful to understand local and recent gene movement, whereas, more distantly related genomes can be used to predict conserved non-coding sequences (CNS). For instance, a fugu-human comparison revealed almost 1,400 CNS and 23/25 tested CNS showed significant enhancer activity (Woolfe et al. 2005).

Defining Gene Models Based on Genes of Different Evolutionary Distance

Annotation of sequenced genomes uses a variety of approaches including *ab initio* prediction using sequence features (Salamov and Solovyev 2000), sequence identity to expressed sequence tags (ESTs) (Wei and Brent 2006) and comparative sequence alignments (Flicek et al. 2003). The assumption is that non-coding sequences should not be selectively constrained across evolutionary time, therefore, conserved sequences between distantly related species should hold some functional significance, often being coding or regulatory. Comparisons of human sequences to various mammals helped to define gene/gene structure and regulatory elements (Woolfe et al. 2005). In addition, using carefully placed species on the evolutionary tree, inferences can be made about conserved non-coding sequences that may play significant roles in gene function (Margulies et al. 2003; Margulies and Green 2003; Margulies et al. 2005). For example, plants species separated by 16–50 MY were useful for such comparisons (Guo and Moose 2003; Inada et al. 2003). The approximate evolutionary distance between *Phaseolus* and soybean is 19.2 Ma (Lavin et al. 2005). Sequence information from *Phaseolus*, combined with sequence from *M. truncatula* and/or *L. japonicus* (~50 Ma) will provide an evolutionary framework in the legumes for determining gene structure, organization and to make hypotheses on gene function.

Assisting with Final Assembly of Sequenced Genome

Very few genomes are likely to be sequenced using a clone-by-clone approach as was done for Arabidopsis (AGI 2000), rice (IRGSP 2005), *C. elegans* (CESC 1998) and a few others. The advent of shotgun sequencing of entire eukaryotic genomes altered the approach by which most future genomes will be sequenced. Shotgun sequences of whole genomes are assembled as scaffolds of sequence contigs with many holes due to repeats and/or lack of sequences coverage and occasional mis-assemblies due to recent duplications (Adams et al. 2000; Myers et al. 2000; Venter et al. 2001; Tuskan et al. 2006). Yet this approach still provides most of the gene models of the species. Most genomes will likely be sequenced using hybrid approaches, most commonly a mixture of shotgun sequencing and limited physical mapping.

In soybean, a physical map is being constructed for the genotype Williams 82, and a shotgun sequence is targeted for completion by 2008 (Jackson et al. 2006). The physical map with its sequence tag connectors (STCs) will be used, inasmuch as possible, to help assemble sequence contigs into sensible scaffolds. In conjunction with this, a physical map of *Phaseolus*, with its own STC database, will be used to (1) check assembly, (2) resolve duplicated regions by comparison to a 'diploid' sister species, and (3) possibly span gaps in the soybean sequence/physical map. In rice, $\sim$ 35 gaps (excluding telomere/centromere gaps) existed in the sequence map at completion (2005), but sister species, in the same genus, were found that contained BAC clones that putatively spanned the gaps (Wing et al. 2005). This is an example of how a closely related out-group such as *Phaseolus* can be used to facilitate genome sequencing/assembly in soybean.

Identifying Adaptation Genes Within the Phaseolaeae

Recently, random genome-wide scans identified genes positively associated with domestication and/or agronomic productivity (Wright et al. 2005; Yamasaki et al. 2005). These scans were costly because they did not have *a priori* information regarding genes that might be undergoing the selection process and necessitated the large scale sequencing of over 1,000 genes from multiple genotypes. Utilizing pairwise sequence data obtained from comparing common bean and soybean, it may be possible to narrow the candidates necessary for this discovery process. Once these domestication or agronomic productivity genes are discovered, breeding populations developed by introgressing wild germplasm with beneficial alleles into an improved variety can be screened using high-throughput technologies to select lines contain a high proportion of essential domestication, agronomic and improvement alleles. An outline of how to leverage soybean and common bean sequences to identify these adaptation loci follows.

By comparing the coding sequence between common bean and soybean orthologous genes, it may be possible to discover genes undergoing purifying or positive selection within the Phaseoleae lineage. The classical method to detect these two forms of selection is to measure the K_a/K_s ratio. K_a is the non-synonymous substitution rate, and K_s is the synonymous substitution rate. If the $K_a/K_s << 1.0$, then the gene is assumed to be undergoing purifying selection to eliminate deleterious mutations, whereas if the K_a/K_s ratio is $>> 1.0$, then the gene is considered to be undergoing positive, and possibly adaptive selection. Key to these comparisons is to ensure that orthologs are compared. Syntenic map data (Choi et al. 2004) provides a reference point from which orhtologs between two legume species can be identified.

Orthologous genes found to have undergone selection by the common bean and soybean pairwise comparison can then be compared with a Galegoid species such as *M. truncatula, L. japonicus*, or *P. sativum*. If a comparison at this level does not indicate selection, then the significant K_a/K_s ratio for the common bean/soybean comparison would mean that the gene may be important in the evolution of the Phaeoleae lineage. Likewise, if the comparison to one of the Galegoid genes is still significant, a similar comparison to an ortholog from an outgroup would allow us to determine if it is important to the legume lineage. Such comparisons, though, would require significant data from a member of the Rosales or Cucurbitales, data which is currently not available in significant depth.

Genes undergoing positive selection deserve a more detailed analysis because they may encode functional changes that drove evolution of a specific taxonomic lineage. With genome-wide sequence data for legumes, similar calculations can be performed to identify genes important to a specific lineage. Additionally, a sliding-window calculation of the K_a/K_s ratio across the gene can identify specific regions of the gene that were strongly affected by selection (Choi and Lahn 2003).

The polymorphism data generated by these K_a/K_s studies can then be used in a manner described by Wright et al. (2005) and Yamasaki et al. (2005). The advantage, though, is that this genome-wide K_a/K_s survey will act as a prefilter to identify candidate genes in the common bean/soybean lineage. For example, a genome-wide scan identified 13,454 human-chimpanzee orthologs, of which 585 had a K_a/K_s ratio greater than 1 (CSAC 2005). These are logical candidate genes for further studies of adaptation in that lineage. By comparison, Yamasaki et al. (2005) prescreened 1,095 sequences and identified eight candidates for maize adaptation. Using rice as a reference for the number of genes (IRGSP 2005), this prescreen only considered 3% of the genes. Clearly, it would be more efficient to use a genome-wide approach than random searches through a subset of the genome to select genes to study adaptation. To apply this approach to the study of adaptation to these two socially and economically important legumes, we simply need the resources and will to collect common bean sequence to the same depth as soybean.

References

Adams, M.D., Celniker, S.E., Holt, R.A., Evans, C.A., Gocayne, J.D. et al. (2000) The genome sequence of the *Drosophila melanogaster*. Science 287, 2185–2195.

AGI (2000) Analysis of the genome sequence of the flowering plant *Arabidopsis thaliana*. Nature 408, 796–815.

Aragão, F., Ribeiro, S., Barros, L., Brasileiro, A., Maxwell, D., Rech, E. and Faria, J. (1998) Transgenic beans (*Phaseolus vulgaris* L.) engineered to express viral antisense RNAs show delayed and attenuated symptoms to bean golden mosaic geminivirus. Mol. Breed. 4, 491–499.

Aragão, F.J.L., Barros, L.M.G., de Sousa, M.V., Grossi de Sa, M.F., Almeida, E.R.P., Gander, E.S. and Rech, E.L. (1999) Expression of a methionine-rich storage albumin from the Brazil nut (*Bertholletia excelsa* H.B.K., Lecythidaceae) in transgenic bean plants (*Phaseolus vulgaris* L., Fabaceae). Genetics & Mol. Biol. 22, 445–449.

Aragão, F.J.L., Vianna, G.R., Albino, M.M.C. and Rech, E.L. (2002) Transgenic dry bean tolerant to the herbicide glufosinate ammonium. Crop Sci. 42, 1298–1302.

Aroca, R., Ferrante, A., Vernieri, P. and Chrispeels, M.J. (2006) Drought, abscisic acid and transpiration rate effects on the regulation of PIP aquaporin gene expression and abundance in *Phaseolus vulgaris* plants. Ann. Bot. 98, 1301–1310.

Arumuganathan, K. and Earle, D.E. (1991) Nuclear DNA content of some important plant species. Plant Mol. Biol. Report. 9, 208–218.

Becerra Velásquez, V.L., and Gepts, P.(1994) RFLP diversity in common bean (*Phaseolus vulgaris* L.). Genome 37, 256–263.

Beebe, S.E., Rojas-Pierce, M., Yan, X., Blair, M.W., Pedraza, F., Muñoz, F., Tohme, J. and Lynch, J.P. (2006) Quantitative trait loci for root architecture traits correlated with phosphorus acquisition in common bean, Crop Sci. 46, 413–423.

Beebe, S.E., Skroch, P.W., Thome, J., Duque, M.C., Pedraza, F., Nienhuis, J. (2000) Structure of genetic diversity among common bean landraces of middle American origin based on correspondance analysis of RAPD. Crop Sci. 40, 264–273.

Bennett, M. and Leitch, I. (1995) Nuclear DNA amounts in angiosperms. Ann. Bot. 76, 113–176.

Bennett, M. and Leitch, I. (2005) Plant DNA C-Values database. Release 4.0 http://www.rbgkew.org.uk/cval/database1.html.

Blanc, G., and Wolfe, K.H. (2004) Widespread paeleopolyploidy in model plant species inferred from age distributions of duplicate genes. Plant Cell 17, 1667–1678.

Blair, M.W., Giraldo, M.C., Buendia, H.F., Tovar, E., Duque, M.C., and Beebe, S.E. (2006) Microsatellite marker diversity in common bean (*Phaseolus vulgaris* L.) Theor. Appl. Genet. 113, 100–109.

Blair, M.W., Pedraza, F., Buendia, H.F., Gaitán-Solís, E., Beebe, S.E., Gepts, P., and Tohme, J. (2003) Development of a genome-wide anchored microsatellite map for common bean (*Phaseolus vulgaris* L.) Theor. Appl. Genet. 107, 1362–1374.

Blyden, E.R., Doerner, P.W., Lamb, C.J. and Dixon, R.A. (1991) Sequence analysis of a chalcone isomerase cDNA of Phaseolus vulgaris L. Plant Mol. Biol. 16, 167–169.

Broughton, W.J., Hernandez, G., Blair, M., Beebe, S., Gepts, P. and Vanderleyden, J. (2003) Beans (*Phaseolus* spp.) - model food legumes. Plant and Soil 252, 55–128.

Cadle-Davidson, M. and Jahn, M. (2006) Differential gene expression in *Phaseolus vulgaris I* locus NILs challenged with bean common mosaic virus. Theor. Appl. Genet. 112, 1452–1457.

CESC (1998) Genome sequence of the nematode *C. elegans*: a platform for investigating biology. Science 282, 2012–2018.

Choi, S.S., and Lahn, B.T. (2003) Adaptive evolution of MRG, a neuron-specific gene family implicated in nociception. Genome Res. 13, 2252–2259.

Choi, H.K., Mun, J.H., Kin, D.J., Zhu, H., Baek, J.M., Mudge, J., Roe, B., Ellis, N., Doyle, J., Kiss, G.B., Young, N.D., and Cook, D.R. (2004) Estimating genome conservation between crop and model legume species. Proc. Natl. Acad. Sci., USA 101, 15289–15294.

Crepet, W.L., Nixon, K.C., and Gandolfo, M.A. (2004) Fossil evidence and phylogeny: the age of major angiosperm clades based on mesofossil and macrofossil evidence from Cretaceous deposits. Amer. J. Bot. 91, 1666–1682.

CSAC (2005) Initial sequence of the chimpanzee genome and comparison with the human genome. Nature 437, 69–87.

Debouck, D.G., Toro, O., Paredes, O.M., Johnson, W.C., and Gepts, P. (1993) Genetic diversity and ecological distribution of *Phaseolus vulgaris* in northwestern South America. Econ. Bot. 47, 408–423.

DeClercq, J., Zambre, M., Van Montagu, M., Dillen, W. and Angenon, G. (2002) An optimized Agrobacterium-mediated transformation procedure for *Phaseolus acutifolius* A. Gray. Pl. Cell Reports 21, 333–340.

Delgado-Salinas, A.., Bibler, R. and Lavin, M. (2006) Phylogeny of the genus *Phaseolus* (Leguminosae): a recent diversification in an ancient landscape. Syst. Bot. 56, in press.

Doyle, J.J., and Luckow, M.A. (2003) The rest of the iceberg. Legume diversity and evolution in a phylogenetic context. Plant Phys. 131, 900–910.

Estrada-Navarrete, G., Alvarado-Affantranger, X., Olivares, J.-E., Díaz-Camino, C., Santana, O., Murillo, E., Guillén, G., Sánchez-Guevara, N., Acosta, J., Quinto, C., Li, D., Gresshoff, P.M,. and Sánchez, F. (2006) *Agrobacterium rhizogenes*-transformation of the *Phaseolus* spp.: A tool for functional genomics. Mol. Plant Microbe Interact. 19, 1385–1393.

Flicek, P., Keibler, E. Hu, P. Korf, I., and Brent, M.R. (2003) Leveraging the mouse genome for gene prediction in human: from whole-genome shotgun reads to a global synteny map. Genome Res. 13, 46–54.

Flint, J., Tufarelli, C., Peden, J., Clark, K., Daniels, R.J., Hardison, R., Miller, W., Philipsen, S., Tan-Un, K.C., McMorrow, T., Frampton, J., Alter, B.P., Frischauf, A.M., and Higgs, D.R. (2001) Comparative genome analysis delimits a chromosomal domain and identifies key regulatory elements in the alpha globin cluster. Hum. Mol. Genet. 10, 371–382.

Freyre, R., Skroch, P.W., Geffroy, V., Adam-Blondon, A.F., Shirmohamadali, A.., Johnson, W.C., Llaca, V., Nodari, R.O., Pereira, P.A.., Tsai, S.M., Tohme, J., Dron, M., Nienhuis, J., Vallejos, C.E., and Gepts.,P. (1998) Towards an integrated linkage map of common bean. 4. Development of a core linkage map and alignment of RFLP maps. Theor. Appl. Genet. 97: 847–856.

Gepts, P. (1990) Biochemical evidence bearing on the domestication of *Phaseolus* (Fabaceae) beans. Econ. Bot. 44, 22–38.

Gepts, P. (1998) Origin and evolution of common bean: past events and recent trends. HortSci. 33, 1124–1130.

Gepts, P., Aragão, F., de Barros, E., Blair, M.W., Brondani, R., Broughton, W.J., Galasso, I., Hernández, G., Kami, J., Lariguet, P., McClean, P., Melotto, M., Miklas, P., Pauls, P., Pedrosa-Harand, A., Porch, T., Sánchez, F., Sparvoli, F. and Yu, K. (2007) Genomics of *Phaseolus* beans, a major source of dietary protein and micronutrients in the Tropics. *In*: P.H. Moore and R. Ming (Eds), *Genomics of Tropical Crop Plants*. Springer, Berlin, in press.

Gepts, P., and Bliss, F.A. (1986) Phaseolin variability among wild and cultivated common beans (*Phaseolus vulgaris*) from Colombia. Econ. Bot. 40, 469–478.

Gepts, P, and Debouck, D. 1991. Origin, domestication, and evolution of common bean (*Phaseolus vulgaris* L.). *In*: A. van Schoonhoven and O. Voyest (Eds.) *Common Beans: Research for Crop Improvement*. C.A.B. Intl., Wallingford, UK and CIAT, Cali, Colombia, pp. 7–53.

González, A., Wong, A., Delgado-Salinas, A., Papa, R., and Gepts, P. (2005) Assessment of inter simple sequence repeat markers to differentiate sympatric wild and domesticated populations of common bean (*Phaseolus vulgaris* L.). Crop Sci. 35, 606–615.

Guo, H., and Moose, S.P. (2003). Conserved noncoding sequences among cultivated cereal genomes identify candidate regulatory sequence elements and patterns of promoter evolution. Plant Cell 15, 1143–1158.

Hoffman, L.M., and Donaldson, D.D. (1985) Characterization of two *Phaseolus vulgaris* phytohemagglutinin genes closely linked on the chromosome. EMBO J. 4, 883–889.

Inada, D.C., Bashir, A., Lee, C., Thomas, B.C., Ko, C., Goff, S.A., and Freeling, M. (2003). Conserved noncoding sequences in the grasses. Genome Res. 13, 2030–2041.

Inoue, S., Konoshita, T., and Simazaki, K. (2005) Possible involvement of phototropins in leaf movement of kidney bean in response to blue light. Plant Physiol. 138, 1994–2004.

IRGSP (2005) The map-based sequence of the rice genome. Nature 436, 793–800.

Islam, F.M.A., Basford, K.E., Jara, C., Redden, R.L., and Beebe, S. (2002) Seed compositional and disease resistance differences among gene pools in cultivated common bean. Genet. Resour. Crop. Evol. 49, 285–293.

Jackson, S.A., Rokshar, D., Stacey, G., Shoemaker, R.C., Schmutz, J., and Grimwood, J. (2006) Toward a reference sequence of the soybean genome: a multiagency effort. Crop Sci. 46, S-55–61.

Kami, J., Becerra Velásquez, B., Debouck, D.G., and Gepts, P. (1995) Identification of presumed ancestral DNA sequences of phaseolin in *Phaseolus vulgaris*. Proc. Natl. Acad. Sci. U.S.A. 92, 1101–1104.

Kami, J., Poncet, V., Geffroy, V. and Gepts, P. (2006) Development of four phylogenetically-arrayed BAC libraries and sequence of the APA locus in *Phaseolus vulgaris*. Theor. Appl. Genet. 112, 987–998.

Kaplan, L., and Lynch, T.F. (1999) *Phaseolus* (Fabaceae) in archaeology: AMS radiocarbon dates and their significance for pre-Columbian agriculture. Econ. Bot. 53, 261–272.

Kellis, M., Patterson, N., Endrizzi, M., Birren, B., and Lander, E.S. (2003) Sequencing and comparison of yeast species to identify genes and regulatory elements. Nature 423, 242–254.

Kelly, J.D., Gepts, P., Miklas, P.N. and Coyne, D.P. (2003) Tagging and mapping of genes and QTL and molecular marker-assisted selection for traits of economic importance in bean and cowpea. Field Crops Res. 82, 135–154.

Koenig, R., and Gepts, P. (1989) Allozyme diversity in wild *Phaseolus vulgaris*: further evidence for two major centers of diversity. Theor. Appl. Genet. 78, 809–817.

Kumar, S., Filipski, A., Swarna, V., Walker A., and Hedges, B.S. (2005) Placing confidence limits on the molecular age of the human–chimpanzee divergence. Proc. Natl, Acad. Sci., U.S.A 102, 18842–18847.

Lai, J., Ma, J., Swigonova, Z., Ramakrishna, W., Linton, E., Llaca, V., Tanyolac, B., Park, Y.J., Jenong, O.Y., Bennetzen, J.L., and Messing, J. (2004) Gene loss and movement in the maize genome. Genome Res. 14, 1924–1931.

Lavin, M., Herendeen, P.S., and Wojciechowski, M.F. (2005) Evolutionary rates analysis of Leguminosae implicates a rapid diversification of lineages during the Tertiary. System. Biol. 54, 530–549.

Lewis, G., Schrire, B., Mackinder, B., and Lock, M. (2005) *Legumes of the World*. Royal Botanic Gardens, Kew Publishing; London.

Lioi, L., Sparvoli, F., Galasso, I., Lanave, C., and Bollini, R. (2003) Lectin-related resistance factors against bruchids evolved through a number of duplication events. Theor. Appl. Genet. 107, 814–822.

Liu, Z.C., Park, B.J., Kanno, A., and Kameya, T. (2005) The novel use of a combination of sonication and vacuum infiltration in *Agrobacterium*-mediated transformation of kidney bean (*Phaseolus vulgaris* L.) with lea gene. Mol. Breed. 16, 189–197.

López, C.E., Acosta, I.F., Jara, C., Pedraza, F., Gaitán-Solís, E., Gallego, G., Beebe, S. and Tohme, J. (2003) Identifying resistance gene analogs associated with resistances to different pathogens in common bean. Phytopathology 93, 88–95.

Ma, J. SanMiguel, P., Lai, J., Messing, J., and Bennetzen, J.L. (2005) DNA rearrangement in orthologous *orp* regions of the maize, rice, and sorghum genomes. Genetics 170, 1209–1220.

Magallón, S., Crane, P.R., and Herendeen, P.S. (1999) Phylogenetic pattern, diversity, and diversification of eudicots. Ann. Mo. Bot. Gard. 86, 297–372.

Magallón, S., and Sanderson, M.J. (2001) Absolute diversification rates in angiosperms. Evolution 55, 1762–1780.

Maréchal, R. (1971) Observations sur quelques hybrides dans le genre *Phaseolus*. II. Les phénomènes méiotiques. Bull. Rech. Agron. Gembloux 6, 461–489.

Margis-Pinheiro, M., Martivet, J. and Burkard, G. (1994) Bean class IV chitinase gene: structure, developmental expression and induction by heat stress Plant Sci. 98, 163–173.

Margulies, E.H., and Green, E.D. (2003) Detecting highly conserved regions of the human genome by multispecies sequence comparison. Cold Spring Harb. Symp. Quant. Biol. 68, 255–263.

Margulies, E.H., Blanchette, M., Haussler, D., and Green, E.D. (2003) Identification and characterization of multi-species conserved sequences. Genome Res. 13, 2507–2518.

Margulies, E.H., Vinson, J.P., Miller, W., Jaffe, D.B., Lindblad-Toh, K., Chang, J.L., Green, E.D., Lander, E.S., Mullikin, J.C., and Clamp, M. (2005) An initial strategy for the systematic identification of functional elements in the human genome by low-redundancy comparative sequencing. Proc. Natl. Acad. Sci., USA. 102, 4795–4800.

Matsuoka, Y., Vigouroux, Y., Goodman, M.M., Sanchez, G.J., Buckler IV, E., and Doebley, J. (2002) A single domestication for maize shown by multilocus microsatellite genotyping. Proc. Natl. Acad. Sci., USA. 99, 6080–6084.

McCallum, C.M., Comai, L., Greene, E.A. and Henikoff, S. (2000) Targeting induced local lesions in genomes (TILLING) for plant functional genomics. Pl. Physiol. 123, 439–442.

McClean, P.E., Kami, J., and Gepts, P. (2004a) Genomics and genetic diversity in common bean. *In*: R.F. Wilson, H.T. Stalker, and E.C. Brummer (Eds.) *Legume Crop Genomics*. AOCS Press, Champaign, Illinois, USA, pp. 60–82.

McClean, P.E., Lee, R.K. and Miklas, P.N. (2004b) Sequence diversity analysis of dihydroflavanol reductase intron 1 in common bean. Genome 47, 266–280.

Melotto, M., Coelho, M.F., Pedrosa-Harand, A., Kelly, J.D. and Camargo, L.E.A. (2004) The anthracnose resistance locus *Co-4* of common bean is located on chromosome 3 and contains putative disease resistance-related genes. Theor. Appl. Genet. 109, 690–699.

Melotto, M., Monteiro-Vitorello, C.B., Bruschi, A.G., and Camargo, L.E. (2005) Comparative bioinformatic analysis of genes expressed in common bean (*Phaseolus vulgaris* L.) seedlings. Genome, 48, 562–570.

Miklas, P.N., Kelly, J.D., Beebe, S.E. and Blair, M.W. (2006) Common bean breeding for resistance against biotic and abiotic stresses: from classical to MAS breeding. Euphytica 147, 106–131.

Moscone, E.A., Klein, F., Lambrou, M., Fuchs, J. and Schweizer, D. (1999) Quantitative karyotyping and dual-color FISH mapping of 5S and 18S–25S rDNA probes in the cultivated *Phaseolus* species (Leguminosae). Genome 42, 1224–1233.

Mutlu, N., Miklas, P.N. and Coyne, D.P. (2005) Resistance gene analog polymorphism (RGAP) markers co-localize with disease resistance genes and QTL in common bean. Mol. Breed. 17, 127–135.

Myers, E.W., Sutton, G.G., Delcher, A.L., Dew, I.M., Fasulo, D.P., *et al.* (2000) A whole-genome assembly of *Drosophila*. Science 287, 2196–2204.

Nagl, W. (1974) The *Phaseolus* suspensor and its polytene chromosomes. Z. Pflanzenphysiol. 73, 1–44.

Nanni, L., Losa, A., Bellucci, E., Kater, M., Gepts, P. and Papa, R. (2005) Identification and molecular diversity of a genomic sequence similar to SHATTERPROOF (SHP1) in Phaseolus vulgaris L. Abstract, Plant and Animal Genome XIII, Jan. 15–19, 2005: http://www.intl-pag.org/pag/13/abstracts/PAG13_P470.html

Nenno, M., Schumann, K. and Nagl, W. (1994) Detection of rRNA and phaseolin genes on polytene chromosomes of *Phaseolus coccineus* by fluorescence in situ hybridization after pepsin treatment. Genome 37, 1018–1021.

Nodari, R.O., Koinange, E.M.K., Kelly, J.D., and Gepts, P. (1992) Towards an integrated linkage map of common bean. I. Development of genomic DNA probes and levels of restriction fragment length polymorphism. Theor. Appl. Genet. 84, 186–192.

Ochoa, I.E., Blair, M.W., and Lynch, J.P. (2006) QTL analysis of adventitious root formation in common bean under contrasting phosphorus availability. Crop Sci. 46, 1609–1621.

Pagel, J., Walling, J.G., Young, N.D., Shoemaker, R.C., and Jackson, S.A. (2004) Segmental duplications within the *Glycine max* genome revealed by fluorescence *in situ* hybrization of bacterial artificial chromosomes. Genome 47, 764–768.

Papa, R., and Gepts, P. (2003) Asymmetry of gene flow and differential geographical structure of molecular diversity in wild and domesticated common bean (*Phaseolus vulgaris* L.) from Mesoamerica. Theor. Appl. Genet. 106, 239–250.

Pedrosa, A., Vallejos, C.E., Bachmair, A., and Schweizer, D. (2003) Integration of common bean (*Phaseolus vulgaris* L.) linkage and chromosomal maps. Theor. Appl. Genet. 106, 205–212.

Pedrosa-Harand, A., Almeida, C.C.S., Mosiolek, M., Blair, M.W., Schweizer, D. and Guerra, M. (2006) Extensive ribosomal DNA amplification during Andean common bean (*Phaseolus vulgaris* L.) evolution. Theor. Appl. Genet. 112, 924–933.

Powell, W., Morgante, M., Doyle, J.J., McNicol, J.W., Tingey, S.V., and Rafalski, A.J. 1996. Genepool variation in genus *Glycine* subgenus soja revealed by polymorphic nucler and chloroplast microsatellites. Genetics 144, 793–803.

Ramirez, M., Graham, M.A., Blanco-Lopez, L., Silvente, S., Medrano-Soto, A., Blair, M.W., Hernandez, G., Vance, C.P. and Lara, M. (2005) Sequencing and analysis of common bean ESTs. Building a foundation for functional genomics. Pl. Physiol. 137, 1211–1227.

Rivkin, M.I., Vallejos, C.E., and McClean, P.E. 1999. Disease-resistance related sequences in common bean. Genome 42, 41–47.

Rodriguez-Uribe, L. and O'Connell, M.A. (2006) A root-specific bZIP transcription factor is responsive to water deficit stress in tepary bean (*Phaseolus acutifolius*) and common bean (*P. vulgaris*). Plant Sci. 171, 300–307.

Ryder, T.B., Hedrick, S.A., Bell, J.N., Liang, X.W., Clouse, S.D. and Lamb, C.J. (1987) Organization and differential activation of a gene family encoding the plant defense enzyme chalcone synthase in Phaseolus vulgaris. Mol. Gen. Genet. 210, 219–233.

Salamov, A.A., and Solovyev, V.V. (2000) *Ab initio* gene finding in *Drosophila* genomic DNA. Genome Res. 10, 516–522.

Schlueter, J.A., Dixon, P., Granger, C., Grant, D., Clark, L., Doyle, J.J., and Shoemaker, R.C. (2004) Mining EST databases to resolve evolutionary events in major crop species. Genome 47, 868–876.

Shoemaker, R.C., Polzin, K., Labate, J., Specht, J., Brummer, E.C., Olson, T.K., Young, N., Concibido, V., Wilcox, J., Tamulonis, J.P., Kochert, G., and Boerma, H.R. (1996) Genome duplications in soybean (*Glycine* subgenus *soja*). Genetics 144, 329–338.

Shultz, J.L., Kurunam, D., Shopinski, K., Iqbal, M.J., Kazi, S., Zobrist, K., Bashir, R., Yaegashi, S., Lavu, N., Afzal, A.J., Yesudas, C.R., Kassem, M.A., Wu, C.C., Zhang, H.B., Town, C.D., Meksem, K. and Lightfoot, D.A. (2006) The Soybean Genome Database (SoyGD): a browser for display of duplicated, polyploid, regions and sequence tagged sites on the integrated physical and genetic maps of *Glycine max*. Nucl. Acids Res. 34, D758–D765.

Singh, S.P., Gepts, P., and DeBouck, D.G. (1991). Races of common bean (*Phaseolus vulgaris*, Fabaceae). Econ.Bot. 45, 379–386.

Stevens, P.F. (2001) Angiosperm Phylogeny Website. Version 7, December 2006 and continuously updated (http://www.mobot.org/MOBOT/research/APweb/).

Sun, S.M., Slightom, J.L., and Hall, T.C. (1981) Intervening sequences in a plant gene – comparison of the partial sequence of the cDNA and genomic DNA of French bean phaseolin. Nature 289, 37–41.

Thomas, J.W., Touchman, J.W., Blakesley, R.W., Bouffard, G.G., Beckstrom-Sternberg, S.M., *et al.* (2003) Comparative analyses of multi-species sequences from targeted genomic regions. Nature 424, 788–793.

Tohme, J., González, D.O., Beebe, S., and Duque, M.C. (1996) AFLP analysis of gene pools of a wild bean core collection. Crop Sci. 36, 1375–1384.

Tuskan, G.A., Difazio, S., Jansson, S., Bohlmann, J., Grigoriev, I., *et al.* (2006) The genome of black cottonwood, *Populus trichocarpa* (Torr. & Gray). Science 313, 1596–1604.

Vallejos, C.E., Astua-Monge, G., Jones, V., Plyler, T.R., Sakiyama, N.S., and Mackenzie, S.A. (2006) Genetic and molecular characterization of the *I* locus of *Phaseolus vulgaris*. Genetics 172, 1229–1242.

Vallejos, C.E., Sakiyama, N.S., and Chase, C.D. (1992) A molecular marker-based linkage map of *Phaseolus vulgaris* L. Genetics 131, 733–740.

Veltcheva, M., Svetleva, D., Petkova, S. and Perl, A. (2005) *In vitro* regeneration and genetic transformation of common bean (*Phaseolus vulgaris* L.) - problems and progress. Scientia Horticulturae 107, 2–10.

Venter, J.C., Adams, M.D., Myers, E.W., Li, P.W., Mural, R.J. *et al.* (2001) The sequence of the human genome. Science 291, 1304–1351.

Walling, J.G., Pires, J.C., and Jackson, S.A. (2005) Preparation of samples for comparative studies of plant chromosomes using *in situ* hybridization methods. Methods in Enzymology 395, 442–459.

Walter, M.H., Liu, J.W., Grand, C., Lamb, C.J. and Hess, D. (1990) Bean pathogenesis-related (PR) proteins deduced from elicitor-induced transcripts are members of a ubiquitous new class of conserved PR proteins including pollen allergens. Mol. Gen. Genet. 222, 353–360.

Wei, C., and Brent, M.R. (2006) Using ESTs to improve the accuracy of *de novo* gene prediction. BMC Bioinformatics 7, 327.

Wikström, N., Savlainen, V., and Chase, M.W. (2001) Evolution of the angiosperms: calibrating the family tree. Proc. Royal Soc. London, B, Biol. Sci. 268, 2211–2220.

Wing, R.A., Ammiraju, J.S., Luo, M., Kim, H. Yu, Y. Kudrna, D., Goicoechea, J.L., Wang, W., Nelson, W., Rao, K., Brar, D., Mackill, D.J., Han, B., Soderlund, C., Stein, L., SanMiguel, P., and Jackson, S. (2005) The *Oryza* map alignment project: the golden path to unlocking the genetic potential of wild rice species. Plant Mol. Biol. 59, 53–62.

Woolfe, A., Goodson, M., Goode, D.K., Snell, P., McEwen, G.K., Vavouri, T., Smith, S.f., North, P., Callaway, H., Kelly, K, Walter, K., Abnizova, I., Gilks, W., Edwards, Y.J., Cooke, J.E., and Elgar, G. (2005) Highly expressed non-coding sequences are associated with vertebrate development. PLoS Biol. 3, e7.

Wright, S.I., Bi, I.V., Schroeder, S.G., Yamasaki, M., Doebley, J.F., McMullen, M.D., Gaut, B.S. (2005). The effects of artificial selection on the maize genome. Science 308, 1310–1314.

Xu, D.H. and Gai, Y. (2003) Genetic diversity of wild and cultivated soybeans growing in China revealed by RAPD analysis. Plant Breeding 122, 503–506.

Yamasaki, M., Tenaillon, M.I., Bi, I.V., Schroeder, S.G,, Sanchez-Villeda, H., Doebley, J.F., Gaut, B.S., McMullen, M.D. (2005) A large-scale screen for artificial selection in maize identifies candidate agronomic loci for domestication and crop improvement. Plant Cell 17,2859–2872.

Yang, S.H., and Zeevaart, J.A. (2006) Expression of ABA 8′-hydroxylases in relation to leaf water relations and seed development in bean. Plant J. 47, 675–686.

Zambre, M., Goossens, A., Cardona, C., Van Montagu, M., Terryn, N. and Angenon, G. (2005) A reproducible genetic transformation system for cultivated *Phaseolus acutifolius* (tepary bean) and its use to assess the role of arcelins in resistance to the Mexican bean weevil. Theor. Appl, Genet. 110, 914–924.

Part II
Tools, Resources and Approaches

Chapter 5
The Soybean Molecular Genetic Linkage Map

Perry B. Cregan

Introduction

Like most plant and animal species, the genesis of molecular genetic mapping in soybean was the landmark work describing the construction of a genetic linkage map using restriction fragment length polymorphisms (RFLPs) by Botstein et al. (1980). It was this approach for the rapid discovery of DNA-based markers that provided the impetus for a revolution in the molecular analyses of plant, animal and other genomes. It made possible the rapid discovery of large numbers of molecular markers and the development of high density genetic linkage maps. Similarly, the advent of polymerase chain reaction (PCR) technology (Mullis et al. 1986) provided a simple method for the detection of DNA sequence polymorphism and led to the development of a number of additional classes of DNA markers. Among these were random amplified polymorphic DNA (RAPD) (Williams et al. 1990) or arbitrary primer PCR (AP-PCR) markers (Welsh and McClelland 1990); SSR or microsatellite markers (Litt and Luty 1989; Tautz 1989; Weber and May 1989); and amplification fragment length polymorphism (AFLP) markers (Vos et al. 1995). Finally, as a result of the availability of high throughput and relatively inexpensive DNA sequencing technology, SNP markers have steadily gained in prominence in numerous plant and animal species.

Genetic Maps Based on RFLP Markers

The first RFLP-based map of the soybean genome was published by Keim et al. (1990). Because of the relatively low level of RFLP in soybean (Apuya et al. 1988), as compared to maize (*Zea maize* L.) in which a large number of RFLP alleles are frequently observed among cultivated genotypes (Helentjaris et al. 1986), Keim

P.B. Cregan
Soybean Genomics and Improvement Laboratory, U.S. Department of Agriculture, Agricultural Research Service, Beltsville, Maryland, MD 20705, USA
e-mail: creganp@ba.ars.usda.gov

G. Stacey (ed.), *Genetics and Genomics of Soybean*,

Table 5.1 A summary of soybean molecular genetic linkage maps and numbers of various types of DNA markers positioned in each

Mapping population and literature citation	Population size and type	Linkage groups - no. -	Map length - cM -	DNA marker type: Total	RFLP	RAPD	SSR	AFLP	SNPs
				....no....					
USDA/Iowa St. (A81-356022 × *G. soja* PI 468.916)	59 F_2 lines								
Keim et al. (1990)		26	1,200	150	150				
Diers et al. (1992)		31	2,147	238	238				
Shoemaker and Olson (1993)		23	3,371	365	355	10			
Shoemaker and Specht (1995)		25	2,473	368	358	10			
Cregan et al. (1999)		23	3,003	1,008	501	10	486	11	
E. I. DuPont Bonus × *G. soja* PI 81762	68 F_2 lines								
Rafalski and Tingey (1993)		21	2,678	600+	600+				
Univ. of Utah (Minsoy × Noir 1)									
Lark et al. (1993)	69 F_3 families	31	1,551	132	132				
Mansur et al. (1996)	240 RILs	35	1,981	265	224		41		
Cregan et al. (1999)	240 RILs	22	2,413	621	209		412		
Univ. of Nebraska (Clark × Harosoy isolines)	57 F_2 lines								
Shoemaker and Specht (1995)		26	1,004	83	80	3			
Akkaya et al. (1995)		29	1,486	118	80	4	34		
Cregan et al. (1999)		28	2,787	514	95	69	339	11	
PI 437654 × BSR301	42 RILs								
Keim et al. (1997)		28	3,441	840	165	25		650	
Ferreira et al. (2000)		35	3,275	356	250	106			
Changnong 4 × Xinmin 6	88 RILs								
Liu et al. (2000)		22	3,713	237	100	62	33	42	
Misuzudaizu × Moshidou Gong 503	190 F_2 plants								
Yamanaka et al. (2001)		20	2,909	498	401	1	96		
Kefeng 1 × Nannong 1138-2	184 RILs								
Wu et al. (2001)		24	2,321	792					
Noir 1 × BARC-2	149 F_2 plants								
Matthews et al. (2001)		35	1,400	186	39	17	25	105	
Essex × Forrest	100 RILs								
Lightfoot et al. (2005)		20	2,823	337	41	90	206		
Consensus Map									
Song et al. (2004)	Multiple	20	2,524	1,803	709	73	1,015	6	
Choi et al. (2007)	populations	20	2,389	2,982	634	4	983		1,361

et al. (1990) constructed their map from a population derived from a cross of cultivated soybean A81-356022 × *G. soja* PI 468916. This map contained a total of 150 RFLP and three classical loci and was 1,200 centiMorgans in length with 26 linkage groups. Subsequent iterations of this map by Diers et al. (1992), Shoemaker and Olson (1993) and Shoemaker and Specht (1995) added more than 200 RFLP loci and resulted in a map 2,473 cM in length with 25 linkage groups (Table 5.1). Researchers at the E.I. DuPont Corporation (Rafalski and Tingey 1993) also used a *G. max* × *G. soja* cross to create a RFLP map with more than 600 loci. The relatively low level of polymorphism in cultivated soybean made the application of RFLP difficult in cultivated soybean populations. However, Lark et al. (1993) developed a small population of F2–3 families from a cross of the *G. max* cultivars Minsoy and Noir 1 and successfully mapped 132 RFLP loci. A population of 240 recombinant inbred lines was subsequently created from a cross of Minsoy × Noir 1 (Mansur et al. 1996) which has been widely used in soybean molecular genetic mapping. Another molecular map that was predominantly build using RFLP markers was that developed by Yamanaka et al. (2001) which was constructed using 190 F2 plants and included 401 RFLP loci. In addition, 250 RFLP loci were positioned on a map constructed from a cross of the PI 437654 × BSR301 which consisted of 42 RILS (Ferreira et al. 2000; Keim et al. 1997).

Genetic Maps Based on RAPD and AFLP Markers

RAPD (Williams et al. 1990) or AP-PCR markers (Welsh and McClelland 1990) are PCR based, require no prior knowledge of DNA sequence and can be analyzed based simply as the presence or absence of an amplicon via agarose gel electrophoresis. RAPDs provided an appealing alternative to RFLP and Southern blot analysis. However, while soybean geneticists used RAPDs extensively in germplasm classification (Brown-Guedira et al. 2000; Li and Nelson 2002; Thompson et al. 1998), only the genetic map developed by Ferreira et al. (2000) contained a large number of RAPD loci. In contrast, an extensive AFLP-based map was constructed by Keim et al. (1997). A total of 650 AFLP loci were mapped in the 42 RILs of the PI 437654 × BSR 101 population described earlier. The authors noted significant clustering of AFLP markers generated using *Eco*RI/*Mse*I restriction enzymes. Thirty-four percent of the loci displayed dense clustering.

Simple Sequence Repeat (SSR) or Microsatellite Markers

The discovery, development and use of microsatellite or SSR markers in human (Litt and Luty 1989; Tautz 1989; Weber and May 1989), prompted soybean researchers to investigate the use SSRs. Indeed, the first demonstration of SSR allelic variation and heritability in a plant species was reported in soybean (Akkaya et al. 1992; Morgante and Olivieri 1993). Akkaya et al. (1992) found up to eight SSR alleles at one locus in a group of 38 *G. max* and five *G. soja* genotypes. Further reports of SSR allelic variation in soybean (Maughan et al. 1995; Morgante et al. 1994; Rongwen et al. 1995) detected very high levels of allelic variation including one locus with 26 alleles among a group of 91 cultivated and five wild soybean genotypes. The first report describing the mapping of SSR loci in soybean (Akkaya et al. 1995) found no evidence of clustering of SSR loci. The high level of polymorphism combined with the random distribution on the genetic map as well as their single locus nature and analysis via PCR suggested that SSRs were an excellent complement to RFLP markers for use in soybean molecular biology, genetics, and plant breeding research.

An SSR-based Soybean Genome Map

With the support of the United Soybean Board, a project was initiated in 1995 to develop and map a large set of soybean SSR markers. A collaborative project between workers at the USDA, Beltsville, MD and Ames, IA; the Univ. of Nebraska and the Univ. of Utah as well as BioGenetic Services Inc. of Brookings, SD resulted in the development and mapping of more than 600 SSR loci in one, two, and occasionally three different mapping populations (Cregan et al. 1999). One population was the USDA/Iowa St. *G. max* × *G. soja* population and the second, the expanded 240 RIL Univ. of Utah population developed from the cultivars Minsoy and Noir 1. The third population was the Univ. of Nebraska Clark × Harosoy isoline population consisting of 57 F_2-derived lines. A total of 187 loci were mapped in each of the three populations while many more loci were common to any two of the populations. Thus, because of the single locus nature of the SSRs used in this mapping project, it was a simple matter to align homologous linkage groups and to create the first soybean linkage map with 20 linkage groups which were assumed to correspond to the 20 soybean chromosomes. The 20 sets of aligned linkage groups included a total of 1,423 unique marker loci including 606 SSRs, 689 RFLP and 26 classical loci. Based upon the mapping of the 20 classical loci and linkage reports in the literature, all but one of the 20 classical linkage groups (Palmer and Shoemaker 1998) were assigned to a corresponding molecular linkage group.

The consensus soybean linkage map with an average of more than 30 SSR markers per linkage group (Cregan et al. 1999) was useful for the alignment of linkage groups in pre-existing or newly created linkage maps with the consensus linkage groups in the SSR-based genome map. A small number of SSRs could be used to associate linkage groups with corresponding linkage groups on the SSR-based consensus map. Wu et al. (2001) mapped 792 markers in a population of 201 RILs from a cross of the genotypes Kefeng 1 × Nannong 1138-2 and aligned their linkage groups with the consensus linkage groups via the analysis of more than 200 SSRs common to the Cregan et al. (1999) map. Similarly, Yamanaka et al. (2001) created a population of 190 F_2 plants from a cross of Misuzudaizu × Moshidou Gong 503 and mapped 401 RFLP loci along with 96 SSRs. The resulting map consisted of 20 major linkage groups and a total length of 2,908.7 cM. Matthews et al. (2001)

successfully positioned cDNA and genomic clones on consensus linkage groups in the SSR-based genome map using a small number of SSR loci to align corresponding linkage groups. Similarly, Lightfoot et al. (2005) developed an F5-derived RIL population from a cross of the U.S. cultivars Essex and Forrest. A total of 337 loci were genetically mapped including 206 SSRs allowing the alignment of linkage groups in the resulting map with the 20 consensus linkage groups.

An updated version of the SSR-based linkage map was published by Song et al. (2004). This map included 420 additional SSR markers that were mapped in one or more of the three populations included in the Cregan et al. (1999) map as well as two additional RIL populations from the Univ. of Utah; Minsoy × Archer and Archer × Noir 1. The resulting consensus map of the five populations was created using JoinMap (Van Ooijen and Voorrips 2001) software. This integrated genetic map spans 2,523.6 cM of Kosambi map distance across 20 linkage groups that contained 1,849 markers, including 1,015 SSRs, 709 RFLPs, 73 RAPDs, 24 classical traits, six AFLPs and ten isozymes.

Single Nucleotide Polymorphisms (SNPs)

Because of a lack of definition of the soybean gene-space that is thought to be clustered in approximately 25% of the genome, Mudge et al. (2004) and Stacey et al. (2004) suggested that 2,000–3,000 cDNA sequences be placed on the physical map. Alternatively, coding sequences could be genetically mapped onto the existing SSR-based map. Such a genetic map would indicate the positions of coding sequences and would also provide information on the relative positions of genes with existing SSR and RFLP markers. The mapping of SSRs from EST (expressed sequence tag) sequence is one means to position genes on the genetic map. This approach has worked successfully in a number of species including *Medicago truncatula* (Eujayl et al. 2004), wheat (*Triticum aestivum* L.) (Yu et al. 2004) and rice (*Oryza sativa* L.) (Eujayl et al. 2004; Gao et al. 2004; Yu et al. 2004) for the genetic mapping of the ESTs. In contrast, only a small proportion of soybean ESTs contain polymorphic SSRs, as indicated by Song et al. (2004) who reported the successful development and mapping of only 24 polymorphic SSR markers from more than 136,000 soybean ESTs. However, the discovery of SNPs (which include single base changes and insertion/deletions) in genic sequence would provide a source of markers to expedite the positioning of genes on the genetic map. SNPs are now the marker of choice in human genetics studies and as of February 2007, 27.8 million human SNPs were catalogued by the National Center for Biotechnology Information SNP database, dbSNP, of which more 5.6 million had been validated. The discovery of SNPs in plant species is now accelerating and reports of SNP discovery and application in species including *Arabidopsis thaliana* (Jander et al. 2002; Nordborg et al. 2005; Schmid et al. 2005, 2003) maize (*Zea mays* L.) (Ching et al. 2002; Tenaillon et al. 2001), rice (Feltus et al. 2004; Nasu et al. 2002), barley (Rostoks et al. 2005), poplar (*Populus trichocarpa* Torr. & Gray) (Tuskan et al. 2006) are now quite common.

SNPs in Soybean

Zhu et al. (2003) reported the frequency of SNPs in a diverse set of 25 soybean genotypes via the analysis of more than 76 kbp of coding sequence, untranslated regions (UTRs), introns, and genomic DNA in close proximity to coding sequence. The frequency of SNPs was reported in terms of nucleotide diversity (θ), the number of polymorphic sites per base in a genotypic sample corrected for sample size (Watterson 1975). θ was equal to 0.00053 in 28.7 kbp of coding sequence (an average of approximately 0.5 SNPs per kbp corrected for sample size) and was more than two-fold higher in UTRs, introns, and genomic DNA ($\theta = 0.00111$). The level of sequence diversity in soybean is low in comparison to species such as maize. For example, Wright et al. (2005) reported nucleotide diversity of $\theta = 0.00627$ in modern maize inbreds. The recent report by Hyten et al. (2006) indicates that the frequency of sequence variants in soybean is low due to historical genetic bottlenecks and low sequence diversity in soybean's wild ancestor, *G. soja*. Nonetheless, the nucleotide diversity in soybean is very similar to that in humans where most estimates indicate nucleotide diversity (θ) values of less than 0.001 (Sachidanandam et al. 2001); (Cargill et al. 1999; Halushka et al. 1999). Indeed, a calculation from Zhu et al. (2003), who reported the discovery of 280 SNPs in 76.4 kbp of sequence, indicated there are likely in excess of 4 million SNPs in cultivated soybean. While this estimate was based mostly on coding sequence and may be biased, it clearly suggests an adequate number of sequence variants for successful SNP discovery in soybean.

A Soybean Transcript Map

The most recent genetic linkage map of the soybean genome (Choi et al. 2007) was constructed via the JoinMap (Van Ooijen and Voorrips 2001) analysis of four mapping populations: the Univ. of Utah Minsoy × Noir 1 and Minsoy × Archer RIL populations, the USDA, ARS, Iowa State Univ. A81-356022 × *G. soja* PI 468916, and a population of 77 RILs derived from a cross of the cultivar Evans × *G. max* PI 209332 (Concibido et al. 1996). This map was built upon the SSR framework map defined by Song et al. (2004) with the objective of discovering and mapping SNPs in, or in close proximity to soybean genes. SNPs were discovered via the re-sequencing of sequence tagged sites (STS) developed by the analysis of 9,459 polymerase chain reaction primer sets, more than 8,500 of which were designed to soybean unigene sequence reported by Vodkin et al. (2004). The analysis of these primer sets resulted in the development of 4,240 STS which were re-sequenced in a set of six diverse *G. max* genotypes: Minsoy, Noir 1, Archer, Evans Peking and PI 209332. In the resulting 2.44 mbp of aligned sequence, a total of 5,551 SNPs were discovered, including 4,712 single base changes and 839 indels for an average nucleotide diversity of $\theta = 0.000997$ (an average of almost 1 SNP per kbp adjusted for the size of the population analyzed). At least one SNP was discovered in 2,032 of the 4,240 STS. A total of 1,158 SNPs derived from 1,141 different genes were positioned in the 20 soybean linkage groups. The SNP detection assays were conducted using single base extension on either the Sequenom MassARRAY^{TM} platform (Griffin et al. 1999) or the Luminex flow cytometer (Chen et al. 2000). The analysis of the observed genetic distances between adjacent genes *versus* the theoretical distribution based upon the assumption of a random distribution of genes across the 20 soybean linkage groups clearly indicated that genes were clustered. One interesting result of the transcript mapping on the SSR framework was the opportunity to better understand the relationship of SSRs and genes. Marek et al. (2001) had reported that end-sequences of soybean BAC-clones identified with *Pst*1-derived RFLP probes had 50% more gene-like sequences and 45% less repetitive sequence than end-sequences of BAC clones identified with SSR markers. This suggested that in relation to SSRs, RFLPs were more closely associated with gene-rich regions. However, an analysis of the relationship of SSR and RFLP loci and mapped transcripts on the new map of Choi et al. (2007) indicated no difference in the proximity of SSRs and RFLP loci to the closest flanking genic sequences. This result was similar to that reported by Morgante et al. (2002) who analyzed DNA sequence data from *Arabidopsis thaliana*, rice, soybean, maize and wheat. They concluded that the frequency of SSRs was significantly higher in transcribed regions and that SSRs are associated with low-copy portions of plant genomes rather than with regions of repetitive DNA.

The new SNP/SSR based transcript map of the soybean genome more than doubles the number of available sequence-based markers and will provide an important resource to soybean geneticists for quantitative trait locus discovery and map-based cloning, as well as to soybean breeders who increasingly depend upon marker assisted selection in cultivar improvement. All information relative to the SNP-containing STS, primer sequences and PCR conditions, SNP positions and SNP alleles, as well as the consensus maps of the 20 soybean linkage groups is available at http://bfgl.anri.barc.usda.gov/soybean/index.html. This map represents the first step in the development of a much denser SNP-based soybean linkage map. As reported by workers at the USDA, Beltsville at the recent Plant Animal Genome XV meeting in San Diego, CA (Hyten et al. 2007) the Illumina Inc. GoldenGate SNP detection assay (http://www.illumina.com/pages.ilmn?ID=11) functions extremely well in soybean. At this time, more than 1,200 additional SNP-containing fragments have been mapped using the GoldenGate assay. An additional set of 1,536 assays were designed and will be mapped in the near future. Thus, in the near term, the number of mapped SNP-containing STS will exceed 4,000. The Department of Energy, Joint Genome Institute sequence of the soybean genome, which will be completed in 2008, followed by re-sequencing of alternative genotypes will make *in silico* SNP discovery a relatively simple task and will greatly accelerate SNP discovery and the creation of a dense SNP-based soybean genome map.

References

Akkaya, M.S., Bhagwat, A.A., and Cregan, P.B. (1992) Length polymorphisms of simple sequence repeat DNA in soybean. Genetics 132, 1131–9.

Akkaya, M.S., Shoemaker, R.C., Specht, J.E., Bhagwat, A.A., and Cregan, P.B. (1995) Integration of simple sequence repeat DNA markers into a soybean linkage map. Crop Sci. 35, 1439–1445.

Apuya, N.R., Frazier, B.L., Keim, P., Roth, E.J., and Lark, K.G. (1988) Restriction fragment length polymorphisms as genetic markers in soybean, Glycine max (L.) Merr. Theor. Appl. Genet. 75, 889–901.

Botstein, D., White, R.L., Skolnick, M., and Davis, R.W. (1980) Construction of a genetic linkage map in man using restriction fragment length polymorphisms. Am. J. Hum. Genet. 32, 314–31.

Brown-Guedira, G.L., Thompson, J.A., Nelson, R.L., and Warburton, M.L. (2000) Evaluation of genetic diversity of soybean introductions and North American ancestors using RAPD and SSR markers. Crop Sci. 40, 815–823.

Cargill, M., Altshuler, D., Ireland, J., Sklar, P., Ardlie, K., Patil, N., Shaw, N., Lane, C.R., Lim, E.P., Kalyanaraman, N., Nemesh, J., Ziaugra, L., Friedland, L., Rolfe, A., Warrington, J., Lipshutz, R., Daley, G.Q., and Lander, E.S. (1999) Characterization of single-nucleotide polymorphisms in coding regions of human genes. Nat. Genet. 22, 231–8.

Chen, J., Iannone, M.A., Li, M.S., Taylor, J.D., Rivers, P., Nelsen, A.J., Slentz-Kesler, K.A., Roses, A., and Weiner, M.P. (2000) A microsphere-based assay for multiplexed single nucleotide polymorphism analysis using single base chain extension. Genome Res. 10, 549–57.

Ching, A., Caldwell, K.S., Jung, M., Dolan, M., Smith, O.S., Tingey, S., Morgante, M., and Rafalski, A.J. (2002) SNP frequency, haplotype structure and linkage disequilibrium in elite maize inbred lines. BMC Genet. 3, 19.

Choi, I.-Y., Hyten, D.L., Matukumalli, L.K., Song, Q.-J., Chaky, J.M., Quigley, C.V., Chase, K., Lark, K.G., Reiter, R.S., Yoon, M.-S., Hwang, E.-Y., Yi, S.-I., Young, N.D., Shoemaker, R.C., van Tassell, C.P., Specht, J.E., and Cregan, P.B. (2007) A soybean transcript map: gene distribution, haplotype and SNP analysis. Genetics (In press).

Concibido, V.C., Denny, R.L., Lange, D.A., Orf, J.H., and Young, N.D. (1996) RFLP mapping and marker-assisted selection of soybean cyst nematode resistance in PI 209332. Crop Sci. 36, 1643–1650.

Cregan, P.B., Jarvik, T., Bush, A.L., Shoemaker, R.C., Lark, K.G., Kahler, A.L., Kaya, N., VanToai, T.T., Lohnes, D.G., and Chung, J. (1999) An integrated genetic linkage map of the soybean genome. Crop Sci. 39, 1464–1490.

Diers, B.W., Keim, P., Fehr, W.R., and Shoemaker, R.C. (1992) RFLP analysis of soybean seed protein and oil content. Theor. Appl. Genet. 83, 608–612.

Eujayl, I., Sledge, M.K., Wang, L., May, G.D., Chekhovskiy, K., Zwonitzer, J.C., and Mian, M.A. (2004) Medicago truncatula EST-SSRs reveal cross-species genetic markers for Medicago spp. Theor. Appl. Genet. 108, 414–22.

Feltus, F.A., Wan, J., Schulze, S.R., Estill, J.C., Jiang, N., and Paterson, A.H. (2004) An SNP resource for rice genetics and breeding based on subspecies indica and japonica genome alignments. Genome Res 14, 1812–9.

Ferreira, A.R., Foutz, K.R., and Keim, P. (2000) Soybean genetic map of RAPD markers assigned to an existing scaffold RFLP map. J. Hered. 91, 392–396.

Gao, L.F., Jing, R.L., Huo, N.X., Li, Y., Li, X.P., Zhou, R.H., Chang, X.P., Tang, J.F., Ma, Z.Y., and Jia, J.Z. (2004) One hundred and one new microsatellite loci derived from ESTs (EST-SSRs) in bread wheat. Theor. Appl. Genet. 108, 1392–1400.

Griffin, T.J., Hall, J.G., Prudent, J.R., and Smith, L.M. (1999) Direct genetic analysis by matrix-assisted laser desorption/ionization mass spectrometry. Proc Natl Acad Sci U S A 96, 6301–6.

Halushka, M.K., Fan, J.B., Bentley, K., Hsie, L., Shen, N., Weder, A., Cooper, R., Lipshutz, R., and Chakravarti, A. (1999) Patterns of single-nucleotide polymorphisms in candidate genes for blood-pressure homeostasis. Nat. Genet. 22, 239–47.

Helentjaris, T., Slocum, M., Wright, S., Schaefer, A., and Nienhuis, J. (1986) Construction of genetic maps in maize and tomato using restriction fragment length polymorprhisms. Theor. Appl. Genet. 72, 761–769.

Hyten, D.L., Song, Q., Zhu, Y., Choi, I.Y., Nelson, R.L., Costa, J.M., Specht, J.E., Shoemaker, R.C., and Cregan, P.B. (2006) Impacts of genetic bottlenecks on soybean genome diversity. Proc. Natl. Acad. Sci. U.S.A. 103, 16666–71.

Hyten, D.L., Choi, I.-Y., Yoon, M.-S., Song, Q.-J., Specht, J.E., Nelson, R.L., Chase, K., Young, N.D., Lark, K.G., Shoemaker, R.C., and Cregan, P.B. (2007) An Assessment of Genome-wide Linkage Disequilibrium in Soybean. Plant & Animal Genome XV, San Diego, CA.

Jander, G., Norris, S.R., Rounsley, S.D., Bush, D.F., Levin, I.M., and Last, R.L. (2002) Arabidopsis map-based cloning in the post-genome era. Plant Physiol 129, 440–50.

Keim, P., Diers, B.W., Olson, T.C., and Shoemaker, R.C. (1990) RFLP mapping in soybean: association between marker loci and variation in quantitative traits. Genetics 126, 735–42.

Keim, P., Schupp, J.M., Travis, S.E., Clayton, K., Zhu, T., Shi, L., Ferreira, A., and Webb, D.M. (1997) A high-density soybean genetic map based on AFLP markers. Crop Sci. 37, 537–543.

Lark, K.G., Weisemann, J.M., Matthews, B.F., Palmer, R., Chase, K., and Macalma, T. (1993) A genetic map of soybean (Glycine max L.) using an intraspecific cross of two cultivars: 'Minosy' and 'Noir 1'. Theor. Appl. Genet. 86, 901–906.

Li, Z., and Nelson, R.L. (2002) RAPD marker diversity among cultivated and wild soybean accessions from four Chinese provinces. Crop Sci. 42, 1737–1744.

Lightfoot, D.A., Njiti, V.N., Gibson, P.T., Kassem, M.A., Iqbal, J.M., and Meksem, K. (2005) Registration of the Essex x Forrest recombinant inbred line mapping population. Crop Sci. 45, 1678–1681.

Litt, M., and Luty, J.A. (1989) A hypervariable microsatellite revealed by in vitro amplification of a dinucleotide repeat within the cardiac muscle actin gene. Am. J. Hum. Genet. 44, 397–401.

Liu, F., F.Y. Dong, J.J. Zou, S.Y. Chen, and B.C. Zhuang. 2000. Soybean germplasm diversity and genetic variation detected by microsatellite markers. (In Chinese.) Acta Genet. Sin. 27, 628–633.

Mansur, L.M., Orf, J.H., Chase, K., Jarvik, T., Cregan, P.B., and Lark, K.G. (1996) Genetic mapping of agronomic traits using recombinant inbred lines of soybean. Crop Sci. 36, 1327–1336.

Marek, L.F., Mudge, J., Darnielle, L., Grant, D., Hanson, N., Paz, M., Huihuang, Y., Denny, R., Larson, K., Foster-Hartnett, D., Cooper, A., Danesh, D., Larsen, D., Schmidt, T., Staggs, R., Crow, J.A., Retzel, E., Young, N.D., and Shoemaker, R.C. (2001) Soybean genomic survey: BAC-end sequences near RFLP and SSR markers. Genome 44, 572–81.

Matthews, B.F., Devine, T.E., Weisemann, J.M., Beard, H.S., Lewers, K.S., MacDonald, M.H., Park, Y.B., Maiti, R., Lin, J.J., and Kuo, J. (2001) Incorporation of sequenced cDNA and genomic markers into the soybean genetic map. Crop Sci. 41, 516–521.

Maughan, P.J., Saghai Maroof, M.A., and Buss, G.R. (1995) Microsatellite and amplified sequence length polymorphisms in cultivated and wild soybean. Genome 38, 715–723.

Morgante, M., and Olivieri, A.M. (1993) PCR-amplified microsatellites as markers in plant genetics. Plant J. 3, 175–82.

Morgante, M., Hanafey, M., and Powell, W. (2002) Microsatellites are preferentially associated with nonrepetitive DNA in plant genomes. Nat Genet 30, 194–200.

Morgante, M., Rafalski, A., Biddle, P., Tingey, S., and Olivieri, A.M. (1994) Genetic mapping and variability of seven soybean simple sequence repeat loci. Genome 37, 763–9.

Mudge, J., Huihuang, Y., Denny, R.L., Howe, D.K., Danesh, D., Marek, L.F., Retzel, E., Shoemaker, R.C., and Young, N.D. (2004) Soybean bacterial artificial chromosome contigs anchored with RFLPs: insights into genome duplication and gene clustering. Genome 47, 361–72.

Mullis, K., Faloona, F., Scharf, S., Saiki, R., Horn, G., and Erlich, H. (1986) Specific enzymatic amplification of DNA in vitro: the polymerase chain reaction. Cold Spring Harb Symp Quant Biol 51 Pt 1, 263–73.

Nasu, S., Suzuki, J., Ohta, R., Hasegawa, K., Yui, R., Kitazawa, N., Monna, L., and Minobe, Y. (2002) Search for and analysis of single nucleotide polymorphisms (SNPs) in rice (Oryza sativa, Oryza rufipogon) and establishment of SNP markers. DNA Res 9, 163–71.

Nordborg, M., Hu, T.T., Ishino, Y., Jhaveri, J., Toomajian, C., Zheng, H., Bakker, E., Calabrese, P., Gladstone, J., Goyal, R., Jakobsson, M., Kim, S., Morozov, Y., Padhukasahasram, B., Plagnol,

V., Rosenberg, N.A., Shah, C., Wall, J.D., Wang, J., Zhao, K., Kalbfleisch, T., Schulz, V., Kreitman, M., and Bergelson, J. (2005) The pattern of polymorphism in Arabidopsis thaliana. PLoS Biol 3, e196.

Palmer, R.G., and Shoemaker, R.C. (1998) Soybean genetics:In: M. V. M. Hrustic, and D. Jockovic (Eds), *Soybean Institute of Field and Vegetable Crops*, Novi Sad, Yugoslavia, pp. 45–82.

Rafalski, A., and Tingey, S. (1993). RFLP map of soybean (Glycine max), S. J. O'Brien (ed.) In:Genetic Maps: Locus maps of complex genomes. pp. 1–6.149–6.156. Cold Spring Harbor Laboratory Press, New York.

Rongwen, J., Akkaya, M.S., Bhagwat, A.A., Lavi, U., and Cregan, P.B. (1995) The use of microsatellite DNA markers for soybean genotype identification. Theor. Appl. Genet. 90, 43–48.

Rostoks, N., Mudie, S., Cardle, L., Russell, J., Ramsay, L., Booth, A., Svensson, J.T., Wanamaker, S.I., Walia, H., Rodriguez, E.M., Hedley, P.E., Liu, H., Morris, J., Close, T.J., Marshall, D.F., and Waugh, R. (2005) Genome-wide SNP discovery and linkage analysis in barley based on genes responsive to abiotic stress. Mol Genet Genomics 274, 515–27.

Sachidanandam, R., Weissman, D., Schmidt, S.C., Kakol, J.M., Stein, L.D., Marth, G., Sherry, S., Mullikin, J.C., Mortimore, B.J., Willey, D.L., Hunt, S.E., Cole, C.G., Coggill, P.C., Rice, C.M., Ning, Z., Rogers, J., Bentley, D.R., Kwok, P.Y., Mardis, E.R., Yeh, R.T., Schultz, B., Cook, L., Davenport, R., Dante, M., Fulton, L., Hillier, L., Waterston, R.H., McPherson, J.D., Gilman, B., Schaffner, S., Van Etten, W.J., Reich, D., Higgins, J., Daly, M.J., Blumenstiel, B., Baldwin, J., Stange-Thomann, N., Zody, M.C., Linton, L., Lander, E.S., and Altshuler, D. (2001) A map of human genome sequence variation containing 1.42 million single nucleotide polymorphisms. Nature 409, 928–33.

Schmid, K.J., Ramos-Onsins, S., Ringys-Beckstein, H., Weisshaar, B., and Mitchell-Olds, T. (2005) A Multilocus Sequence Survey in Arabidopsis thaliana Reveals a Genome-Wide Departure From a Neutral Model of DNA Sequence Polymorphism. Genetics 169, 1601–15.

Schmid, K.J., Sorensen, T.R., Stracke, R., Torjek, O., Altmann, T., Mitchell-Olds, T., and Weisshaar, B. (2003) Large-scale identification and analysis of genome-wide single-nucleotide polymorphisms for mapping in Arabidopsis thaliana. Genome Res. 13, 1250–7.

Shoemaker, R.C., and Olson, T.C. (1993). Molecular linkage map of soybean (Glycine max L. Merr.), S. J. O'Brien (ed.) In:Genetic Maps: Locus maps of complex genomes. pp. 1–6. 131–6.138. Cold Spring Harbor Laboratory Press, New York.

Shoemaker, R.C., and Specht, J.E. (1995) Integration of the soybean molecular and classical genetic linkage groups. Crop Sci. 35, 436–446.

Song, Q.J., Marek, L.F., Shoemaker, R.C., Lark, K.G., Concibido, V.C., Delannay, X., Specht, J.E., and Cregan, P.B. (2004) A new integrated genetic linkage map of the soybean. Theor. Appl. Genet. 109, 122–128.

Stacey, G., Vodkin, L., Parrott, W.A., and Shoemaker, R.C. (2004) National Science Foundation-sponsored workshop report. Draft plan for soybean genomics. Plant Physiol. 135, 59–70.

Tautz, D. (1989) Hypervariability of simple sequences as a general source for polymorphic DNA markers. Nucleic Acids Res. 17, 6463–71.

Tenaillon, M.I., Sawkins, M.C., Long, A.D., Gaut, R.L., Doebley, J.F., and Gaut, B.S. (2001) Patterns of DNA sequence polymorphism along chromosome 1 of maize (Zea mays ssp. mays L.). Proc. Natl. Acad. Sci. U.S.A. 98, 9161–6.

Thompson, J.A., Nelson, R.L., and Vodkin, L.O. (1998) Identification of diverse soybean germplasm using RAPD markers. Crop Sci. 38, 1348–1355.

Tuskan, G.A., Difazio, S., Jansson, S., Bohlmann, J., Grigoriev, I., Hellsten, U., Putnam, N., Ralph, S., Rombauts, S., Salamov, A., Schein, J., Sterck, L., Aerts, A., Bhalerao, R.R., Bhalerao, R.P., Blaudez, D., Boerjan, W., Brun, A., Brunner, A., Busov, V., Campbell, M., Carlson, J., Chalot, M., Chapman, J., Chen, G.L., Cooper, D., Coutinho, P.M., Couturier, J., Covert, S., Cronk, Q., Cunningham, R., Davis, J., Degroeve, S., Dejardin, A., Depamphilis, C., Detter, J., Dirks, B., Dubchak, I., Duplessis, S., Ehlting, J., Ellis, B., Gendler, K., Goodstein, D., Gribskov, M., Grimwood, J., Groover, A., Gunter, L., Hamberger, B., Heinze, B., Helariutta, Y., Henrissat, B., Holligan, D., Holt, R., Huang, W., Islam-Faridi, N., Jones, S., Jones-Rhoades, M.,

Jorgensen, R., Joshi, C., Kangasjarvi, J., Karlsson, J., Kelleher, C., Kirkpatrick, R., Kirst, M., Kohler, A., Kalluri, U., Larimer, F., Leebens-Mack, J., Leple, J.C., Locascio, P., Lou, Y., Lucas, S., Martin, F., Montanini, B., Napoli, C., Nelson, D.R., Nelson, C., Nieminen, K., Nilsson, O., Pereda, V., Peter, G., Philippe, R., Pilate, G., Poliakov, A., Razumovskaya, J., Richardson, P., Rinaldi, C., Ritland, K., Rouze, P., Ryaboy, D., Schmutz, J., Schrader, J., Segerman, B., Shin, H., Siddiqui, A., Sterky, F., Terry, A., Tsai, C.J., Uberbacher, E., Unneberg, P., *et al.* (2006) The genome of black cottonwood, Populus trichocarpa (Torr. & Gray). Science 313, 1596–604.

Van Ooijen, J.W., and Voorrips, R.E. (2001) JoinMap 3.0 software. Plant Research Internation, Wageningen, the Netherlands.

Vodkin, L.O., Khanna, A., Shealy, R., Clough, S.J., Gonzalez, D.O., Philip, R., Zabala, G., Thibaud-Nissen, F., Sidarous, M., Stromvik, M.V., Shoop, E., Schmidt, C., Retzel, E., Erpelding, J., Shoemaker, R.C., Rodriguez-Huete, A.M., Polacco, J.C., Coryell, V., Keim, P., Gong, G., Liu, L., Pardinas, J., and Schweitzer, P. (2004) Microarrays for global expression constructed with a low redundancy set of 27,500 sequenced cDNAs representing an array of developmental stages and physiological conditions of the soybean plant. BMC Genomics 5, 73.

Vos, P., Hogers, R., Bleeker, M., Reijans, M., van de Lee, T., Hornes, M., Frijters, A., Pot, J., Peleman, J., Kuiper, M., and Zabeau, M. (1995) AFLP: a new technique for DNA fingerprinting. Nucleic Acids Res. 23, 4407–14.

Watterson, G.A. (1975) On the number of segregating sites in genetical models without recombination. Theor. Popul. Biol. 7, 256–76.

Weber, J.L., and May, P.E. (1989) Abundant class of human DNA polymorphisms which can be typed using the polymerase chain reaction. Am. J. Hum. Genet. 44, 388–96.

Welsh, J., and McClelland, M. (1990) Fingerprinting genomes using PCR with arbitrary primers. Nucleic Acids Res. 18, 7213–8.

Williams, J.G., Kubelik, A.R., Livak, K.J., Rafalski, J.A., and Tingey, S.V. (1990) DNA polymorphisms amplified by arbitrary primers are useful as genetic markers. Nucleic Acids Res. 18, 6531–5.

Wright, S.I., Bi, I.V., Schroeder, S.G., Yamasaki, M., Doebley, J.F., McMullen, M.D., and Gaut, B.S. (2005) The effects of artificial selection on the maize genome. Science 308, 1310–4.

Wu, X.L., He, C.Y., Wang, Y.J., Zhang, Z.Y., Dongfang, Y., Zhang, J.S., Chen, S.Y., and Gai, J.Y. (2001) Construction and analysis of a genetic linkage map of soybean. Yi Chuan Xue Bao 28, 1051–1061.

Yamanaka, N., Ninomiya, S., Hoshi, M., Tsubokura, Y., Yano, M., Nagamura, Y., Sasaki, T., and Harada, K. (2001) An informative linkage map of soybean reveals QTLs for flowering time, leaflet morphology and regions of segregation distortion. DNA Res. 8, 61–72.

Yu, J.K., Dake, T.M., Singh, S., Benscher, D., Li, W., Gill, B., and Sorrells, M.E. (2004) Development and mapping of EST-derived simple sequence repeat markers for hexaploid wheat. Genome 47, 805–818.

Zhu, Y.L., Song, Q.J., Hyten, D.L., Van Tassell, C.P., Matukumalli, L.K., Grimm, D.R., Hyatt, S.M., Fickus, E.W., Young, N.D., and Cregan, P.B. (2003) Single-nucleotide polymorphisms in soybean. Genetics 163, 1123–34.

Chapter 6
Soybean Genome Structure and Organization

Randy C. Shoemaker, Jessica A. Schlueter, and Scott A. Jackson

Introduction

Phylogenetic Relationships Among the Legumes

The three subfamilies that comprise the legumes include the Mimosoideae, Caesalpinioideae, and Papilionoideae (Doyle and Luckow 2003). Nearly all of the economically important crop legumes belong to the Papilionoideae subfamily. This is a very diverse subfamily that includes soybean (*Glycine max*), common bean (*Phaseolus vulgaris*), peanut (*Arachis hypogaea*), mungbean (*Vigna radiata*), chick-pea (*Cicer arietinum*), lentil (*Lens culinaris*), pea (*Pisum sativum*), and alfalfa (*Medicago sativa*). Most of these important crop legumes fall into two clades; Galegoid (cool season legumes) and Phaseoloid (tropical season legumes).

Macrosynteny

A compelling rationale for conducting genomic studies in multiple legumes is the promise of translating findings in one genus directly to another, thus speeding advancement. Success of this strategy relies upon the general conservation of gene sequence, structure and order among the genera.

Early studies based on comparative mapping between members of the subfamily suggested that members possessed high levels of genome conservation (Weeden et al. 1992; Menacio-Hautea et al. 1993). Boutin et al. (1995) reported that synteny between soybean (*Glycine max*) and mungbean (*Vigna radiata*) and common bean (*Phaseolus vulgaris*) was limited to short linkage blocks. However, a later study by Lee et al. (2001), showed that soybean linkage groups were more syntenic with those of mungbean and common bean than previously thought.

Using soybean RFLP sequences and a bioinformatics approach, Grant et al. (2000) reported observations that suggested substantial macrosynteny between

R.C. Shoemaker
USDA-ARS CICGRU, Department of Agronomy, Iowa State University, Ames, IA 50011, USA
e-mail: rcsshoe@iastate.edu

G. Stacey (ed.), *Genetics and Genomics of Soybean*,

soybean and Arabidopsis, while comparison between *M. truncatula* and Arabidopsis revealed a lack of extended macrosynteny between the two genomes (Zhu et al. 2003). Although it is unlikely that chromosomal level conservation will be observed among any of the legumes and Arabidopsis, it is clear that synteny is frequently maintained over small chromosomal segments. Genetically linked loci in *M. truncatula* are often collinear with several segments of Arabidopsis, consistent with the fact that the Arabidopsis genome has experienced post-duplication segmental reshuffling accompanied by selective gene loss (Vision et al. 2000; Bowers et al. 2003). Sequence analyses reveals networks of microsynteny that are often highly degenerate (Schlueter et al. unpublished).

Choi et al. (2004a,b) mapped 60 markers that were homologous between soybean and *Medicago truncatula*. Twenty-three of the 60 mapped markers identified 11 syntenic blocks between *M. truncatula* and soybean. Their results showed that synteny was limited to small genetic intervals, probably reflecting chromosomal rearrangements coincident with times of divergence. A complicating factor in establishing macrosyntenic relationship between soybean and other legumes is the ancient polyploidy, or paleopolyploidy of the soybean genome. Genome duplication is often followed by gene loss and reshuffling of chromosomal segments thus making it difficult to identify lengthy stretches of syntenic chromosome segments between soybean and related legumes.

Microsynteny

Although the picture that has emerged with macrosynteny studies is one that predicts limited large-scale colinear relationships between the soybean genome and that of other legumes, the story with microsynteny is just emerging. Yan et al. (2003) estimated the level of microsynteny between *M. truncatula* and soybean using a hybridization strategy involving bacterial artificial chromosome (BAC) contigs. Fifty-four percent (27 of 50 soybean contigs) possessed some level of microsynteny with *M. truncatula*. Cannon et al. (2003) showed conserved gene order between *M. truncatula* and soybean, with at least 6 genes in common, across a 70 kb region around putatively orthologous apyrase genes. A similar analysis was conducted between a soybean genomic region and what is thought to be the orthologous region in *M. truncatula* (Choi et al. 2004b). In that study, 48% (14 of 29) of the genes identified between the two genomes were conserved. Even more recently, an analysis of three megabases of soybean sequence showed unusually high synteny to two Medicago chromosomes with upwards of 75% gene colinearity (Mudge et al. 2005).

Duplication Shapes the Soybean Genome

It was estimated that 80–100% of all angiosperms are likely to have a polyploid origin (Masterson 1994; Lockton and Gaut 2005; Adams and Wendel 2005). It is not surprising that soybean is also an ancient polyploid. Early map-based studies showed that the soybean genome was likely an ancient polyploid with putative homoeologous chromosomal regions readily identifiable through hybridization-based mapping (Shoemaker et al. 1996; Lee et al. 1999, 2001). The presence of nested duplications observed on the genetic maps suggested that the genome may have undergone two or more round of large-scale duplication (Shoemaker et al. 1996). Hybridization-based analyses of homoeologous BAC (bacterial artificial chromosome) clones further supported the ancient polyploidy of the genome (Foster-Hartnett et al. 2002; Yan et al. 2003). More recently, fluorescence in situ hybridization of BACs was used to visualize segmental duplications on soybean chromosomes (Pagel et al. 2004). More surprisingly, FISH analyses revealed nearly chromosomal-level homeology between chromosomes, with only limited disruption of colinearity (Walling et al. 2006).

Transcript Evidence for Ancient Polyploidy

The soybean research community possesses an extensive EST (expressed sequence tag) collection containing more than 350,000 sequences. These sequences reflect transcripts of duplicated genes and can provide important information on the history of the genome. Compelling evidence of large-scale genome duplication events was obtained through the analysis of duplicated genes from ESTs. Schlueter et al. (2004) and Blanc and Wolfe (2004) identified duplicate transcripts from the EST collections of both of soybean and Medicago and estimated genetic distances of the pairs using synonymous substitution measurements. Large numbers of paralog gene pairs with similar levels of divergence from each suggested at least two rounds of duplication had occurred. The estimated coalescence times placed these two events at approximately 14.5 MYA and 41.6 MYA (Schlueter et al. 2004). Interestingly, it was also estimated that *M. truncatula*, a cool season legume, underwent a genome duplication event at approximately 58 MYA (Schlueter et al. 2004). An analysis using a phylogenetic multigene approach concluded that the more ancient duplication events probably represent a single event that occurred before soybean and Medicago diverged (Pfeil et al. 2005).

Homology and Gene Density of Duplicated Regions

The first detailed analysis between paralogous genomic regions in soybean involved regions containing a tandemly duplicated gene family of hydroxycinnamoyl/ benzoyltransferase (HCBT) (Schlueter et al. 2006). This region corresponded to a duplication event that occurred at approximately 14 MYA (Schlueter et al. 2004). Nine out of ten genes were mutually retained between the two BACs. All shared genes were colinear in both order and orientation; results that are very different from what was previously observed in maize (Illic et al. 2003; Ma et al. 2005). Differences in the number of tandemly duplicated genes were implicated in genome expansion of one region over the other. The average nucleotide identity between homeologous coding regions is 89.8%, with some regions showing upwards of 95% identity. The authors also noted striking similarities in some intergenic sequence suggesting conservation of regulatory elements in the upstream regions between genes (Schlueter et al. 2006).

The average gene density of one BAC was one gene every 9.1 kb and for the other, one gene every 9.9 kb. This density was somewhat less than a previous estimate of one gene every 8 kb (Young et al. 2003) and significantly less dense than the most recent estimate of one gene every 5.8–6.7 kb (Mudge et al. 2005). Another group of homeologous BACs containing FAD2 genes were sequenced and compared (Schlueter et al. 2007b). Two of the BACs showed high levels of conservation of genes and gene order with only an inversion being the exception. Nucleotide identity between the coding regions of these two BACs was 90.7%. The other two BACs were connected only through members of the FAD2 gene family. Gene density showed a wide range with three BACs possessing genes every 6.70–7.95 kb, while the fourth had one gene every 19.2 kb. The latter BAC possessed a large amount of repetitive sequences (Schlueter et al. 2007b).

A caveat to gene density analyses conducted so far is that differences in the software and the criteria used to call genes may result in differences in predicted gene numbers with the same regions. Although the existing data supports a range of gene densities along the genome (sparse to very dense), on a whole, soybean gene density seems to be less than that of Arabidopsis (AGI 2000), *Oriza sativa* (TIGR rice Annotation webpage), *Medicago truncatula* (Young et al. 2005) or *Lotus japonicus* (Young et al. 2005).

Genome Organization is Affected by Duplication

Genome duplications have the potential to create a complex genome structure. The chromosomal regions resulting from large-scale duplication events are subject to a wide range of structural changes that include accumulation of indels (Petrov 2001; Illic et al. 2003), illegitimate recombination (Devos et al. 2002; Ma et al. 2005), gene loss, rearrangements, gene duplications and nucleotide divergence (Schlueter et al. 2006). All of these events are involved in the process of 'diploidization' (Ohno 1970) and complicate the interpretation of comparative genomic data.

The analysis of available information suggests that the soybean contains a mixture of genomic structural models ranging from that of high conservation between homeologous regions (the cotton model; Grover et al. 2004) and regions with a high degree of gene loss and rearrangement (the maize model: Illic et al. 2003; Langham et al. 2004; Ma et al. 2005) (see Fig. 6.1 for examples). In most cases, assembly of anciently duplicated regions from whole-genome sequence should prove to be little problem. However, it is likely that some regions, particularly recent tandem duplications, may remain so identical in sequence as to prove problematic in genome assembly (Schlueter et al. 2007a).

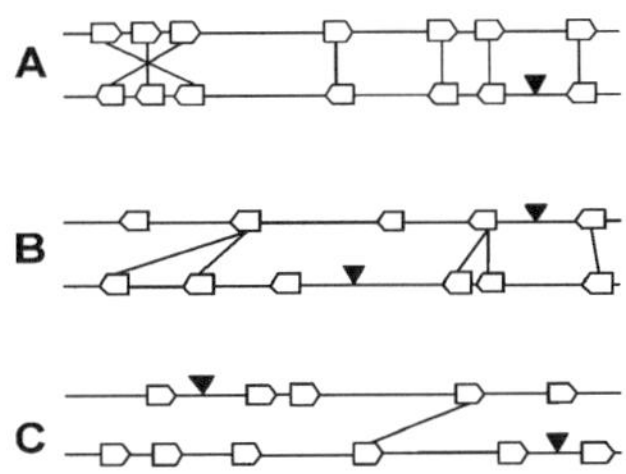

Fig. 6.1 Examples of the structures of homeologous regions in the soybean genome. (**A**) Homeologous regions showing gene for gene similarity with only an inversion and a transposable element (inverted black pyramid) as major structural distinctions. (**B**) Homeologous regions showing examples of fractionation (gene addition or deletion). Tandem duplications and transposable elements are distinguishing features. (**C**) Putative homeologous regions with only a single gene as the connection

Repeats Also Shape the Soybean Genome

Repetitive DNA sequences are important factors shaping plant genomes. Proliferation of repeats can drive genome size expansion (Bennetzen and Kellogg 1997); though, there are counteracting forces that can effectively remove repeats, thereby stabilizing genome size (Devos et al. 2002). Two basic models exist for genome organization relative to repeats: (1) the Arabidopsis model where there are few dispersed high copy repeats in the genic regions (The Arabidopsis Genome 2000); and (2) the maize model where there are high copy repeats throughout the genome and the genes, or gene islands, are distributed throughout (Messing et al. 2004).

Computational studies of BAC end sequences (BES) showed that BACs anchored with SSR (simple sequence repeat) markers, as opposed to those anchored with RFLP (restriction fragment length fragment polymorphism) markers, tended to contain more sequences with homology to repeats preliminarily suggestive of clustering of repeats within the soybean genome (Marek et al. 2001).

Only a few repeats were characterized in any depth, but the initial observations seem to indicate that soybean is more similar to Arabidopsis in organization than to maize. Most of the repeats that were mapped to chromosomes do localize to discreet locations such as the pericentromeric and centric regions (Vahedian et al. 1995; Lin et al. 2005) or ribosomal clusters (Shi et al. 1996) but no truly dispersed high-copy sequences. Further bolstering this conclusion is that more than 700 kb of DNA in the form of BAC clones were hybridized in a single experiment to soybean chromosomes with no blocking DNA and no cross-hybridization was observed except to discrete duplicated loci (Walling et al. 2006). Thus, the genic regions of soybean, which may comprise ~343 or more Mb of DNA, seem to be organized discretely from the major genome component, the repetitive DNA (Nunberg et al. 2006). Those interested in further pursuing soybean repeats can find information at the following URLs: http://www.tigr.org/tdb/e2k1/plant.repeats/; http://www.soymap.org/data/misc/soy_repeats.fasta; and http://digbio.rnet.missouri.edu/soybeangenome/NovelRepeat_BacEnd_Combine.txt

Gene Space Estimates

DNA:DNA re-association studies indicated that nearly 40–60% of the soybean genome consists of repetitive DNA sequences (Goldberg 1978; Gurley et al. 1979). Repetitive sequences are often methylated (Bender 2004), so sequencing of hypomethylated sequences can give an indication of the 'gene-space'. Using this approach, the gene space of soybean was estimated to be $\sim$ 343 Mb (Nunberg et al. 2006), approximately 1/3 of the total genome and concordant with the DNA:DNA re-association studies. This estimate is roughly in line with that of Mudge et al. (2004). Using a Poisson distribution model and redundant hits (multiple RFLP probes hybridized to the same BAC), Mudge et al. (2004) estimated gene space to be approximately 24% of the genome.

Summary Statements

The organization and structure of this complex genome promises to provide challenges as the whole genome sequence is assembled. At this writing, the initial 4 X draft is already completed. When the final 8 X draft is completed, the structure of the genome and the organization of the genic and non-genic spaces of homeologous regions will no longer be only inferred from snapshots of the genome.

The soybean is in a unique position in the crop world in that it possesses a very detailed and saturated genetic map. That map is salted with more than 900 QTL collected on more than 50 important agronomic traits over the last 15 years. The whole-genome sequence overlaid onto this map will provide a wealth of candidates from which to develop markers for marker-assisted selection and from which to expand into functional analyses and determinations through various gene-knockout technologies, e.g. VIGS, RNAi, TILLING, etc. Moreover, the whole-genome sequence of the soybean will used alongside that of the model legumes Medicago and Lotus to facilitate resequencing for marker development, as well as sequencing of a growing number of legume species. Comparisons of these sequences will provide fascinating stories on the evolutionary history of the legumes.

References

Adams, K. and Wendel, J. (2005) Polyploidy and genome evolution in plants. Current Opinion in Plant Biology 8,135–141.

Arabidopsis Genome Initiative. (2000) Analysis of the genome sequence of the flowering plant Arabidopsis thaliana. Nature 408, 796–815.

Bender, J. (2004) DNA methylation and epigenetics. Annu Rev Plant Biol 55, 41–68.

Bennetzen, J. L. and Kellogg, E. A. (1997) Do plants have a one-way ticket to genomic obesity? Plant Cell 9, 1509–1514.

Blanc, G., and Wolfe, K.H., (2004) Widespread paleopolyploidy in model plant species inferred from age distributions of duploicate genes. Plant Cell 16, 1667–1678.

Boutin, S.R., Young, N.D., Olson, Yu, Z.-H., Shoemaker, R.C., and Vallejos C.E. (1995) Genome conservation among three legume genera detected with DNA markers. Genome 38, 928–937.

Bowers, J.E., Chapman, B.A., Rong, J., Paterson, A.H. (2003) Unravelling angiosperm genome evolution by phylogenetic analysis of chromosomal duplication events. Nature 422, 433–438.

Cannon, S.B., McCombie, W.R., Sato, S., Tabata, S., Denny, R., Palmer, L., Katari, M., Young, N.D., Stacey, G. (2003) Evolution and microsynteny of the apyrase gene family in three legume genomes. Mol Genet Genomics 270, 347–361.

Choi, H.K., Kim, D., Uhm, T., Limpens, E., Lim, H., Mun, J.H., Kalo, P., Penmetsa, R.V., Seres, A., Kulikova, O., Roe, B.A., Bisseling, T., Kiss, G.B., Cook, D.R. (2004a) A sequence-based genetic map of Medicago truncatula and comparison of marker colinearity with M. sativa. Genetics 166, 1463–1502.

Choi, H.K., Mun, J.H., Kim, D.J., Zhu, H., Baek, J.M., Mudge, J., Roe, B., Ellis, N., Doyle, J., Kiss, G.B., Young, N.D., Cook, D.R. (2004b) Estimating genome conservation between crop and model legume species. Proc Natl Acad Sci USA 101, 15289–15294.

Devos, K.M., Brown, J.K.M., and Bennetzen, J.L. (2002) Genome size reduction through illegitimate recombination counteracts genome expansion in Arabidopsis. Genome Res. 12, 1075–1079.

Doyle, J.J., Luckow, M.A. (2003) The rest of the iceberg. Legume diversity and evolution in a phylogenetic context. Plant Physiol 131, 900–910.

Foster-Hartnett, D., Mudge, J., Danesh, D., Yan, H., Larsen, D., Denny, R., and Young N.D. (2002) Comparative genomic analysis of sequences sampled from a small region on soybean molecular linkage group 'G'. Genome 45, 634–645.

Goldberg, R. B. (1978) DNA sequence organization in the soybean plant. Biochem Genet 16, 45–68.

Grant, D. Cregan, P., and Shoemaker, R.C. (2000) Genome organization in dicots: genome duplication in Arabidopsis and synteny between soybean and Arabidopsis. Proc. Natl. Acad. Sci. U.S.A. 94, 4168–4173.

Grover, C.E., Kim, H, Wing, R., Paterson A.H., and Wendel, J.F. (2004) Incongruent patterns of local and global genome size evolution in cotton. Genome Res. 14, 1474–1482.

Gurley, W. B., Hepburn, A. G., Key, J.L. (1979) Sequence organization of the soybean genome. Biochem Biophys Acta 561, 167–183.

Illic, K., SanMiguel, P.J., and Bennetzen, J.L. (2003) A complex history of rearrangement in an orthologous region of the maize, sorghum, and rice genomes. Proc. Natl. Acad. Sci. U.S.A. 100, 12265–12270.

Langham, R.J., Walsh, J., Dunn, M., Ko, C., Goff, S.A., and Freeling, M. (2004) Genome duplication, fractionation and the origin of regulatory novelty. Genetics 166, 935–945.

Lee, J.M., Bush A., Specht J.E., and Shoemaker R. (1999) Mapping duplicate genes in soybean. Genome 42, 829–836.

Lee, J.M., Grant, D.,Vallejos C.E., and Shoemaker R. (2001) Genome organization in dicots. II. *Arabidopsis* as a 'bridging species' to resolve genome evolution events among legumes. Theor. Appl. Genet. 103, 765–773.

Lin, J. Y., Jacobus, B. H., SanMiguel, P., Walling, J.G., Yuan, Y., Shoemaker, R.C., Young, N.D., Jackson, S.A. (2005) Pericentromeric regions of soybean (Glycine max L. Merr.) chromosomes consist of retroelements and tandemly repeated DNA and are structurally and evolutionarily labile. Genetics 170, 1221–1230.

Lockton, S., and Gaut, B.S. (2005) Plant conserved non-coding sequences and paralogue evolution. Trends in Genet. 21, 80–86.

Ma, J., SanMiguel, P., Lai, J., Messing, J. and Bennetizen, J.L. (2005) DNA rearrangement in orthologous Orp regions of the maize, rice and sorghum genomes. Genetics 170, 1209–1220.

Marek, L.F., Mudge, J., Darnielle, L., Grant, D., Hanson, N., Paz, M., Huihuang, Y., Denny, R., Larson, K., Foster-Hartnett, D., Cooper, A., Danesh, D., Larsen, D., Schmidt, T., Staggs, R., Crow, J.A., Retzel, E., Young, N.D., and Shoemaker, R.C. (2001) Soybean genomic survey: BAC-end sequences near RFLP and SSR markers. Genome 44, 572–581.

Masterson, J. (1994) Stomatal size in fossil plants: evidence for polyploidy in majority of angiosperms. Science (Washington, D.C.), 264, 421–424.

Menacio-Hautea, D., Fatokum, C.A., Kumar, L., Danesh, D., Young, N.D. (1993) Comparative genome analysis of mungbean (Vigna radiata (L.) Wilczek) and cowpea (V. unguiculata (L.) Walpers) using RFLP mapping data. Theor Appl Genet 86, 797–810.

Messing, J., Bharti, A. K., Karlowski, W.M., Gundlach, H., Kinm H.R., Yu, Y., Wei, F., Fuks, G., Soderlund, C.A., Mayer, K.F., Wing, R.A. (2004) Sequence composition and genome organization of maize. PNAS 101, 14349–14354.

Mudge, J., Huihuang, Yan, Denny, R. L., Howe, D. K., Danesh, D., Marek, L. F., Retzel, E., Shoemaker, R. C., and Young, N. D. (2004) Soybean bacterial artificial chromosome contigs anchored with RFLPs: insights into genome duplication and gene clustering. Genome 47, 361–372.

Mudge, J., Cannon, S.B., Kalo, P., Oldroyd, G.E.D., Roe, B.A., Town, C.D., and Young, N.D. (2005) Highly syntenic regions in the genomes of soybean, Medicago truncatula and Arabidopsis thaliana. BMC Plant Biology 5, 15.

Nunberg, A., Bedell, J. A.,Budiman, M.A., Citek, R., Clifton, S.W., Fulton, L., Pape, D., Cai, Z., Joshi, T., Nguyen, H., Xu, D., Stacey, G. (2006) Survey sequencing of soybean elucidates the genome structure, composition and identifies novel repeats. Functional Plant Biology 33, 765–773.

Ohno, S. (1970) Evolution by Gene Duplication, Springer-Verlag, New York.

Pagel, J., Walling, J.G., Young, N.D., Shoemaker, R.C., Jackson, S.A. (2004) Segmental duplications within the *Glycine max* genome revealed by fluorescence in situ hybridization of bacterial artificial chromosomes. Genome 47, 764–768.

Petrov, D.A. (2001) Evolution of genome size: new approaches to an old problem. Trends in Genetics 17, 23–28.

Pfeil, B.E., Schlueter, J.A., Shoemaker, R.C., Doyle, J.J. (2005) Placing paleopolyploidy in relation to taxon divergence: a phylogenetic analysis in legumes using 39 gene families. Systematic Biology 54(3), 441–454.

Schlueter, J.A., Dixon, P., Granger, C. Grant, D., Clark, L., Doyle, J.J., and Shoemaker, R.C. (2004) Mining EST databases to resolve evolutionary events in major crop species. Genome 47, 868–876.

Schlueter, J.A., Scheffler, B.E.,, Schlueter, S.D., and Shoemaker, R.C. (2006) Sequence conservation of homeologous Bacterial Artificial Chromosomes and transcription of homeologous genes in soybean (*Glycine max* L. Merr.). Genetics 174, 1017–1028.

Schlueter, J., Lin, J.-Y., Schlueter, S., Vasylenko-Sanders, I., Deshpandem, S., Yi, J., O'Bleness, M., Roe, B., Nelson R., Scheffler, B., Jackson, S. and Shoemaker, R., (2007a) Gene duplication and paleopolypoloidy in soybean and the implications for whole genome sequencing. BMC Genomics 8, 330.

Schlueter, J.A., Vasylendo-Sanders, I.F., Deshpande, S., Yi, J., Siegfried, M., Roe, B.A., Schlueter, S.D., Scheffler, B.E. and R.C. Shoemaker. (2007b) The FAD2 gene family of soybean: insights into the structural and functional divergence of a paleopolyploid genome. The Plant Genome 1, 14–26.

Shi, L., Zhu, T., Keim, P. (1996) Ribosomal RNA genes in soybean and common bean : chromosomal organization, expression, and evolution. Theor Appl Genet 93, 136–141.

Shoemaker, R., Polzin K., Labate J., Specht J., Brummer E.C., Olson T., Young N., Concibido V., Wilcox J., Tamulonis J., Kochert, G., and Boerma H.R. (1996) Genome duplication in soybean (*Glycine* subgenus *soja*). Genetics 144, 329–338.

Vahedian, M., Shi, L., Zhu, T., Okimoto, R., Danna, K., Keim, P. (1995) Genomic organization and evolution of the soybean SB92 satellite sequence. Plant Molecular Biology 29, 857–862.

Vision, T.J, Brown, D.G., and Tanksley, S.D. (2000) The origins of genomic duplications in Arabidopsis. Science 290, 2114–2117.

Walling, J. G., Shoemaker, R. C., Young, N., Mudge, J., Jackson, S. (2006) Chromosome level homeology in paleopolyploid soybean (Glycine max) revealed through integration of genetic and chromosome maps. Genetics 172, 1893–1900.

Weeden, N.F., Muehlbauer, F.J., Ladizinsky, G. (1992) Extensive conservation of linkage relationships between pea and lentil genetic maps. J Hered 83, 123–129.

Yan, H.H., Mudge, J., Kim, D.-J., Shoemaker, R.C., Cook, D.R., Young, N.D. (2003) Estimates of conserved microsynteny among the genomes of Glycine max, Medicago truncatula, and Arabidopsis thaliana. Theor Appl Genet 106, 1256–1265.

Young, N.D., Mudge, J., Ellis, T.N. (2003) Legume genomes: more than peas in a pod. Curr. Opin. Plant Biol. 6, 199–204.

Young, N.D., Cannon, S.B., Sato, S., Kim, K., Cook, D.R., Town, C.D., Roe, B.A., Tabata, S. (2005) Sequencing the Genespaces of Medicago truncatula and Lotus japonicus. Plant Phys 137, 1174–1181.

Zhu, H., Kim, D.J., Baek, J.M., Choi, H.K., Ellis, L.C., Kuester, H., McCombie, W.R., Peng, H.M., Cook, D.R. (2003) Syntenic relationships between Medicago truncatula and Arabidopsis reveal extensive divergence of genome organization. Plant Physiol 131, 1018–1026.

Chapter 7
Sequence and Assembly of the Soybean Genome

Jeremy Schmutz, Jarrod Chapman, Uffe Hellsten, and Daniel Rokhsar

Introduction

In January 2006, the U.S. Department of Energy (DOE) announced an agreement to coordinate the sequencing and share resources with the U.S. Department of Agriculture (USDA) for agency relevant plant and microbial genome projects. The first genome to be sequenced under this cooperative agreement is the soybean (*Glycine max*) genome with the DOE Joint Genome Institute (JGI) providing the sequence and analysis of the genome and the USDA providing genomic resources and integration with the soybean genomics and breeding communities (http://www.doe.gov/news/2979.htm).

Soybean is the most valuable legume crop, with numerous nutritional and industrial uses because of its unique seed chemical position. The soybean seed is the world's main source of vegetable protein and oil, accounting for over 55% of all oilseed production and 80% of the edible consumption of fats and oils in the US. Industrial applications of soybean oil include lubricants, emulsifiers, coatings, and biodiesel. Over 3.1 billion bushels of soybeans were produced in the US on nearly 75 million acres in 2004, with an estimated annual economic value exceeding $16 billion, second only to maize and approximately twice that of wheat and ten times that of rice (National Agricultural Statistics Service http://www.usda.gov/nass). For these reasons, a proposal was made in late 2005 to the Department of Energy Joint Genome Institute (JGI) to sequence the soybean genome to provide a resource for the research community and support rational improvements in soybean quality and yield for biodiesel and other uses.

Following the model of other JGI sequencing projects, the whole genome shotgun strategy (WGS) was adopted as being the most rapid and cost-effective path to a draft genome, with a target coverage of ~8-fold redundant coverage (8x). Sequencing began in summer 2006, and as of February 2007 over 6.8 million shotgun reads

J. Schmutz
Department of Energy Joint Genome Institute, Stanford Human Genome Center, 975 California Avenue, Palo Alto, CA 94304, USA
e-mail: jeremy@shgc.stanford.edu

G. Stacey (ed.), *Genetics and Genomics of Soybean*,

(an estimated 3.7x coverage of the genome) were deposited in the NCBI Trace Archive (http://www.ncbi.nlm.nih.gov/Traces), where they are available for downloads and/or BLAST searches (http://www.ncbi.nlm.nih.gov/blast/mmtrace.shtml). The remaining ~ 4x of shotgun sequence will be produced in 2007 and early 2008. Here we describe the basic plans for the sequencing and annotation of the soybean genome and give a brief overview of the initial assembly and analysis of the first assembled soybean genome.

Choice of Strategy for Sequencing Soybean

Although the first microbial, plant, and animal genomes were sequenced by a clone-by-clone approach, almost all current animal and fungal genome sequencing projects use the whole genome shotgun strategy in which the entire genome is randomly sheared, sub-cloned, and redundantly sequenced (Fleischmann et al. 1995). The ease, cost-efficiency, and speed of whole genome shotgun approach made it the method of choice in many cases. The principal advantages are (a) obviating the need for a physical map at the start of the project, (b) avoiding labor and time-intensive sub-cloning of thousands of BACs, and (c) ease of high throughput sequencing of a small number of high quality shotgun libraries. While there are lingering concerns about its effectiveness for large repeat-rich plant genomes, mammalian genomes of apparent comparable complexity are now routinely sequenced with the shotgun method.

Whole Genome Shotgun Methodology

In a modern whole genome shotgun project, the genome is redundantly sampled by a collection of high fidelity shotgun "reads" that each span more than ~800 bp of genome at an accuracy of better than 99%. The redundancy, or "depth," ensures that most genomic regions will be sampled by multiple shotgun reads. A whole genome shotgun strategy is feasible if it is possible to reliably distinguish the "true" overlaps between reads (i.e., between reads that align because they are derived from the same genomic locus) out of the total pool of computed read-read alignments (which include confounding alignments between reads derived from similar repeats found at distinct genomic loci). If these "true" overlaps can be identified, they can be followed computationally to "walk" across the genome, reconstructing contiguous stretches of sequence ("contigs"). Thus, while 70% identity is a useful criterion for classifying repeats by superfamily, the figure of merit for WGS success is much more stringent ~99.8% identity ($= 0.999 \times 0.999$, allowing for errors in both aligning reads).

Soybean is polyploid, formed from the hybridization of two diploid ancestors that has undergone several cycles of genome duplication (Arumuganathan and Earle 1991; Shoemaker et al. 1996; Schlueter et al. 2004). It might be expected that distinguishing between paralogous regions would be a challenge for the project. The

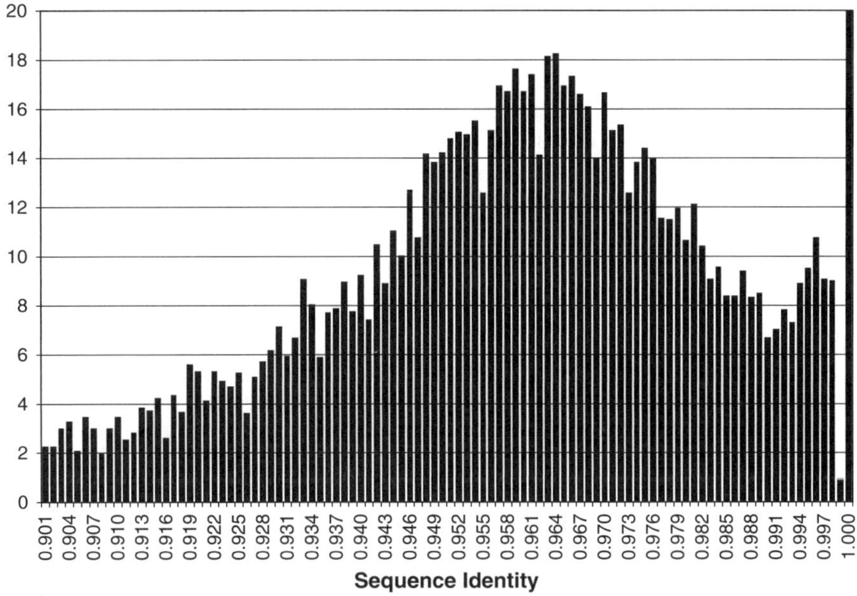

Fig. 7.1 Sequence identity comparison of nearest paralogs of soybean EST clusters

divergence of the two diploid progenitors can be estimated using EST analysis. Pairwise comparison of open reading frames from soybean ESTs demonstrates a clear peak at ~ 96% nucleotide identity (Fig. 7.1). From this analysis, we conclude that with 99.9% accurate sequence, paralogs from this epoch will be easily distinguished relative to each other, even in the more highly conserved coding regions.

A second key to the WGS approach is that the vast majority (> 85%) of reads are paired with a "sister" read that represents a nearby ~800 bp segment on the opposite strand whose relative position is approximately known, several kb downstream of the original read. Pairing between these sister reads can be used to string contigs together to reconstruct longer genomic regions ("scaffolds"). The linked contigs are separated by gaps whose approximate size is known (from the known distribution of clone sizes). Once a provisional scaffold is established, pairing data can be used to help place additional reads into these gaps. Thus, a hypothesis of true overlaps can be corroborated by this sister pair information and, conversely, knowledge of the reliable placement of reads within contigs constrain the placement of their sisters. Specifically, the position of an easy-to-place (e.g., non-repetitive) read can often help distinguish correct overlaps of its repetitive sister read from "false" alignments. Such false alignments typically show up in the context of "deep" reads that have an unusually large number of high fidelity alignments – more than expected at the given shotgun depth. Based on the fraction of high fidelity deep reads whose sister read is not deep, we expect that more than half of the repetitive reads excluded based on the number of high fidelity alignments can be placed. WGS will not be able to recover tandem arrays of high fidelity repeats, but these are problematic even for a clone-by-clone approach.

In outline, the approach is to randomly clone genomic DNA into plasmid and fosmid vectors, producing several high titer libraries with insert sizes ranging from a few kilobases for plasmids to ~35 kb for fosmids. End-sequencing these libraries (as well as available BAC clones) provides reads of ~7–800 bp of high quality (>99% accurate) genomic sequence at each end. The order and orientation (i.e., relative strand placement) of the two end sequences are known, as well as their approximate separation on the genome. Sequences are generated to a target depth, which measures the average number of shotgun reads that cover a typical base pair in the genome. This redundancy of coverage, and the accuracy of the reads, allows the computational identification of sequence reads that are derived from the same genomic locus. These overlapping reads can then be assembled together to reconstruct contiguous stretches of genome sequence (contigs). Contigs can be further linked into longer genomic spans (scaffolds or supercontigs) by using the order and orientation constraints of the paired end sequences. Scaffolds can contain gaps of unknown sequence that arise from some combination of (a) genomic positions that are not sampled at the targeted depth, (b) difficult to sequence or clone regions that are not sampled, or (c) highly repetitive regions whose covering reads cannot be confidently assembled.

There are several potential pitfalls of a whole genome shotgun approach, which follow from the requirement that shotgun reads derived from the same genomic locus be confidently identified during the assembly process. First, it is essential that the shotgun sequences be high quality, so that very high percent identity alignments between reads can be interpreted as evidence that two reads belong together. Conversely, we must be able to reject alignments of reads that are derived from distinct locations on the genome that have similar sequence. This situation can occur if (a) there are families of highly similar repetitive elements that are longer than a typical read length, or (b) recent tandem or larger-scale segmental duplication has occurred. Note that the presence of polymorphism can be confounding, since one must then also recognize and assemble together reads from different alleles, which means accepting lower percent identity alignments. For highly inbred and/or low polymorphism genomes such as soybean, we can take advantage of the high quality of the shotgun data and reject alignments that are not nearly identical which decreases false joins across nearly identical duplication regions.

Applying Whole Genome Shotgun Sequencing to Soybean

For the sequencing of soybean, purified DNA of the cultivar "Williams 82" was provided by J. Schlueter and S. Jackson, Purdue University (at the request of the Soybean Genetics Executive Committee). Three sized insert libraries were constructed from this purified DNA. The smallest library, in the high copy number vector pUC18, averages 3.3 kb in length and provides small-scale definitive linking in the WGS assembly due to the very narrow size range and low chimeric rate. The second library, in the lower copy number pMCL200 vector, is larger at an average insert size of 6.8 kb. These larger inserts have a slightly broader size range but the larger paired end links allow transposon elements and most common genome repeats to be spanned, allowing backfilling with repetitive pairs into 5 kb or less repetitive elements, with one of the pair anchored in unique sequence outside of the repeat. The final library created by the JGI is a fosmid library that averages 36 kb in length. The vector for this library is pCC1FOS, which uses a bacteriophage lambda packaging method and generally only accepts DNA between 28 kb and 45 kb. While a fosmid library is more challenging to create than the smaller subclone libraries, the paired links from the library ensure longer-range contiguity across the genome assembly and bridge common larger genomic repeats up to 30 kb in length. This is especially important for soybean because of the recent (8–10 MYA) large genome duplication (i.e., allopolyploidization), which is thought to contain large, very similar regions of sequence.

Finally, to provide very long range linking in the assembly, we used three BAC libraries (Marek and Shoemaker 1997; Marek et al. 2001) that are available from Clemson University (http://www.genome.clemson.edu) and Arizona Genomics Institute (http://www.genome.arizona.edu/) and that were end-sequenced by the Arizona Genomics Institute. These three libraries GM_WBa, GM_WBb and GM_WBc average 115 kb, 135 kb, and 133 kb. The BAC end sequences (BESs) are very valuable for a WGS project as they provide a framework that spans the assembly to ensure megabase scale contiguity across the WGS assembly. In addition the BAC clones provide a clone-based resource that can be sequenced to address problem areas in the WGS assembly of the genome and as an important resource for the community. Indeed, as part of the soybean genome project, the JGI is shotgun sequencing and finishing 500 BAC clones, selected by the community to cover genes or genomic regions of interest such as QTL trait regions. These finished BAC clones will be integrated into the final soybean assembly.

Initial Glimpse of the Soybean Genome at the Sequencing Midpoint

To build the 4x coverage assembly, we combined the available data from the 3 subclone libraries and the three BAC libraries (Table 7.1) and used the Arachne assembler (Batzoglou et al. 2002) to construct a WGS assembly of the genome. This genome assembly represents a first pass at building the soybean genome which does not include ongoing BAC based sequencing projects and was not screened for contamination.

We find the results of the initial 4x assembly of soybean encouraging, with large stretches of assembled contiguous sequence. The assembly information is shown in Table 7.2. The majority of the sequence is assembled in 1,063 scaffolds greater than 50 kb in size with 72% of the assembly in scaffolds greater than 1 megabase.

Table 7.1 Sequence data included in the 4x coverage genome assembly

Library designation	Insert size	Number of reads	Genome sequence coverage
BIWS	3,300	3,207,506	1.68x
BXCB	6,800	3,204,240	1.68x
BIWU	36,300	1,139,802	0.55x
GM_WBa	115,000	59,286	0.03x
GM_WBb	135,000	121,680	0.06x
GM_WBc	133,000	83,031	0.04x

Table 7.2 Initial assembly statistics for the 4x soybean assembly. The statistics are cumulative as the rows increase so there are a total of 1,285 scaffolds greater than 25,000 bp in length

Scaffold size in base pairs	No. of scaffolds > size	Contigs	Total scaffold size including gaps	Total base pairs	Percent base pairs
> 5, 000, 000	36	18,244	302,009,125	293,366,340	97.14%
> 2, 500, 000	87	30,014	469,466,219	454,555,990	96.82%
> 1, 000, 000	211	45,148	667,406,054	644,071,020	96.50%
> 500, 000	399	56,634	800,457,065	768,808,479	96.05%
> 250, 000	624	63,960	880,188,598	842,197,925	95.68%
> 100, 000	875	68,184	921,957,940	878,059,110	95.24%
> 50, 000	1, 063	69,700	935,578,969	888,941,075	95.02%
> 25, 000	1, 285	70,685	944,408,014	893,676,215	94.63%
> 10, 000	1, 678	71,678	950,130,186	898,920,546	94.61%
> 5, 000	2, 547	73,631	956,122,992	903,977,849	94.55%
> 2, 500	3, 852	75,777	961,056,892	908,291,978	94.51%
> 1, 000	5, 181	77,169	963,279,339	910,472,951	94.52%

Between 95 and 97% of the assembled scaffolds are covered by sequence, leaving less than 5% of the scaffolds as gaps. Based on past experience, this is very good for a 4x assembly and suggests that most of the assembled sequence of soybean is not highly repetitive. Note that the high number of contigs is due to gaps caused by both repetitive sequences as well as regions of low sequence coverage. We anticipate that at the completion of the project, the majority (>90%) of the remaining gaps will be the result of repetitive sequence rather than from low sequence coverage.

While we showed above that the soybean polyploidization events should be distinguishable from each other within the WGS sequence, with the 4x assembly we can look in depth at the recent large-scale duplication and verify that it is separating during the assembly phase. To verify that there is indeed enough variation accumulated between the duplicated copies, we examined regions of the WGS assembly surrounding sequenced BAC clones that had originally been selected from duplicate gene pairs. The clone placements for 5 pairs of clones show they have both a primary, near identical placement, and one or more secondary placements that align to the duplicated region(s). Figure 7.2 shows a dot plot of one such 600 kb region of scaffold 183 and scaffold 354, two of these duplicated regions in the soybean

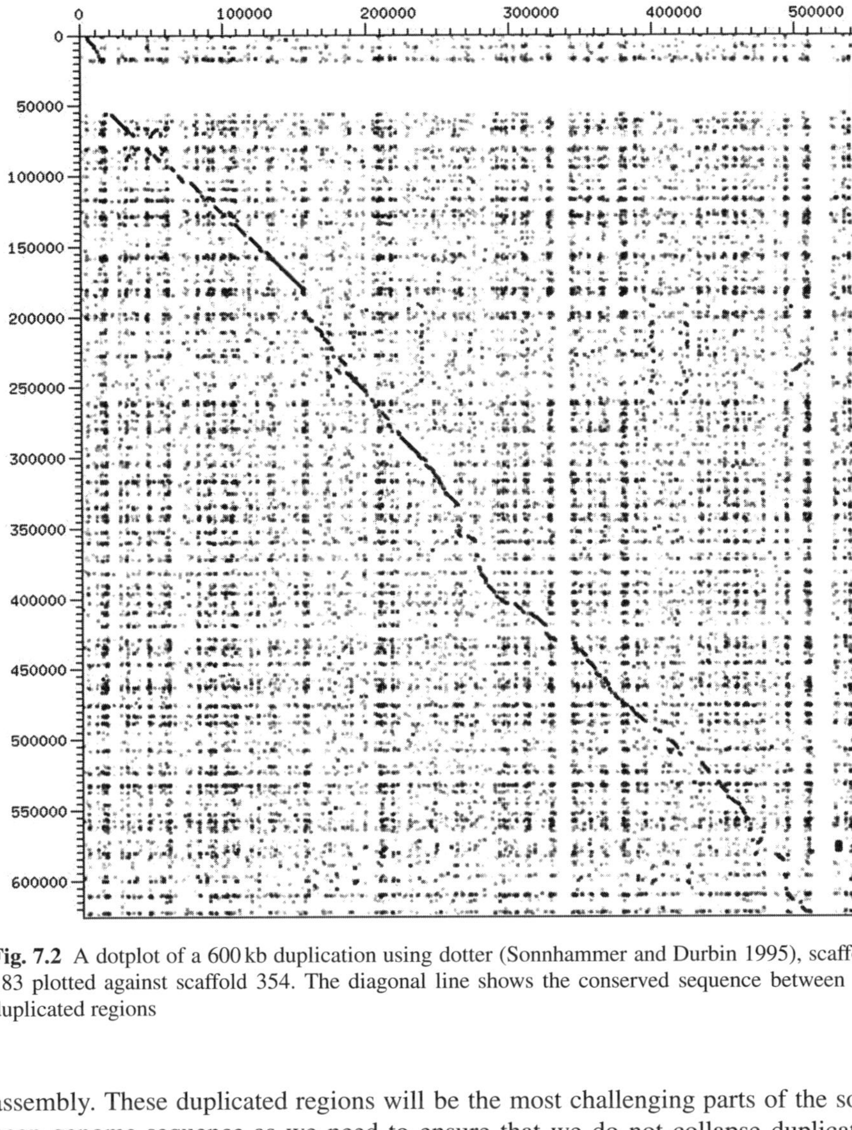

Fig. 7.2 A dotplot of a 600 kb duplication using dotter (Sonnhammer and Durbin 1995), scaffold 183 plotted against scaffold 354. The diagonal line shows the conserved sequence between the duplicated regions

assembly. These duplicated regions will be the most challenging parts of the soybean genome sequence as we need to ensure that we do not collapse duplicated regions, nor wander back and forth between duplicated genomic segments during the assembly process.

To assess the completeness of the 4x assembly, we aligned 50 finished BAC clone sequences to the genomic assembly. For each clone we counted the matching bases in the best-placed genomic region of the WGS. Of the 50 clones 40 were 93% identical to the assembly, with 29 clones matching 95% or more to the assembly (Fig. 7.3). These are very encouraging results from only a 4x coverage data set and show that we are sampling the majority of the euchromatic genome sequence. Currently, we are producing additional 4x shotgun sequence coverage for the three sized

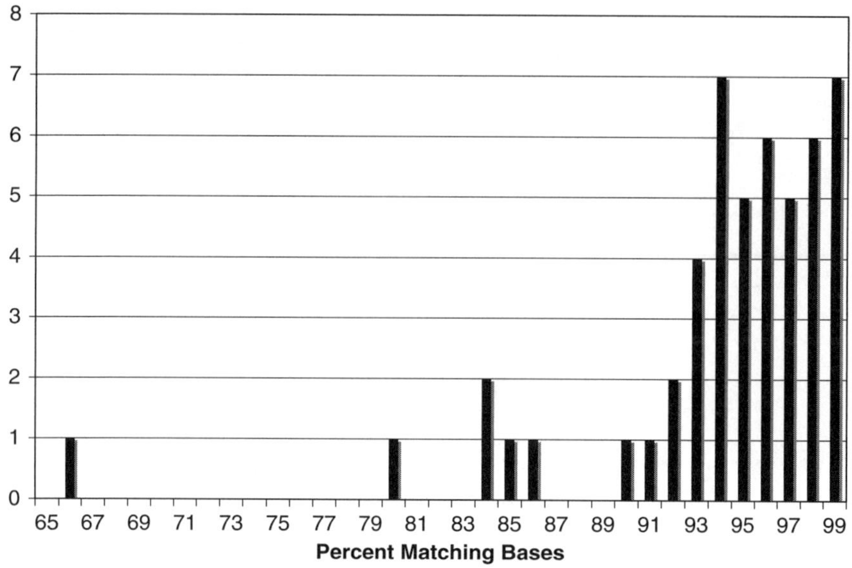

Fig. 7.3 Finished clone placements against the 4x coverage assembly

libraries and adding additional BAC end sequences that will result in a final genome of 8x sequence coverage and 20x clone coverage in BACs. Once the sequencing has been completed, a new shotgun assembly will be constructed that incorporates the 8x shotgun coverage, 20x BAC end sequence coverage and any finished soybean sequenced BAC clone sequence. This assembly will then be integrated with the available soybean genetic and physical mapping data to produce a chromosome scale assembly. This will be denoted as the final draft soybean assembly. Subsequent improvements will require future targeted efforts.

Annotating the Final Draft Assembly

To maximize the usefulness of the soybean sequence, the assembled sequence will be annotated at JGI using state-of-the-art methods and this information distributed on the Web in both downloadable and web-navigable formats and to major public databases. The annotation plan is modeled after the JGI's experience with *Populus* and *Sorghum*, which can be found at www.jgi.doe.gov/poplar and www.phytozome.net/sorghum, respectively. The JGI has extensive experience with genome annotation of both finished and draft genome sequences, and tailors its pipeline to the needs of each project.

Methods developed for annotating human, poplar, and other genomes will be used to predict protein-coding genes, including intron/exon structure and 5′ and 3′ UTRs if supported by cDNA or well-aligned EST evidence. Gene prediction will be based on all available evidence, starting from known full-length soybean genes and ESTs that align well to the genome. We will also align the public cDNAs and ESTs from other legumes, including *Medicago truncatula, M. sativa*, and *Lotus japonica*, as evidence for gene structures in soybean. Based on neutral substitution rates, these legumes bear approximately the same relationship to soybean as other mammals do to human, so we can take advantage of the extensive experience and computational toolkit for comparative gene modeling developed while annotating and curating the human genome (Schmutz et al. 2004; Martin et al. 2004; Grimwood et al. 2004). In total, over half a million legume ESTs are available for this purpose.

Comparison with the gene sets of rice and the sequenced eudicots (Arabidopsis, Populus, Medicago, and any others) will augment *ab initio* modeling based on FgenesH and other available tools, with the parameterization of FgenesH (Salamov and Solovyev 2000) to be optimized for soybean. One challenge in annotating genomes is to avoid propagating errors.

The automated JGI functional annotation pipeline incorporates standard methodologies including:

- InterProScan (Zdobnov and Apweiler 2001), a suite of pattern detection algorithms for domain recognition that subsumes Pfam (Bateman et al. 2004), Prosite (Letunic et al. 2004), Smart (Letunic et al. 2004), and other standard domain finding algorithms, including InterPro-to-GeneOntology mapping;
- KOG [euKaryotic Orthologous Groups (Koonin et al. 2004)] assignment based on RPS-BLAST of position specific scoring matrices for core conserved eukaryotic genes (Altschul et al. 1997), mapped to associated GeneOntology terms using JGI tools. Note that this set does not include plant specific genes;
- KEGG [Kyoto Encyclopedia of Genes and Genomes (Kanehisa et al. 2004)] Enzyme Classification mapping, focusing on rice and Arabidopsis enzymes but including other eukaryotes, also translated to GeneOntology terms.

In addition, we developed phylogenetic methods for identifying orthologs based on nucleotide and protein alignments that are being applied and tested in *Populus* (Tuskan et al. 2006). We note that a molecular clock at four-fold synonymous sites appears to hold (and remain unsaturated) within the legumes and eudicots (Hellsten, Dirks, Rokhsar, unpublished) and will use available legumes genes and ESTs to create gene families descended from a single ancestral gene in the common legume and eudicot ancestora to facilitate accurate transfer of molecular functional annotations between legumes and suggest possibly conserved biological roles for experimental testing. At a deeper phylogenetic level, this will be repeated with all angiosperms. This phylogenomic approach will leverage the intensive efforts to functionally annotate *Arabidopsis, Medicago*, rice, and other species, and suggest putative transfers of Gene Ontology terms based on assumed conservation of core ancestral molecular function, and produce hypotheses for organismal roles that can be subjected to experimental test.

End Goal of the Sequencing Project

We expect to produce a final draft sequence of soybean by the fall of 2008. We anticipate that this genome sequence will include more than 98% of the non-repetitive sequence regions with the majority of the sequence scaffolds ordered and oriented in a chromosome-scale assembly. This assembly, is expected to span about 900 Mb of the soybean genome, and will also include 60–100 Mb of finished BAC clone sequences, derived during our on-going clone sequencing effort. This high quality final draft will be annotated and released with a public gene and genome browser hosted at the JGI. The availability of a high quality draft assembly and annotation for *Glycine max* Williams 82 will serve as a reference sequence for genome-enabled studies of soybean and related crops.

References

Altschul SF, Madden TL, Schaffer AA, Zhang J, Zhang Z, Miller W, Lipman DJ. (1997) Gapped BLAST and PSI-BLAST: a new generation of protein database search programs. *Nucleic Acids Res*. **25**: 3389–3402.

Arumuganathan K, Earle ED. (1991) Nuclear DNA content of some important plant species. *Plant Mol. Biol. Rep*. **9**:211–215.

Bateman A, Coin L, Durbin R, Finn RD, Hollich V, Griffiths-Jones S, Khanna A, Marshall M, Moxon S, Sonnhammer EL, Studholme DJ, Yeats C, Eddy SR. (2004). "The Pfam protein families database." *Nucleic Acids Res*. **32**:D138–41.

Batzoglou S, Jaffe DB, Stanley K, Butler J, Gnerre S, Mauceli E, Berger B, Mesirov JP, Lander ES. (2002) ARACHNE: a whole-genome shotgun assembler. *Genome Research* **12**:177–189.

Fleischmann RD, Adams MD, White O, Clayton RA, Kirkness EF, Kerlavage AR, Bult CJ, Tomb JF, Dougherty BA, Merrick JM. (1995) Whole-genome random sequencing and assembly of *Haemophilus influenzae*. *Science* **269**: 496–512.

Grimwood J, Gordon LA, Olsen A, Terry A, Schmutz J, Lamerdin J, Hellsten U, Goodstein D, Couronne O, Tran-Gyamfi M, Aerts A, Altherr M, Ashworth L, Bajorek E, Black S, Branscomb E, Caenepeel S, Carrano A, Caoile C, Chan YM, Christensen M, Cleland CA, Copeland A, Dalin E, Dehal P, Denys M, Detter JC, Escobar J, Flowers D, Fotopulos D, Garcia C, Georgescu AM, Glavina T, Gomez M, Gonzales E, Groza M, Hammon N, Hawkins T, Haydu L, Ho I, Huang W, Israni S, Jett J, Kadner K, Kimball H, Kobayashi A, Larionov V, Leem SH, Lopez F, Lou Y, Lowry S, Malfatti S, Martinez D, McCready P, Medina C, Morgan J, Nelson K, Nolan M, Ovcharenko I, Pitluck S, Pollard M, Popkie AP, Predki P, Quan G, Ramirez L, Rash S, Retterer J, Rodriguez A, Rogers S, Salamov A, Salazar A, She X, Smith D, Slezak T, Solovyev V, Thayer N, Tice H, Tsai M, Ustaszewska A, Vo N, Wagner M, Wheeler J, Wu K, Xie G, Yang J, Dubchak I, Furey TS, DeJong P, Dickson M, Gordon D, Eichler EE, Pennacchio LA, Richardson P, Stubbs L, Rokhsar DS, Myers RM, Rubin EM, Lucas SM (2004). "The DNA sequence and biology of human chromosome 19." *Nature*. **428**:529–35.

Kanehisa M, Goto S, Kawashima S, Okuno Y, Hattori M. (2004) "The KEGG resource for deciphering the genome." *Nucleic Acids Res*. **32**:D277–80.

Koonin EV, Fedorova ND, Jackson JD, Jacobs AR, Krylov DM, Makarova KS, Mazumder R, Mekhedov SL, Nikolskaya AN, Rao BS, Rogozin IB, Smirnov S, Sorokin AV, Sverdlov AV, Vasudevan S, Wolf YI, Yin JJ, Natale DA. (2004) "A comprehensive evolutionary classification of proteins encoded in complete eukaryotic genomes." *Genome Biol*. **5**(2):R7.

Letunic I, Copley RR, Schmidt S, Ciccarelli FD, Doerks T, Schultz J, Ponting CP, Bork P. (2004). "SMART 4.0: towards genomic data integration." *Nucleic Acids Res*. **32**:D142–4.

Marek LF, Mudge J, Darnielle L, Grant D, Hanson N, Paz M, Huihuang Y, Denny R, Larson K, Foster-Hartnett D, Cooper A, Danesh D, Larsen D, Schmidt T, Staggs R, Crow JA, Retzel E, Young ND, Shoemaker RC. (2001) Soybean genomic survey: BAC-end sequences near RFLP and SSR markers. *Genome* **44**:572–581.

Marek LF, Shoemaker RC. (1997) BAC contig development by fingerprint analysis in soybean. *Genome* **40**:420–427.

Martin J, Han C, Gordon LA, Terry A, Prabhakar S, She X, Xie G, Hellsten U, Chan YM, Altherr M, Couronne O, Aerts A, Bajorek E, Black S, Blumer H, Branscomb E, Brown NC, Bruno WJ, Buckingham JM, Callen DF, Campbell CS, Campbell ML, Campbell EW, Caoile C, Challacombe JF, Chasteen LA, Chertkov O, Chi HC, Christensen M, Clark LM, Cohn JD, Denys M, Detter JC, Dickson M, Dimitrijevic-Bussod M, Escobar J, Fawcett JJ, Flowers D, Fotopulos D, Glavina T, Gomez M, Gonzales E, Goodstein D, Goodwin LA, Grady DL, Grigoriev I, Groza M, Hammon N, Hawkins T, Haydu L, Hildebrand CE, Huang W, Israni S, Jett J, Jewett PB, Kadner K, Kimball H, Kobayashi A, Krawczyk MC, Leyba T, Longmire JL, Lopez F, Lou Y, Lowry S, Ludeman T, Manohar CF, Mark GA, McMurray KL, Meincke LJ, Morgan J, Moyzis RK, Mundt MO, Munk AC, Nandkeshwar RD, Pitluck S, Pollard M, Predki P, Parson Quintana B, Ramirez L, Rash S, Retterer J, Ricke DO, Robinson DL, Rodriguez A, Salamov A, Saunders EH, Scott D, Shough T, Stallings RL, Stalvey M,Sutherland RD, Tapia R, Tesmer JG, Thayer N, Thompson LS, Tice H, Torney DC, Tran-Gyamfi M, Tsai M, Ulanovsky LE, Ustaszewska A, Vo N, White PS, Williams AL, Wills PL, Wu JR, Wu K, Yang J, Dejong P, Bruce D, Doggett NA, Deaven L, Schmutz J, Grimwood J, Richardson P, Rokhsar DS, Eichler EE, Gilna P, Lucas SM, Myers RM, Rubin EM, Pennacchio LA. "The sequence and analysis of duplication rich human chromosome 16." (2004) *Nature* **432**:988–94.

Salamov AA, Solovyev VV. (2000) "Ab initio gene finding in Drosophila genomic DNA." *Genome Res.* **10**:516–22.

Sonnhammer EL, Durbin R. (1995) A dot-matrix program with dynamic threshold control suited for genomic DNA and protein sequence analysis. Gene **167**:1–2.

Schlueter JA, Dixon P, Granger C, Grant D, Clark L, Doyle JJ, Shoemaker RC. (2004) Mining EST databases to resolve evolutionary events in major crop species. *Genome* **47**:868–876.

Schmutz J, Martin J, Terry A, Couronne O, Grimwood J, Lowry S, Gordon LA, Scott D, Xie G, Huang W, Hellsten U, Tran-Gyamfi M, She X, Prabhakar S, Aerts A, Altherr M, Bajorek E, Black S, Branscomb E, Caoile C, Challacombe JF, Chan YM, Denys M, Detter JC, Escobar J, Flowers D, Fotopulos D, Glavina T, Gomez M, Gonzales E, Goodstein D, Grigoriev I, Groza M, Hammon N, Hawkins T, Haydu L, Israni S, Jett J, Kadner K, Kimball H, Kobayashi A, Lopez F, Lou Y, Martinez D, Medina C, Morgan J, Nandkeshwar R, Noonan JP, Pitluck S, Pollard M, Predki P, Priest J, Ramirez L, Retterer J, Rodriguez A, Rogers S, Salamov A, Salazar A, Thayer N, Tice H, Tsai M, Ustaszewska A, Vo N, Wheeler J, Wu K, Yang J, Dickson M, Cheng JF, Eichler EE, Olsen A, Pennacchio LA, Rokhsar DS, Richardson P, Lucas SM, Myers RM, Rubin EM. "The DNA sequence and comparative analysis of human chromosome 5" (2004). *Nature* **431**:268–74.

Shoemaker RC, Polzin K, Labate J, Specht J, Brummer EC, Olson T, Young N, Concibido V, Wilcox J, Tamulonis JP, Kochert G, Boerma HR. (1996) Genome duplication in soybean (Glycine subgenus soja). *Genetics* **144**:329–338.

Tuskan GA, Difazio S, Jansson S, Bohlmann J, Grigoriev I, Hellsten U, Putnam N, Ralph S, Rombauts S, Salamov A, Schein J, Sterck L, Aerts A, Bhalerao RR, Bhalerao RP, Blaudez D, Boerjan W, Brun A, Brunner A, Busov V, Campbell M, Carlson J, Chalot M, Chapman J, Chen GL, Cooper D, Coutinho PM, Couturier J, Covert S, Cronk Q, Cunningham R, Davis J, Degroeve S, Dejardin A, Depamphilis C, Detter J, Dirks B, Dubchak I, Duplessis S, Ehlting J, Ellis B, Gendler K, Goodstein D, Gribskov M, Grimwood J, Groover A, Gunter L, Hamberger B, Heinze B, Helariutta Y, Henrissat B, Holligan D, Holt R, Huang W, Islam-Faridi N, Jones S, Jones-Rhoades M, Jorgensen R, Joshi C, Kangasjarvi J, Karlsson J, Kelleher C, Kirkpatrick R, Kirst M, Kohler A, Kalluri U, Larimer F, Leebens-Mack J, Leple JC, Locascio P, Lou Y, Lucas S, Martin F, Montanini B, Napoli C, Nelson DR, Nelson C, Nieminen K,

Nilsson O, Pereda V, Peter G, Philippe R, Pilate G, Poliakov A, Razumovskaya J, Richardson P, Rinaldi C, Ritland K, Rouze P, Ryaboy D, Schmutz J, Schrader J, Segerman B, Shin H, Siddiqui A, Sterky F, Terry A, Tsai CJ, Uberbacher E, Unneberg P, Vahala J, Wall K, Wessler S, Yang G, Yin T, Douglas C, Marra M, Sandberg G, Van de Peer Y, Rokhsar D. (2006) The genome of black cottonwood, *Populus trichocarpa. Science* **313**:1596–604.

Zdobnov EM, Apweiler R. (2001). "InterProScan–an integration platform for the signaturerecognition methods in InterPro." *Bioinformatics*. 17(9):847–8.

Chapter 8
Advances in Soybean Breeding

M.S. Pathan and David A. Sleper

Introduction

Soybean center of origin is China and was domesticated during 1500–1100 B.C. Soybean was introduced in Europe during 16th and 17th centuries and was brought into North America in 1765 and then into Central and South America during the mid 1900s (Hymowitz 1990, 2004). Soybean is a member of the genus *Glycine* and genus *Glycine* is divided into two subgenera; *Glycine* (perennials) and *Soja* (Moench) F. J. Herm. (annuals). The subgenus *Soja* includes the cultivated soybean, *Glycine max* and the wild annual soybean, *Glycine soja*. The haploid soybean genome consists of 1,100 million base pairs (Mbp), which is relatively larger than the model plant, *Arabidopsis* (120 Mbp) (Arumuganathan and Earle 1991). Both the cultivated and wild soybean are paleopoloid, $2n = 40$ with base chromosome number of 20 and perfectly cross compatible (Hymowitz 2004). The subgenus *Glycine* contains 22 species including important species like *G. tabacina* and *G. tomentella*. Soybean genome evolved by two rounds of polyploidization or duplication (Shoemaker et al. 1996; Blanc and Wolfe 2004; Schlueter et al. 2004) and 35 % of the soybean genome is diploidized (Shultz et al. 2006). About 170,000 *G. max* accessions are maintained in 70 countries. China has the largest collection of soybean germplasm in the world with nearly 26,000 accessions of *G. max* (Chang et al. 1999; Carter et al. 2004) and followed by the United States with about 19,000 accessions of *G. max* (USDA soybean germplasm collection). The genetic base of North American soybean cultivars is narrow (Gizlice et al. 1994; Singh and Hymowitz 1999).

World soybean production has increased steadily in the last decade, rising from 133 million metric tons in 1996 to 228 million metric tons in 2006 (www.soystats.com; www.stratsoy.uiuc.edu), due to higher economic values of protein and oil contents, industrial uses and medicinal importance. In the United States, soybean production was 65 million tons in 1996 and production rose to 87 million metric

M.S. Pathan
National Center for Soybean Biotechnology, Division of Plant Sciences, University of Missouri-Columbia, MO 65211, USA
e-mail: Pathanm@missouri.edu

G. Stacey (ed.), *Genetics and Genomics of Soybean*,

tons in 2006. At the same time, about 82 % of the world's soybean was produced in only 3 countries, United States, Brazil and Argentina. United States is the leading soybean producer and produced 87 million tons (38 % of the total world production) and Brazil and Argentina produced about 56 and 44 million metric tons (25 % and 19 % of the total), respectively. Although soybean is native to China, China produced only 16 million metric tons (7 % of the total) and India produced about 7 million metric tons (3 % of the total) (Fig. 8.1a). During the same time, total oilseed production was 397 million metric tons, out of this, 57 % (228 million metric tons) was covered by soybean alone, hence, it makes soybean the world's number one oil seed crop followed by rapeseed at 12 % and cotton seed at 11 % (47 and 44 million metric tons, respectively) (Fig. 8.1b). In 2006, US exported grain soybean, soybean oil and soybean meal totaling about 9 billion dollars, and China, Japan, European Union and Mexico were the main importers. Generally, soybean seed is composed of 40 % protein, 35 % carbohydrate, 20 % oil, and 5 % ash (Liu 1997) and a 60-pound bushel of soybean yields about 48 pounds of protein-rich meal and 11 pounds of

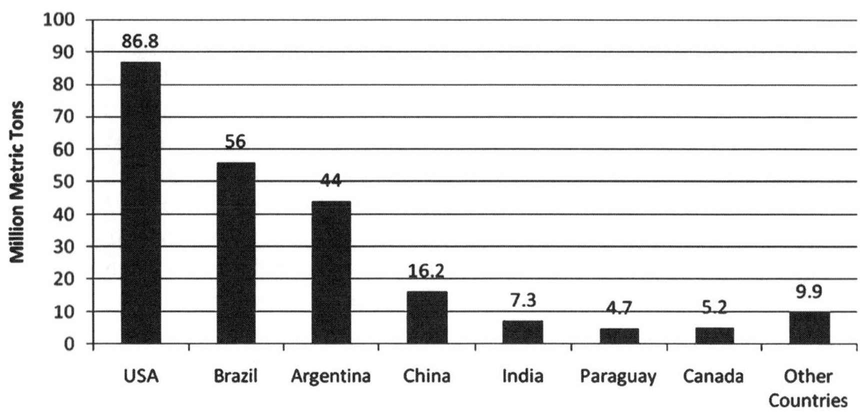

Fig. 8.1a World soybean production in 2006 (www.soystats.com)

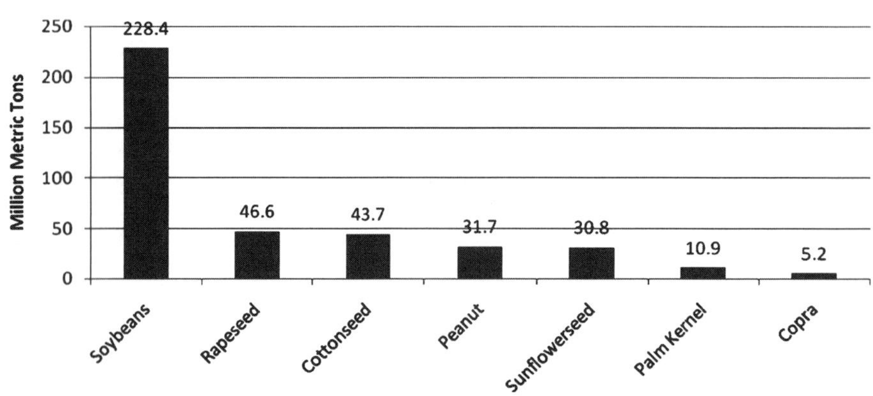

Fig. 8.1b World oilseed production in 2006 (www.soystats.com)

oil (www.soystats.com). Soybean is the world's primary source of protein and oil for human and also an important source of protein feed supplement for livestock. Soy-based nutritious popular foods such as, tofu, miso, soy milk, soy sauce etc have been developed. Soybean is also used as raw materials for industrial and pharmaceutical uses, biodiesel production and a large number of soybean-based industries have been established for the production of building and construction materials, plastics, papers, printing inks, paints, engine oils, lubricants, hydraulic fluids, pesticides, cosmetics, pharmaceuticals, and biodiesel. List of hundreds of other soy products are available on the web (www.soystats.com). Soybean consumption reduces cancer, blood cholesterol, osteoporosis and heart diseases in human (Birt et al. 2004) and is also a good source of minerals, vitamin B, folic acid and isoflavones attributed to slowing down cancer development, heart diseases and osteoporosis (Wilson 2004).

Advances in soybean breeding have been discussed below in two sections, pre-1996 era and post-1996 era. Until 1996, soybean breeding was mostly performed through conventional breeding, and more recently conventional breeding was enhanced through the support of molecular and other techniques. Today, molecular breeding and other techniques are playing important roles along with conventional breeding methods for genetic improvement of soybean.

Advances Through Conventional Breeding

Conventional breeding is based on phenotypic selection of superior individuals from segregating populations. It takes about 8–10 years to complete the cycle starting with making crosses to release of variety/germplasm. Phenotypic selection is often difficult and unreliable due to significant genotype-environment interactions. In spite of these, farmers have been doing selection since the domestication of soybean similar to other crops, by saving seeds from selected plants to grow in the following years (Gai et al. 1997; Qiu et al. 1999). Farmer's selections thus established a global reservoir of soybean germplasm resources, much of which is available to modern soybean breeders. Early soybean varieties were Asian introduced lines, selections from these lines, and/or natural crosses from these introduced lines. Before 1940s, soybean was mostly grown as a forage crop in the USA because of lodging and seed-shattering. Intensive soybean breeding programs initiated with the establishment of the United States Regional Soybean Industrial Products Laboratory at Urbana, Illinois in 1936 and in cooperation with other north central State Agricultural Experiment Stations (SAES) (Hartwig 1973). Initially, soybean breeding was mainly confined to SAES or United States Department of Agriculture's Agriculture Research Service (USDA-ARS). Private industries came forward with soybean breeding at the later part of the 20th century. Conventional breeding strategies have made significant improvements in soybean yield, quality, disease and insect resistance, and consequently, soybean became a major food from a minor crop in about 60 years. Wilcox (2001) estimated that seed yield of northern soybean varieties have increased about 60 % over the past 60 years. Today, soybean ranks second in production to corn and the first in oil production in the USA. Soybean is a self pollinated legume with natural out-crossing of less than 1 % (Carlson and Lersten 1987) and soybean breeding follows the common procedures for varietal development like other self pollinating crops, similar to wheat and rice.

Overall goal of plant breeding is to increase yield with improved qualities and increased resistance to biotic and abiotic stresses. Based on the objectives of the breeding program, soybean breeders generally consider, (a) maturity groups (MG) based on latitude, starting from MG 000 for Canada and extreme northern regions of the USA to MGX for southern Florida, (b) growth habit- indeterminate (MG 00–IV) and determinate (MG V–IX) based on soybean plant stem growth and flowering, and (c) seed size (larger seed size for tofu, and small seed for misu and natto). A number of studies have been conducted in United States to evaluate genetic improvement for yield potential of soybean with a number of plant introduction (PI) lines and cultivars developed by hybridizations at different periods (Boerma 1979; Boyer et al. 1980; Luedders 1977; Specht et al. 1999; Wilcox 2001; Wilcox et al. 1979) and reported that increase in yield potential ranged from 0.5 to 1.0 % per year. Luedders (1977) evaluated 21 soybean cultivars of MG I–MG IV developed over a period of 50 years (1920–1970) and found that yield increased about 1.0 % per year, partially due to increased lodging resistance. Wilcox et al. (1979) assessed yield improvement of five cultivars of MG II and MG III released from 1921 to 1974 and found yield increased consistently at a rate of 0.5 % per year with successive release of new cultivars because newer cultivars were more stable across a wide range of environments. Boerma (1979) evaluated 18 cultivars of MG VI, VII and VIII, released during the period 1914–1973 and reported yield increases of 0.7 % per year without a consistent relationship with any of the traits studied. Boyer et al. (1980) compared new and old cultivars and observed that new cultivars were more tolerant to water stress than older cultivars and yield increased about 0.6 % per year for new cultivars. Specht et al. (1999) reported that during last 70 years (1924–1998), USA soybean yield increased at a linear rate of 22.6 kg ha^{-1} $year^{-1}$ due to adaptation of improved cultivars and production methods, increases in atmospheric CO_2 concentrations and greater nitrogen fixation. The authors also suggested that recently released cultivars are able to supply more assimilates during seed filling period (SFP), have increased N_2 fixation capacity and better tolerance to the stress of high plant populations than older cultivars. Wilcox (2001) has evaluated soybean lines (MG 00–IV) across the northern soybean production areas of the USA and Canada, and reported that soybean production has increased about 60 % over the last 60 years (1941–2000) through the use of elite soybean lines developed through the use of conventional breeding programs by public breeders. Genetic improvement was possible due to development and cultivation of soybean cultivars with increased lodging resistance and increased resistance to major pathogens.

Similar studies were also conducted in Canada with short-seasoned (MG 000, 00, and 0) soybean cultivars released from 1943 to 1992 (Morrison et al. 1999, 2000; Voldeng et al. 1997) and found that yield of short-season soybean increased by approximately 0.5 % per year since they were first cultivated in the 1930s. Morrison et al. (1999, 2000) reported that modern soybean cultivars were more efficient in production and allocation of carbon resources to seeds, lodging resistance, and phenotypically stable in plant height than their predecessors, increase in seed yield was correlated with an increase in harvest index, photosynthesis, stomatal conductance and decrease in leaf area index. There was a significant decrease of seed protein with year of release with an associated increase in seed oil concentration. Voldeng et al. (1997) also pointed out that genetic improvement in short-season soybean is associated with lodging resistance and decreases in seed protein levels with an increase in seed oil. Similar results were also reported by Wilcox (2001). Recently, the USDA reported that during the last 30 years (1975–2005), soybean production has increased about 0.4 Bu per acre per year (http://www.usda.gov/nass/aggraphs/soyyld.htm).

Advances Through the Incorporation of Molecular and Other Techniques

Molecular Breeding

Although conventional breeding has made a significant contribution to soybean yield increases in the last 65 years (1930–1995), but progress is slow specially for improving quantitative traits, due to significant environmental effects, time-consuming long selection processes and problems associated with selecting appropriate genotypes. Molecular technologies are capable to dissect quantitative traits into their individual components, known as quantitative trait loci (QTL) (Tanksley 1993; Quarrie 1996; Tuberosa et al. 2002). New genomic technology, marker-assisted selection (MAS) uses molecular markers closely linked to a target gene/QTL as a molecular tag that can be used for quick indirect selection of the target QTL/gene. Selection can be done at an early stage of plant growth, even before maturation is possible, thus enhancing conventional breeding process and aiding genetic improvement of the traits of interest (MacKill 2003; Varshney et al. 2005; Wang et al. 2005). MAS can enhance conventional breeding in various ways, such as, (a) early selection of traits in the lab, using seed DNA even before planting or using leaf DNA from young seedlings well before phenotypic expression of traits in the field. This reduces number of samples to be planted and reduces time to wait until maturity for evaluation, subsequently saving time, space and cost, (b) MAS is independent to season and location consideration for the trait whose expression is dependent upon certain environmental conditions, allows for screening of important trait any time anywhere, (c) gene pyramiding by incorporating different alleles for multiple traits or multiple alleles related to a single trait, (d) selection of parental lines with wider genetic base, as genetic variance of a population for a metric trait, such as yield, will increase as the parents differ for genes that affect the trait (Falconer 1981), (e) recovery of recurrent parent genome after repeated backcrossing and selection, (f) discovery of new beneficial alleles from wild germplasm, (g) monitoring seed purity and germplasm identity.

Molecular markers and genetic maps are two important components of molecular marker technology. An integrated genetic linkage map of soybean has been constructed with 1,849 markers, including 1,015 simple sequence repeats (SSRs), 709 Restriction Fragment Length Polymorphisms (RFLPs), 73 Random Amplification of Polymorphic DNAs (RAPDs), 6 Amplified Fragment Length Polymorphisms (AFLPs), 24 classical, 10 isozymes, and 12 others (Song et al. 2004). Among different markers, RFLP, SSR (microsatellite) and single nucleotide polymorphism (SNP) markers are widely and frequently used for genetic map construction, QTL analysis and MAS, while SSR and SNP markers are commonly used for molecular breeding. Both SSR and SNP markers are PCR-based; they are abundant and easy to use in high throughput genotyping systems. Recently, Choi et al. (2007) developed the first genetic transcript map of soybean genome by mapping one SNP in each of 1,141 genes on the integrated genetic map. The new map consists of 2,989 markers, (1,141 SNPs and 1,848 markers from Song et al., 2004 integrated soybean map). Preparation of the high density genetic linkage map by adding more markers around the region harboring the gene of interest will more precisely facilitate dissection of genetic loci. An integrated genetic-physical map which includes anchoring molecular markers of the genetic map to the physical map will provide breeders 'mile markers' allowing us to access the soybean genome (Wu et al. 2006). Recently, soybean genome sequencing was in progress which will facilitate gene function analysis in soybean. Genome or gene space or EST sequence information facilitates in identification of candidate genes either by *in silico* approaches, or transcript profiling. Identification and validation of candidate genes for molecular breeding depends on expression patterns, chromosomal positions, biological functions and performance of alleles under phenotypic selection (Wang et al. 2005).

MAS is considered as one of the first direct benefits that breeders obtained from genomics (Xu et al. 2005) and being applied successfully in molecular breeding for genetic improvement of many major crops including soybean. Recently, Varshney et al. (2006) highlighted some important contributions of MAS in the area of cereal genomics (barley, maize, pearl millet, rice, sorghum, and wheat). Ribaut and Ragot (2007) also discussed the application of MAS to improve drought tolerance in maize. So far, a total of 1,174 QTLs for more than 70 traits of soybean have been reported in the database (http//:soybeanbreederstoolbox.org) and Lee et al. (2006) summarized the reported QTLs. Scientists mostly identified QTLs for traits with high heritability and importance to producers such as SCN resistance, protein and oil content, and yield (Orf et al. 2004). Some important genes have been well-documented for different traits of soybean (Table 8.1).

Among different soybean insects and diseases, SCN is economically the most destructive pest in USA. So far, more than 90 QTLs associated with different races of SCN were reported in the database but most of them need confirmation. Only two QTLs, namely *rhg1* (located on LG G) and R*hg4* (located on LG A2) are confirmed across the different populations, time and locations and commonly used in MAS for

Table 8.1 Important genes mapped in soybean (modified from Lee et al. 2006)

Trait	Gene	Reference
Aphid resistance	*Rag1*	Hill et al. 2006
Bacterial blight	*Rpg1, Rpg4*	Ashfield et al. 1998
Bacterial pustule	*Rxp*	Narvel et al. 2001
Brown stem rot	*Rbs1, Rbs2, Rbs3*	Bachman et al. 2001; Lewers et al. 1999
Chlorophyll deficiency	*Y9, Y10, Y13, G1, G2*	Cregan et al. 1999; Zou et al. 2003; Luquez and Guiamet 2001
Drought and salt tolerance	*GmDREBa, GmDREBb, GmDREBc*	Li et al. 2005
Fatty acid (Linolenic acid)	*GmFAD3, Fan Fas*	Bilyeu et al. 2003, 2005; Spencer et al. 2003
Flower color	*W1, Wp, gmfls1*	Karakaya et al. 2002; Hegstad et al. 2000; Takahashi et al. 2007
Frog leaf spot	*Rcs3*	Mian et al. 1999
Herbicide resistance- metribuzin	*Hm*	Palmer et al. 2004
Leaf form	*Lf1, Ln*	Cregan et al. 1999
Lipoxygenase	*Lx2*	Kim et al. 2005
Nodulation	*Rj1, rj2, rj4*	Devine and Kuykendall 1996; Polzin et al. 1994; Matthews et al. 2001
Phytophthora root and stem rot	*Rps1 to Rps8*	Burnham et al. 2003; Demirbas et al. 2001; Lohnes and Schmitthenner 1997
Powdery mildew	*Rmd*	Polzin et al. 1994
Salt stress	*GmPAP3, GmCAX1*	Liao et al. 2005; Luo et al. 2005
Seed coat and hilum color	*I, K, R*	Cregan et al. 1999; Senda et al. 2002; Karakaya et al. 2002
Seed shape and seed-coat structure	*B1, Shr*	Chen and Shoemaker 1998
Soybean cyst nematode	*rhg1, Rhg4*	Meksem et al. 1999; Cregan et al. 1999
Soybean mosaic virus	*Rsv1, Rsv4*	Yu et al. 1994; Hayes et al. 2000
Soybean rust	*Rpp1*	Hyten et al. 2007
Stem, petiole, plant growth	*Dt1, F*	Karakaya et al. 2002; Kiang 1990
Sudden death syndrome	*Rfs*	Meksem et al. 1999

SCN screening. Recently, several SNP markers have been developed for *rhg1* and *Rhg4* for more efficient and high-throughput genotyping for SCN resistance (Hyten and Cregan 2006; Wu et al. 2004). A number of QTLs also have been identified for other pests including soybean white mold, sclerotinia stem rot (Arahana et al. 2001), sudden death syndrome (SDS) (Meksem et al. 1999), soybean corn earworm (Rector et al. 1998, 1999, 2000), brown stem rot (Bachman et al. 2001; Cregan et al. 1999; Patzoldt et al. 2005b), root knot nematode (Tamulonis et al. 1997).

Soybean seed is a major source of protein for animal feed and oil for human consumption. It is difficult to improve seed protein and oil simultaneously, as they are negatively correlated (Burton 1987). There are a number of important QTL studies for soybean seed protein and oil content (Diers et al. 1992; Lee et al. 1996; Mansur et al. 1993, 1996; Brummer et al. 1997; Sebolt et al. 2000; Chung et al. 2003; Hyten et al. 2004; Panthee et al. 2005) and as of 13 June 2007, about 140 QTLs reported in soybase for these traits. Confirmation of these reported QTLs across the different environment and genetic backgrounds are important for practical use of QTL in molecular breeding programs. But there are a few reports available on confirmed and validated QTLs related to protein and oil. QTL region of Satt239 of LG I for protein and oil was found to be in common in different genetic backgrounds (Brummer et al. 1997; Chung et al. 2003; Sebolt et al. 2000) and recently, Diers and his group also confirmed this (Nichols et al. 2006). Fasoula et al. (2005) have confirmed seed protein, oil and seed weight QTLs using two populations. Although there is a lot of reports on QTL mapping for oil in soybean, but there are only a few reports on QTL mapping on fatty acid composition. Panthee et al. (2006) identified several QTLs for palmitic, stearic, oleic, linoleic and linolenic acids in soybean. From human nutritional viewpoint, edible soybean oil containing low linolenic acid will increase storage life, reduces oily odor and avoid partial hydrogenation process, this process increase tras-fat. Spencer et al. (2004) identified two SSR markers Satt534 and Satt560 on LG B2. Recently, molecular markers have been developed that identify specific mutations in three genes associated with low-linolenic acid content in soybean (Bilyeu personnel communication). These markers will improve the identification of soybean lines with low linolenic acid content and finally development of soybean varieties with low linolenic acid. United Soybean Board (USB) has initiated a project to develop soybean varieties with improved oil content (low saturates high stearic, high oleic and low linolenic acid content) using marker assisted backcrossing.

More than 80 QTLs have been reported in SoyBase for seed yield and yield related traits like, lodging, plant height and seed weight. Narrow genetic base is one of the reasons for slow yield increases of soybean in North America (Gizlice et al. 1994) and greater genetic diversity may increase the rate of yield improvement (Kisha et al. 1997). QTL mapping of yield from soybean PI's and exotic germplasm demonstrated the possibility to incorporate yield enhancing alleles from PI's and exotic germplasm in to soybean elite lines (Orf et al. 1999; Specht et al. 2001; Concibido et al. 2003; Kabelka et al. 2005; Wang et al. 2004; Guzman et al. 2007; Winter et al. 2007). Two QTLs for lodging on LG C2 and L were detected in three independent studies (Mansur et al. 1996; Lee et al. 1996; Orf et al. 1999). Yield and yield related QTLs with larger effects need confirmation in different genetic backgrounds, locations and time. Although a notable success has been made to map QTLs/genes for other important quantitative traits, but only a few reports are available on abiotic stress (drought, salinity and submergence/waterlogging) related traits like, water use efficiency (Mian et al. 1996, 1998), leaf wilting (Bhatnagar et al. 2005), genetic basis of beta and carbon isotope discrimination (Specht et al. 2001), salt stress (Lee et al. 2004), and flooding/waterlogging tolerance (Githiri et al. 2006; VanToai et al. 2001; Reyna et al. 2003). Lee et al. (2004) identified a major QTL for salt stress linked to SSR marker sat_091 on LG N explaining more than 40 % phenotypic variation both in the greenhouse and field. These results need to be confirmed in different genetic backgrounds and time.

North American soybean cultivars are composed from a narrow genetic base and about 85 % of the genes present in modern soybean cultivars could be trace back to 18 ancestors and their progenies (Gizlice et al. 1994). Recently, Diers (2006) reported that soybean PI88788 is the sole source of 93 % of SCN resistance in modern soybean cultivars. Introduction of new sources of genetic variability into soybean breeding programs is very important and wild/unadapted soybean germplasm is a new source of variability. Tanksley and his group have demonstrated the concept of mining beneficial alleles from wild tomato and transfer into cultivated species (Tanksley et al. 1996; Fulton et al. 2000). Beneficial alleles have been uncovered from unadapted and exotic germplasm for quantitative trait improvement in rice (Xiao et al. 1998; Moncada et al. 2001), in barley (von Korff et al. 2005; Yun et al. 2005) and in other crops. In soybean, using molecular markers, a number of beneficial alleles have been extracted from wild or unadapted soybean germplasm for the improvement of cultivated soybean. Beneficial alleles include traits such as SCN resistance (Concibido et al. 1997; Riggs et al. 1998), soybean rust (Patzoldt et al. 2007), Brown stem rot (Lewers et al. 1999; Patzoldt et al. 2005a,b), Phytophthora root rot (Hegstad et al. 1998), soybean mosaic virus (Hayes et al. 2000), yield and yield related traits (Chakraborthy et al. 2006; Concibido et al. 2003; Li and Pfeiffer 2006). Molecular markers like SSR and SNP may play a significant role in full exploitation of exotic soybean germplasm to identify beneficial alleles for the improvement of elite soybean lines by widening the genetic base.

Pioneer® used molecular marker technology to develop soybean variety 94M80, the soybean variety that broke the world soybean yield record in 2006, by achieving 139 bushels per acre. Pioneer soybean varieties developed by MAS in the last seven years had average yearly increases of 1.4 bushels per acre per year while non-MAS Pioneer® varieties had increases over 10 years of 0.5 bushels per acre per year. At the same time, non-MAS USDA varieties achieved yield increases of 0.4 bushels per acre per year (http://www.prnewswire.com/mnr/pioneer/26118/docs/26118a- SoybeanDecadeStudy_whitepaper.doc). Development of MAS screening facilities and availability of suitable markers, confirmed genes/QTLs for the trait of interest are prerequisites for integration of MAS into modern plant breeding programs. Only a few public funded high throughput MAS facilities have been developed and on the other hand, private industries have established MAS facilities for their routine screening. Pioneer celebrated its 10th anniversary of molecular breeding in 2006. Monsanto, Syngenta and other companies are also routinely using MAS for their varietal development programs. Important genes/QTLs related to most of the agronomic traits have been identified and more is in progress using high through-put SSR and SNP markers. Easy access to screening facilities, availability of useful markers and reasonable costs for sample analysis will help in integration of MAS in public breeding programs. High throughput MAS operation at the Soybean Genome

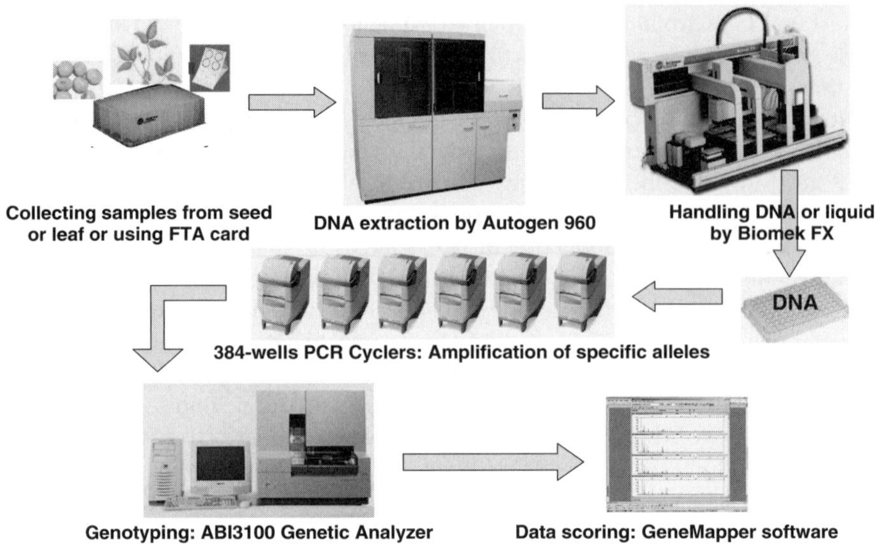

Fig. 8.2 High throughput MAS operation at the Soybean Genome Mapping Lab of the University of Missouri-Columbia (*See* also Color Insert)

Mapping Lab of the University of Missouri-Columbia and suggested general MAS steps are presented in Figs. 8.2 and 8.3, respectively.

Genetic Engineering

Crops containing foreign genes are known as transgenic or genetically modified (GM) or more often known as biotech crops. Conventional, transgenic and molecular plant breeding techniques depend on identification, production and use of genetic variation for varietal improvement. Genetic engineering, in combination with conventional plant breeding has provided additional tools to enhance efficiency of plant breeding by helping to alleviate the narrow genetic base of soybean for varietal improvement. Nearly 100 transgenic plant species are ready to be released or are presently grown in the field commercially. Among them, GM cotton, maize, rapeseed and soybean are major crops grown globally (Wenzel 2006). Both public and private soybean breeders are involved in transgenic work but public breeders are mostly involved in conducting basic research and private breeders are engaged in varietal development. Sharing of information on genetic engineering is limited between public and private soybean breeders.

James (2006) reported that global area of biotech crops has increased more than 50-fold in the last ten years (1996–2005), from 1.7 million hectares (M ha) to 90 M ha. Herbicide tolerant soybean and canola, insect resistance by the *Bacillus*

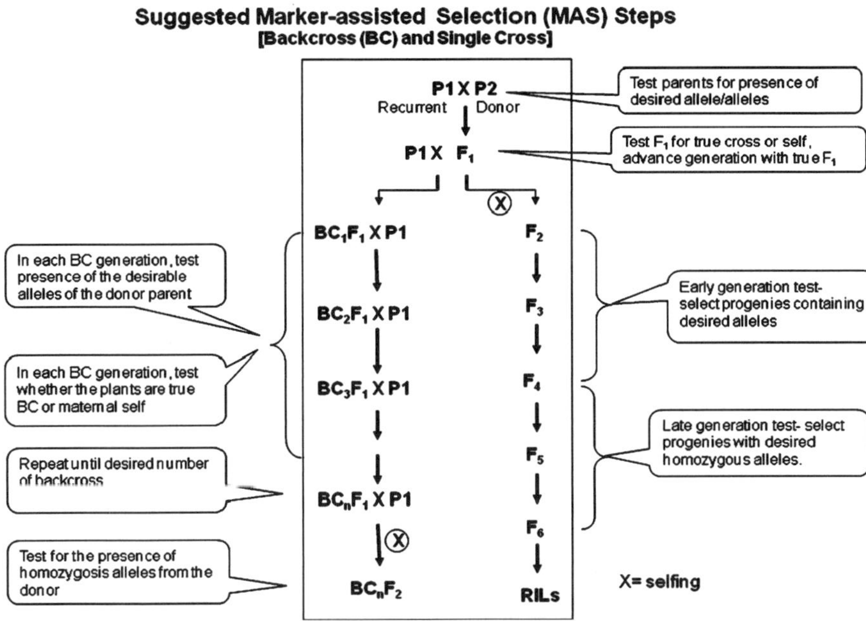

Fig. 8.3 Suggested Marker-assisted Selection (MAS) steps in backcross (BC) and single cross operations

thuringiensis T-toxin protein (Bt) maize and cotton, and Bt/herbicide tolerant maize and cotton are major biotech crops. In 2005, herbicide tolerant soybean was the major biotech crop cultivated on 54.4 M ha, 60 % of the world biotech crop areas, followed by maize 21.2 M ha, 24 % of biotech crop areas. Farmers of the United States have widely adopted GM crops including soybean since 1996. In 2005, US farmers planted 49.8 M ha of transgenic crops, 55 % of the total world biotech crop area. Herbicide tolerant 'Roundup Ready' soybean patented and marketed by Monsanto is the most rapidly and widely adopted transgenic crop in the United States followed by herbicide tolerant and insect resistant cotton and corn. According to USDA-NASS, in 2005, 89–90 % of US soybean areas were covered by GM soybean and 'Roundup Ready' varieties were cultivated in almost all of the GM soybean fields. Acreages of GM crops indicate that farmers are adopting GM crops very rapidly and widely due to substantial and consistent improvement, social and economic benefits to both large and small farmers, both in the industrial and developing countries. Transgenic soybean is also grown in Argentina, Brazil, Paraguay, Canada, South Africa, Uruguay, Romania and Mexico. Genetic engineering deals with important traits that are not easily changeable by conventional breeding, to shorten varietal development time and to widen the genetic base. But, genetic engineering is costly from both the invention and regulatory aspects and moreover, a large number of people are not willing to accept GM crops. GM crop industries are mostly under the control of private industries, and they invest huge amount of money, patenting traits and earning profits.

Hybrid Breeding

Use of heterosis or hybrid vigor in first generation seed (F_1), generated by crossing genetically distinct lines, has been exploited in many crops to increase yield potential. Hybrid maize yields about 15 % (Tollenaar 1994) and hybrid rice 10–15 % (Khush 2005) more than non-hybrid cultivars. China has made tremendous progress in hybrid rice production, now, half of the total rice growing area in China is planted with hybrid rice cultivars (Wang et al. 2005). Major drawbacks of hybrid breeding programs are identification of appropriate germplasm resources for selection and development of parental lines are very time consuming in conventional breeding programs. Like rice, soybean is also a self-pollinated species; researchers have made crosses manually to produce hybrid seed in limited quantities for breeding and genetic studies. It is difficult to produce large amounts of hybrid seed for commercial purposes. With the identification of male sterile and female fertile mutant lines, it is possible to produce hybrid soybean seed (Orf et al. 2004). Male sterile mutant lines may be used for random mating in a recurrent selection program (Burton 1987), facilitating crossing in a population improvement and cultivar development program (Specht and Graef 1990), and introgression of exotic germplasm into elite lines (Lewers et al. 1999). Palmer et al. (2001, 2004) mentioned five important components for developing hybrid soybean; (1) stable male-sterile and female-fertile lines, (2) efficient pollen transfer mechanisms, (3) parents of superior level of heterosis, (4) higher percentage of normal seed-seed in male sterile-female fertile plants, and (5) production of large quantities of hybrid seed. First two points are very important, once male-sterile and female-fertile lines become available, the next most critical factor is efficient pollen transfer. Recently, Ortiz-Perez et al. (2006a) evaluated 21 male-sterile soybean lines for efficient transfer of pollen from fertile male parents to male-sterile female-fertile parents using *Megachile rotundata* as a pollinator vector at Ames, Iowa in 2001–2003. They reported that preferential cross-pollination among male-sterile lines was significantly influenced by maturity group, pubescence color, and male-sterile loci. Ortiz-Perez et al. (2006b) also evaluated insect-mediated seed-set in different soybean lines segregating for male sterility at the *ms6* locus. They found significant differences in seed set among lines, suggesting preferential attraction of pollination. Both of the above groups suggested that environmental conditions that favors plant-pollinator relationships for seed set need to be identified for efficient hybrid seed production systems in soybean. In 2003, Chinese scientists bred the first hybrid soybean with more than two decades of unremitting efforts (Beijing Time, January 17, 2003). They have developed and used a cytoplasmic–nuclear male sterility system in soybean to produce large amounts of hybrid seed for commercial production (Zhao and Gai 2006). So far, limited success has been made towards development of hybrid soybean in the USA and more research is needed from both the private and public sectors.

Genomics Assisted Breeding

Significant progress has been made in development of soybean genetic and genomics resources, such as genetic maps, molecular markers, integrated genetic physical map, genome sequencing, ESTs, proteomics, transcriptomics, metabolomics, and bioinformatics. Genomics-assisted breeding (GAB) is a newly introduced breeding strategy based on integration and exploitation of genetic and genomic tools and resources used for crop improvement (Varshney et al. 2005). Varshney et al. (2006) have reviewed the recent advances made in cereal genomics and their applications in crop breeding. They described how various available genetic and genomics tools and resources are used to understand related components between genotype and pheno-type for crop improvement. With the success of GAB, MAS and molecular breeding will evolve into genomic-assisted breeding to enhance conventional plant breeding. Recently, Tuberosa and Salvi (2006) have discussed genomics-based approaches to improve drought tolerance of crops.

Economic Implication

Several groups have conducted cost analysis of marker-assisted breeding (Dreher et al. 2003; Kuchel et al. 2005; Moreau et al. 2000; Morris et al. 2003). Moreau and his group (2000) concluded that cost efficiency of MAS depends on the stage at which MAS is applied and number of genes controlling the trait under study. At the International Maize and Wheat Improvement Center (CIMMYT) at Mexico, Morris et al. (2003) found that if operating funds are abundant, MAS maximizes the net present values (time can be saved by expenditure of more funds with trade offs occurring between money spent and time) and for this reason private industries invest more money than many public breeding programs. Pioneer Hi-bred has used MAS for the last 10 years as a cost-effective breeding tool, frequently used for the development of soybean cyst nematode (SCN) resistant varieties (Cahill and Schmidt 2004). In a simulated analysis in wheat, Kuchel et al. (2005) found that incorporation of MAS at BC_1F_1 and haploid stages increased genetic gain over phenotypic selection and reduced overall cost by 40 %. It is clear from the studies that MAS is beneficial in support of conventional plant breeding strategies, however, economic success of MAS depends on many factors, for example, importance of the trait, genetic nature of traits under selection, that is number of gene/genes controlling the trait, population type (F_1, backcross or doubled haploid), size of the sample, and availability of funds, equipment and facilities. There are certain traits for which suitable molecular markers have not yet been identified and conventional breeding strategies are cost-effective and efficient method for selection of these traits.

Conclusion and Future Perspectives

It is clear that genomic tools are being exploited by soybean breeders for the development of superior varieties. One of the greatest advancements with these genetic tools is the use of molecular markers for MAS. This technology was not available to soybean breeders or breeders of other important crops until recently. MAS have been proven to improve upon the efficiency of soybean breeding. It must be pointed out that these new genomic tools do not take the place of conventional soybean breeding, but rather, they are tools used to improve upon the efficiency and development of superior soybean varieties.

It must be reiterated that the use of genomic tools in plant breeding can be expensive. Development and use of genomic tools, such as DNA markers come with a price and are not within the reach of most public soybean breeding programs. As molecular breeding becomes more popular and the use of it increases, costs are likely to become more favorable, but it will always remain an added expense to the breeding program. Whether a soybean breeder decides to use genomic tools in plant breeding depends on many factors including: (a) costs, (b) ease of using the technology, (c) consumer and producer acceptance, (d) plant breeding objectives, (e) estimated improvement in efficiency of selection, and others dependent upon individual circumstances associated with the breeding effort.

As we look towards the future, it is evident that the number of genomic tools is increasing in soybean. It is our hope that as soybean becomes an increasing desirable species to work with and because it has an escalating array of genomic possibilities, that it will attract more scientists to use soybean in their basic genomic studies. As the application of genomics increases in soybean, the soybean breeder stands to be one of the important recipients of these new technologies. The future of soybean breeding will certainly be advanced with the development of more genomic tools that can only be imagined today. Soybean breeders have outstanding challenges, and at the same time, exciting possibilities for the future as much is yet to be accomplished.

References

Arahana, V.S., Garef, G.L., Specht, J.E., Steadman, J.R., and Eskridge, K.M. (2001) Identification of QTLs for resistance to *Sclerotinia sclerotiorum* in soybean. Crop Sci. 41, 180–188.

Arumuganathan, K. and Earle, E.D. (1991) Nuclear DNA content of some important plant species. Plant Mol. Biol. Rep. 9, 208–219.

Ashfield T., Saghai Maroof, M.A., Webb, D.M., Innes, R.W., Keim, P., Danzer, J.R., Held, D., and Clayton, K. (1998) *Rpg1*, a soybean gene effective against races of bacterial blight, maps to a cluster of previously identified disease resistance genes. Theor. Appl. Genet. 96, 1013–1021.

Bachman, M.S., Tamulonis, J.P., Nickell, C.D., and Bent, A.F. (2001) Molecular markers linked to brown stem rot resistance genes, *Rbs1* and *Rbs2* in soybean. Crop Sci. 41, 527–535.

Bhatnagar, S., King, C.A., Purcell, L., Ray, J.D. (2005) Identification and mapping of quantitative trait loci associated with crop responses to water-deficit stress in soybean [*Glycine max* (L.) Merr.]. The ASA-CSSA-SSSA International Annual Meeting (Abstract), November 6–10, 2005. p 9.

Bilyeu, K., Palavalli, L., Sleper, D., and Beuselinck, P. (2003) Three microsomal omega-3 fatty acid desaturase genes contribute to soybean linolenic acid levels. Crop Sci. 43, 1833–1838.

Bilyeu, K., Palavalli, L., Sleper, D., and Beuselinck, P. (2005) Mutations in soybean microsomal omega-3 fatty acid desaturase genes reduce linolenic acid concentration in soybean seeds. Crop Sci. 45, 1830–1836.

Birt, D.F., Hendrick, S. and Alekel, D.L. (2004) Soybean and the prevention of chronic human disease. In: H.R. Boerma and J. E. Specht (Eds), *Soybeans: Improvement, Production, and Uses*. Agronomy Monographs 3rd ed. No. 16, ASA-CSSA-SSSA, Madison, WI, USA, pp.1047–1117.

Blanc, G. and Wolfe, K.H. (2004) Wide spread paleoploidy in model plant species inferred from age distributions of duplicate genes. Plant Cell. 16, 1667–1678.

Boerma, H. R. (1979) Comparison of past and recently developed soybean cultivars in maturity groups VI, VII and VIII. Crop Sci. 19, 611–613.

Boyer, J.S., Johnson, R.R., and Saupe, S.G. (1980) Afternoon water deficits and grain yields in old and new soybean cultivars. Agron. J. 72, 981–986.

Brummer, E.C., Greaf, G.L., Orf, J., Wilcox, J.R., and Shoemaker, R.C. (1997) Mapping QTL for seed protein and oil content in eight soybean populations. Crop Sci. 37, 370–378.

Burnham, K.D., Dorrance, A.E., Francis, D.M., Fioritto, R.J., and St. Martin, S.K. (2003) A new locus in soybean for resistance to *Phytophthora sojae*. Crop Sci. 43, 101–105.

Burton, J.W. (1987) Soybean [(*Glycine max* (L.) Merr.)]. Field Crop Res. 53, 171–186.

Cahill, D.J. and Schmidt, D.H. (2004) Use of marker assistant selection in a product development breeding program. Proceedings of the 4th International Crop Science Congress, 26 Sept- 1 Oct 2004, Brisbane, Australia. (www.cropscience.org.au).

Carter Jr., T. E., Nelson, R. L., Sneller, C. H. and Cui, Z. (2004) Genetic diversity in soybean. In: H.R. Boerma and J. E. Specht (Eds), *Soybeans: Improvement, Production, and Uses*. Agronomy Monographs 3rd ed. No. 16, ASA-CSSA-SSSA, Madison, WI, USA, pp. 303–416.

Carlson, J.B., and Lersten, N.R. (1987) Reproductive morphology. In: J.R. Wilcox (Ed.), *Soybeans: Improvement, Production, and Uses*. Agronomy Monographs 2nd ed. No. 16, American Society of Agronomy (ASA), Madison, WI, USA, pp. 95–134.

Chakraborthy, N., Curley, J., Neece, D., Diers, B. and Nelson, R. L. (2006) QTLs for soybean seed yield and agronomic traits in populations derived from exotic lines. Poster abstract. Biennial conference on the molecular and cellular biology of the soybean, August 5–8, 2006, Lincoln, Nebraska.

Chang, R, Qiu, L., Sun, J., Chen, Y., Li, X. and Xu, Z. (1999) Collection and conservation of soybean germplasm in China. In. H. E. Kauffman (Ed). *Proceedings of the World Soybean Conference VI*. Superior Printing, Champange, Illinois, USA. pp. 172–176.

Chen, Z., and Shoemaker, R.C. (1998) Four genes affecting seed traits in soybeans map to linkage group F. J. Heredity. 89, 211–215.

Choi, I-Y., Hyten, D.L., Matukumalli, L.K., Song, Q., Chaky, J.M., Quigley, C.V., Chase, K., Lark, K.G., Reiter, R.S., Yoon, Hwang, E-Y., Yi, S-I., Young, N.D., Shoe,amker, R.C., Tassell, C.P., Specht, J.E., and Cregan, P. B. (2007) A soybean transcript map: gene distribution, haplotype and Single-nucleotide polymorphism (SNP) analysis. Genetics. 176, 685–696.

Chung, J., Babka, H.L., Graef,G.L., Staswick, P.E., Lee, D.J., Cregan, P.B., Shoemaker, R.C., and Specht, J.E. (2003) The seed protein, oil, and yield QTL on soybean linkage group I. Crop Sci. 43, 1053–1067.

Concibido, V.C., La Vallee, B., McLaird, P., Pineda, N., Meyer, J., Hummel, L., Yang, J., Wu, K., and Delannay, X. (2003) Introgression of a quantitative trait locus for yield from Glycine soja in to commercial soybean cultivars. Theor. Appl. Genet. 106, 575–582.

Concibido, V.C., Denny, R., Lange, D., Danesh, D., Orf, J., and Young, N. (1997) Genome mapping on soybean cyst nematode resistance genes in Peking, PI90763, and PI88788 using DNA markers. Crop Sci. 37, 258–264.

Cregan, P.B., Mudge, J., Fickus, E.W., Danesh, D., Denny, R. and Young, N.D. (1999) Two simple sequence repeat markers to select for soybean cyst nematode resistance conditioned by the *rhg1* locus. Theor. Appl. Genet. 99, 811–818.

Demirbas, A., Cregan, P.B., Shoemaker, R.C., Specht, J.E., Graef, G.L., Rector, B.G., Lohnes, D.G., and Fioritto, R.J. (2001) Simple sequence repeat markers linked to the soybean *Rps* genes for phytophthora resistance. Crop Sci. 41, 1220–1227.

Devine, T.E., and Kuykendall, L.D. (1996) Host genetic control of symbiosis in soybean (*Glycine max* L.). Plant Soil. 186, 173–187.

Diers, B. (2006) Staying ahead of SCN, alternative sources of resistance. Field Day 2006, University of Illinois, found at http://agronomyday.cropsci.uiuc.edu/2006/tour a/SCN/ (verified on 10-03-07).

Diers, B.W., Keim, P., Fehr, W.R., and Shoemaker, R.C. (1992) RFLP analysis of soybean seed protein and oil content. Theor. Appl. Genet. 83, 608–612.

Dreher, K., Khairallah, M., Ribaut, J-M. and Morris, M. (2003) Money matters (I) costs of field and laboratory procedures associated with conventional and marker-assisted maize breeding at CIMMYT. Mol. Breeding. 11, 221–234.

Falconer, D.S. (1981) Introduction to quantitative genetics. 2nd ed. Longman Inc., New York.

Fasoula, V.S., Harris, D.K., and Boerma, H.R. (2005) Validation and designation of quantitative trait loci for seed protein, seed oil, and seed weight from two soybean populations. Crop Sci. 44, 1218–1225.

Fulton, T.M., Grandillo, S., Beck-Bunn, T., Friedman, E., Frampton, A., Lopez, J., Petiard, J., Uhlig, J., Zamir, D., and Tanksley, S.D. (2000) Advanced backcross QTL analysis of a *Lycopersicon* × *Lycopersicon parviflorum* cross. Theor. Appl. Genet. 100, 1025–1042.

Gai, J., Zhao, T., and Qiu, J. (1997) A review on the advances of soybean breeding since 19981 in China. In: Seed industry and agricultural development, CAASS. China Agric. Press, Beijing, China, pp. 168–174.

Githiri, S.M., Watanabe, S., Harada, K., and Takahashi, R. (2006) QTL analysis of flooding tolerance in soybean at an early vegetative growth stage. Plant Breed. 125, 613–618.

Gizlice, Z., Carter, T.E., and Burton, J.W. (1994) Genetic base for North American public soybean cultivars released between 1947 and 1998. Crop Sci. 34, 1143–1151.

Guzman, P.S., Diers, B.W., Neece, D.J., St.Martin, S.K., LeRoy, A.R., Grau, C.R., Hughes, T.J., and Nelson, R.L. (2007) QTL associated with yield in three backcross-derived populations of soybean. Crop Sci. 47, 111–122.

Hartwig, E.E. (1973) Varietal development. In: B.E. Caldwell (Ed.), *Soybeans: improvement, production, and uses*. Publication No. 16, American Society of Agronomy (ASA), Madison, WI, USA, pp. 187–210.

Hayes, A.J., Ma, G.R., Buss, G.R., and Maroof, M.A.S. (2000) Molecular marker mapping of *RSV4*, a gene conferring resistance to all known strains of soybean mosaic virus. Crop Sci. 40, 1434–1437.

Hegstad, J.M., Nickell, C.D., and Vodkin. L.O. (1998) Identifying resistance to phytophthora sojae in selected soybean accessions using RFLP techniques. Crop Sci. 38, 50–55.

Hegstad, J.M., Tarter, J.A., Vodkin, L.O., and Nickell, C.D. (2000) Positioning the *wp* flower color locus on the soybean genome map. Crop Sci. 40, 534–537.

Hill, C.B., Li, Y., and Hartman, G.L. (2006) A single dominant gene for resistance to the soybean aphid in the soybean cultivar Dowling. Crop Sci. 46, 1601–1605.

Hymowitz, T. (2004) Speciation and cytogenetics. In: H.R. Boerma and J. E. Specht (Eds), *Soybeans: Improvement, Production, and Uses*. Agronomy Monographs 3rd ed. No. 16, ASA-CSSA-SSSA, Madison, WI, USA, pp. 97–136.

Hymowitz, T. (1990) Soybean: The success story. In: J. Janick and J. E. Simon (Eds), Advances in new crops. Timber Press, Portland, OR, USA, pp. 159–163.

Hyten, D.L., and Cregan, P.B. (2006) Saturation of the rhg1 genomic region with SNP markers to determine linkage drag in resistant soybean cultivars and to demonstrate association analysis in soybean. International Plant and Animal Genome (PAG) Conference. January 13–17, 2006, San Diego, CA. p. 414.

Hyten, D. L., Pantalone, V.R., Sams, C.E., Saxton, A.M., Landau-Ellis, D., Stefaniak, T.R., and Schmidt, M.E. (2004) Seed quality QTL in a prominent soybean population. Theor. Appl. Genet.109:552–561.

Hyten, D. L., Hartman, G. L., Nelson, R. L., Frederick, R. D., Concibido, V. C., Narvel, J. M., and Cregan, P. B. (2007) Map Location of the *Rpp1* Locus That Confers Resistance to Soybean Rust in Soybean. Crop Sci. 47, 837–838.

James, C. (2006) Global status of commercialized biotech/GM crops- 2005, ISAAA Briefs no 34–2005, Ithaca, New York, USA.

Kabelka, E.A., Carlson, S.R., and Diers, B.W. (2005) Localization of two loci that confer resistance to soybean cyst nematode from *Glycine soja* PI 468916. Crop Sci. 45, 2473–2481.

Karakaya, H.C., tang, Y., Cregan, P.B. and Knap, H.T. (2002) Molecular mapping of the fasciation mutation in soybean in soybean, *Glycine max* (Leguminose). Am. J. Bot. 89, 559–565.

Khush, G. S. (2005). What it will take to feed 5.0 billion rice consumers in 2030. Plant Mol. Biol. 59, 1–6.

Kiang, Y.T. (1990) Linkage analysis of *Pgd 1*, *Pgi 1*, pod color (*L1*), and determinate stem (*dt1*) loci on soybean linkage group 5. J. Heredity. 81, 401–404.

Kim, M. Y., Van, K., Lestari, P., Moon, J. K., and Lee, S. H. (2005) SNP identification and SNAP marker development for a GmNARK gene controlling supernodulation in soybean, Theor. Appl. Genet. 110, 1003–1010.

Kisha, T.J., Sneller, C.H., and Diers, B.W. (1997) Relationship between genetic distance among parents and genetic variance in populations of soybean. Crop Sci. 37, 1317–1325.

Kuchel, H., Ye, G., Fox, R. and Jefferies, S. (2005) Genetic and economic analysis Sof a targeted marker-assisted wheat breeding strategy. Mol. Breeding. 16, 67–78.

Lee, G.J., Carter, T.E. Jr., Li, Z., Gibbs, M.O., Boerma, H.R., Villagarcia, M.R., and Zhou, X. (2004) A major QTL conditioning salt tolerance in S-100 soybean and descendent cultivars. Theor. Appl. Genet. 109, 1610–1619.

Lee, S.H., Bailey, M.A., Mian, M.A.R., Carter, Jr., T.E., Ashley, D.A., Hussey, R.S., Parrott, W.A., and Boerma, H.R. (1996) Identification of quantitative trait loci for plant height, lodging, and maturity in a soybean population segregating for growth habit. Theor. Appl. Genet. 92, 516–523.

Lee, G-J., X. Wu, Shannon, J.G., Sleper, D.A. and Nguyen, H.T. (2006). Genome mapping and molecular breeding in plants: soybean. In: C. Kole (ed). *Genome mapping and molecular breeding in plants*, Vol. 2 (oilseeds), Springer, USA.

Lewers, K.S., Crane, E.H., Bronson, C.R., Schupp, J.M., Keim, P, and Shoemaker, R.C. (1999) Detection of linked QTL for soybean brown stem rot resistance in 'BSR101' as expressed in a growth chamber environment. Mol. Breed. 5, 33–42.

Li, D., and Pfeiffer, T. (2006) Soybean QTLs for yield and yield components associated with Glycine soja alleles. Abstract. The 11th Biennial conference on the molecular and cellular biology of the soybean, August 5–8, 2006, Lincoln, Nebraska.

Li, X-P., Tian, A-G., Luo, G-Z., Gong, Z-Z., Zhang, J-S., and Chen, S-Y. (2005) Soybean DRE-binding transcription factors that are responsive to abiotic stresses. Theor. Appl. Genet. 110, 1355–1362.

Liu, K. (1997) Soybeans: Chemistry, Technology and Utilization. Aspen Publishers, Gaithersburg, Maryland, USA. pp. 532.

Lohnes, D.G., and Schmitthenner, A.F. (1997) Position of the phytophthora gene *Rps7* on the soybean molecular map. Crop Sci. 37, 555–556.

Luedders, V.D. (1977) Genetic improvement in yield of soybean. Crop Sci. 17, 971–972.

Luo, G-Z., Wang, H-W., Huang, J., Tian, A-G., Wang, Y-J., Zhang, J-S., and Chen, S-Y. (2005) A putative plasma membrane cation/protein antiporter from soybean confers salt tolerance in Arabidopsis. Plant Mol. Biol. 59, 809–820.

Luquez, V.M., and Guiamet, J.J. (2001) Effects of the 'stay green' genotype *GGd1d1d2d2* on leaf gas exchange, dry matter accumulation and seed yield in soybean (*Glycine max* L. Merr.). Ann. Bot. 87, 313–318.

Mackill, D.J. (2003) Applications of genomics to rice breeding. Int. Rice Res. Note. 28, 9–15.

Mansur, L.M., Lark, K.G. Kross, H., and Oliveira, A. (1993) Internal mapping of quantitative trait loci for reproductive, morphological, and seed traits of soybean (*Glycine max* L.). Theor. Appl. Genet. 86, 907–913.

Mansur, L.M., Orf, J.H., Chase, K., Jarvik, T., Cregan, P.B., and Lark, K.G. (1996) Genetic mapping of agronomic traits using recombinant inbred lines of soybean. Crop Sci. 36, 1327–1336.

Matthews, B.F., Devine, T.E., Weisemann, J.M., Beard, H.S., Lewers, K.S., McDonald, M.H., Park, Y.B., Maiti, R., Lin, J.J., Kuo, J., Pedroni, M.J., Cregan, P.B., and Saunders, J.A. (2001) Incorporation of sequenced cDNA and genomic markers into the soybean genetic map. Crop Sci. 41, 516–521.

Meksem, K., Doubler, T.W., Chancharoenchai, K., Njiti, V.N., Chang, S.J.C., Arelli, A.P.R., Cregan, P.E., Gray, L.E., Gibson, P.T., and Lightfoot, D.A. (1999) Clustering among loci underlying soybean resistance to *Fusarum solani*, SDS and SCN in near-isogenic lines. Theor. Appl. Genet. 99, 1131–1142.

Mian, M.A.R., Ashley, D.A., and Boerma, H.R. (1998) An additional QTL for water use efficiency in soybean. Crop Sci. 38, 390–393.

Mian, M.A.R., Mailey, M.A., Ashley, D.A., Wells, R., Carter, T.E., Parrot, W.A., and Boerma, H.R. (1996) Molecular markers associated with water use efficiency and leaf ash in soybean. Crop Sci. 36, 1252–1257.

Mian, M.A.R., Wang, T., Phillips, D.V., Alvernaz, J. and Boerma, R.R. (1999) Molecular mapping of the *Rcs3* gene for resistance to frogeye leaf spot in soybean. Crop Sci. 39, 1687–1691.

Moncada, P., Martinez, C.P., Borrero, J., Chatel, M., Gauch, H.Jr., Guimaraes, E., Tohme, J., and McCouch, R. (2001) Quantitative trait loci for yield and yield components in an Oryza sativa × Oryza rufipogon BC_2F_2 population evaluated in an upland environment. Theor. Appl. Genet. 102, 41–52.

Moreau, L., Lamarie, S., Charcosset, A. and Gallais. A. (2000) Economic efficiency of one cycle of marker-assisted selection. Crop Sci. 40, 329–337.

Morris, M., Dreher, K.. Ribaut, J-M. and Khairallah, M. (2003) Money matters (II): costs of maize inbred line conversion schemes at CIMMYT using conventional and marker-assisted selection. Mol. Breed. 11, 235–247.

Morrison, M. J., Voldeng, H.D. and Cober, R.R. (2000) Agronomic changes from 58 years of genetic improvement of short-season soybean cultivars in Canada. Agron. J. 92, 780–784.

Morrison, M. J., Voldeng, H.D. and Cober, R.R. (1999) Physiological changes from 58 years of genetic improvement of short-season soybean cultivars in Canada. Agron. J. 91, 685–689.

Narvel, J.M., Lee, S.H., Boerma, H.R., Wang, T., Jakkula, L.R., and Philips, D.V. (2001) Molecular mapping of *Rxp* conditioning reaction to bacterial pustule in soybean. J. Hered. 92, 267–270.

Nichols, D.M., Glover, K.D., Carlson, S.R., Specht, J.E., and Diers, B.W. (2006) Fine mapping of a seed protein QTL on soybean linkage map I and its correlated effects on agronomic traits. Crop Sci. 46, 834–839.

Orf, J.H., Chase, K., Jarvik, T., Mansur, L.M., Cregan, P.B., Adler, F.R., Lark, K.G. (1999) Genetics of soybean agronomic traits. I. Comparison of three related recombinant inbred populations. Crop Sci. 39, 1642–1651.

Orf, J. H., Diers, B. W. and Boerma, H. R. (2004) Genetic improvement: conventional and molecular-based strategies. In: H.R. Boerma and J. E. Specht (Eds), *Soybeans: Improvement, Production, and Uses*. Agronomy Monographs 3rd ed. No. 16, ASA-CSSA-SSSA, Madison, WI, USA, pp. 417–450.

Ortiz-Perez, E., Horner, H. T., Hanlin, S. J. and Palmer, R. G. (2006a) Insect-mediated seed-set evaluation of 21 soybean lines segregating for male sterility at different loci. Euphytica 152, 351–360.

Ortiz-Perez, E., Horner, H. T., Hanlin, S. J. and Palmer, R. G. (2006b) Evaluation of insect-mediated seed set among soybean lines segregating for male sterility at the *ms6* locus. Field Crops Res. 97, 353–362.

Palmer, R.G., Gai, J., Sun, H., and Burton, J.W. (2001) Production and evaluation of hybrid soybean. Plant Breed. Rev. 21, 263–307.

Palmer, R.G., Pfeiffer, T.W., Buss, G.R., and Kilen, T.C. (2004) Qualitative genetics. In: Boerma HR, Specht JE (Eds) *Soybeans: Improvement, Production, and Uses*. Agronomy Monographs 3rd ed. No. 16, ASA-CSSA-SSSA, Madison, WI, USA, pp. 133–233.

Panthee, D.R., Pantalone, V.R., West, D.R., Saxton, A.M., and Sams, C.E. (2005) Quantitative trait loci for seed protein and oil concentration, and seed size in soybean. Crop Sci. 45, 2015–2022.

Panthee, D.R., Pantalone, V.R., and Saxton, A.M. (2006) Modifier QTL for fatty acid composition in soybean oil. Euphytica 152, 67–73.

Patzoldt, M.E., Tyagi, R.K., Hymowitz, T., Miles, M.R., Hartman, G.L., and Frederick, R.D. (2007) Soybean rust resistance derived from Glycine tomentella in amphiploid hybrid lines. Crop Sci. 47, 158–161.

Patzoldt, M.E., Carlson, S.R., and Diers, B.W. (2005a) Characterization of resistance to brown stem rot of soybean in five accessions from central China. Crop Sci. 45, 1092–1095

Patzoldt, M.E., Grau, C.R., Stephens, P.A., Kurtzweil, N.C., Carlson, S.R. and Diers, B.W. (2005b) Localization of a quantitative trait locus providing brown stem rot resistance in the soybean cultivar Bell. Crop Sci. 45, 1241–1248.

Polzin, K.M., Shoemaker, R.C., Nickell, C.D., and Lohnes, D.G. (1994) Integration of *Rps2*, *Rmd*, and *Rj2* into linkage group J of the soybean molecular map. J. Hered. 85, 300–303.

Qiu, B. X., Arelli, P.R. and Sleper, D.A. (1999) RFLP markers associated with soybean cyst nematode resistance and seed composition in a Peking × Essex population. Theor. Appl. Genet. 96, 786–790.

Quarrie, S. A. (1996). New molecular tools to improve the efficiency of breeding for increased drought resistance, Plant Growth Regul. 20, 167–178.

Rector, B.G., All, J.N., Parrott, W.A., and Boerma, H.R. (1998) Identification of molecular markers associated with quantitative trait loci for soybean resistance to corn earworm. Theor. Appl. Genet. 96, 786–790.

Rector, B.G., All, J.N., Parrott, W.A., and Boerma, H.R. (1999) Quantitative trait loci for antixenosis resistance to corn earworm in soybean. Crop Sci. 39, 531–538.

Rector, B.G., All, J.N., Parrott, W.A., and Boerma, H.R. (2000) Quantitative trait loci for antibiosis resistance to corn earworm in soybean. Crop Sci. 40, 233–238.

Reyna, N., Cornelious, B., Shannon, J.G., and Sneller, C.H. (2003) Evaluation of a QTL for waterlogging tolerance in southern soybean germplasm. Crop Sci. 43, 2077–2082.

Ribaut, J.-M., and Ragot, M. (2007) Marker-assisted selection to improve drought adaptation in maize: the backcross approach, perspectives, limitations and alternatives. J. Expt. Bot. 58, 351–360.

Riggs, R.D., Wang, S., Singh, R.J., Hymowitz, T. (1998) Possible transfer of resistance to *Heterodera glycines* from *Glycine tomentella* to *Glycine max*. J. Nematol. 30, 547–552.

Sebolt, A.M., Shoemaker, R.C., and Diers, B.W. (2000) Analysis of a quantitative trait locus allele from wild soybean that increases seed protein concentration in soybean. Crop Sci. 40, 1438–1444.

Schlueter, J. A., Dixon, P., Granger, C., Grant, D., Clark, L., Doylee, J. J. and Shoemaker, R. C. (2004) Mining EST databases to resolve evolutionary events in major crop species. Genome. 47, 868–876.

Schmidt, D.H. and Cahill, D. J. (2006) Marker-assisted selection and its contribution to soybean product development- glancing back, looking forward. Abstract. The 11th Biennial conference on the molecular and cellular biology of the soybean, August 5–8, 2006, Lincoln, Nebraska.

Senda, M., Jumonji, A., Yumoto, S., Ishikawa, R., Harada, T., Niizeki, M and Akada, S. (2002) Analysis of the duplicated CHS1 gene related to the suppression of the seed coat pigmentation in yellow soybeans. Theor. Appl. Genet. 104, 1086–1091.

Shoemaker, R. C., lzin, K., Labate, J., Specht, J., Brummer, E. C., Olson, T., Young, N., Concibido, V., Wilcox, J., Tamulonis, J. P., Kochert, G. and Boerma, H. R. (1996) Genome duplication in soybean (*Glycine* subgenus *soja*). Genetics. 144, 329–338.

Shultz, J. L., and others. (2006) The soybean genome database (SoyGD): a browser for display a duplicated, polyploidy, *regions* and sequence tagged sites on integrated physical and genetic maps of Glycine max. Nucleic Acid Res. 34, D758–D765.

Singh, R. J. and Hymowitz, T. (1999) Soybean genetic resources and crop improvement. Genome, 42, 605–616.

Song, Q. J., Marek, L. F., Shoemaker, R. C., Lark, K. G., Concibido, V. C., Delannay, X., Specht, J. E. and Cregan, P. B. (2004) A new integrated genetic linkage map of the soybean. Theor. Appl. Genet. 109, 122–128.

Specht, J. E., Hume, D. J. and Kumudini, S. V. (1999) Soybean yield potential- A genetic and physiological perspective. Crop Sci. 39, 1560–1570. Specht, J.E., Germann, M., Markwell, J.P., Lark, K.G., Orf, J.H., Macrander, M., Chase, K., Chung, J., Graef, G.L. (2001) Soybean response to water: a QTL analysis of drought tolerance. Crop Sci. 41, 493–509.

Specht, J.E., and Graef, G.L. (1990) Breeding methodologies for chickpea: new avenues to greater productivity. In: B.J. Walby and S.d. Hall (rds), *Chickpea in the nineties.* Proc. 2nd International Worhshop on Chickpea Improvement. ICRISAT, India. pp. 217–223.

Spencer, M.M., Pantalone, V.R., Meyer, E.J., Landau-Ellis, D., and Hyten, D.L. (2003) Mapping the *Fas* locus controlling stearic acid content in soybean. Theor. Appl. Genet. 106, 615–619.

Spencer, M.M., Landau-Ellis, D., Meyer, E.J., and Pantalone, V.R. (2004) Molecular marker associated with linolenic acid content in soybean. JAOCS. 81, 559–562.

Takahashi, R,, Githiri, S., Hatayama, K., Dubouzet, E., Shimada, N., Aoki, T., Ayabe, S., Iwashina, T., Toda, K., and Matsumura, H. (2007) A single-base deletion in soybean flavonol synthase gene is associated with magenta flower color. Plant Mol. Biol. 63, 125–135.

Tamulonis, J.P., Luzzi, B.M., Hussey, R.S., Parrott, W.A., and Boerma, H.R. (1997) RFLP mapping of resistance to southern root-knot nematode in soybean. Crop Sci. 37, 1903–1909.

Tanksley, S. D. (1993) Mapping polygenes. Ann. Rev. Genet. 27, 205–233.

Tanksley, S.D., Grandillo, S., Fulton, T.M., Zamir, D., Eshed, Y., Petiard, V., Lopez, J., and Beck-Bunn, T. (1996) Advanced backcross QTL analysis in a cross between an elite processing line of tomato and its wild relative *pimpinellifolium.* Theor. Appl. Genet. 92, 213–224.

Tollenaar, M. (1994) Yield potential of maize: impact of stress tolerance. In: K.G. Cassman (Ed.), Breaking the yield barrier. Proceedings of workshop on rice yield potential in favorable environment. International Rice Research Institute, Manila, Philippines, pp. 103–109.

Tuberosa, R., Gill, B. S. and Quarrie, S. A. (2002) Cereal genomics: ushering in a brave new world. Plant Mol. Biol. 48, 445–449.

Tuberosa, R. and Salvi, S. (2006) Genomics-based approaches to improve drought tolerance of crops. Trends Plant Sci. 11, 405–412.

VanToai, T.T., Martin, S.K., Chase, K., Boru, G., Schnipke, V., Schmitthenner, A.F., and Lark, K.G. (2001) Identification of a QTL associated with tolerance of soybean to soil waterlogging. Crop Sci. 41, 1247–1252.

Varshney, R. K., Graner, A. Sorrells, M.E. (2005) Genomics-assisted breeding for crop improvement. Trends Plant Sci. 10, 621–630.

Varshney, R. K., Hoisington, D.A. and Tyagi, A.K. (2006) Advances in cereal genomics and application in crop breeding. Trends Biotech. 24, 11.

Voldeng, H.D., Cober, E.R., Hume, D.J., Gillard, C. and Morrison, M.J. (1997) Fifty-eight years of genetic improvement of short-season soybean cultivars. Crop Sci. 37, 428–431.

von Korff, M., Wang, h., leon, J., and Pillen, K. (2005) AB-QTL analysis in spring barley. I. Detection of resistance genes against powdery mildew, leaf rust and scald introgressed from wild barley. Theor. Appl. Genet. 111, 583–590.

Wang, D., Graef, G.L., Procopiuk, A.M., and Diers, B.W. (2004) Identification of putative QTL that underline yield in interspecific soybean backcross populations. Theor. Appl. Genet. 108, 458–467.

Wang, Y., Xue, Y. and Li, J. (2005) Towards molecular breeding and improvement of rice in China. Trends Plant Sci. 10(12), 610–614.

Wenzel, G. (2006) Molecular plant breeding: achievements in green biotechnology and future perspectives. Appl. Microbiol. Biotechnol. 70, 642–650.

Wilcox, J. R. (2001) Sixty years of improvement in publicly developed elite soybean lines. Crop Sci. 49, 1711–1716.

Wilcox, J.R., Schapaugh Jr, W.T., Bernard, R. L., Copper, R. L., Fehr, W.R. and Niehaus, M.H. (1979) Genetic improvement of soybeans in the Midwest. Crop Sci. 19, 803–805.

Wilson, R. F. (2004) Seed Composition. In: H.R. Boerma and J. E. Specht (Eds), *Soybeans: Improvement, Production, and Uses*. Agronomy Monographs 3rd ed. No. 16, ASA-CSSA-SSSA, Madison, WI, USA, pp. 621–677.

Winter, S.M., Shelp,B.J., Anderson, T.R., Welacky, T.W., and Rajcan, I. (2007) QTL associated with horizontal resistance to soybean cyst nematode in *Glycine soja* PI464925B. Theor. Appl. Genet. 114, 461–472.

Wu, C., Sun, S., Nimmakayala, P., Santos, F. A., Meksem, K., Springman, R., Ding, K., Lightfoot, D. A. and Zhang, H-B. (2004) A BAC- and BIBAC-based physical map of the soybean genome. Genome Res. 14, 319–326.

Xiao, J., Li, J., Grandillo, S., Ahn, S.N., Yuan, L., Tanksley, S.D., and McCouch, S.R. (1998) Identification of trait-improving quantitative trait loci alleles from a wild rice relatives, *Oryza rufipogon*. Genetics 150, 899–909.

Xu, Y., McCouch, S. R. and Zhang, Q. (2005) How can we use genomics to improve cereals with rice as a reference genome. Plant Mol. Biol. 59, 7–26.

Yu, Y.G., Saghai Maroof, M.A., Buss, G.R., Maughan, P.J., and Tolin, S.A. (1994) RFLP and microsatellite mapping of a gene for soybean mosaic virus resistance. Phytopathology 84, 60–64.

Yun, S.J., Gyenis, L., Hayes, P.M., Matus, I., Smith, K.P., Steffenson, B.J., and Muehlbauer, G.J. (2005) Quantitative trait loci for multiple disease resistance in wild barley. Crop Sci. 45, 2563–2572.

Zhao, T. J., and Gai, J. Y. (2006) Discovery of new male-sterile cytoplasm sources and development of a new cytoplasmic-nuclear male-sterile line in NJCMS3A in soybean. Euphytica 152, 387–396.

Zou, J.J., Singh, R.J., and Hymowitz, T. (2003) Association of the yellow leaf (*y10*) mutant to soybean chromosome 3. J. Heredity. 94, 352–355.

Chapter 9
Forward and Reverse Genetics in Soybean

Kristin D. Bilyeu

Forward Genetics

Efforts towards soybean improvement have relied heavily on what is now termed "forward" genetics. Forward genetics is the process of identifying a specific phenotype of interest within a population of randomly induced mutants. Forward genetics is a common technique in many facets of biology, but has been especially effective in delivering desired traits for soybeans. Identification of the specific trait can be by selection, where restraints are put on the system so that only individuals of the desired phenotype are capable of survival. Alternately, identification of the specific trait can be accomplished by screening, where all individuals in a population are tested for the desired phenotype. Because of the complexity of soybean as a biological system, most forward genetic work on soybeans has been by screening.

Forward genetics research projects in soybean have utilized both chemical mutagens and radiation to induce genetic changes. The chemical mutagens ethylmethane sulfonate (EMS) and N-nitroso-N-methylurea (NMU) are alkylating agents that typically induce single nucleotide polymorphisms (SNPs), which are the result of G-C to A-T transitions. Chemical mutagens were used very successfully to induce mutations in soybean populations that were found to contain a variety of plant type traits, many changes in fatty acid composition, and meal quality traits (reviewed in Bhatia et al. 1999; Palmer et al. 2004; Rajcan et al. 2005; also Wilcox et al. 2000; Hitz et al. 2002).

Radiation has also been a powerful soybean mutagenesis tool. Radiation exposure in the form of X-rays, fast neutrons, and gamma rays induces genomic deletions. The deletions can result in the loss of parts of single genes or can extend to large regions encompassing many genes. Many early mutant soybean releases were the result of screening for plant type after gamma ray exposure (reviewed in Bhatia et al. 1999). More recently, X-ray and fast neutron exposure have been valuable

K.D. Bilyeu
USDA-ARS, Plant Genetics Research Unit, University of Missouri-Columbia, MO 65211, USA
e-mail: bilyeuk@missouri.edu

G. Stacey (ed.), *Genetics and Genomics of Soybean*,

induced mutation sources for screening soybeans for seed fatty acid composition traits and investigations into soybean nodulation (Takagi et al. 1990; Rahman et al. 1994, 1995; Takagi et al. 1995; Men et al. 2002).

In forward genetics research the desired phenotype is identified in a line that can be characterized genetically to determine the qualitative or quantitative nature of the trait, as well as allelism to other identified mutant lines. Unfortunately, the underlying molecular genetic basis for the trait is not easily elucidated in soybeans. Despite significant research investment in physical and genetic soybean maps, map-based positional cloning in soybean has proven to be exceedingly difficult with relatively few success stories (Searle et al. 2003; Ashfield et al. 2004; Gao et al. 2005). It should be noted, however, that anonymous DNA markers linked to desired phenotypes can be used by soybean breeders to incorporate valuable traits without the need to determine the molecular basis for the phenotype.

The candidate gene approach is an alternative to map-based cloning methods to identify the underlying molecular genetic basis for soybean traits identified using forward genetics. Candidate genes are chosen based on molecular genetic or biochemical information in soybeans or other model organisms that have been characterized for the phenotype. The soybean homologues of genes thought to be involved in the phenotype are then characterized for mutations that associate with the trait. The candidate gene approach was applied to a number of soybean lines containing fatty acid modifications, and the research provided not only the causal mutation in the candidate gene but also "perfect" molecular markers that allow soybean breeders to directly select for the trait in early generations by identifying plants with the appropriate genotype (Bilyeu et al. 2003; Alt et al. 2005; Anai et al. 2005; Bilyeu et al. 2005; Aghoram et al. 2006; Bilyeu et al. 2006; Chappell and Bilyeu 2006).

Because many soybean traits are additively qualitative or quantitative in nature, forward genetics research has also included re-mutagenesis of existing mutant soybean lines. The production of linolenic acid in the oil of soybean seeds was shown to be the result of microsomal omega-three fatty acid desaturase enzymes encoded by at least three soybean genes (Yadav et al. 1993; Bilyeu et al. 2003; Anai et al. 2005; Bilyeu et al. 2005, 2006; Chappell and Bilyeu 2006). Mutations in at least two of these soybean genes are necessary to reduce linolenic acid levels to produce oil that is oxidatively stable without *trans* fats (Stojsin et al. 1998; Ross et al. 2000; Anai et al. 2005; Bilyeu et al. 2005, 2006). In two cases, re-mutagenesis of existing low linolenic acid lines led to the identification of the desired phenotype (Stojsin et al. 1998; Takagi et al. 1999).

The expectation for a forward genetics project utilizing chemical or radiation-based mutagenesis is the identification of single locus that contributes to the phenotype. However, when selecting EMS mutants for low linolenic acid and separately for low phytate phenotypes, mutants lines were recovered with two independent loci controlling both of these phenotypes (Wilcox et al. 2000; Bilyeu et al. 2005; Walker et al. 2006). An explanation for this unexpected result will require further research.

Reverse Genetics

In the "omics" era, whole genome, transcriptome, proteome, and metabolome data for soybean are actively being generated. With all of this new information comes a desire to identify the function of previously uncharacterized genes. Soybean breeding programs must capitalize on this wealth of information to target important traits. Genetics projects that start with an identified gene sequence and aim to determine the functional role of that gene are termed "reverse" genetics because the strategy is the reverse of the more common forward genetics project. Typically, soybean lines containing mutations in the target genes are identified and then characterized for the resulting phenotype. Confirmation of gene function can under some circumstances be established by the generation and characterization of transgenic soybean lines overexpressing or suppressing the target gene.

Model plants such as Arabidopsis and rice have a community resource of insertional mutagenesis lines in the form of T-DNA insertions with characterized border sequences (Krysan et al. 1999; Sallaud et al. 2004). The low transformation efficiency of soybean prohibits the development of a similar resource for soybean. Populations of insertional mutants can also be achieved through transposon insertion (Walbot 2000; Muskett et al. 2003; Kolkman et al. 2005) Efforts are now underway to generate a community resource of transposon-based insertional mutations in soybean with a goal to target every soybean gene (G. Stacey personal communication).

Recently, an efficient reverse genetics method for detection of mutant genes in plant populations was described [Targeted Induced Local Lesions IN Genomes, TILLING, (Colbert et al. 2001)]. The advantages to this reverse genetics approach include the ability to obtain an allelic series of mutations and the ability to utilize identified lines to develop varieties for commercial release unencumbered by regulatory restraints. For TILLING, seeds are heavily dosed with mutagen and grown for two generations. During this time, tissue is sampled from individual plants for DNA extraction, and the seeds produced from each plant are banked. Isolated DNA from individual plants is pooled eightfold to develop template for querying target genes.

For TILLING, target genes are chosen based on prior knowledge of putative gene function or other factors, and genomic sequence is determined for the gene to facilitate selection of regions for amplification. Genomic DNA corresponding to a characterized target gene is amplified and subjected to mismatch cleavage to detect pools containing putative mutations in the target gene. Positive pools are further screened and confirmed by DNA sequencing to determine the individual plant containing the mutation. Seeds from the identified line are then made available for phenotyping. Because of the heavy mutagen load, a number of backcrosses are necessary to reduce genetic drag from the mutant line.

Several soybean populations were developed for use in TILLING. Currently, two soybean TILLING libraries are available for screening, one from the cultivar Forrest and one from Williams 82 (K. Meksem personal communication). Other groups are in the population development phase for soybean. One of the drawbacks to TILLING in soybean is the difficulty in identifying sequences of candidate genes that allow specific amplification of only one locus in the soybean genome. However, the characterization of the sequence of the whole soybean genome will undoubtedly resolve some of the specific amplification issues. It seems clear that reverse genetics will be a powerful tool for future advances in soybean research that will ultimately lead to the release of more valuable soybean varieties.

References

Aghoram, K., Wilson, R.F., Burton, J.W. and Dewey, R.E. (2006) A mutation in a 3-keto-acyl-ACP synthase II gene is associated with elevated palmitic acid levels in soybean seeds. Crop Sci. 46, 2453–2459.

Alt, J.L., Fehr, W.R., Welke, G.A. and Sandhu, D. (2005) Phenotypic and molecular analysis of oleate content in the mutant soybean line M23. Crop Sci. 45, 1997–2000.

Anai, T., Yamada, T., Kinoshita, T., Rahman, S.M. and Takagi, Y. (2005) Identification of corresponding genes for three low-*alpha*-linolenic acid mutants and elucidation of their contribution to fatty acid biosynthesis in soybean seed. Plant Sci. 168, 1615–1623.

Ashfield, T., Ong, L.E., Nobuta, K., Schneider, C.M. and Innes, R.W. (2004) Convergent evolution of disease resistance gene specificity in two flowering plant families. Plant Cell 16, 309–318.

Bhatia, C.R., Nichterlein, K. and Maluszynski, M. (1999) Oilseed cultivars developed from induced mutations and mutations altering fatty acid composition. Mutation Breeding Review 11, 1–36.

Bilyeu, K.D., Palavalli, L., Sleper, D.A. and Beuselinck, P.R. (2003) Three microsomal omega-3 fatty-acid desaturase genes contribute to soybean linolenic acid levels. Crop Sci. 43, 1833–1838.

Bilyeu, K., Palavalli, L., Sleper, D. and Beuselinck, P. (2005) Mutations in soybean microsomal omega-3 fatty acid desaturase genes reduce linolenic acid concentration in soybean seeds. Crop Sci. 45, 1830–1836.

Bilyeu, K., Palavalli, L., Sleper, D.A. and Beuselinck, P. (2006) Molecular genetic resources for development of 1% linolenic acid soybeans. Crop Sci. 46, 1913–1918.

Chappell, A.S. and Bilyeu, K.D. (2006) A *GmFAD3A* mutation in the low linolenic acid soybean mutant C1640. Plant Breeding 125, 535–536.

Colbert, T., Till, B.J., Tompa, R., Reynolds, S., Steine, M.N., Yeung, A.T., McCallum, C.M., Comai, L. and Henikoff, S. (2001) High-throughput screening for induced point mutations. Plant Phys. 126, 480–484.

Gao, H., Narayanan, N.N., Ellison, L. and Bhattacharyya, M.K. (2005) Two classes of highly similar coiled coil-nucleotide binding-leucine rich repeat genes isolated from the rps1-k locus encode phytophthora resistance in soybean. Mol. Plant. Micr. Int. 18, 1035–1045.

Hitz, W.D., Carlson, T.J., Kerr, P.S. and Sebastian, S.A. (2002) Biochemical and molecular characterization of a mutation that confers a decreased raffinosaccharide and phytic acid phenotype on soybean seeds. Plant Physiol. 128, 650–660.

Kolkman, J.M., Conrad, L.J., Farmer, P.R., Hardeman, K., Ahern, K.R., Lewis, P.E., Sawers, R.J.H., Lebejko, S., Chomet, P. and Brutnell, T.P. (2005) Distribution of activator (ac) throughout the maize genome for use in regional mutagenesis. Genetics 169, 981–995.

Krysan, P.J., Young, J.C. and Sussman, M.R. (1999) T-DNA as an insertional mutagen in Arabidopsis. Plant Cell 11, 2283–2290.

Men, A.E., Laniya, T.S., Searle, I.R., Iturbe-Ormaetxe, I., Gresshoff, I., Jiang, Q., Carroll, B.J. and Gresshoff, P.M. (2002) Fast neutron mutagenesis of soybean (*Glycine soja* l.) produces a supernodulating mutant containing a large deletion in linkage group H. Genome Letters 1, 147–155.

Muskett, P.R., Clissold, L., Marocco, A., Springer, P.S., Martienssen, R. and Dean, C. (2003) A resource of mapped dissociation launch pads for targeted insertional mutagenesis in the Arabidopsis genome. Plant Physiol. 132, 506–516.

Palmer, R.G., Pfeiffer, T.W., Buss, G.R. and Kilen, T.C. (2004) Qualitative genetics. In: J.E. Specht and H.R. Boerma (Eds.), *Soybeans: Improvement, Production, and Uses*, Crop Sci. Society of America Monograph, Madison, WI, pp. 137–234.

Rahman, S.M., Takagi, Y., Kubota, K., Miyamoto, K. and Kawakita, T. (1994) High oleic acid mutant in soybean induced by X-ray irradiation. Biosci. Biotech. Biochem. 58, 1070–1072.

Rahman, S.M., Takagi, Y., Miyamoto, K. and Kawakita, T. (1995) High stearic acid soybean mutant induced by X-ray irradiation. Biosci. Biotech. Biochem.59, 922–926.

Rajcan, I., Hou, G.Y. and Weir, A.D. (2005) Advances in breeding of seed-quality traits in soybean J. Crop Improvement 14, 145–174.

Ross, A.J., Fehr, W.R., Welke, G.A. and Cianzio, S.R. (2000) Agronomic and seed traits of 1%-linolenate soybean genotypes. Crop Sci. 40, 383–386.

Sallaud, C., Gay, C., Larmande, P., Bes, M., Piffanelli, P., Piegu, B., Droc, G., Regad, F., Bourgeois, E., Meynard, D., Perin, C., Sabau, X., Ghesquiere, A., Glaszmann, J.C., Delseny, M. and Guiderdoni, E. (2004) High throughput T-DNA insertion mutagenesis in rice: A first step towards *in silico* reverse genetics. The Plant J. 39, 450–464.

Searle, I.R., Men, A.E., Laniya, T.S., Buzas, D.M., Iturbe-Ormaetxe, I., Carroll, B.J. and Gresshoff, P.M. (2003) Long-distance signaling in nodulation directed by a clavata1-like receptor kinase. Science 299, 109–112.

Stojsin, D., Luzzi, B.M., Ablett, G.R. and Tanner, J.W. (1998) Inheritance of low linolenic acid level in the soybean line RG10. Crop Sci. 38, 1441–1444.

Takagi, Y., Hossain, A.B.M.M., Yanagita, T., Matsueda, T. and Murayama, A. (1990) Linolenic acid content in soybean improved by X-ray irradiation. Agric. Biol. Chem 54, 1735–1738.

Takagi, Y., Rahman, S.M., Anai, T., Wasala, S.K., Kinoshita, T. and Khalekuzzaman, M. (1999) Development of a reduced linolenate soy mutant by re-irradiation and its genetic analysis. Breeding Science 49, 1–5.

Takagi, Y., Rahman, S.M., Joo, H. and Kawakita, T. (1995) Reduced and elevated palmitic acid mutants in soybean developed by X-ray irradiation. Biosci. Biotech. Biochem.59, 1778–1779.

Walbot, V. (2000) Saturation mutagenesis using maize transposons. Curr. Op. Plant Bio. 3, 103–107.

Walker, D.R., Scaboo, A.M., Pantalone, V.R., Wilcox, J.R. and Boerma, H.R. (2006) Genetic mapping of loci associated with seed phytic acid content in CX1834-1-2 soybean. Crop Sci. 46, 390–397.

Wilcox, J.R., Premachandra, G.S., Young, K.A. and Raboy, V. (2000) Isolation of high seed inorganic P, low-phytate soybean mutants. Crop Sci. 40, 1601–1605.

Yadav, N.S., Wierzbicki, A., Aegerter, M., Caster, C.S., Perez-Grau, L., Kinney, A.J., Hitz, W.D., Booth, J.R., Jr., Schweiger, B., Stecca, K.L., Allen, S.M., Blackwell, M., Reiter, R.S., Carlson, T.J., Russell, S.H., Feldmann, K.A., Pierce, J. and Browse, J. (1993) Cloning of higher plant omega-3 fatty acid desaturases. Plant Physiol. 103, 467–76.

Chapter 10
Bioinformatic Resources for Soybean Genetic and Genomic Research

David Grant, Rex T. Nelson, Michelle A. Graham, and Randy C. Shoemaker

Introduction

The last decade has seen an explosion in soybean [*Glycine max* (L.) Merr.] research. The molecular genetic map has grown from only a few hundred RFLP markers to over 2000 loci encompassing RFLP, RAPD, SSR and most recently, SNP markers. Over a thousand quantitative trait loci (QTL) have been mapped in soybean, representing ~90 agronomically important traits. More than 650,000 nucleotide and expressed sequence tag (EST) sequences are available. Soybean researchers around the world are using soybean macro- and microarrays to generate expression data for thousands of genes under different experimental conditions. Two complementary physical maps of the soybean genome were developed and the complete genomic sequence of soybean is expected to be available in 2008. Researchers can use these data resources to address many important issues. For example, the data described above can be used to identify genes expressed during defense responses, QTLs associated with disease resistance, or homologs of known disease resistance genes from other species. Clearly, asking these sorts of questions and analyzing such large data sets requires the use of bioinformatic tools. In this chapter, we will present a brief summary of the currently available data sets and the bioinformatic resources available for the analysis of soybean genetic and genomic data.

Genetic Resources

Formal soybean genetic research began in 1921 when Woodworth (1921) reported the first linkage data. As was the case for many other species, growth of the genetic map was slow and intermittent due to the relatively small number of Mendelian

---

D. Grant
USDA-ARS CICGRU, Department of Agronomy, Iowa State University, Ames, IA 50011, USA
e-mail: dgrant@iastate.edu

G. Stacey (ed.), *Genetics and Genomics of Soybean*,

inherited genes that controlled easily discernable phenotypes. In addition, these easily scorable traits would likely never be sufficient for breeders and researchers to identify and characterize the genes controlling many important agronomic traits. Therefore, several groups began to add Restriction Fragment Length Polymorphism (RFLP, Apuya et al. 1998; Keim et al. 1990), Rapid Amplification of Polymorphic DNA (RAPD, Shoemaker and Olson 1993), Single Sequence Repeat (SSR, Akkaya et al. 1992, 1995) and Single Nucleotide Polymorphism (SNP, Zhu et al. 2003) molecular markers to the genetic map. These efforts resulted in a rapid expansion of the genetic map and in 2004 a new integrated genetic linkage map of soybean was released (Song et al. 2004). Already this map has been used to examine germplasm diversity (Hyten et al. 2006), disease resistance (Sandhu et al. 2004; Kopisch-Obuch et al. 2005), seed quality (Hyten et al. 2004) and many other projects.

SoyBase

Based on the ongoing progress and needs of the soybean research community, the USDA-ARS initiated a central repository for genetics data and related resources. This effort, directed by Randy Shoemaker in Ames, Iowa, resulted in the release of the Soybase (http://soybase.org) database in 1993. Originally, SoyBase was implemented using the ACeDB database (Durbin and Thierry-Mieg 1994). The first release contained all of the available information on soybean genetics, including the classical genetic map, four molecular marker maps, and QTL studies on more than 30 traits. In addition, related information covering phenotypic data for selected germplasm, details on 15 soybean diseases and their causative agents, and diagrams of metabolic pathways identified as important to soybean agronomic traits were included. As shown in Table 10.1, SoyBase has been expanded to include many more data types and now continues to be updated with the published genetics and associated data on soybean.

Most of the original molecular markers on the soybean genetic map were RFLPs. Because of the low level of sequence polymorphism for RFLP alleles in the adapted US germplasm, most RFLP loci were monomorphic between the parents in any given cross (Cregan et al. 1999). In practical terms, this meant that genetic maps developed in different labs had few markers in common and thus comparisons across experiments were difficult. Two approaches were used to overcome this difficulty. First, SSR loci were developed that showed more variability in the soybean germplasm pool (Akkaya et al. 1995). Second a composite genetic map for soybean was constructed by generating a genetic map in three populations using a common set of markers. JoinMap (Stam 1993) was then used to construct a dense composite map that included most of the genetic markers that had been mapped in soybean (Cregan et al. 1999). In 2003, an updated version of the soybean composite map was prepared (Song et al. 2004). The SoyBase curators have used this map collection as the framework onto which all of the published loci and QTL data have been overlaid.

Table 10.1 Data types available in the Soybase database

Data class	Short description
Map Collection	Classical genetic map and molecular marker maps, including the composite genetic/molecular map
Map	Unique linkage groups for each Map Collection, with QTL displays
Locus	Morphological, biochemical and molecular (RFLP, AFLP, RAPD, SSR, PCR) markers
Gene	Data include gene class, locus, alleles, phenotypes and 2-point data
QTL	QTL information for agronomic traits such as yield, seed quality, maturity, and disease resistance
Picture	Autoradiograms for RFLP probes and SSR markers, graphical genotypes of selected cultivars, and additional QTL mapping diagrams
Reaction or Pathway	Clickable diagrams of >900 metabolic pathways showing kinetics, enzymes, metabolites and regulation
Enzyme	Data include EC number, purification, clones, physical properties and tissues and cultivars studied
Misc_Protein	Proteins mapped, sequenced, or purified from soybean or other plants but not easily categorized as enzymes or storage proteins
Trait	Traits associated with entries in the GRIN and PVP databases, with links to genes, QTL, and pathology
Pathology	Information on soybean diseases including causative organism, symptoms, differentials, distribution, and resistance mechanisms
Insect pests	Data concerning the pathologies and genetics of resistance to insects
Storage Protein	Data on vegetative and seed storage proteins
Nodulin	Data on the nodulins of soybean including gene and protein information, and probe and antibody availability
Nodulation	Plant and microbial processes and genes involved in nodulation of soybean
Transformation	Summary of research done on transformation in soybean, including methodology, transgenes, cultivar, and regeneration of plants
Paper	References for all data in SoyBase and to other papers relevant to SoyBase

The Soybean Breeder's Toolbox

SoyBase was initially conceived as a portal for soybean geneticists to easily access data relevant to their work, and as such, has been quite successful. To better accommodate the needs of soybean breeders, the Soybean Breeder's Toolbox (SBT, http://soybeanbreederstoolbox.org) was developed. The SBT is an alternate interface to most of the data in SoyBase, but the querying and data display tools are aligned more closely to the needs of soybean breeders. The map displays in the SBT are identical to those described for SoyBase above.

One important feature of the SBT is that genetic map displays were changed to use CMap (Stein et al. 2002, http://www.gmod.org/cmap/). CMap allows multiple maps to be simultaneously displayed. In Fig. 10.1A, linkage groups D1b, C2 and J are displayed. Loci in common between the maps are highlighted and connected with blue lines. QTL for related traits (i.e. pathology, plant architecture, seed composition, etc.) are displayed in the same color to allow the user to quickly recognize commonalities between maps. Users can control the visibility of each QTL class or

(A)

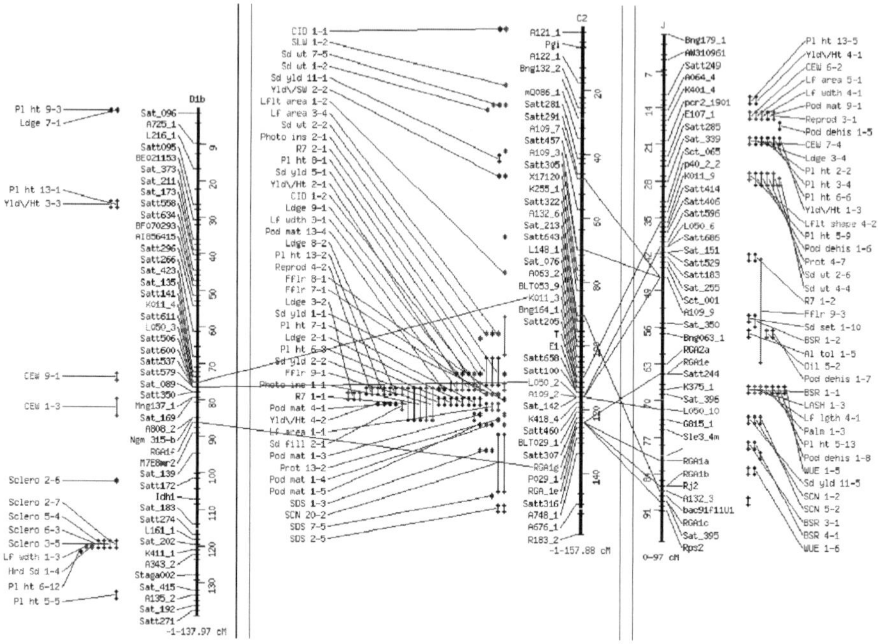

Fig. 10.1 Comparative Map Display Capabilities of the Soybean Breeder's Toolbox. **Panel A**. Comparison of molecular linkage groups D1b, C2 and J. Thick black vertical lines represent the linkage group genetic maps. The positions of molecular markers are designated by black horizontal hashes with their identifiers linked by black lines. Molecular markers that are in common between any two linkage groups are highlighted in red with blue lines connecting them. The positions of various QTL are identified by colored vertical dumbbell shapes. The colors used for QTL group QTL of similar phenotype thus all QTL for fungal resistance are in brick red, resistance to bacterial pathogens are in orange, oil related QTL are in gold etc. A listing of all color codes is included in each CMap legend. The comparison of linkage groups with many mapped QTL can be quite difficult to interpret. CMap allows the user to display individual types of markers in order to make the interpretation of maps with dense markers easier. **Panel B** depicts the same comparison with only QTL for insect, bacterial and fungal visible for ease of interpretation (*See* also Color Insert)

marker type (RFLP, SSR, etc.). Maps can be individually scaled or cropped to allow detailed comparisons of specific regions. Figure 10.1B shows the results of cropping the linkage groups from Fig. 10.1A while limiting QTLs to those that are pathology related. Clicking on a locus or QTL in any map display opens a new window giving the details about that feature. In the future, CMap will allow us to make comparisons between cultivars and related legumes.

(B)

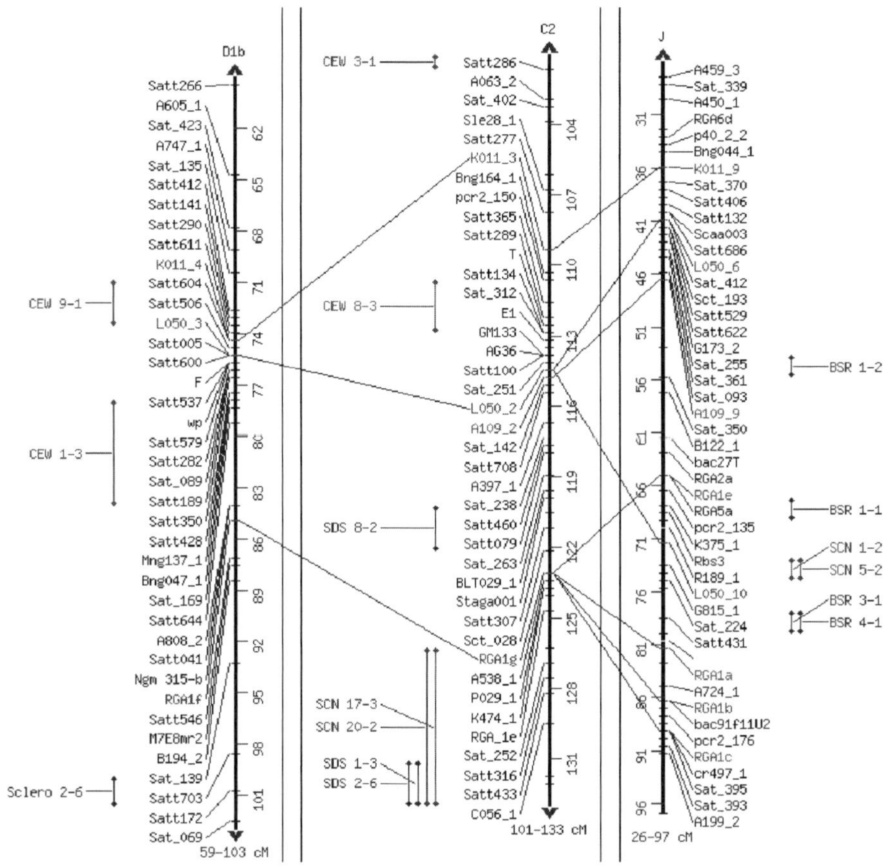

Fig. 10.1 (Continued) (*See* also Color Insert)

Physical Map Resources

One of the important areas of soybean research is the isolation of the genes controlling important agronomic traits. A key tool is a robust physical map that is well anchored to the genetic map. Two projects have been initiated to develop a physical map for soybean, each of which use different visualization programs.

Forrest Physical Map

The first soybean physical map was developed using the cultivar Forrest (Wu et al. 2004; Shultz et al. 2006a). The Forrest physical map is displayed using GBrowse, a component of the GMOD program (Stein et al. 2002). Figure 10.2 shows a

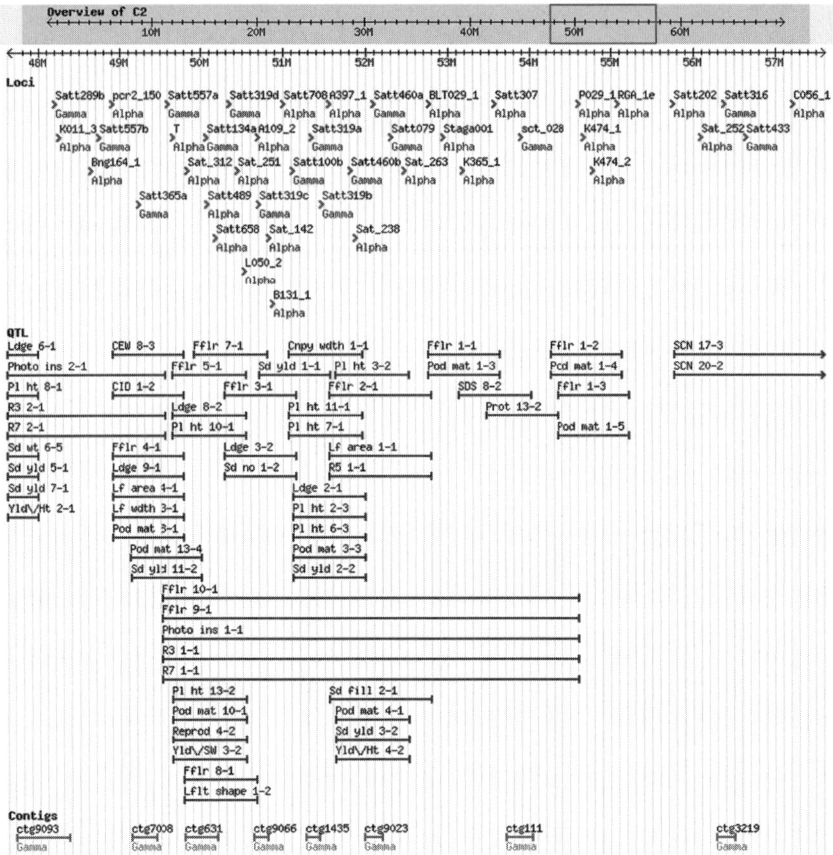

Fig. 10.2 Example of the Forrest Physical Map Display and Gbrowse. An example of the Forrest Physical Map displayed using the Gbrowse display engine. Pictured is a 9.9 Mbp section of molecularlinkage group C2 around the *T* locus. The black horizontal line at the top of the display in the gray box represents the whole linkage group. The region in the red box represents the segment displayed in the light box below it. In the light box with light blue rules, we can see the representation of the linkage group as a black horizontal line ruled in 100 kbp segments. Under the horizontal lines are different "tracks" containing different types of features. Displayed are loci, QTL and BAC contigs (Contigs). More types were available but not displayed for ease of interpretation. As indicated, the Gbrowse engine was designed as a map display and annotation tool thus it allows the presentation of a single map at a time. Features in common with other maps or species could be displayed as a new set of "tracks". Interpretation of synteny or genomic context between the comparisons would be difficult because the different maps could not be directly visualized together as they could be in a CMap display (*See* also Color Insert)

sample of some of the kinds of data available at the Forrest physical map web site (http://soybeangenome.siu.edu/). In a typical Gbrowse display, only one reference sequence (chromosome/linkage group) can be displayed at a time. Features of that sequence can be overlaid using tracks as seen in Fig. 10.2. Some of these tracks represent EST or other sequences that were identified and mapped onto the reference sequence regardless of their physical arrangement on other genomes or linkage groups.

Williams 82 Physical Map

In 2005, a project was initiated to develop a physical map using the soybean cultivar Williams 82. This cultivar was chosen as it (1) was the source of over half of the ESTs sequenced from soybean and (2) it allows comparisons between the Northern soybean germplasm pool represented by cultivars Williams/Williams 82 and the Southern germplasm pool (cultivar Forrest). The Williams 82 physical map can be viewed on the web using CMap from the SoyBase home page (http://soybase.org – > Williams Physical Map) or at http://soybeanphysicalmap.org. Figure 10.3 shows an example of how a user can drill down from a genetic map to an individual BAC. In this example, a section of linkage group C2 near the E1 and T genes is shown on the right. The middle panel shows a version of this genetic map with only those genetic markers that have been associated with BACs in the physical map. The contigs containing those BACs and their relative sizes and positions are also shown. The green and orange highlighted markers are both associated with BACs in

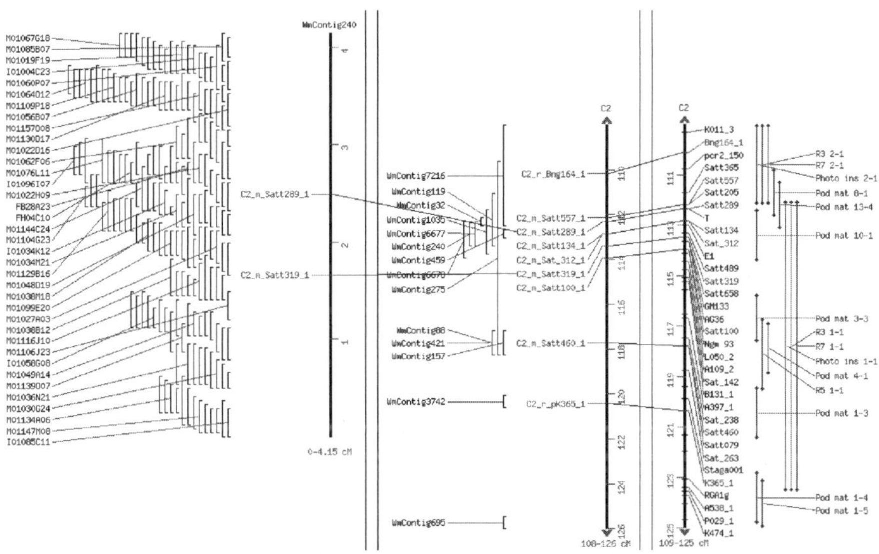

Fig. 10.3 Example of the Williams Physical Map Display using CMap. Pictured is a view of molecular linkage group C2 similar to that depicted in Fig. 10.2 in the Forrest Physical Map. The genetic map is drawn on the right with only QTL for pod maturity visible. The physical map positions and approximate sizes of the Williams physical map contigs are displayed in the middle. On the left is an explosion of a physical map contig that may contain the *T* locus (WmContig240) based on the presence of markers Satt289 and Satt319. Individual BAC clones mapped with those markers are indicated in green (Satt289) and light orange (Satt319). Thus these BACs could serve as a starting point for the fine scale mapping of the *T* locus (*See* also Color Insert)

WmContig240, and the BACs are similarly colored in the expanded representation of this BAC contig.

Expressed Sequence Tag Resources

Prior to the arrival of high-throughput DNA genome sequencing projects, the sequence data available for many species was largely the result of small to large-scale expressed sequence tag (EST) sequencing projects. EST sequencing has long been looked at as a way to derive sequence data on genes without the overhead of sequencing non-coding DNA. It has also been referred to as a "poor man's genome" (Rudd 2003). EST libraries represent the messenger RNA present in a particular tissue under study. The consequence of this is that messages from highly or constitutively expressed genes are over-represented in the library. The over-representation of these messages can be mitigated using various methods of library construction (Soares et al. 1994; Bonaldo et al. 1996; Scheetz et al. 2004). However most of the soybean EST libraries were not normalized and thus represent a biased sampling of the soybean transcriptome.

The National Center for Biotechnology Information (NCBI) separated sequences from single-pass cDNA sequencing projects into a database called dbEST (Boguski et al. 1993). Currently, this section of GenBank consists of ESTs collected from many plant species including soybean (*G. max)* and its wild relative, *G. soja*. At the time of this writing, more than 25 plant species were represented in the database with total EST counts for each species ranging from a few sequences to more than 1.2 million. Currently, *G. max* has the 6th largest plant EST collection with over 350,000 sequences. The Public Soybean EST Project (Shoemaker et al. 2002) produced the majority of these sequences ($> 297,000$ ESTs). This collection of ESTs was derived from more than 25 cultivars and from various tissues ranging from roots/root nodules, stems and/or leaves, flowers, seeds, pods and somatic embryos. Some of the tissues also were collected from plants challenged with various biotic and abiotic stresses.

Some EST libraries are derived from RNA from multiple individuals or cultivars. It was hoped that ESTs from different cultivars could be used to generate SNP markers. However, this approach also introduces allele specific sequence differences into later steps of reproducing the original gene sequences, sometimes called a "unigene" sequence. A collection of all unigenes produced from an EST dataset has been termed a "Gene Index". Depending on the representation of a specific message in the library and the assembly method used, a unigene sequence could incorporate sequence differences from more than one allele. If very similar paralogous sequences are expressed and at different rates, the presence of the weakly expressed paralog could also be missed altogether.

The gene indices described below were created using the NCBI EST collection. However, there are large variances in the total number of genes reported by the various projects. This underscores the impact that different assembly procedures have on the final gene totals for each index.

NCBI Unigene Sets

The NCBI utilizes species-specific large EST collections to produce UniGene sets. These sets represent collections of EST and curated gene sequences that appear to be highly related at the level of primary sequence similarity. EST collections are first purged of vector and known repetitive element contaminants. The sequences remaining are then associated with each other based on overall sequence similarity as previously described (Wheeler et al. 2003). As described above, identification of collections of EST sequences representing whole or partial transcript sequences is complicated in part by the heterogeneous nature of some EST collections. EST collections derived from multiple individuals of the same species may produce unigene sequence collections that contain or miss allele specific gene differences, as well as differences between recent gene duplicates (paralogs) unless specific measures are taken to partition the different sequences. Currently, the *Glycine max* UniGene set (Build 27) contains 18,907 unigene clusters, most of which ($\sim$18,000) are supported by EST sequences alone. Most of the unigene clusters are composed of two or more sequences (89%) with 64% of the unigenes being composed of eight or fewer sequences. Soybean has the second largest set of unigenes in the eucotyledons next to Arabidopsis (29,215). Soybean has the largest legume unigene set with *Medicago truncatula* (Build 27) second with 13,904 and *Lotus japonicus* (Build 1) third with 13,218 unigenes. The composition of the unigene sets, in terms of the number of sequences in each cluster, for the included legume species appears to be roughly similar with 89% of *M. truncatula* unigenes (91%, *L. japonicus*) composed of two or more sequences and 72% composed of eight or fewer sequences (78%, *L. japonicus*).

TIGR TCs

The Institute for Genomic Research (TIGR) has been producing gene indices termed "TCs" which represent "Tentative Consensus", or TCs, of sequences (Quackenbush et al. 2001). TCs represent EST sequences that were clustered using a "modified" version of MegaBLAST (Zhang et al. 2000). These clusters were then used as input into Cap3 (Huang and Madan 1999) to produce a final consensus sequence for each cluster. Currently, the TIGR gene indices are being transferred to the Dana Farber Institute Cancer Institute (http://compbio.dfci.harvard.edu/tgi/plant.html) where they will be available in the future. The gene indices are composed of plant, animal, protist and fungal sequences. The plant gene indices were assembled for 34 plant species including soybean. Other legume species with gene indices include common bean *(Phaseolus vulgaris), M. truncatula* and *L. japonicus*. The soybean gene index (Release 12.0) is composed of 31,928 TC assemblies, 31,636 singleton ESTs and 112 singleton mRNAs for a total of 63,676 unique sequences and is by far the largest legume gene index. For comparison, the gene indices of *P. vulgaris* (Release 1.0), *L. japonicus* (Release 3.0) and *M. truncatula* (Release 8.0) contain

9,484, 28,460 and 36,878 unigene sequences, respectively. In all cases, the original ESTs used to create the TCs came from the EST collection (dbEST) of GenBank.

PlantGDB PUTs

A variety of plant EST collections, including soybean, obtained from NCBI have been clustered into unigene sets by the Plant Genome Database project (PlantGDB, www.plantgdb.org) including soybean. These unigenes are termed "PUTs" for plant unique transcripts. EST sequences used in the construction of PUTs are first masked for vector contamination and poly-A tails. Redundant ESTs are removed in a data reduction step and the remaining sequences are then clustered using PaCE (Kalyanaraman et al. 2003). Cap3 (Huang and Madan 1999) produces a consensus sequence for the cluster. If the initial clustering method does not distinguish between highly similar sequences (recent paralogs), then some of the clusters could contain a mixture of gene sequences. Cap3 was designed to produce a multiple sequence alignment and consensus sequence of the aligned sequences, thus the consensus sequence would be assumed to represent the most numerous gene sequence at any one point along the multiple sequence alignment. A detailed description of the clustering and assembly methods used to construct PlantGDB PUTs is available at the PlantGDB website (http://www.plantgdb.org/prj/ESTCluster/PUT_procedure.php). Currently, there are 102,305 PUTs assembled for soybean (Build 157a) representing assemblies and singleton sequences. PlantGDB also provides a number of analysis resources for its PUT collection such as the ability to BLAST query sequences against selected parts of the PUT collection, align PUT and other sequences to an input genomic sequence and search for a specified pattern in an input sequence using VMATCH (Abouelhoda et al. 2004). Other resources available include the ability to BLAST query sequences against the Affymetrix® GeneChip® Soybean Genome array probe sequences provided by the Plant Expression Database, PLEXdb (www.plexdb.org).

Soybean Potential Gene Sequences (pHaps)

An EST clustering procedure, ESTminer (Nelson et al. 2005) was developed that attempted to assign EST sequences to clusters based on shared redundant sequence features. This procedure was applied to a subset of the ESTs produced by the Public Soybean EST project. The ESTminer suite of programs seeks to assemble ESTs into clusters based on repeated and shared sequence similarities. This procedure treats all sequences that have high similarities but are not identical as separate sequences without regard to their representation in the library. The procedure also requires the use of ESTs derived from a single genotype thus reducing the number of alleles of a gene identified as a separate gene. These clusters of identical sequences are termed potential gene haplotypes (pHaps). For soybean, the ESTs were derived from the Williams and Williams 82 cultivars. This clustering procedure produced 45,255 pHaps and 49,895 singleton sequences for a total gene index of 95,150. The pHaps form 12,702 separate gene families based on overall sequence similarity. Each family is composed of closely related genes and gene paralogs. These sequences are available through the Soybean Breeders Toolbox web site (www.soybeanbreederstoolbox.org/pHapDB/). At this site, users can BLAST input sequences against the pHap/singleton sequences and can retrieve detailed descriptions of each of the pHaps.

Microarray Technologies

While sequencing of ESTs helped to identify the minimally redundant set of expressed genes in soybean, not all biological conditions were represented and comparing expression in different conditions was often difficult. The advent of micro/macro array technology allowed researchers to examine the expression of identified genes in different tissues, in different genotypes and in response to particular treatments including abiotic/biotic stresses. Several different macro/micro array platforms were developed for soybean including cDNA arrays, long oligo arrays and the Affymetrix® GeneChip® Soybean Genome array.

cDNA and Oligo Arrays

Several different research groups developed custom macro/microarrays to examine genes of interest, most often those involved in defense. Iqbal et al. (2002) used a 135 cDNA macroarray to examine gene expression in response to Sudden Death Syndrome caused by *Fusarium solani* f.sp. *glycines* in two different genotypes. Alkharouf et al. (2006) developed a custom microarray from cDNAs expressed following soybean cyst nematode (*Heterodera glycines)* infection and from cDNAs with known roles in defense. Similarly, Moy et al. (2004) developed a custom 4,896 cDNA array from publicly available soybean and *Phytophthora sojae* ESTs to examine host and pathogen expression in susceptible soybean inoculated with *P. sojae*.

While custom macroarrays allowed users to examine particular genes of interest, large-scale microarrays were needed to identify genes involved in broad pathways. Vodkin et al. (2004) developed a 27,513 low redundancy cDNA microarray using unigenes identified from the public EST project (Shoemaker et al. 2002). This array included cDNAs from a variety of tissues, development stages, and stress or pathogen treated tissues. In addition to assembling the array, Vodkin et al. (2004) also provided 3′ sequences for the majority of cDNAs on the array. This array has now been expanded to include 36,864 low redundancy cDNAs. A 38,000 long oligo (70mer) array was also developed. Together, these arrays were used to examine gene expression in response to pathogens (Zou et al. 2005; Zabala et al. 2006), elevated CO_2 concentration (Ainsworth et al. 2006), somatic embrogenesis (Thibaud-Nissen et al. 2003) and iron chlorosis deficiency in near isogenic lines (O'Rourke et al. 2007). Further details on these arrays can be found at the NSF Soybean Functional Genomics website at the University of Illinois (http://soybeangenomics.cropsci.uiuc.edu).

Affymetrix® GeneChip® Soybean Genome Array

Following the advent of cDNA based microarrays, there was also a need for short oligo arrays that might better differentiate between homologous sequences. This was especially important in soybean given the polyploid nature of the soybean genome (Hymowitz and Singh 1987). In addition, researchers were interested in assaying host and microbe gene expression in a single chip. Therefore, the Affymetrix® GeneChip® Soybean Genome array was developed by Affymetrix® in close collaboration with the soybean research community (http://www.affymetrix.com/products/arrays/specific/soybean.affx). It represents 35,611, 15,421 and 7,431 unigenes from soybean, *P. sojae* and *H. glycines*, respectively. Affymetrix® developed the unigenes represented on the array using publicly available ESTs (prior to November 2003). Each unigene is represented on the chip as eleven 25mer oligo pairs, spread along the length of the transcript. For each intended oligo, a mismatch oligo also is included that differs by a single nucleotide (Wosik 2006). Researchers can use the hybridization signal from the mismatch oligo to identify and subtract non-specific hybridization.

Bioinformatic Resources for Microarray Analyses

While microarray technology has allowed users to identify thousands of biologically interesting genes, determining the function of these genes remains difficult.

The NetAffx analysis center (http://www.affymetrix.com/analysis/index.affx), developed by Affymetrix®, allows Affymetrix® users to upload a list of identifiers and get probe sequences, consensus sequence information and annotation data. In addition, users can blast a query sequence against the Affymetrix probes to see if their gene or transcript is represented on the array.

Scientists at the USDA-ARS also developed a web annotation tool for the Affymetrix® GeneChip® Soybean Genome array. This website, delivered as part of the soybean breeder's toolbox, allows users to assign their own annotation based on information provided by the site (http://soybase.org/affychip/). The consensus sequences, provided by Affymetrix, were compared to the Uniprot protein database (Apweiler et al. 2004), the predicted coding sequences from the *A. thaliana* genome [The Arabidopsis Information Resource (TAIR), TAIR6_cds_20051108], and the Pfam protein database (Bateman et al. 2004) using BLASTX (Altschul et al. 1997) with an E-value cutoff of $E < 10^{-4}$. For BLAST against Uniprot database, the top three hits are reported including the E-value and the percent overlap and percent identity between query and subject sequences. For BLAST analyses against PFAM and the TAIR coding sequences, only the top hit and E-value are reported. The TAIR Gene Ontology (GO) and GO slim annotations also are provided for the top *A. thaliana* coding sequence (Berardini et al. 2004). All of the data can be downloaded as text in a tab delimited file or can be viewed online with corresponding hyperlinks. In addition, users can upload a file of sequence identifiers of interest and download the corresponding annotation. Currently, only soybean sequences on the chip are annotated, but the next version will include annotations for *P. sojae* and *H. glycines* sequences.

As mentioned earlier, the duplicated nature of the soybean genome suggests that some gene transcripts may hybridize to unintended targets. This makes validation of microarray results by reverse-transcriptase polymerase chain reaction (RT-PCR) or other methods extremely difficult. While the use of mismatch probes will identify highly similar transcripts, it cannot distinguish transcripts that are identical to the target probe but come from different genes. To help identify probes on the Soybean GeneChip® array that might cross hybridize, the sequences of all "perfect match" 25mer probes for all *G. max* consensus sequences were compared to various EST assemblies including the soybean TIGR gene index (Quackenbush et al. 2001), the PlantGDB PUTs (www.plantgdb.org) and the Soybean Potential Gene Sequences (Nelson et al. 2005, www.soybeanbreederstoolbox.org/pHapDB/). Analyses were performed using BLASTN and required a perfect (100% nucleotide identity) match over the entire length of the probe (25 nucleotides). As expected, probes could be identified that hybridized to a single gene transcript or to multiple gene transcripts. The complete results of this analysis can be found at http://www.soybase.org/affyprobe/. This information can be used following RT-PCR to examine unexpected results. However, it is better used prior to RT-PCR to identify sequences that are likely to cross-amplify and may allow users to design better oligonucleotide primers that will distinguish closely related genes.

Expression Databases

GEO

The NCBI maintains an expression database called GEO for gene expression omnibus (Edgar et al. 2002; Barrett et al. 2006). GEO is a MIAME compliant expression database. Access to donated microarray experiments from the GEO web site (www.ncbi.nlm.nih.gov/geo/) is accomplished by either a database lookup using keywords used to describe an individual experiment or by BLAST analysis using a starting nucleotide sequence to identify potentially homologous exemplars from the stored chip sets. The database currently contains 16 experiments using various soybean micro- and macro-arrays.

PLEXdb

PLEXdb (www.plexdb.org) is a MIAME compliant gene expression database for plants and plant pathogens. PLEXdb (Shen et al. 2005) contains expression experiments from a wide variety of plants ranging from Arabidopsis to sugar cane. The database also includes expression experiments from soybean and *M. truncatula*. The *M. truncatula* exemplar data is further subdivided into exemplars from *M. truncatula, M. sativa* and the endosymbiont *Sinorhizobium meliloti*. Similarly, the soybean exemplars are divided into their constituent parts, exemplars from *G. max*, the soybean pathogen *P. sojae* and the nematode pest *H. glycines*. Currently, there are no soybean or *M. truncatula* experimental datasets available but they are setup to assimilate experiments using the Affymetrix® GeneChip® Soybean and Medicago Genome Arrays.

A number of analysis tools are available through the PLEXdb interface including tools for the functional annotation of exemplars, the ability to compare gene lists within a group of experiments to draw correlations base on globally normalized data and the ability to find potential homologs to different chip exemplars by BLAST analysis.

Genome Sequence Resources

RefSeq Database

The NCBI provides a highly curated database representing annotated genomic sequence from a number of organisms including soybean (Pruitt et al. 2005, http://www.ncbi.nlm.nih.gov/RefSeq/). These sequences are prepared from the data available from the NCBI GenBank sequence databases. Because of the paucity of genomic data related to soybean, only 83 genes sequences have been assembled as of the time of this writing. In addition, the chloroplast genome for the Chinese cultivar Er-hej-jan (PI 437654) also is available. The RefSeq database will be augmented with many more soybean sequences in the near future given the soybean genome sequencing effort being undertaken by the Joint Genome Initiative of the DOE and the Agricultural Research Service of the USDA.

Sequenced BAC Clones

Currently, there are 33 fully sequenced bacterial artificial chromosome clones for *Glycine max* in GenBank. This collection totals approximately 3.9 million bases (Mb) of largely un-annotated genomic sequence or 0.3% of the estimated 1,100 Mb soybean genome.

BAC-End Sequences

Another source of soybean genomic sequence comes from a large bacterial artificial chromosome (BAC) collection and end sequences of their genomic DNA inserts.

Currently, there are approximately 238,000 soybean BAC-end sequences (BES). Of this total, 21,360 BES were derived from the Forrest physical map project (Wu et al. 2004; Shultz et al. 2006b) and 217,000 from the Williams physical map project. The Forrest BES can be retrieved using "LargeInsertSoybeanGenLib" and "LargeInsert-SoybeanGenLib Build4" as the library identifiers. The Williams sequences can be retrieved using the following library identifiers: GMW1, GMW2, ISb001, UMb001, GM_WBb and GM_WBa.

Methyl-Filtered Genomic Clones

One of the strategies to sample the gene-space of an organism involves the cloning of only those genomic sequences which may be expressed, thus ignoring "junk" DNA and instead spending sequencing resources and through-put on only clones containing genic sequences. Methylation filtration cloning (Rabinowicz et al. 1999; Whitelaw et al. 2003) is a cloning strategy that relies on the hypo-methylation of expressed (genic) sequences in the target genome. This strategy utilizes a methylation-restriction competent *E. coli* host. When random genomic sequences are cloned and inserted in to a methylation-restriction competent *E. coli* cell, the host cell will selectively degrade the inserts of plasmids containing DNA that is methylated. Thus, only cells containing plasmids with hypo-methylated inserts will be retained and those with hyper-methylated inserts will be selectively degraded. This step will selectively enrich all transformants for those with potentially expressed (genic) sequences. This step greatly improves the efficiency of a random sequencing effort in sequencing the gene-space of an organism. At the present time, there are 13,449 methylation-filtered soybean sequences in GenBank.

Assembled Genomic Sequence

Currently, the only assembled genomic sequence available for soybean is obtained by sequencing BAC clones. In the future, assembled chromosome or near-chomosome-sized genomic sequences may be available as a result of the USDA-Agricultural Research Service and Joint Genome Initiative soybean sequencing project. In the near-term, in excess of 6 million shotgun sequencing reads are or will be available through the TraceDB database of the NCBI.

Comparative Legume Genomics

The Legume Information System (LIS) is a service of the National Center for Genome Resources (NCGR.org) in Santa Fe, NM. (Gonzales et al. 2005) LIS (www.comparative-legumes.org) contains genetic and sequence information for a variety of legume species, including peanut (*Arachis hypogaea*), chick pea (*Cicer arietinum*), soybean, barrel medic (*M. truncatula*), common bean (*P. vulgaris*) and pea (*Pisum sativum*). Using the EST resources of NCBI, LIS also prepares gene indices for these and other legume species using an analysis pipeline called XGI (http://www.ncgr.org/xgi/components). Currently, LIS has assembled 63,676 consensus gene sequences for soybean as well as 28,460 *L. japonicus* and 35,726 *M. truncatula* gene sequences. Their assemblies are available for download from their web site (www.comparative-legumes.org/fastas/lis). Their assemblies and raw EST sequences are also available for BLAST analysis. Genomic sequences available from NCBI in the form of partial or complete BAC sequences also are curated by LIS and annotated using the XGI pipeline. The XGI analysis pipeline attempts to annotate all sequences using a variety of data sources such as prints, pfam, and GO. All sequences and assemblies potentially can be searched using keywords such as "kinase" or by GO number to retrieve sequences with those annotations using the LIS web interface thus facilitating the identification of potential legume homologs. Genetic and physical map data for selected legume species is also available from LIS.

LIS contains genetic maps for 5 legume species including soybean, barrel medic, peanut, pea and common bean. Copies of the two physical maps for soybean and one for *M. truncatula* also are available through LIS. Genetic and physical maps are displayed using the comparative map viewer CMap (gmod.org/cmap). This allows the comparison of all physical or genetic maps to the included legume species where a common marker exists. Currently, it is possible to compare soybean genetic and physical maps with a limited number of homologous markers from *M. truncatula* and *P. vulgaris*.

Access to metabolic or pathway information for soybean is also available from the LIS interface. The metabolic data was transferred to LIS from SoyBase (www.soybase.org). The pathway information is queried and retrieved from both sites using a modified AceDB display.

Conclusions

It is an exciting time in soybean genetics given the recent advances in microarray technologies and the forthcoming whole genome sequence. These technologies will allow an explosion of soybean research in basic biology leading to cultivar improvement. In addition, these advances herald soybean as an emerging model for crop legumes. Because of its phylogenetic position within the Phaseoleae, soybean will be a better scaffold for comparisons among crop legumes than the model legumes *M. truncatula* and *L. japonicus*, both members of the Trifolieae. In this review, we describe data and resources currently available for soybean researchers, in the hope of underscoring soybean's research potential.

References

Abouelhoda, M.I., Kurtz, S. and Ohlebusch, E. (2004) Replacing suffix trees with enhanced suffix arrays. J. Discrete Algorithms 2, 53–86.

Ainsworth, E.A., Rogers A., Vodkin, L.O., Walter, A. and Schurr, U. (2006.) The effects of elevated CO_2 on soybean gene expression: An analysis of growing and mature leaves. Plant Physiol. 142, 135–147.

Akkaya, M.S., Shoemaker, R.C., Specht, J.E., Bhagwat, A.A. and Cregan, P.B. (1995) Integration of simple sequence repeat DNA markers into a soybean linkage map. Crop Sci. 35, 1439–1445.

Akkaya, M.S., Bhagwat, A.A. and Cregan, P.B. (1992) Length polymorphisms of simple sequence repeat DNA in soybean. Genetics 132, 1131–1139.

Alkharouf, N.W., Klink, V.P., Chouikha, I.B., Beard, H.S., MacDonald, M.H., Meyer, S., Knap, H.T., Khan, R. and Matthews, B.F. (2006) Timecourse microarray analyses reveal global changes in gene expression of susceptible *Glycine max* (soybean) roots during infection by *Heterodera glycines* (soybean cyst nematode). Planta 224, 838–852.

Altschul, S.F., Madden, T.L., Schaffer, A.A., Zhang, J., Zhang, Z., Miller, W. and Lipman, D.J. (1997) Gapped BLAST and PSI-BLAST: a new generation of protein database search programs. Nucl. Acids Res. 25, 3389–3402.

Apuya, N.R., Frazier, B.L., Keim, P., Roth, E.J. and Lark, K.G. (1998) Restriction fragment length polymorphisms as genetic markers in soybean, *Glycine max* (L.) Merrill. Theor. Appl. Genet. 75, 889–901.

Apweiler, R., Bairoch, A., Wu, C.H., Barker, W.C., Boeckmann, B., Ferro, S., Gasteiger, E., Huang, H., Lopez, R., Magrane, M., Martin, M.J., Natale, D.A., O'Donovan, C., Redaschi, N. and Yeh, L.S. (2004) UniProt: the universal protein knowledgebase. Nucl. Acids Res. 32, D115–D119.

Barrett, T., Troup, D.B., Wilhite, S.E., Ledoux, P., Rudnev, D., Evangelista, C., Kim, I.F., Soboleva, A., Tomashevsky, M. and Edgar, R. 2006. NCBI GEO: mining tens of millions of expression profiles—database and tools update. Nucl. Acids Res. 35, D760–D765.

Bateman, A., Coin, L., Durbin, R., Finn, R.D., Hollich,V., Griffiths-Jones, S., Khanna, A., Marshall, M., Moxon, S., Sonnhammer, E.L.L., Studholme, D.J., Yeats, C. and Eddy, S.R. (2004) The Pfam protein families database. Nucl. Acids Res. 32, D138–D141.

Berardini, T.Z., Mundodi, S., Reiser, R., Huala, E., Garcia-Hernandez, M., Zhang, P., Mueller, L.M., Yoon, J., Doyle, A., Lander, G., Moseyko, N., Yoo, D., Xu, I., Zoeckler, B., Montoya, M., Miller, N., Weems, D. and Rhee, S.Y. (2004) Functional annotation of the Arabidopsis genome using controlled vocabularies. Plant Physiol. 135, 1–11.

Boguski, M.S., Lowe, T.M.J. and Tolstoshev, C.M. (1993) dbEST — database for "expressed sequence tags". Nature Genetics 4, 332–333.

Bonaldo, M.F., Lennon, G. and Soares, M.B. (1996) Normalization and subtraction: two approaches to facilitate gene discovery. Genome Res. 6, 791–806.

Cregan, P.B., Jarvik, T., Bush, A.L., Shoemaker, R.C., Lark, K.G., Kahler, A.L., Kaya, N., VanToai, T.T., Lohnes, D.G., Chung, J. and Specht, J.E. (1999) An integrated genetic linkage map of the soybean genome. Crop Sci. 39, 1464–1490.

Durbin, R., Thierry-Mieg, J. (1994) The AceDB genome database. In *Computational Methods in Genome Research* (ed. S. Suhai), pp. 7821–7825. Plenum Press, New York.

Edgar, R., Domrachev, M., and Lash, A.E. (2002) Gene Expression Omnibus: NCBI gene expression and hybridization array data repository. Nucl. Acids Res. 30, 207–210.

Gonzales, M.D., Archuleta, E., Farmer, A., Gajendran, K. Grant, D., Shoemaker, R., Beavis, W.D. and Waugh M.E. (2005) The Legume Information System (LIS): an integrated information resource for comparative legume biology. Nucl. Acids Res. 33, D660–D665.

Huang, X. and Madan, A. (1999) CAP3: a DNA sequence assemblyprogram. Genome Res. 9, 868–877.

Hymowitz, T. and Singh, R.J. (1987) Taxomony and speciation. In *Soybeans: Improvement, production, and uses* (ed. RJ Wilcox), pp. 23–48. ASA, CSSA, SSSA, Madison, Wisconsin.

Hyten, D.L., Pantalone, V.R., Sams, C.E., Saxton, A.M., Landau-Ellis, D., Stefaniak, T.R. and Schmidt, M.E. (2004) Seed quality QTL in a prominent soybean population. Theor. Appl. Genet. 109, 552–561.

Hyten, D.L., Song, Q., Zhu, Y., Choi, I.Y., Nelson, R.L., Costa, J.M., Specht, J.E., Shoemaker, R.C. and Cregan, P.B. (2006) Impacts of genetic bottlenecks on soybean genome diversity. Proc. Natl. Acad. Sci. USA 103, 16666–16671.

Iqbal, M.J., Yaegashi, S., Njiti, V.N., Ahsan, R., Cryder, K.L. and Lightfoot, D.A. (2002) Resistance locus pyramids alter transcript abundance in soybean roots inoculated with *Fusarium solani* f.sp. *glycines*. Mol. Genet. Genomics 268, 407–417.

Kalyanaraman, A., Aluru, S., Kothari, S. and Brendel, V. (2003) Efficient clustering of large EST data sets on parallel computers. Nucl. Acids Res. 31, 2963–2974.

Keim, P., Diers, B.W., Olson, T.C. and Shoemaker, R.C. (1990) RFLP mapping in soybean: association between marker loci and variation in quantitative traits. Genetics 126, 735–742.

Kopisch-Obuch, F.J., McBroom, R.L. and Diers, B.W. (2005) Association between soybean cyst nematode resistance loci and yield in soybean. Crop Sci. 45, 956–965.

Moy, P., Qutob, D., Chapman, B.P., Atkinson, I., Gijzen, M. (2004) Patterns of gene expression upon infection of soybean plants by *Phytophthora sojae*. Mol. Plant-Microbe Interact. 17, 1051–1062.

Nelson, R.T., Grant, D. and Shoemaker, R.C. (2005) ESTminer: a suite of programs for gene and allele identification. Bioinformatics 21, 691–693.

O'Rourke, J.A., Graham, M.A., Vodkin, L.O., Gonzales, D.O., Cianzio, S.R. and Shoemaker, R.C. (2007) Recovering from iron deficiency chlorosis in near isogenic soybeans: a microarray study. Plant Physiology and Biochemistry: in press.

Pruitt, K.D., Tatusova, T. and Maglott, D.R. (2005) NCBI Reference sequence (RefSeq): a curated non-redundant sequence database of genomes, transcripts and proteins. Nucl. Acids Res. 33, D501–D504.

Quackenbush, J., Cho, J., Lee, D., Liang, F., Holt, I., Karamycheva, S., Parvizi, B., Pertea, G., Sultana, R. and White, J. (2001) The TIGR Gene Indices: analysis of gene transcript sequences in highly sampled eukaryotic species. Nucl. Acids Res. 29: 159–164.

Rabinowicz, P.D., Schutz, K., Dedhia, N., Yordan, C., Parnell, L.D., Stein, L., McCombie, R.W. and Martienssen, R.A. (1999) Differential methylation of genes and retrotransposons facilitates shotgun sequencing of the maize genome. Nature Genetics 23, 305–308.

Rudd, S. (2003) Expressed sequence tags: alternative or complement to whole genome sequence? Trends Plant Sci. 8, 321–329.

Sandhu, D., Gao, H., Cianzio, S. and Bhattacharyya, M.K. (2004) Deletion of a disease resistance nucleotide-binding-site leucine-rich-repeat-like sequence is associated with the loss of the Phytophthora resistance gene *Rps4* in soybean. Genetics 168, 2157–2167.

Scheetz, T.E., Laffin, J.J., Berger, B., Holte, S., Baumes, S.A., Brown, R. II, Chang, S., Coco, J., Conklin, J., Crouch, K., Donohue, M., Doonan, G., Estes, C., Eyestone, M., Fishler, K., Gardineer, J., Guo, L., Johnson, B., Keppel, C., Kreger, R., Lebeck, M., Marcelin, R., Miljkovich, V., Perdue, M., Qui, L., Rehmann, J., Reiter, R.S., Rhoads, B., Schaefer, K., Smith, C., Sunjevaric, I., Trout, K., Wu, N., Birkett, C., Bischof. J., Gackle, B., Gavin, A., Grundstad, A.J., Mokrzycki, B., Moressi, C., O'Leary, B., Pedretti, K., Roberts, C., Robinson, N.L., Smith, M., Tack, D., Trivedi, N., Kucaba, T., Freeman, T., Lin, J.J., Bonaldo, M.F., Casavant T.L., Sheffields, V.C. and Soares, M.B. (2004) High-throughput gene discovery in the rat. Genome Res. 14, 733–741.

Shen, L., Gong, J., Caldo, R.C., Nettleton, D., Cook, D., Wise, R.P. and Dickerson, J.A. (2005) BarleyBase — An expression profiling database for plant genomics. Nucl. Acids Res. 33, D614–D618.

Shoemaker R.C. and Olson T.C. (1993) Molecular linkage map of soybean (*Glycine max* L). In *Genetic maps: Locus maps of complex genomes* (ed. SJ O'Brien), pp. 6–131. Cold Spring Harbor Laboratory Press, New York.

Shoemaker, R., Keim, P., Vodkin, L., Retzel, E., Clifton, S.W., Waterson, R., Smoller, D., Coryell, V., Khanna, A., Erplding, J., Gai, X., Brendel, V., Raph-Schmidt, C., Shoop, E.G., Vielweber, C.V., Schimatz, M., Pape, D., Bowers, Y., Theising, B., Martin, J., Dante, M., Wylie, T. and Granger, C. (2002) A compilation of soybean ESTs: generation and analysis. Genome 45, 329–338.

Shultz. J.L., Kurunam, D., Shopinski, K., Iqbal, M.J., Kazi, S., Zobrist, K., Bashir, R., Yaegashi, S., Lavu, N., Afzal, A.J., Yesudas, C.R., Kassem, M.A., Wu, C., Zhang, H.B., Town, C.D., Meksem, K. and Lightfoot, D.A. (2006a) The Soybean Genome Database (SoyGD): a browser for display of duplicated, polyploid, regions and sequence tagged sites on the integrated physical and genetic maps of *Glycine max*. Nucl. Acids Res. 34, D758–D765.

Shultz, J.L., Yesudas, C., Yaegashi, S., Afzal, A.J., Kazi, S. and Lightfoot, D.A. (2006b) Three minimum tile paths from bacterial artificial chromosome libraries of the soybean (*Glycine max* cv. 'Forrest'): tools for structural and functional genomics. Plant Methods. 2, 9.

Soares, M.B., Bonaldo, M.D.F., Jelene, P., Su, L., Lawton, L. and Efstratiadis, A. (1994) Construction and characterization of a normalized cDNA library. Proc. Natl. Acad. Sci. USA 91, 9228–9232.

Song, Q.J., Marek, L.F., Shoemaker, R.C., Lark, K.G., Concibido, V.C., Delannay, X., Spect, J.E. and Cregan, P.B. (2004) A new integrated genetic linkage map of the soybean. Theor. Appl. Genet. 109, 122–128.

Stam, P. (1993) Construction of integrated genetic linkage maps by means of a new computer package: Join Map. Plant Journal 3, 739–744.

Stein, L.D., Mungall, C., Shu, S., Caudy, M., Mangone, M., Day, A., Nickerson, E., Stajich, J.E., Harris, T.W., Arva, A. and Lewis, S. (2002) The generic genome browser: a building block for a model organism system database. Genome Res. 12, 1599–610.

Thibaud-Nissen, F., Shealy, R.T., Khanna, A. and Vodkin, L.O. (2003) Clustering of microarray data reveals transcript patterns associated with somatic embryogenesis in soybean. Plant Physiol 132, 118–136.

Vodkin, L.O., Khanna, A., Shealy, R., Clough, S.J., Gonzalez, D.O., Philip, R., Zabala, G., Thibaud-Nissen, F., Sidarous, M., Strömvik, M.V., Shoop, E., Schmidt, C., Retzel, E., Erpelding, E., Shoemaker, R.C., Rodriguez-Huete, A.M., Polacco, J.C., Coryell, V., Keim, P., Gong, G., Lui, L., Pardinas, J. and Schweitzer, P. (2004) Microarrays for global expression constructed with a low redundancy set of 27,500 sequenced cDNAs representing an array of developmental stages and physiological conditions of the soybean plant. BMC Genomics 5, 73.

Wheeler, D.L., Church, D.M., Federhen, S., Lash, A.E., Madden, T.L., Pontius, J.U., Schuler, G.D., Schriml, L.M., Sequeira, E., Tatusova, T.A. and Wagner, L. (2003) Database resources of the National Center for Biotechnology. Nucl. Acids Res. 31, 28–33.

Whitelaw, C.A., Barbazuk, W.B., Pertea, G., Chan, A.P., Cheung, F., Lee, Y., Zheng, L., van Heeringen, S., Karamycheva, S., Bennetzen, J.L., SanMiguel, P., Lakey, N., Bedell, J., Yuan, Y., Budiman, M.A., Resnick, A., Van Aken, S., Utterback, T., Riedmuller, S., Williams, M., Feldblyum, T., Schubert, K., Beachy, R., Fraser, C.M. and Quackenbush, J. (2003) Enrichment of gene-coding sequences in maize by genome filtration. Science 302, 2118–2120.

Woodworth, C.M. (1921) Inheritance of cotyledon, seed-coat, hilum, and pubescence colors in soybeans. Genetics 6, 487–555.

Wosik, E. (2006) Oligonucleotide Arrays-Detailed Information. Connexions, March 31, 2006. http://cnx.org/content/m12388/1.4/

Wu, C.C., Nimmakayala, P., Santos, F.A., Springman, R., Scheuring, C., Meksem, K., Lightfoot, D.A. and Zhang, H.B. (2004) Construction and characterization of a soybean bacterial artificial chromosome library and use of multiple complementary libraries for genome physical mapping. Theor. Appl. Genet. 109, 1041–1050.

Zabala, G., Zou, J.,Tuteja, J., Gonzales, D.O., Clough, S.J. and Vodkin, L.O. (2006) Transcriptome changes in the phenylpropanoid pathway of *Glycine max* in response to *Pseudomonas syringae* infection. BMC Plant Biol 6, 26.

Zhang, Z., Schwartz, S., Wagner, L. and Miller, W. (2000), "A greedy algorithm for aligning DNA sequences", J. Comput. Biol. 7, 203–214.

Zhu, Y.L., Song, Q.J., Hyten, D.L., Van Tassell, C.P., Matukumalli, L.K., Grimm, D.R., Hyatt, S.M., Fickus, E.W., Young, N.D. and Cregan, P.B. (2003) Single-nucleotide polymorphisms in soybean. Genetics 163 1123–1134.

Zou, J., Rodriguez-Zas, S., Aldea, M., Li, M., Zhu, J., Gonzalez, D.O., Vodkin, L.O., DeLucia, E. and Clough, S.J. (2005) Expression profiling soybean response to *Pseudomonas syringae* reveals new defense-related genes and rapid HR-specific downregulation of photosynthesis. Mol Plant-Microbe Interact 18, 1161–1174.

Part III
Investigations of Soybean Biology

Chapter 11
Genomics of Soybean Seed Development

Lila Vodkin, Sarah Jones, Delkin Orlando Gonzalez, Françoise Thibaud-Nissen, Gracia Zabala, and Jigyasa Tuteja

Generation of cDNA Microarray Resources for Soybean

The "Public EST Project for Soybean" was a multi-university and team-oriented research project funded by soybean grower check-off funds. This project stimulated development of a large public database of soybean Expressed Sequence Tags (ESTs). The number of ESTs in this database rose from less than 100 in 1998 to over 300,000 in 2004. As part of this project, a large number of cDNA libraries were made from the mRNAs extracted from numerous tissue and organ systems of the soybean plant (Shoemaker et al. 2002, 2004; Vodkin et al. 2004). Over 80 different cDNA libraries were constructed from which the 300,000 ESTs were generated. The soybean EST resource data reside in public databases such as Genbank, maintained by the National Center for Biotechnology Information (NCBI), and also in the databases of The Institute of Genomic Research (TIGR), as do EST resources developed in recent years for other plant species (Walbot 1999; Cullis 2004).

One of the goals of the NSF-sponsored "A Functional Genomics Program for Soybean" was to select 36,000 unique ESTs from the collection of over 300,000 primary ESTs that represented a wide array of cDNAs made from mRNAs expressed in various tissue and organ systems, and physiological stages under pathogen and stress challenges (Vodkin et al. 2004). The ESTs were compared by computer programs such as PHRAP (Green 2001) or CAP3 (Huang and Madan 1999) and assembled into overlapping clones that have sequence similarity. These assemblies are known as contigs (contiguous segments). In this way, longer sequences representing expressed genes were assembled and identical sequences representing redundant clones were recognized. The number of sequences in the contigs in a non-normalized cDNA library is a rough approximation of the relative abundance of the mRNAs within that tissue, and this information can be used as a "virtual" RNA blot (Stromvik et al. 2004).

L. Vodkin
Department of Crop Sciences, University of Illinois, Urbana, Illinois, IL 61801, USA
e-mail: l-vodkin@uiuc.edu

G. Stacey (ed.), *Genetics and Genomics of Soybean*,

The EST with the furthest 5′ end sequence was chosen to represent a contig and its identity was verified by sequencing, generally at the opposite or 3′ end. This created a tentative "unigene" set in which each contig was represented by a member of the cluster and many singletons in the EST collection were also identified. The complex process of preparing the soybean cDNA microarrays (Vodkin et al. 2004) involved physically picking or "reracking" the 36,000 selected soybean cDNAs that represented the "unigene" set from among the 300,000 cDNA clones that are maintained as individual recombinant *E. coli* cultures stored in 384-well plates. The plasmid DNA templates were then purified and the 3′ end of each cDNA was sequenced to verify the identity of each clone and to obtain additional sequence information for each of the unique genes. Afterwards, the inserts from the 36,000 selected cDNA clones were amplified by PCR (polymerase chain reaction), the PCR products were purified, and nanoliter volumes of each were spotted onto hundreds of glass microscope slides.

As a community resource, our laboratory prints two microarray slide sets each consisting of 18,432 single-spotted PCR products derived from the low redundancy cDNA sets. The GmcDNA18kA set (representing sequence-driven unigene clone libraries Gm-r1021, Gm-r1083, and Gm-r1070) is highly representative of genes expressed in the developing flowers and buds, young pods, developing seed coats, and immature cotyledons, as well as from roots of seedlings and adult plants, including roots infected with the nodulating bacterium, *Bradyrhizobium japonicum.* The GmcDNA18kB microarray set (unigene clone libraries Gm-r1088 and Gm-r1089) is highly representative of clones selected from libraries derived from tissue-culture embryos, germinating cotyledons, and seedlings subjected to various stresses including some challenged by pathogens. Completion of both sets brings the total number of cDNAs represented to 36,864. The cDNA platforms are entered in the Gene Expression Omnibus database at http://www.ncbi.nlm.nih.gov/geo.

Also on the arrays, are a set of 64 control or "choice" clones that are printed eight times repetitively throughout each array. Some represent constitutively expressed genes (such as ubiquitin and EF1). Some are cDNAs whose expression is restricted to a subset of specific plant tissues (such as Rubisco or seed storage proteins). Some are clones of enzymes representing commonly used antibiotic resistance markers in transgenic plants (as hygromycin or kanamycin resistance), and 32 are cDNAs that represent at least 13 different enzymes of the flavonoid pathway. The flavonoid pathway was chosen because the corresponding genes often respond to many biotic and abiotic stress conditions and it has been widely studied in plant systems. The pathway is also very important in soybean for the synthesis of isoflavones in the developing cotyledons.

Development of Oligo Arrays for Soybean

As recommended by the Soybean Genomics Executive Committee (SoyGEC) and several workshops sponsored by the United Soybean Board or National Science Foundation (Stacey et al. 2004), a proposal for the synthesis of 70-mer long oligos representing the soybean EST collection was funded by the United Soybean Board. The long oligos are uniformly of 70 bases and were preferentially chosen to represent the 3′ region of the cDNAs where possible. The 3′ region generally has more sequence variability and can be used to design oligonucleotides that distinguish among gene family members. Clustering analysis and oligo design and synthesis were performed by Illumina, Inc., San Diego, CA. A total of 38,000 unique oligos were synthesized and printed by our laboratory at the University of Illinois on two sets of slides (GmOLIGO19KA and GmOLIGO19KB) containing 19,200 spots each. Further information about the availability of the soybean oligo or cDNA microarrays can be found at the following web site http://soybeangenomics.cropsci.uiuc.edu. One of the first uses of the oligo arrays is to examine the transition of the cotyledons from a storage organ to the photosynthetic "first leaves" of the germinating seedling (Gonzalez and Vodkin 2006, 2007).

Other oligo array platforms available for soybean include Affymetrix (Santa Clara, CA) arrays of 25-mer oligos synthesized by photolithography that represent 37,500 of the unigenes from the soybean EST collection. They also contain 7,500 ESTs from the soybean cyst nematode pathogen (*Heterodera glycines*) and over 15,800 from the fungal pathogen *Phytophtora sojae*.

Global Expression Analyses Using Soybean Microarrays

A number of studies using the soybean microarrays for global expression analyses have been published including examination of somatic embryo development (Thibaud-Nissen et al. 2003), zygotic development (Dhaubhadel et al. 2007), the germinating cotyledons (Gonzalez and Vodkin 2006), the early response to challenge by the pathogen *Pseudomonas syringae* (Zou et al. 2005; Zabala et al. 2006), and the effect of elevated CO_2 atmospheric conditions (Ainsworth et al. 2006). In this chapter, we concentrate specifically on several studies in progress or recently reported on seed development in soybean and we briefly review several microarray investigations of seed development in other plant systems.

Transcript Profiles of the Induction of Somatic Embryos During Tissue Culture

Somatic embryos are the tissue of choice for transformation by particle bombardment in several crop species, including soybean (Finer and McMullen 1991), due to their ability to regenerate into entire plants. Somatic and zygotic embryos follow the same general pattern of development (Zimmerman 1993; Goldberg et al. 1994). However, large quantities of somatic embryos can be produced in vitro, making them more amenable to experimentation than their zygotic counterparts, which are protected by fruit structures and less accessible. Therefore, somatic embryos constitute a model system to study basic aspects of embryogenesis, in addition to their practical use as a tool for efficient transformation.

Somatic embryos follow the same general pattern of development as zygotic embryos, but the progression from one stage to the next is induced externally by changes in the culture medium. In soybean, somatic embryos are initiated from immature cotyledons on high levels of the synthetic auxin 2,4-dichlorophenoxyacetic acid (2,4-D, 40 mg/L) (Finer 1988). Within 30 days, embryos appear from the epidermal or subepidermal layers of the upper side (adaxial side, away from the medium), while the rest of the cotyledon (abaxial side) degenerates into a brown callus mass (Finer 1988). Auxin inhibits the differentiation of embryo cells beyond the globular stage in the soybean system while in many species, such as the carrot, auxin inhibits the organization of the callus into embryos. In soybean, embryos can be maintained indefinitely at the globular stage on 20 mg/L 2,4-D (Wright et al. 1991). The heart through the cotyledon stages occur on MS medium free of auxin (Finer and McMullen 1991) and are followed by several days of desiccation. The mature embryos can then be placed on a germination medium and grown into plants.

Two critical differences with zygotic embryos are that somatic embryos are never surrounded by an endosperm and that they do not develop a suspensor (Zimmerman 1993). In addition, somatic embryos are often larger than zygotic embryos. The morphological diversity and the lower rates of germination seen in somatic embryos compared to zygotic embryos are generally attributed to in vitro culture. However, the extent to which the zygotic and somatic developmental routes are molecularly similar is unclear.

A detailed analysis of the global expression patterns during somatic embryogenesis in soybean revealed many aspects of the events that occur during reprogramming of the cotyledon cells during the induction process (Thibaud-Nissen et al. 2003). Somatic embryos were induced on the adaxial (upper) side of soybean cotyledons, and RNA from this side was compared to RNA from the abaxial (bottom, seed coat) side at five time points using soybean cDNA microarrays. RNA from just the adaxial side at four time points was also compared in a loop design using these arrays (Fig. 11.1). The results indicated that about 500 genes (5.3% of the over 9,000 cDNAs on the array) had at least a two-fold expression level change in one or more of the comparisons. These 500 genes were clustered into 11 k-means sets by the similarity of their expression profiles. Data suggested that auxin causes the cells of both sides of the cotyledon to dedifferentiate during the first seven days. However, when the auxin levels have gone down after 14 days, cells on both sides of the cotyledon differentiate again, leading to the formation of somatic embryos (adaxial side) and callus (abaxial side). The transcripts for gibberellic acid and storage protein synthesis are also transcribed at this time in the embryo tissue. During these first two weeks of development, an oxidative burst initiated by auxin may cause programmed cell death, allowing plants to recycle molecules from structures that are no longer needed. A summary of the major pathways in which transcripts fluctuated during the 28 days of somatic embryo induction is shown in Fig. 11.2. These results give a global picture of the likely molecular events unfolding in the cotyledons during their reprogramming.

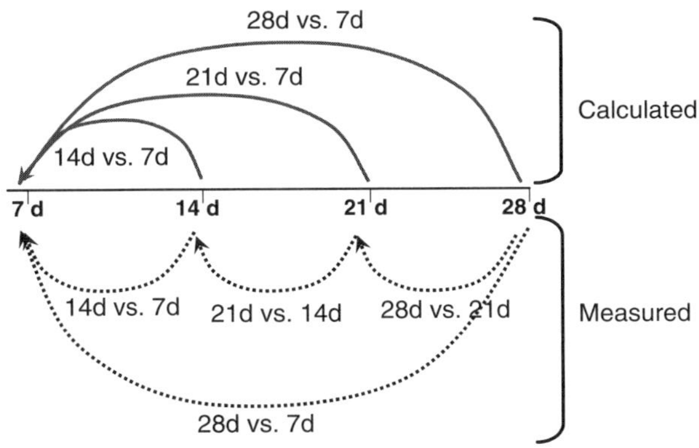

Fig. 11.1 Illustration of sampling protocol for the induction of somatic embryos on auxin medium over a 28 day period using a loop design. The bottom half with the dotted lines illustrates the actual sampled time points measured in the two color microarray hybridizations. The solid line indicates the calculated transformation of the data to illustrate each time point with reference to the 7 day time point. (From Thibaud-Nissen et al. 2003 ©American Society Plant Biologists)

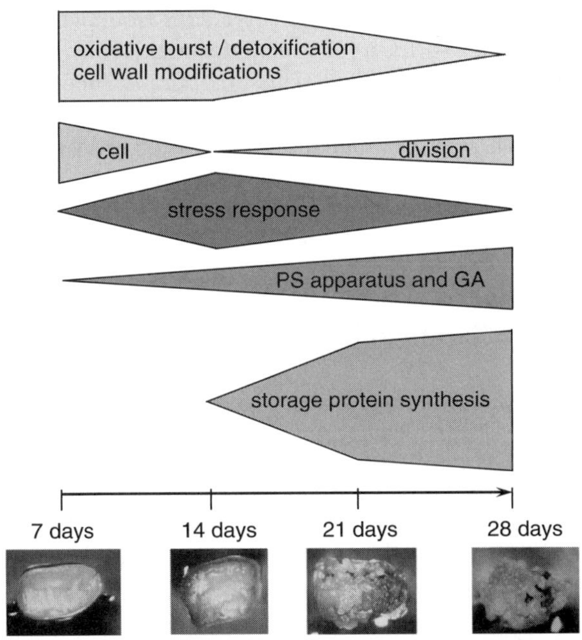

Fig. 11.2 Transcript patterns associated with the early induction phase of somatic embryogenesis in the adaxial sides. The timescale represents the number of days the cotyledons were on high auxin media. The figures underneath illustrate the appearance of the sampled tissue at each timepoint. The heights of the blocks represent the relative levels of the transcript expression (From Thibaud-Nissen et al. 2003 ©American Society Plant Biologists) (*See* also Color Insert)

Another intriguing observation from this study was the contrast in gene expression between the adaxial and abaxial side of the cotyledon. In this study, the cotyledon adaxial side was placed face up on auxin medium. The orientation of the explant is critical for successful induction of somatic embryos in several species including alfalfa (Chen et al. 1987) and soybean (Santarem et al. 1997) and is consistent with the fact that shoot apical meristems form from cells with adaxial cell fate (McConnell and Barton 1998). The explanation for this requirement most likely derives from the polar expression of one or several factors in plant lateral organs.

Even at the initiation of the study just after dissection from the pods and at the 0-time point of the cotyledons on the medium, homologues of the transcription factors *YABBY2* and *FIL/YABBY1* showed higher expression in the abaxial side of the cotyledons than in the adaxial side (Thibaud-Nissen et al. 2003). The YABBY family was shown to specify abaxial cell fate in *Arabidopsis* leaves, cotyledons and ovules (Sawa et al. 1999; Siegfried et al. 1999), probably in conjunction with other proteins. In our study, the polarity of *YABBY2* mRNA persisted up to 14 days after the beginning of the 2,4-D treatment, while that of other indicators such as seed proteins did not. This observation supports the hypothesis that, like in *Arabidopsis*, YABBY is a determinant of abaxial cell fate in soybean and that its low abundance in the adaxial cells allows the formation of shoot apical meristems and of somatic embryos from these cells. Also at the initiation of the study at the 0-time point after dissection of the cotyledons from the pods, transcripts for storage proteins (Bowman-Birk trypsin inhibitor and lectin) were more abundant in the abaxial side of the cotyledons. This observation supports previous in situ hybridizations performed with Kunitz trypsin inhibitor, beta-conglycinin, lectin and glycinin probes that show progression of the expression in a wave-like pattern from the abaxial to the adaxial side during the development of the cotyledon (Goldberg et al. 1989).

Zygotic Seed Development in Soybean

Flowering occurs at a stage known as reproductive stage R1 and early pod development at R3 occurs when the plant typically has up to 16 vegetative nodes (Ritchie et al. 1996). Early seed development consists of cell division and tissue differentiation, with the embryo surrounded by hexose resulting from sucrose cleaved by invertase in the seed coat (Bewley et al. 2000; Hills 2004). The maternal seed coat has many important structural and developmental influences on the seed (Moise et al. 2005). The seeds grow rapidly between the stages of R4 and R7. Hexose is supplanted by sucrose in the seed, and storage proteins like globulin, 2S albumin, alpha- and beta-conglycinin, and legumin are synthesized (Hills 2004; Bewley et al. 2000). Ureide-N makes up 10–15% of the nitrogen entering soybean seeds, though seed coat enzymes convert this to the more common forms of seed nitrogen, asparagine and glutamine (Bewley et al. 2000). Seed coat enzymes also convert amino acids entering the seeds from one composition to another (Bewley et al. 2000).

During the R6 stage, the vegetative parts of the parent plant begin to turn yellow, and the older, lower leaves start to senesce and fall from the plant (Ritchie et al. 1996). Soon after this stage, the accumulation of nutrients begins to slow down in the seeds, and by R7, virtually all the seed's dry weight has been acquired (Ritchie et al. 1996). In the cultivar Williams, the total fresh weight of the seed peaks around 400–500 mg. At this point, the seed contains about 60% moisture, and its tissue begins to turn yellow (Ritchie et al. 1996). The seeds are capable of germinating now, although only a small percentage of the seeds are able to sustain seedling growth (Rosenberg and Rinne 1986). Around this time, the developmental processes in the seed end and the embryo prepares for desiccation. The total fresh weight of the seed decreases as water is lost. Raffinose and other nonreducing oligosaccharides are produced to keep the sucrose in the seeds from crystallizing during desiccation, which would destabilize the cytoplasm in the cells (Bewley et al. 2000). Heat shock proteins, late embryogenesis abundant (LEA) proteins, and other molecules that protect against dehydration damage are synthesized in the seed as well (Bewley et al. 2000).

By stage R8, almost all of the pods and seeds have turned brown and dry (Ritchie et al. 1996). At about 55% moisture, most seeds are able to germinate and sustain seedling growth, though they may take several more days before they reach the best moisture content for harvest, around 15% (Rosenberg and Rinne 1986; Ritchie et al. 1996). Later, when the seed imbibes water and begins germination, metabolic activity can restart immediately using enzymes, ribosomes, initiation and elongation factors, and other compounds that were produced during development and stored (Bewley et al. 2000).

The developing soybean seed was among the first model systems used at the dawn of plant molecular biology in the early 1980s because of the large seed size and the abundance of the seed protein transcripts. Isolation and characterization of the major storage protein genes in soybean and their expression yielded many insights into the regulation of plant gene expression (Goldberg et al. 1981a,b; Meinke et al. 1981; Vodkin 1981; Fisher and Goldberg 1982; reviewed in Goldberg 1988; Goldberg et al. 1989; Le et al. 2007).

Transcriptomics of Zygotic Soybean Seed Development

Since the recent development of genomic technologies for soybean including EST and cDNA and synthetic oligonucleotide microarray resources, it is possible to gain a global view of gene expression in soybean in a way not previously available with single gene-by-gene studies. A soybean cDNA array with about 27,600 cDNAs from various tissues and conditions was used to examine gene expression profiles over five stages of development in soybean cotyledons, including the stage of largest fresh weight and dry whole seeds, as well as three stages of seed coat development (Jones 2004; Jones et al. 2006). Cotyledons and seed coats were dissected and examined independently. In the design of this experiment, cotyledons or seed coats taken from seeds in the 100–200 mg fresh weight range were used as the reference tissue for the cotyledon and seed coat time course studies, respectively. The weight range of these reference tissues falls roughly in the middle of each developmental series. The data was normalized using an in-house analysis program then further refined with commercially-available software (GeneSpring) to select genes with measurements for at least two replicates at every stage, a standard error ≤ 0.5 at every stage, and two-fold differential expression in at least one stage compared to the reference. In total, over 3,700 genes met these requirements in cotyledons and over 2,500 in seed coats. This data was then divided into k-means clusters by the similarity of their expression profiles, with 11 clusters used for each tissue. The genes in these clusters were functionally categorized on the basis of their 5′ and 3′ annotations.

A few of the general trends gleaned from this project (Jones 2004) showed that genes related to cell growth and maintenance processes such as cell wall expansion and preservation, protein folding, and fatty acid metabolism, were found to be decreasing in expression levels as the cotyledons approached the mature, dry stage. Likewise, genes related to energy processes like photosynthesis and chlorophyll binding had very low expression levels in the oldest seeds, compared to the reference. These genes showed their highest expression levels in the youngest stage in the cotyledon study and in the second-youngest stage in the seed coat study. Both findings illustrate the shutdown of metabolic processes that occurs as the seed prepares for desiccation. Genes involved with seed maturation and storage proteins were observed to be most highly expressed at the stage in development that represented seeds at their largest fresh weight before the onset of desiccation. Interestingly, a significant number of genes were found to be highly expressed at the final stage of development studied in cotyledons, that of the whole dry seed. Transcription factors were especially prominent among the genes with this expression pattern, similar to the findings of Duan et al. (2005) in rice. These transcription factors could be expressed in the mature dry seed in preparation for imbibition and germination, when a wide variety of products will need to be synthesized within a short time period.

In addition to obtaining the global view, specific gene profiles can be gleaned from the microarray data. Profiles for many enzymes involved in compositional traits such as oil can be discerned from the microarray data of cotyledon development. An example is shown in Fig. 11.3 of the expression changes in mRNAs encoding omega-3 versus omega-6 soybean desaturases. Clearly, the transcripts encoding these enzymes have different expression profiles at the later stages of seed development with the omega-3 transcripts tending to increase late in development while the omega-6 transcripts decline.

Examination of genes relating to flavonoid synthesis in this data revealed that genes encoding the same enzyme could have highly variable expression patterns, suggesting that isoforms of these enzymes may function to subtly adjust the flavonoid synthesis process. Other researchers (Dhaubhadel et al. 2007) found that flavonoid-related genes increase in expression at these older stages including chalcone synthase (CHS). They found that the CHS7 and CHS8 cDNAs are more highly

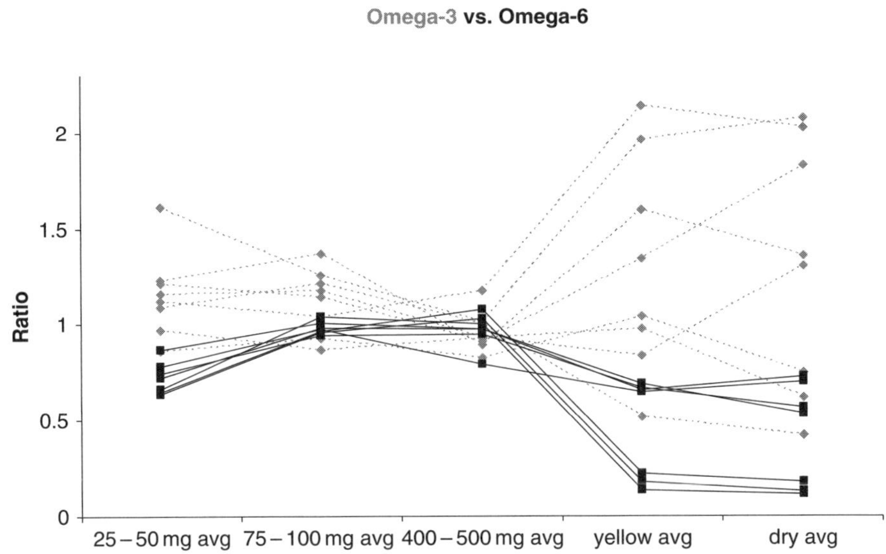

Fig. 11.3 Expression profiles of omega-3 versus omega-6 fatty acid desaturase mRNAs in five stages of the developing cotyledons. Each line represents an individual cDNA annotated as either an omega-3 (*dotted lines*, diamonds) or omega-6 desaturase (*solid lines*, squares). Change in gene expression ratios of each fresh weight stage shown on the x axis relative to the 100–200 mg stage are shown

expressed in a high isoflavone cultivar relative to a low isoflavone cultivar. These data suggest a critical role for CHS in controlling the level of the final isoflavone metabolites.

Genomic Exploration of Early Stages of Seed Development

The early developmental stages of seeds are a time of tremendous growth and tissue differentiation, as the body plan of the mature plant is set within the zygotic embryo and the organs that will later accumulate food reserves begin to form. Defects in morphology at these early stages can severely alter the phenotype and function of the mature plant or even prevent it from forming. Unfortunately, global expression studies of the gene activity during these early seed stages are relatively rare, due to the technical difficulties involved in obtaining tissue and RNA from the very small seeds and embryos. However, new technologies such as RNA amplification and laser capture microdissection are beginning to allow more exploration of how gene expression changes during early seed development.

Presently, we are expanding the global view of transcript expression in developing seed to these much earlier stages of soybean seed development, starting at the globular stage just a few days after fertilization. Using a technique to amplify RNA from very small initial quantities, data on 36,864 soybean cDNAs was obtained over five stages of early seed development and are currently being analyzed (Jones et al. 2006). In these experiments, the whole seed (seed coat + developing cotyledon and axis) from 4 DAF (days after flowering), 8–10 DAF, 12–14 DAF, 17–19 DAF, and 22–24 DAF were compared to the whole seed from an older reference point (5–6 mg fresh wt). An amplification step of total RNA was used and shown to correlate well with the results from non-amplified RNA. Again, the flux in many transcription factors can be seen in these early stages of development as well as the expression of many enzymes for structural and compositional components of the seed.

An ambitious series of experiments using laser capture microscopy (LCM) is currently underway in the laboratory Robert Goldberg as part of a recently funded NSF project to determine the genes required to "make a seed". One of the objectives is to examine in detail the tissues of the globular stage of soybean development just several cell divisions after fertilization (Goldberg et al. 2006; Le et al. 2007). In this study, soybean Affymetrix chips are hybridized with amplified biotinylated cRNAs generated from RNAs that were isolated from different tissues of the soybean seed using laser capture microscopy. Results from array data indicate approximately 20,000 diverse transcripts are expressed in the globular stage embryos, a value that close to that obtained originally by Rot curve hybridization results (Goldberg et al. 1981a). The tissues being examined by LCM include the embryo proper, suspensor, epidermis, endothelium, inner integument, outer integument, endosperm, and suspensor. Transcription factors that are localized in the specific parts of the seed are being identified. Some of these are likely critical to setting in motion the differentiation of the embryo as well as the compositional content of the seed. More information about this project can be found at the following web site: http://estdb.biology.ucla.edu/seed as well as in the Gene Expression Omnibus databases at http://www.ncbi.nlm.nih.gov/geo.

Gene Expression Studies of Embryogenesis in Other Plant Systems

Comparison of the soybean studies to those in other plant systems will likely provide a broader view of seed development. There are several recent investigations that examined the very early seed development stages, primarly in *Arabidopsis*, barley, and rice. Casson et al. (2005) used laser capture microdissection and the 24,000-gene Affymetrix *Arabidopsis* array to analyze gene expression in apical and basal tissues from globular- and heart-stage *Arabidopsis* embryos. This project was continued in Spencer et al. (2007) with the addition of a third early embryo developmental stage, the torpedo stage, with three new tissues (the shoot apical meristem, root, and cotyledon). An amplification procedure was used to obtain sufficient amounts of aRNA (amplified RNA) from limited tissue supplies, with their tests showing that the amplified and unamplified RNA gave similar results. For the one-color Affymetrix microarray experiment, cotyledon and root RNA from a 7-day-old seedling hybridized to the same kind of chip were used as an older comparative time point, somewhat analogous to the reference tissue in a two-color experiment.

A minimum signal value was established to estimate the number of genes that were considered expressed at each tissue and time point. Using this technique, approximately 8,000–11,000 genes were considered expressed, at some level, in the various tissues studied. Several genes with expression patterns of interest were selected for promoter-GUS fusion experiments whose results supported those of the microarray data, and semi-quantitative RT-PCR was performed on a number of genes as well.

For the most part in these studies (Casson et al. 2005; Spencer et al. 2007), the time course data from the apical and basal tissues of the embryos were considered separately, but a variety of analyses were performed, including k-means clustering and differential expression from a set value based on Student's t-test results. Additionally, fold changes were calculated between the expression levels at different stages of development and lists of the 50 or 100 most over-expressed genes in one stage compared to another were produced. These genes were then functionally categorized to reveal what types of gene products were the most highly expressed in, for example, the heart basal tissue compared to the torpedo basal tissue. In this case the data show that cell wall growth genes, such as hydroxyproline-rich glycoproteins, are often over-expressed, corresponding to the embryo elongation occurring at this time. Between the heart and torpedo stages in the apical tissues, in contrast, many protein synthesis-related genes increase in expression levels, which may reflect the maturation of the embryo to a protein-accumulating stage of development. For the most part, their analyses revealed that the difference in gene expression between the apical and basal regions is not as important as the difference between developmental stages.

Nielsen et al. (2006) also used whole genome Affymetrix chips in their study of three stages of barley embryo development, from the middle of differentiation to late maturation. Using more traditional dissection techniques, these researchers explicitly noted the difficulties in obtaining tissue from very early stages and mentioned that their youngest stage, 12 days after flowering, was the earliest from which they could collect embryos with efficiency. The data from this study was analyzed using one-way ANOVA, with lists of the 1,000 and 5,000 most significantly differentially expressed genes being generated. The top 1,000 significant genes were then divided into 9 clusters based on the similarity of their expression profiles using the k-means method, and these genes were also divided into 29 functional categories based on sequence similarity to *Arabidopsis* proteins. The expression profiles of selected genes were confirmed with quantitative RT-PCR.

Analysis of these functional categories revealed that a high percentage of genes involved with cell rescue, defense, and virulence; cellular environment; and protein fate were up-regulated during development. In contrast, genes related to cell cycle and DNA processing; cell fate; and the control of cellular organization were generally down-regulated as development progressed. For most of the analysis presented here, however, Nielsen et al. focused on the defense processes occurring in young seeds, specifically examining the expression profiles of genes encoding enzymes in the entire lipoxygenase-related pathway from stearic acid to oxylipins. The data suggest that there are at least two bursts of defense-related gene activation during development, activation which is *not* associated with increased amounts of endogenous jasmonic acid or salicylic acid in the tissues. Therefore, the researchers conclude that these "developmental defense activations" (DDA) must be induced by developmental signals rather than actual pathogen attacks, perhaps with the goal of protecting the embryo against such attacks in the future by preparing gene products like PR (pathogenesis-related) proteins, chitinases, thionins, and phytoalexins in advance.

In contrast to the global gene expression project of Spencer et al. (2007), Duan et al. (2005) created a thematic cDNA array spotted with 325 known or putative transcription factor genes from twelve families such as WRKY, YABBY, MADS-box, and Myb, hoping to learn more about the transcriptional networks involved with seed development in rice (*Oryza sativa*). The expression levels of the transcription factor clones were studied over eight stages of seed development, beginning with the flower and continuing through mature seed, using mature leaf RNA as the reference tissue. A two-fold expression level change was used to designate clones as differentially expressed, with 135 of the 325 transcription factor clones meeting this criterium in at least one stage studied. Eight of these 135 clones were selected for expression level confirmation by semi-quantitative RT-PCR, which supported the results of the microarrays. These seed-preferentially expressed genes were clustered by the similarity of their expression patterns using a self-organizing map algorithm. This clustering revealed that different genes from the same family often have different expression patterns throughout development, but that every stage of development had highly-expressed transcription factors, suggesting the coordinated expression of these genes to ensure normal seed development. Interestingly, this study noted that highly-expressed transcription factors are found even in mature, fully desiccated seeds, a phenomenon also observed in soybean (Jones 2004).

The 135 seed-preferentially expressed transcription factors were then further studied to discover their expression levels in stems, seedlings, and roots, and to reveal changes in their expression levels in seedlings due to various hormones and abiotic stresses. The results indicated that a large number of transcription factors that were preferentially expressed during seed development were also highly up- or down-regulated by hormones or environmental stresses, perhaps indicating the versatility of their functions. For example, a gene that is highly expressed *both* in a mature normal seed *and* in a seedling under drought conditions may encode a product related to the regulation of desiccation tolerance, which is able to function in both situations. The promoter regions of some of the 135 clones were also analyzed for common *cis*-elements, with the results suggesting that Dof transcription factors may be important in regulating other transcription factors downstream, forming a complex but finely-tuned network of gene expression regulation.

Becerra et al. (2006) took a different approach to the study of young seed gene expression. For the most part, they did not perform new gene expression experiments in their search for genes specific to early seed development in *Arabidopsis*. Instead, they utilized existing data from other researchers' experiments available in public databases. This analysis led to 49 genes that may be specifically expressed in young seeds.

The legume *Phaseolus coccineus* (scarlet runner bean) was also explored as a model system for early embryo development because of the large size of the suspensor region compared to other legumes, which enabled hand dissection of the early stages. Using the soybean Affymetrix chips, cross species hybridizations compared the suspensor and embryo proper regions of the two legumes, forming the foundation of comparative expression studies between seed legumes that will likely be expanded rapidly in the future to show key aspects of seed development in many different legumes (Le et al. 2007).

Gene Identification in Isolines of Developing Soybean Seed Coats Using Microarrays

In another application of microarrays, we used them successfully to examine gene mutations employing isogenic lines in order to define a small list of candidate genes that may cause the mutant phenotype. For example, as a test of this approach, we compared RNA from seed coats of two isogenic lines differing at the *T* (tawny) locus, which encodes a flavonoid 3′-hydroxylase (F3′H) enzyme, and found that the levels of the F3′H transcripts varied repeatedly by more than two-fold among the seed coats of the two isolines (Vodkin et al. 2004). The *T* locus is responsible for the tawny or gray color of the soybean trichome hairs on the plant and was the first genetic locus to be defined in soybean by crossing and segregation analysis (Woodworth 1921). The encoded F3′H gene is also highly expressed in the seed coat of *T* genotypes. The microarray data agree with RNA blots comparing the isogenic lines at the *T* locus, one carrying an unstable allele of the *T* locus (Zabala and Vodkin 2003). Other criteria, including cosegregation data and sequencing of alleles, showed previously that F3′H is encoded by the *T* locus (Toda et al. 2002; Zabala and Vodkin 2003).

Recently, the use of soybean microarrays was essential to the discovery that the molecular basis of the pink flower (*wp*) locus in soybean is a mutation in the *F3H1* gene that encodes a flavanone 3-hydroxlyase (Zabala and Vodkin 2005). We found that flavanone 3-hydroxylase (F3H) cDNAs from soybean hybridized more strongly to RNAs from either immature seed coats or young flower buds of a purple flower line (*WpWp*) than to the corresponding RNAs from a pink flower (*wpwp*) isoline using soybean cDNA microarrays. A novel gene fragment rich insertion of the CACTA family of elements (designated *Tgm-Express1*) that interrupts the *F3H1* gene is the reason for the reduced mRNA expression in pink flowers and in seed coats with the homozygous *wpwp* genotype compared to the normal purple flowers that carry the standard *Wp1* allele.

Molecular Genetic and Genomic Exploration of Unique Genetic Mutants in Soybean Seed Pigmentation

In the future, genomic resources including microarrays, genomic sequences, SNPs, physical maps, and proteomic tools will be enormously useful for examining the nearly 600 near-isogenic lines in the USDA collection maintained at the University of Illinois at Urbana-Champaign.

These isolines have already proved to be an important research tool and were used to study many morphological and physiological traits (reviewed in Carter et al. 2004; Palmer et al. 2004), including pigmentation of the soybean tissues, namely the seed coats, flowers, trichomes, pods, and hypocotyls. While some of the loci controlling pigmentation appear to be organ-specific, others affect pigmentation in more than one tissue. The molecular and genetic control of pigmentation in soybean has been the focus of our laboratory for a number of years, and we elucidated the molecular basis of several of these classical loci as shown in Fig. 11.4 (Lindstrom and Vodkin 1991; Nicholas et al. 1993; Todd and Vodkin 1993; Schmidt et al. 1994; Wang et al. 1994; Fasoula et al. 1995; Todd and Vodkin 1996; Johnson et al. 1998; Percy et al. 1999; Hegstad et al. 2000; Tuteja et al. 2004; Clough et al. 2004; Zabala and Vodkin 2003. 2005).

The recessive *i* allele of the *I* locus (Inhibitor of color) results in a completely pigmented seed coat and the color depends upon the particular alleles present at the *R* and *T* loci. Black (*i, R, T*) and brown (*i, r, T*) pigments are found when the

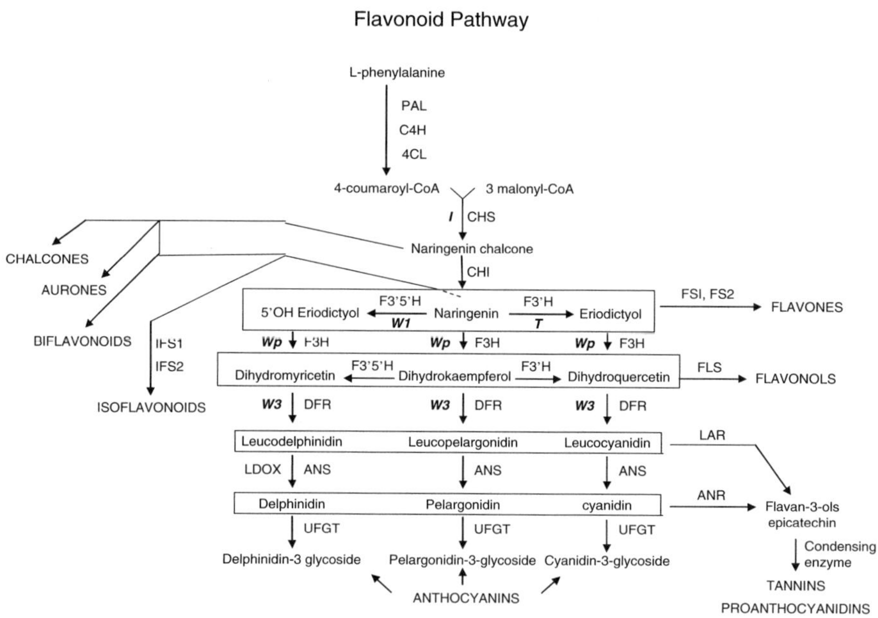

Fig. 11.4 General flavonoid pathway with classical genetic loci in soybean marked. In soybean, the following classical loci have been defined at the molecular level. The *I* (inhibition of seed color) locus is CHS, chalcone synthase (Todd and Vodkin 1996); the *T* locus (trichome color) encodes F3′H, flavonoid 3′ hydroxylase (Zabala and Vodkin 2003); the *W1* (purple flower) locus is F3′5′H, flavonoid 3′5′ hydroxylase (Zabala and Vodkin, 2007a); *W3* is DFR, dihydroflavonol-4-reductase (Fasoula et al. 1995); and *Wp* (light pink flower) is F3H, flavanone 3-hydroxylase (Zabala and Vodkin 2005). The *W1* and *Wp* loci also affect seed color. (Adapted from Zabala and Vodkin 2005)

dominant form of the *T* gene is present, whereas imperfect black color (*i, R, t*) and buff (*i, r, t*) are found when the homozygous, recessive *t* allele is present. The nature of the *R* gene is unknown, but we isolated the *T* locus and showed that it encodes a flavonoid 3′ hydroxylase (Zabala and Vodkin 2003), which adds a 3′ hydroxyl group to the phenolic, flavonoid ring (Buzzell et al. 1987; Todd and Vodkin 1993).

The *T* gene appears to be active in many plant tissues. The flavonoid quercetin, which is hydroxylated at the 3′ position, accumulates in the leaves of tawny plants. Plants with homozygous *t* genotypes have gray pubescence and kaempherol in the leaves. The imperfect black (*i, R, t*) and buff (*i, r, t*) seed coats have reduced amounts of cyanidin glucoside, a 3′ hydroxylated anthocyanin. Also, they do not contain procyanidin, a long chain proanthocyanidin that is capable of precipitating proteins and nucleic acids when released from the cell vacuole through injury or breakage (Todd and Vodkin 1993; Wang et al. 1994). Thus, the biochemical evidence also indicated that the *T* locus encodes F3′H and this was demonstrated at the molecular level (Zabala and Vodkin 2003).

The regulation of both the *T* (F3′H) and *Wp* (F3H) loci has consequences for the channeling of naringenin to the flavonols and anthocyanins, which are found in pigmented seed coats, and away from the isoflavones that are found in high abundance in the cotyledons. Interestingly, transcripts for both F3′H and F3H are very strongly expressed in the developing seed coats but not in the cotyledons (Zabala and Vodkin 2003, 2005). Thus, the plant may achieve differential regulation of the types of flavonoids in the different tissues based on the expression of these transcripts and enzymes.

The *wp* Allele Also Affects Seed Protein Content

In addition to its affect on seed coat pigmentation and flower color, the recessive *wpwp* genotype is also associated with 4% higher protein and 22% greater seed weight (Stephens and Nickell 1992; Stephens et al. 1993; Johnson et al. 1998). Protein content is generally a complex trait, and the loci that influence the trait are unknown at the molecular level, as are the molecular mechanisms that shift the ratio between protein and oil within the seed. With identification of the *wp* allele as the insertion of a novel transposon (described below) that carries gene fragments into the *F3H* (flavanone 3-hydroxylase) gene (Zabala and Vodkin 2005), two possibilities are opened as to the mechanism of the epistatic effect on protein content in this line. One possibility is that the decreased expression of the *F3H* gene is the primary cause mediated through an interaction of the flavonoid pathway with seed protein content. Alternatively, the gene fragments carried into the *F3H* gene somehow affect the metabolic pathways in which they are found to mediate an effect on protein content. Elucidation of the mechanism could lead to biotechnological approaches to manipulation of seed protein and oil content.

An Example of Transposable-Element Induced Chimeric Transcripts During Soybean Seed Development

Recently, we showed that a member of the soybean *Tgm* family of elements is capable of carrying host gene fragments (Zabala and Vodkin 2005). *Tgm* elements were first found to interrupt the soybean seed lectin gene that is expressed in developing cotyledons (Goldberg et al. 1983; Vodkin et al. 1983; Rhodes and Vodkin 1985, 1988). Pink flowered plants (*wpwp*) were first observed in 1989 (Stephens and Nickell 1992; Stephens et al. 1993) and were derived from a mutable, chimeric plant having purple and pink flowers on the same plant. The inheritance of the mutable phenotype and derived pink and purple lines showed a high rate of instability (Johnson et al. 1998). As described above, we identified a candidate gene for the *Wp* locus as F3H using microarrays to screen the isogenic lines (Zabala and Vodkin 2005). The DNA sequence of a 5.7 kb insertion in the mutant *wp* allele revealed a transposable element member of the CACTA family of transposons (*Tgm, Spm,* and *Tam*). However, the element in the *wp* pink flower mutation differs from the other *Tgm* family members previously characterized in that it lacks the sub-terminal repeats and carries five genic fragments picked up from the host genome. Thus, the ability to acquire and transport host DNA segments as found for the Helitrons (Lal et al. 2003; Gupta et al. 2005) in maize and the Pack Mule elements in rice (Jiang et al. 2004) was extended to the CACTA family of elements to which both *Tgm* and the prototypical maize *Spm/En* elements belong.

Recently, we examined whether chimeric transcripts could be generated from the string of gene fragments present in *Tgm-Express 1* that represent enzymes in basic metabolism. Cloning and sequencing of the transcripts isolated by RT-PCR from RNA of the *wp/wp* mutant line carrying the *Tgm-Express 1* element showed variable degrees of splicing of the intronic regions between the gene fragments carried by the *Tgm-Express 1* element. These events resulted in the fusion of exons from coding regions of unrelated genes as shown for the *wp-4* product; for example, that fuses exons of the *F3H* gene and three of the other coding region fragments, including those from a cell division cycle kinase (CDC2), a fructose-6-phosphate 2-kinase/fructose-2-6-bisphosphatase, and an unknown protein (Zabala and Vodkin, 2007b).

The Dominant *I* Locus is a Naturally Occurring Example of RNAi Mediated Gene Silencing

Another unusual genetic locus in soybean is the *I* (inhibitor) locus. The *I* locus controls the presence and absence of, as well as the spatial distribution of, anthocyanins and proanthocyanidins in the seed coat. The dominant *I* allele inhibits pigmentation over the entire seed coat resulting in a light or yellow color on mature harvested seeds. Though all wild *Glycine* accessions have black or brown seed coats due to the recessive *i* allele, thereby permitting accumulation of pigments, most commercial soybean varieties were selected for a yellow seed coat (*I* or i^i alleles) (Palmer and Kilen 1987; Buzzell et al. 1987; Palmer et al. 2004) to mitigate the undesirable effects of the pigments on protein and oil extractions for use in commercial soybean products.

Infrequently, spontaneous mutations in inbred cultivars having either the *I* or i^i alleles give rise to recessive *i* alleles (pigmented seed coats when homozygous). These were preserved as isogenic lines, differing only with respect to the alleles at the *I* locus (Wilcox 1988). We documented quantitative differences in *CHS* transcripts in the seed coats of both the pigmented and non-pigmented isolines (Todd and Vodkin 1996; Tuteja et al. 2004), leading us to propose a post-transcriptional mechanism of homology-dependent gene silencing operating in a *trans*-dominant manner among alleles of the *I* locus. In a study by another group (Senda et al. 2004), on another isogenic pair of the *I* locus alleles, evidence establishing a post-transcriptional mechanism of *CHS* silencing was provided by nuclear run-on assays.

We determined that the *I* locus corresponds to a 27 kb long chalcone synthase (CHS) gene cluster that exhibits a unique tissue specific gene silencing mechanism in the seed coats (Wang et al. 1994; Todd and Vodkin 1996; Tuteja et al. 2004; Clough et al. 2004). Senda et al. (2004) showed that CHS siRNAs are involved in mediating the silencing of the CHS in the yellow seed coats and that the silencing is released by transformation with viral HcPro sequences, as is found for other genes that are silenced by siRNAs (Baulcombe 2004).

Gene silencing by co-suppression was first discovered in plants as co-suppression of both the endogenous and transgenic CHS genes in transgenic petunia lines (Napoli et al. 1990; Van der Krol et al. 1990). Now known as RNA interference (RNAi), silencing is an evolutionarily conserved mechanism that protects genomes from exogenous (viral) and endogenous (transposon) invasion, and greatly impacts cellular programs of gene expression and development (Baulcombe 2004; Matzke and Matzke 2004). The research on the complexity and importance of the silencing pathway is growing rapidly, and the hallmark feature is the production of the 20–30 nucleotide small RNAs (sRNAs), one kind being the short interfering RNAs (siRNAs) (Hamilton and Baulcombe 1999) that are initially derived from long double stranded RNA (Fire et al. 1998). The microRNAs (miRNAs) form another class of small RNAs that are encoded within the genome. These small non-coding RNAs have a hairpin structure that triggers the action of the enzyme Dicer to produce the 21–30 nt small RNAs, which in turn are loaded onto the RNA silencing complex (RISC) where they target destruction of mRNAs containing the matching sequences. It was postulated that miRNAs are evolved from inverted gene duplications of the target genes (Allen et al. 2004).

Though our system bears similarities with two other cases (in rice and maize) in which the dominant loci composed of inverted repeats induce RNA silencing of an endogenous gene (Kusaba et al. 2003; Della Vedova et al. 2005), ours is unique in that the endogenous inverted repeat drives the RNA silencing in a tissue-specific manner (Tuteja et al. 2004).

In a recent study, we examined the response of the flavonoid pathway to infection by *Pseudomonas syringae (avrB)* (Zabala et al. 2006). Eight hours post-infection during the plant's hypersensitive response, the leaves of the yellow seeded soybean cultivar Williams (which harbors the i^i allele at the *I* locus that suppresses expression of *CHS7/CHS8* in the seed coats) accumulated high levels of transcripts of all *CHS* gene family members, including two of the three *CHS* genes constituting the *I* locus (*CHS1* and *CHS3*) that barely accumulated in the control leaves. Thus, most of the CHS genes (including CHS7 and CHS8) are capable of producing abundant transcripts even in varieties in which CHS expression is silenced in the seed coats. These results reinforced the tissue specific nature of silencing seen in the dominant *I* and i^i alleles of the *I* locus.

There are still many unanswered questions about the naturally occurring RNA interference mechanism exhibited by the *I* locus. The cotyledons of dominant *I* genotype express CHS transcripts as this enzyme is necessary for synthesis of isoflavones in these varieties. If we could understand the mechanism by which the dominant *I* alleles are silenced in such a tissue specific manner in the seed coats only, we may be able to design more specific vectors for RNAi downregulation by transgenics. In addition, the phenotype of the i^k allele of the *I* locus that results in a saddle-shaped pattern of pigmentation on the seed coat is very intriguing as it suggests that silencing is present or absent between adjacent cells of the same cell type within the epidermal cells of the seed coat.

Future Prospects

In summary, the elucidation of seed development using the power of genomic tools is still in its infancy. The application of genomic tools to this important developmental stage is poised to make substantial breakthroughs in the next few years. Many EST libraries representing seed tissues have been, and microarray resources are available and being used to explore the transcript profiles during seed development, as well as in tissue and cell types combined with laser capture detection. The pathways for composition and the transcriptional regulatory factors that control these pathways will likely be elucidated. In addition, genomic tools including microarrays will be applied to a wide array of genotypic lines that display compositional and morphological variants. In conjunction with the expanding ways to conduct reverse genetics including RNAi transgenics, researchers will be able to directly test the effect of candidate genes in soybean that influence morphological and compositional traits.

References

Allen, E., Zie. Z., Gustafson, A.M., Sung, G-H., Spatafora, J.W. and Carrington, J.C. (2004) Evolution of microRNA genes by inverted duplication of target gene sequences in *Arabidopsis thaliana*. Nat. Genet. 36, 1282–1290.

Ainsworth, E.A., Rogers, A., Vodkin, L.O., Walter, A. and Schurr, U. (2006) The effects of elevated CO_2 on soybean gene expression: An analysis of growing and mature leaves. Plant Physiol. 142, 135–147.

Baulcombe, D. (2004) RNA silencing in plants. Nature 431, 356–363.

Becerra, C., Puigdomenech, P. and Vicient, C.M. (2006) Computational and experimental analysis identifies Arabidopsis genes specifically expressed during early seed development. BMC Genomics, **7**, 38.

Bewley, J.D., Hempel, F.D., McCormick, S. and Zambryski, P. (2000) Reproductive development. In: *Biochemistry and Molecular Biology of Plants* (Buchanan, B.B., Gruissem, W. and Jones, R.L., eds), pp. 988–1043. Rockville, MD: American Society of Plant Physiologists.

Buzzell, R.I., Buttery, R.B. and MacTavish, D.C. (1987) Biochemical genetics of black pigmentation of soybean seed. J. Hered. 78, 53–54.

Carter, T.E., Nelsone, R.L., Sneller, C.H. and Cru, Z. (2004) In J.R. Wilcox, H.R. Boerma, and J.E. Specht (ed.) *Soybeans: Improvement, Production and Uses*. Agronomy No. 16. 3rd edition. ASA, CSSA, SSSA, Madison, WI.

Casson, S., Spencer, M., Walker, K. and Lindsey, K. (2005) Laser capture microdissection for the analysis of gene expression during embryogenesis of Arabidopsis. Plant Jour. 42, 111–123.

Chen, T.H.H,, Marowitch, J., and Thompson, B.G. (1987) Genotypic effects on somatic embryogenesis and plant regeneration from callus cultures of alfalfa. Plant Cell Tissue Organ Cult 8, 73–81.

Clough, S.J., Tutcja, J.H., Li, M., Marek, L.F., Shoemaker, R.C. and Vodkin, L.O. (2004). Features of a 103-kb gene-rich region in soybean include an inverted perfect repeat cluster of *CHS* genes comprising the *I* locus. Genome 47, 819–831.

Cullis, C.A. (2004) *Plant Genomics and Proteomics*. Wiley-Liss, Hoboken, NJ.

Duan, K., Luo, Y-H., Luo, D., Xu, Z-H. and Xue, H-W. (2005) New insights into the complex and coordinated transcriptional regulation networks underlying rice seed development through cDNA chip-based analysis. Plant Mol. Biol. 57, 785–804.

Della Vedova, C.B., Lorbiecke, R., Kirsch, H., Schulte, M.B., Scheets, K., Borchert, L.M., Scheffler, B.E., Wienand, U., Cone, K.C. and Birchler, J.A. (2005) The dominant inhibitory chalcone synthase allele *C2-Idf (Inhibitor diffuse)* from *Zea mays* (L.) acts via an endogenous RNA silencing mechanism. Genet. 170, 1989–2002.

Dhaubhadel, S., Gijzen, M., Moy, P. and Farhangkhoee, M. (2007) Transcriptome analysis reveals a critical role of *CHS7* and *CHS8* genes for isoflavonoid synthesis in soybean seeds. Plant Physiol. 143, 326–338.

Fasoula, D.A., Stephens, P.A., Nickell, C.D. and Vodkin, L.O. (1995) Cosegregation of purple-throat flower color with a DNA polymorphism in soybean. Crop Science 35, 1028–1031.

Finer, J.J. (1988) Apical proliferation of embryonic tissue of soybean [*Glycine max* (L.) Merrill]. Plant Cell Rep **7**, 238–241

Finer, J.J. and McMullen, M.D. (1991) Transformation of soybean *via* particle bombardment of embryogenic suspension culture tissue. In Vitro Cell Dev Biol **27**: 175–182

Fisher, R. L. and Goldberg, R.B. (1982) Structure and flanking regions of soybean seed protein genes. Cell 29, 651–660.

Fire, A. Xu, S., Montgomery, S., Kostas, S., Driver, S.E. and Mello, C.C (1998) Potent and specific genetic interference by double stranded RNA in *Caenorhabditis elegans*. Nature 391, 806–811.

Goldberg, R.B. (1988) Plants: novel developmental processes. Science 240, 1460–1467.

Goldberg, R.B., Hoschek, G., Tam, S. H. Ditta, G.S. and Breidenbach, R.W. (1981a) Abundance, diversity, and regulation of mRNA sequence sets in soybean embryogenesis. Develop Biol 83, 201–217.

Goldberg, R.B., Hoschek, G., Ditta, G.S., and Breidenbach, R.W. (1981b) Developmental regulation of cloned superabundant embryo mRNAs in soybean. Develop Biol 83, 218–231.

Goldberg, R.B., Hoschek, G. and Vodkin, L.O. (1983) An insertion sequence blocks the expression of a soybean lectin gene. Cell 33, 465–475.

Goldberg, R.G., Barker, S.J., and Perez-Grau, L. (1989) Regulation of gene expression during plant embryogenesis. Cell 56, 149–160

Goldberg, R.B., Depaiva, G. and Yadegari, R. (1994) Plant embryogenesis - Zygote to seed. Science 266, 605–614

Goldberg, R.B., Bui, A.Q., Le, B., Kawashima, R., Wang, X., Wagmaister, J, Fisher, R.L., and Harada, J.J. (2006) Using genomics to dissect seed development. 8th International Congress of Plant Molecular Biology, p. 69.

Gonzalez, D.O. and Vodkin, L.O. (2006) Clustering analysis of transcript abundance in soybean cotyledons during germination and emergence. Plant and Animal Genome XIV, San Diego, CA, p. 295.

Gonzalez, D.O. and Vodkin, L.O. (2007) Specific elements of the glyoxylate pathway play a significant role in the functional transition of the soybean cotyledon during seedling development. BMC Genomics 8, 468.

Green P., 2001. Documentation for Phrap. http;//bozeman.mbt.washington.edu, Genome Center, University of Washington.

Gupta, S., Gallavotti, A., Stryker, G.A., Schmidt, R.J. and Lal, S.K. (2005) A novel class of *Helitron*-related transposable elements in maize contain portions of multiple pseudogenes. Plant Mol. Biol. 57, 115–127.

Hamilton, A.J. and Baulcombe D.C. (1999) A species of small antisense RNA in posttranscriptional gene silencing in plants. Science 286, 950–952.

Hegstad, J.M., Tarter, J.A., Vodkin, L.O. and Nickell, C.D. (2000) Positioning the *wp* flower color locus on the soybean genome map. Crop Sci. 40, 534–537.

Hills, M.J. (2004) Control of storage-product synthesis in seeds. Curr. Opin. Plant Biol. **7**, 302–308.

Huang X. and Madan A. (1999) CAP3: A DNA sequence assembly program. Genome Research 9, 868–877.

Jiang, N., Bao, Z., Zhang, X., Eddy, S.R. and Wessler, S.R. (2004) Pack-MULE transposable elements mediate gene evolution in plants. Nature 431, 569–573.

Johnson, E.O., Stephens, P.A., Fasoula, D.A., Nickell, C.D. and Vodkin, L.O. (1998) Instability of a novel multicolored flower trait in inbred and outcrossed soybean lines. J. Hered. 89, 508–515.

Jones, S. (2004) A global view of gene expression during soybean seed development using microarrays. Masters thesis. University of Illinois at Urbana-Champaign.

Jones, S., Gonzalez, D.O. and Vodkin, L.O. (2006) Global gene expression profiles of early soybean seed development using microarrays. Am. Soc. Plant Biologists, Boston, MA, p. 255.

Kusaba, M., Miyahara, K., Iida, S., Fukuoka, H., Takano, T., Sassa, H., Nishimura, M. and Nishio, T. (2003). Low glutelin content1: a dominant mutation that suppresses the glutelin multigene family via RNA silencing in rice. Plant Cell 15, 1455–1467.

Lal, S.K., Giroux, M.J., Brendel, V., Vallejos, E. and Hannah, L.C. (2003). The maize genome contains a *Helitron* Insertion. Plant Cell 15, 381–391.

Le, B.H., Wagmaister, J.A., Kawashima, T., Bui, A.Q., Harada, J.J., and Goldberg, R.B. (2007) Using genomics to study legume seed development. Plant Physiol. 144, 562–574.

Lindstrom, J.T. and Vodkin, L.O. (1991) A soybean cell wall protein is affected by seed color genotype. Plant Cell 3, 561–571.

Matzke, M.A. and Matzke, A.J.M. (2004) Planting the seeds of a new paradigm. Plos Biol. 2, 582–585.

McConnell J. and Barton, M. (1998) Leaf polarity and meristem formation in *Arabidopsis*. Development 125, 2935–2942.

Meinke, D.W., Chen, J., and Beachy, R.N. (1981) Expression of storage protein genes during soybean seed development. Planta 153, 130–139.

Moise, J., Han, S., Gudynaite-Savitch, L., Johnson, D.A., and Miki, B.L. (2005) Seed coats: structure, development, composition, and biotechnology. In Vitro Cell. Dev. Biol. Plant 41, 620–644.

Napoli, C., Lemieux, C. and Jorgensen, R. (1990) Introduction of a chimeric chalcone synthase gene into petunia results in reversible co-suppression of homologous genes in trans. Plant Cell 2, 279–289.

Nicholas C.D., Lindstrom J.T. and Vodkin, L.O. (1993) Variation in proline rich cell wall proteins in soybean lines with anthocyanin mutations. Plant Molecular Biol. 21: 145–156.

Nielsen, M.E., Lok, F. and Nielsen, H.B. (2006) Distinct developmental defense activations in barley embryos identified by transcriptome profiling. Plant Mol. Biol. 61, 589–601.

Palmer, R.G. and Kilen, T.C. (1987) Qualitative Genetics. In Wilcox, J.R. (ed.), *Soybean: Improvement, Production, and Uses*, 2nd Edition, American Society of Agronomy, Madison, Wisconsin, pp.135–209.

Palmer, R.G., Pfeiffer, T.W., Buss, G.R. and Kilen, T.C. (2004) Qualitative genetics. p. 137–233. In J.R. Wilcox, H.R. Boerma, and J.E. Specht (ed.) *Soybeans: Improvement, Production and Uses*. Agronomy No. 16. 3rd edition. ASA, CSSA, SSSA, Madison, WI.

Percy, J. D., Philip R. and Vodkin, L.O. (1999) Defective seed coat mutations that affect the post-transcriptional abundance of soluble proline rich cell wall proteins. Plant Molecular Biol. 40, 603–613.

Rhodes, P.R. and Vodkin, L.O. (1985). Highly structured sequence homology between an insertion element and the gene in which it resides. Proc. Natl. Acad. Sci. 82, 493–497.

Rhodes, P.R. and Vodkin, L.O. (1988). Organiztion of the Tgm family of transposable elements in soybean. Genet. 120, 597–604.

Ritchie, S.W., Hanway, J.J., Thompson, H.E. and Benson, G.O. (1996) How a soybean plant develops. Spec. Rep. No. 53. Ames, IA: Iowa State Univ. of Science and Tech. Coop. Ext. Serv.

Rosenberg, L.A. and Rinne, R.W. (1986) Moisture loss as a prerequisite for seedling growth in soybean seeds (Glycine max L. Merr.). J. Exp. Bot. 37, 1663–1674.

Santarem, E.R., Pelissier B. and Finer, J.J. (1997) Effect of explant orientation, pH, solidifying agent and wounding on initiation of soybean somatic embryos. In Vitro Cell Dev Biol 33, 13–19.

Sawa, S., Watanabe, K., Goto, K., Kanaya, E., Morita, E.H. and Okada, K. (1999) FILAMENTOUS FLOWER, a meristem and organ identity gene of Arabidopsis, encodes a protein with a zinc finger and HMG-related domains. Genes Dev. 13, 1079–1088.

Schmidt, J.S., Lindstrom, J.T. and Vodkin, L.O. (1994) Genetic length polymorphisms create size variation in proline rich proteins of the cell wall. Plant Journal 6: 177–186.

Senda, M., Masuta, C., Ohnishi, S., Goto, K., Kasai, A., Sano, T., Hong, J-S. and MacFarlane, S. (2004) Patterning of virus-infected soybean seed coat is associated with suppression of endogenous silencing of chalcone synthase genes. Plant Cell 16, 807–818.

Shoemaker, R., Keim, P., Vodkin, L., Retzel, E., Clifton, S.W., Waterston, R., Smoller, D., Coryell, V., Khanna, A., Erpelding, J., Gai, X., Brendel, V., Raph-Schmidt, C., Shoop, EG., Vielweber, C.J., Schmatz, M., Pape, D., Bowers, Y, Theising, B., Martin, J., Dante, M., Wylie, T. and Granger, C. (2002) A compilation of soybean ESTs: generation and analysis. Genome 45, 329–338.

Shoemaker, R.C., Cregan, P.B. and Vodkin, L.O. (2004) Soybean Genomics. Pages 235–263 in *Soybeans: Improvement, Production and Uses*, 3rd edition, ASA monograph, J. Specht (ed). American Society of Agronomy, Madison, WI.

Siegfried, K., Eshed, Y., Baum, S., Otsuga, D., Drews, G. and Bowman, J. (1999) Members of the YABBY gene family specify abaxial cell fate in Arabidopsis. Development 126, 4117–4128.

Spencer, M.W.B., Casson, S.A. and Lindsey, K. (2007) Transcriptional profiling of the Arabidopsis embryo. Plant Physiol. 143, 924–940.

Stacey, G., Vodkin, L. O., Parrott, W. and Shoemaker, R.C. (2004) Draft plant for soybean Genomics. Plant Physiol. 135, 59–70.

Stephens, P.A. and Nickell, C.D. (1992) Inheritance of pink flower color in soybean. Crop Sci. 32, 1131–1132.

Stephens, P.A, Nickell, C.D. and Vodkin, L.O. (1993) Pink flower color associated with increased protein and seed size in soybean. Crop Sci. 33, 1135–1137.

Stromvik, M.V., Thibaud-Nissen, F., and Vodkin, L.O. (2004) Mining soybean expressed sequence tag and microarray data. Ann Rev. Phytochemistry 38, 177–195.

Thibaud-Nissen, F., Shealy, R.T., Khanna, A., and Vodkin, L.O. (2003) Clustering of microarray data reveals transcript patterns associated with somatic embryogenesis in soybean. Plant Physiol. 132, 118–136.

Toda, K., Yang, D., Yamanaka, N., Watanabe, S., Harada, K. and Takahashi, R. (2002) A single-base deletion in soybean flavonoid 3′-hydroxylase gene is associated with gray pubescence color. Plant Mol. Biol. 50, 187–196.

Todd, J.J. and Vodkin, L.O. (1993) Pigmented soybean seed coats accumulate proanthocyanidins during development. Plant Physiol. 102, 663–670.

Todd, J.J. and Vodkin, L.O. (1996) Duplications that suppress and deletions that restore expression from a CHS multigene family. Plant Cell 8, 687–699.

Tuteja, J.H., Clough, S.J., Chan, W.C. and Vodkin, L.O. (2004). Tissue-specific gene silencing mediated by a naturally occurring chalcone synthase gene cluster in *Glycine max*. Plant Cell 16, 819–835.

Van der Krol, A.R., Mur, L.A., Beld, M., Mol, J.N.M and Stuitje, A.R. (1990) Flavonoid genes in petunia: addition of a limited number of gene copies may lead to a suppression of gene expression. Plant Cell 2, 291–299.

Vodkin, L.O. (1981) Isolation and characterization of messenger RNAs for seed lectin and Kunitz trypsin inhibitor in soybean. Plant Physiology 68, 766–771.

Vodkin, L.O., Rhodes, P.R. and Goldberg, R.B. (1983) A lectin gene insertion has the structural features of a transposable element. Cell, 34, 1023–1031.

Vodkin, L.O., Khanna, A., Shealy, R., Clough, S.J., Gonzalez, D.O., Philip, P., Zabala, G., Thibaud-Nissen, F., Sidarous, M., Strömvik, M.V., Shoop, E., Schmidt, C., Retzel, E., Erpelding, J., Shoemaker, R.C., Rodriguez-Huete, A.M., Polacco, J.C., Coryell, V., Keim, P., Gong, G., Liu, L., Pardinas, J., and Schweitzer, P. (2004) Microarrays for global expression constructed with a low redundancy set of 27,500 sequenced cDNAs representing an array of developmental stages and physiological conditions of the soybean plant. BMC Genomics 5, 73.

Wang, C.S., Todd, J.J. and Vodkin, L.O. (1994). Chalcone synthase mRNA and activity are reduced in yellow soybean seed coats with dominant I alleles. Plant Physiol. 105, 739–748.

Walbot, V. (1999) Genes, genomes, genomics. What can plant biologists expect from the 1998 National Science Foundation Plant Genome Research Program. Plant Physiol. 119, 1151–1155.

Wilcox, J.R. (1988) Performance and use of seed coat mutants in soybean. Crop Sci 28, 30–32.

Wright M.S., Launis, K.L., Novitzky, R., Duesing, J.H. and Harms, C.T. (1991) A simple method for the recovery of multiple fertile plants from individual somatic embryos of soybean [Glycine max (L.) Merrill]. In Vitro Cell Dev Biol 27, 153–157.

Woodworth, C. M. (1921) Inheritance of cotyledon, seed-coat, hilum, and pubescence colors in soy-beans. Genet. 6, 487–553.

Zabala, G.C. and Vodkin, L.O. (2003) Cloning of the pleiotropic *T* locus in soybean and two recessive alleles that differentially affect structure and expression of the encoded flavonoid 3′ hydroxylase. Genet. 163, 295–309.

Zabala, G. and Vodkin, L.O. (2005) The *wp* mutation of *Glycine max* carries a gene-fragment-rich transposon of the CACTA superfamily. Plant Cell 17, 2619–2632.

Zabala, G., Zou, J., Tuteja, J., Gonzalez, D.O., Clough, S. J. and Vodkin, L.O. (2006) Transcriptome changes in the phenylpropanoid pathway of Glycine max in response to *Pseudomonas syringae* infection. BMC Plant Biology 6, 26.

Zabala, G. and Vodkin, L.O. (2007a) A rearrangement resulting in small tandem repeats in the F3′5′ H gene of white flower genotypes is associated with the soybean W1 locus. Plant Genome 47, 113–124.

Zabala, G. and Vodkin, L.O. (2007b) Novel exon combinations generated by alternative splicing of gene fragments mobilized by a CACTA transposon in *Glycine max*. BMC Plant Biology 7, 38.

Zimmerman, J. L. (1993) Somatic embryogenesis: a model for early development in higher plants. Plant Cell 5, 1411–1423.

Zou, J., Rodriguez-Zas, S., Aldea, M., Li, M., Zhu, J., Gonzalez, D.O., Vodkin, L.O., DeLucia, E. and Clough, S.J. (2005) Expression profiling soybean response to *Pseudomonas syringae* reveals new defense-related genes and rapid down regulation of photosynthesis. Mol Plant Microbe Interact. 18, 1161–1174.

Chapter 12
Genomics of Soybean Oil Traits

David F. Hildebrand, Runzhi Li, and Tomoko Hatanaka

Introduction

Plant oils including soybean, *Glycine max*, oil are mainly triacylglycerol (TAG) which represents an important edible and industrial resource. TAG also comprises a major part of the value of soybeans and soybeans are the most important source of renewable oil in the US. World-wide production of soybean oil is about the same as that of palm oil. Biodiesel is made from plant TAG and represents an important and growing renewable fuel resource. Surprisingly, the contribution of all the important enzymes to TAG accumulation in plant seeds (or any other plant tissue) even in Arabidopsis is currently unknown. We know even less about TAG biosynthesis in soybeans.

Most plant seeds accumulate storage products during seed development to provide nutrients and energy for seedlings to grow competitively for light and nutrients, especially nitrogen. Soybean seed protein provides the nitrogen and soybean oil provides most of the energy for seedling establishment. Most seeds either accumulate starch or TAG as an energy store for seedling establishment. Many seed crops such as corn, wheat, rice, peas and common beans (*Phaseolus vulgaris*) accumulate starch as the main form of energy storage in the seeds although the embryo or germ portion is often high in oil. Oilseeds such as soybeans, canola (*Brassica napus*), sunflowers, cotton seed, peanuts and many other oilseeds accumulate oil instead of starch. Like many tiny seeds *Arabidopsis thaliana* and tobacco (*Nicotiana tabacum*) seeds accumulate oil as an energy store with Arabidopsis seeds usually being ~42% oil (O'Neill et al., 2003; Zhang et al., 2005). Commercial soybean seeds are about 40% protein and 20% oil on a dry weight basis. Soybean seeds are very high in protein among seed crops but relatively low in oil among oilseeds. Macadamia nut seeds can be as much as 76% oil and palm fruit as much as 79% oil on a dry weight basis (Fig. 12.1). Many oilseeds such as canola, peanuts and sunflower are 40–50%

D.F. Hildebrand
University of Kentucky, Lexington, KY 40546-0312, USA
e-mail: dhild@uky.edu

G. Stacey (ed.), *Genetics and Genomics of Soybean*,

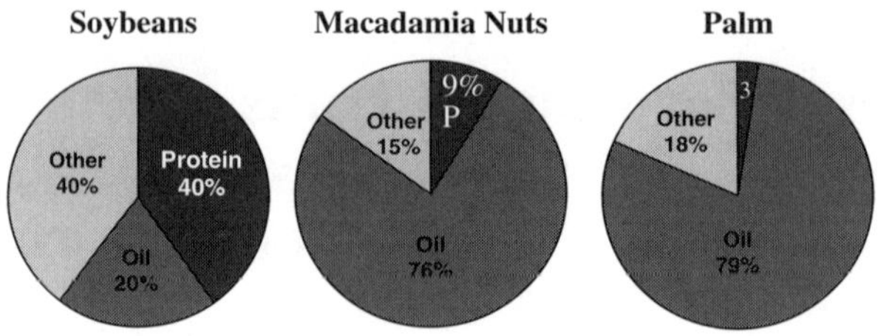

Fig. 12.1 Oil and protein content of soybeans compared with macadamia nuts and palm fruit

oil. The protein + oil content in macadamia nuts is ~85% as opposed to 60% for soybeans. As such soybeans have relatively low oil yields per unit land area among oilseeds with palm being the highest after the tree plantations are well established (Fig. 12.2). Palm oil and olive oil are from the fruit of these trees although palm seeds or kernels also accumulate oil which is different in fatty acid composition from that of the fruit oil.

The genomics of TAG biosynthesis in Arabidopsis has been studied rather well (although much remains to be elucidated) by Ohlrogge and colleagues (Beisson et al., 2003; Ruuska et al., 2004). They maintain an Arabidopsis lipid gene database, http://www.plantbiology.msu.edu/lipids/genesurvey/index.htm, which is regularly updated (e.g. as of January 2007 at the writing of this chapter). To date they have identified more than 620 genes in Arabidopsis involved in acyl-lipid metabolism. This is about 2.4% of the total number of predicted genes in the Arabidopsis genome. They are classified into eight groups plus a miscellaneous class (Table 12.1). Interestingly the largest group in terms of numbers of genes is the lipid signaling group. The groups of greatest importance in the synthesis of seed oil and, therefore, for this chapter are the synthesis of plastid fatty acids, endomembrane lipid synthesis and TAG synthesis and storage.

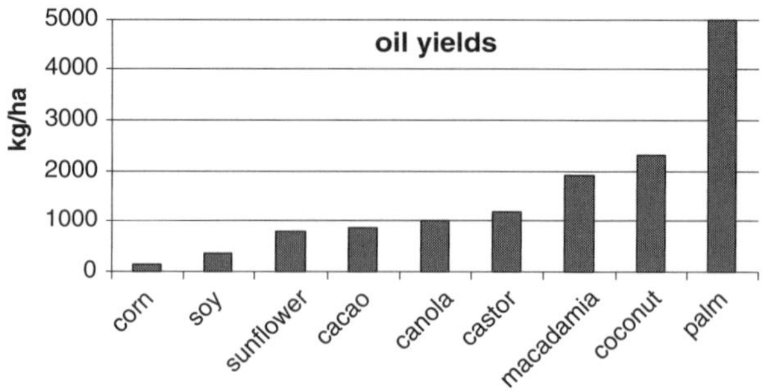

Fig. 12.2 Relative oil yields per unit land area of soybeans compared to some other oil crops

Table 12.1 Genes in Arabidopsis involved in acyl-lipid metabolism (Beisson et al., 2003; Ruuska et al., 2004); http://www.plantbiology.msu.edu/lipids/genesurvey/index.htm

Function	# genes
Synthesis of plastid FA	47
Synthesis of plastid membranes	20
Endomembrane synthesis	59
Mitochondrial acyl-lipid metabolism	29
Oil synthesis and storage	20
Lipid catabolism	43
Lipid signaling	153
Fatty acid elongation, wax & cutin metabolism	75
Miscellaneous	178
Total	624

Fatty Acid Synthesis

As mentioned above the main storage products that accumulate in soybean seeds are protein and oil. Protein mostly accumulates in the form of storage proteins in protein bodies. Oil accumulates as TAG in oil bodies. The amino acids asparagine and glutamine provide the nitrogen and initial carbon skeletons for protein deposition and sucrose provides the energy for protein biosynthesis and the energy and hydrocarbon precursors for TAG accumulation. The key processes and enzymatic steps involved in TAG biosynthesis in seeds such as soybeans are summarized in Fig. 12.3, Table 12.2. This is fueled by sucrose made in the leaves delivered to the apoplasts of the developing seeds. Sucrose is converted into hexose phosphates and then to fructose 1,6 bis- phosphate, which is cleaved into triosphosphates. Triosphosphates, such as dihydroxy acetone phosphate, are reduced to glycerol-3-P that provides the glycerol backbone for membrane lipids and TAGs and oxidized to 3-phosphoglycerate. 3-Phosphoglycerate is isomerized to phosphoenol pyruvate (PEP) and then to pyruvate. Pyruvate and possibly PEP enter the plastids and the

Table 12.2 Enzymes of oil or triacylglycerol (TAG) biosynthesis (Fig. 12.3)

Enzyme abbreviation	Enzyme name
PDH	Pyruvate dehydrogenase
ACC	Acetyl-CoA carboxylase
FAS	Fatty acid synthase complex
KASII	Keto-acyl-ACP synthase II
D9D	Δ-9 desaturase
TE	Thioesterase
ACS	Acyl-CoA synthetase
AT	Acyltransferase
*Fad*2-1	Δ-12 desaturase
*Fad*3	ω-3 desaturase
GPAT	Glycerol-3 phosphate acyltransferase
LPAT	Lysophosphatidic acid acyltransferase
PAP	Phosphatidic acid phosphatase
DGAT	Diacylglycerol acyltransferase

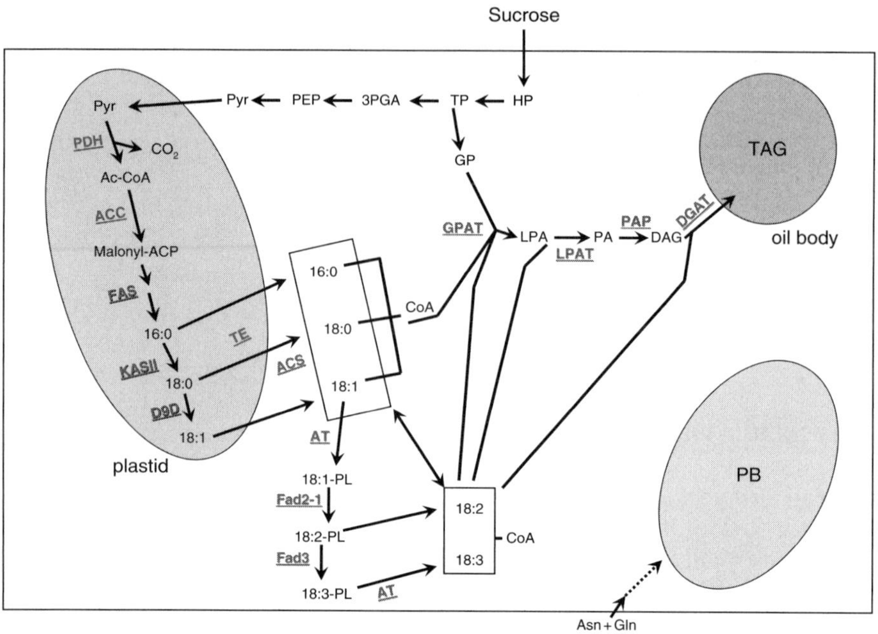

Fig. 12.3 Summary of some of the major steps of oil or triacylglycerol (TAG) biosynthesis in plant cells including developing soybean seeds. PB = protein body. The enzymes discussed are underlined and described in Table 12.2 (*See* also Color Insert)

pyruvate is converted to acetyl-CoA by pyruvate dehydrogenase. Pyruvate synthesis and/or transport to plastids might be a limiting step in fatty acid biosynthesis and accumulation in oil. Acetyl-CoA is a precursor to many molecules in plants and other organisms in multiple organelles. The first committed step of fatty acid biosynthesis is the conversion of acetyl-CoA into malonyl-CoA by acetyl-CoA carboxylase and then to malonyl-ACP by a transacylase. In most tissues of most eukaryotic organisms including plants, malonyl-ACP (malonyl-CoA in eukaryotes without plastids) is elongated in 8 cycles, two carbon units at a time, via the fatty acid synthase complex, to palmitoyl (16:0)-ACP (or -CoA). The fatty acid synthase complex involves four different enzymatic reactions with each cycle starting with a condensation, followed by a reduction, a dehydration and a second reduction (Ohlrogge and Jaworski, 1997). The condensation reactions are catalyzed by enzymes known as **3-ketoacyl-ACP synthases** or KASs. The 1st condensation reaction going from acetyl-CoA to 3-ketobutyrate is catalyzed by KAS III, the reaction from butyryl-ACP (C4) to palmitoyl-ACP (C16) by KAS I and from palmitoyl-ACP to stearoyl-ACP (C18) by KAS II. The reaction stops at C16 and C18 fatty acids not only by virtue of the specificity of the KAS enzymes but also by the action of thioesterases (TEs), which hydrolyze the acyl-S-ACP thioester bonds. Some plants such as coconuts have unusual TEs, known as medium chain TEs, which stop the reaction at C8, C10, C12 or C14 fatty acid chain lengths and these plants can accumulate medium chain fatty acids in their seed oil.

Oil is biosynthesized during the second stage of seed maturation (Harwood and Page, 1994) at which time the relevant biosynthetic enzymes are highly expressed. The major fatty acids of plants (and most other eukaryotic organisms) have a chain length of 16 or 18 carbons and contain from zero to three cis-double bonds. Five fatty acids (18:1, 18:2, 18:3, 16:0 and in some species 16:3) make up over 90% of acyl chains of structural glycerolipids of almost all plant membranes (Ohlrogge and Browse, 1995). The nature of the acyl composition of the TAG is dependent on the availability of the fatty acids from the acyl-CoA substrate pool, as well as the selectivity of the acyltransferases of the Kennedy pathway (Harwood, 1998) and possibly transacylases. These same five fatty acids are the main fatty acids present in soybean oil. The fatty acid composition of the oil of normal soybean cultivars (Fig. 12.4) is:

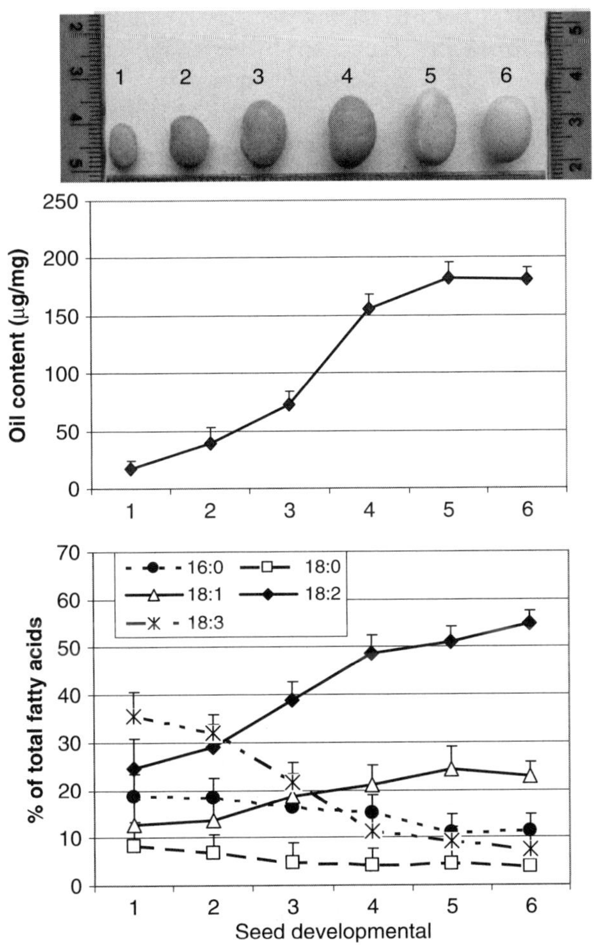

Fig. 12.4 Oil and fatty acid accumulation in soybean seed development (*See* also Color Insert)

Fatty acid	%
16:0	11
18:0	4
18:1	25
18:2	52
18:3	8

Reactions of fatty acid synthesis are terminated by hydrolysis or transfer of the acyl chains from the ACP by ACP-hydrolase or an acyltransferase consecutively. The 'competition' for substrate is thus a competition between the termination of synthesis, a function of thioesterase and transferase activity, and extension, a function of KASI and KASII isoforms (Voelker et al., 1992; Budziszewski et al., 1996). ACP-thioesterases are one of two main types (Klaus et al., 2004). One thioesterase is relatively specific for 18:1 ACP, encoded by Fat A, and a second more specific for saturated acyl-ACPs encoded by Fat B. FA molecules formed in the chloroplast stroma are released from ACPs by thioesterases and cross the membrane by an unknown mechanism. As the FA cross the membrane, they are converted to acyl-CoA esters through the activity of an acyl-CoA synthase (ACS) located on the outer membrane. Plants have multiple ACSs that participate in lipid metabolism (Schnurr et al., 2002; Shockey et al., 2002). ACS enzymes encoded by different genes have differential specificities for particular fatty acids (McKeon et al., 2006).

Fatty Acid Desaturases

In most plant tissues, over 75% of the fatty acids are unsaturated. Two types of desaturases have been identified, one soluble and the other membrane bound, that have different consensus motifs. Database searching for these motifs reveals that these enzymes belong to two distinct multifunctional classes, each of which includes desaturases, hydrolases and epoxygenases that act on FA or other substrates (Shanklin and Cahoon, 1998). Free FA are not thought to be desaturated in vivo, rather they are esterified to acyl carrier protein (ACP) for the soluble plastid desaturase or to coenzyme-A (CoA) or to phospholipids for integral membrane desaturases.

Δ-9 Desaturases

The first double bond in unsaturated FAs is introduced by the soluble enzyme stearoyl-ACP desaturase. This fatty acid desaturase is special to the plant kingdom in that only few other known desaturases are soluble. Soluble Δ-9 stearoyl-ACP desaturases are found in all plant cells and are essential for the biosynthesis of unsaturated membrane lipids (Shanklin and Somerville, 1991; Kaup et al., 2002; Cahoon et al., 1997; Cahoon et al., 1994). Desaturases that convert saturated fatty acid to mono-unsaturated FA share several common characteristics. They perform stereospecific Δ-9 desaturation of an 18:0/16:0 substrate with the removal of the 9-D and 10-D hydrogens (Bloomfield and Bloch, 1960; Mudd and Stumpf, 1961). Protein crystallographic studies on the purified desaturase from castor bean showed that it contains a diiron cluster (Fox et al., 1993). The protein is active as a homodimer and consists of a single domain of 11 helices. This diiron center is the active site of the desaturase (Lindqvist et al., 1996).

Expression and regulation of Δ-9 desaturase in plants have been studied extensively (Fawcett et al., 1994; Slocombe et al., 1994). The expression of the promoter of the *Brassica napus* stearoyl desaturase gene in tobacco was found to be temporally regulated in developing seed tissues. However, the promoter was also particularly active in other oleogenic tissues such as tapetum and pollen grains, raising the interesting question of whether seed expressed lipid synthesis genes are regulated by separate tissue specific determinants or by a single factor common to all oleogenic tissues (Slocombe et al., 1994). In *Saccharomyces cerevisiae*, addition of saturated fatty acids induced Δ-9 fatty acid desaturase mRNA (Ole1 mRNA) by 1.6-fold, whereas a large family of unsaturated fatty acids repress Ole1 transcription by 60-fold. A 111 bp fatty acid regulation region (FAR), approximately 580 bp upstream of the start codon, was identified that is essential for the transcription activation and unsaturated fatty acid repression (Quittnat et al., 2004). In addition to transcriptional regulation, unsaturated fatty acids mediate changes in the half-life of the Ole1 mRNA (Gonzalez and Martin, 1996).

Currently, industries that manufacture shortening, margarine, and confectionery products use considerable amounts of stearate (18:0) produced mainly from partially hydrogenated plant oils (Facciotti et al., 1999). Hydrogenation not only generates extra cost but also is a generator of significant amounts of trans-fatty acids that were associated with an elevated risk of heart disease (Facciotti et al., 1999; Katan et al., 1995; Nelson, 1998). Industries manufacturing shortenings and confectionery products could benefit from an oil crops capable of accumulating high levels of stearate. However, stearate (18:0) does not naturally accumulate to abundant levels in most cultivated oil crops including soybeans and efforts to produce a high-stearates phenotype through conventional breeding and mutagenesis techniques has had only modest success (Facciotti et al., 1999). Although, stearic acid (18:0) is one of the major saturated fatty acids in most seed oils, its percentages vary among the different oilseed crops from 1.0% in rape seed oil to 3.6% in sesame and corn seed oils and 4.0% in soybean oil with a range from 2.2–7.2% for the genotypes available in the world germplasm collection (Downey and McGregor, 1975; Hymowitz et al., 1972; Rahman et al., 1997; Yasumoto et al., 1993). The fatty acid composition of soybeans has been improved by using selective breeding techniques utilizing natural variants or induced mutagenesis (Graef et al., 1985a,b; Ladd and Knowles, 1970). Hammond and Fehr (1983) were able to increase the amount of stearate (C18:0) produced in the soybean oil to levels up to about 28.1% of the total fatty acid content using mutagenesis. Rahman et al. (2003) reported novel soybean germplasm with high stearic levels. This novel soybean was obtained as a consequence of the combination of the loci of high palmitic and stearic acids leading to alterations in other fatty acids. As a result, two lines (M25 and HPS) with a 5-fold increase in stearic acid (from 34 to 181 and 171 g kg^{-1}) were developed. This increase in stearic acid was also found to be associated with a change in oleic and linoleic acids content. Furthermore, these authors reported that when both palmitic and stearic acids were considered together in the oil of HPS, this line had a saturated fatty acid content of >380 g kg^{-1}. Thus, such oil might have the potential to increase the utility and also to improve the quality of soybean oil for specific purposes (Rahman et al., 2003).

Vegetable oils rich in monounsaturated fatty acids (MUFA) are not only important in human nutrition but also can be used as renewable sources of industrial chemicals (Cahoon et al., 1997). One particular output trait of current interest is the use of transgenic soybean plans to produce palmitoleic acid fatty acids that have either nutraceutical or pharmaceutical and even industrial properties. Macadamia nut is another source of palmitoleic acid. Its oil is unique in that monounsaturated fatty acids are the predominant component (about 80%) and a considerable portion (17–21%) of this is palmitoleic acid (a component not present in substantial amounts in olive oil; Curb et al., 2000). Grayburn and Hildebrand (1995) and Wang et al. (1996) reported large increases in palmitoleic acid (16:1 Δ-7) after expressing a mammalian or yeast Δ-9 deasturase gene in tobacco or tomato. Since soybeans are an important oil source that is high in linoleic and saturated fatty acids (mostly linoleic and palmitic acid; about 55% and 15%, respectively), conversion of all or part of these saturated fatty acids into palmitoleic acid would be a great benefit, not only for health, since converting much of the remaining PUFAs into palmitoleic acid could have industrial value. Liu et al. (1996) reported converting $\sim$ half of the palmitic acid of soybean somatic embryos into palmitoleic acid with good expression of a Δ9-CoA desaturase. The transformed embryos had 16:1 levels from 0% to over 10% of total fatty acids, while the levels of 16:0 dropped from 25% to approximately 5% of total fatty acids.

A number of studies demonstrated apparently beneficial effects of diets based on high monosaturated fatty acid content primarily derived from olive oil (Curb et al., 2000; Hegsted et al., 1993; Kris-Etherton et al., 1988; Spiller et al., 1992). The health implications of palmitoleic acid were first addressed by Yamori et al. (1986) and Abraham et al. (1989). Curb et al. (2000) compared the effects of a typical American diet (TAD) (diet high in saturated fat '37% energy from fat', the AHA (American Heart Association) 'step 1' diet '30% energy from fat' (half the SFA's, normal amounts of MUFA's and PUFA's, and high levels of carbohydrates), and a macadamia nut-based monounsaturated fat diet (MND) (37% energy from fat). When compared to the typical diet, step 1 and macadamia nut diets both had potentially beneficial effects on cholesterol and LDL cholesterol levels. These results are consistent with previously-reported lipid altering benefits of MUFA-rich diets particularly those involving macadamia nut oil (Ako and Okuda, 1995). Palmitoleic acid was also reported to protect rats from stroke (Yamori et al., 1986), apparently by increasing cell membrane fluidity, clearing lipids from the blood, and altering the activity of important cell membrane transport systems particularly through inhibition of the Na+, K(+)-ATPase activity within a narrow range (Swarts et al., 1990). In men and women, elevated blood/tissue levels of palmitoleic acid were found to be correlated with protection from ventricular arrhythmias (Abraham et al., 1989) and negatively correlated with markers of artherosclerosis (Theret et al., 1993). Palmitoleic acid was also found to reportedly inhibit mutagenesis in animals (Hayatsu et al., 1988) and was found to be negatively correlated with breast cancer incidence in women (Simonsen et al., 1998).

Many vegetable oils are partially hydrogenated to increase the stability of cooking oils and hydrogenated further for use as margarines and shortenings. The goal of plant geneticists has mainly to develop high stearate oils in order to reduce or eliminate the need for hydrogenation of vegetable oils used for margarines and shortenings. As described in the section above on high stearate oils, many groups were successful in achieving this goal with a variety of vegetable oils using different approaches, including genetic engineering soybeans to a 53% stearic acid content of oil (Knutzon et al., 1992; Kridl, 2002; Martinez-Force et al., 2002). With the advent of genetic engineering, several strategies for increasing stearic acid levels in oilseed crops have been possible and the increase in levels of stearic acid is usually at the expense of oleic (18:1) and linoleic (18:2) acids. Among other strategies, anti-sense suppression or co-suppression to reduce or knock out the activity of stearoyl-ACP desaturase, which is responsible for converting stearoyl-ACP (saturated) to oleoyl-ACP (unsaturated) (Budziszewski et al., 1996) was used routinely. Also, the stearoyl-ACP thioesterase is another possible metabolic target. Thus, up-regulation of this enzyme by sense-oriented reintroduction of the stearoyl-ACP thioesterase was found to increase free stearate release. Kridl (2002) reported transgenic soybeans with stearate levels of as high as approximately 53%, while levels of approximately 4% were observed in non-transformed control plants. This line is low in linoleic and linolenic acids and high in oleic acid in addition to stearic acid. It is important that these large increases in stearate are seed-specific and more so in triacylglycerol than in membrane lipids because high stearate in membranes can reduce membrane fluidity and result in relatively poor germination rates (Kaup et al., 2002; Voelker and Kinney, 2001; Wiberg et al., 2000).

Changes in palmitate levels of soybean oil has been another long time goal of soybean breeders and geneticists and development of genotypes with levels of <4 and >40% have been achieved (Stoltzfus et al., 2000). Because saturated fatty acids, especially mid-chain saturated fatty acids such as 16:0, are dietary health-risk factors, particularly cardiovascular health (www.americanheart.org), reduced palmitate has been the main goal. Alleles for altered palmitate in soybean oil are known as *fap* alleles. A low palmitate mutant, A22 with 6.8% 16:0, has a single recessive *fap3* allele (Schnebly et al., 1994). A mutant soybean line with elevated 16:0 containing the *fap2* allele had a single base-pair substitution in codon 299 of the *GmKASIIA* gene with TGG → TAG converting a tryptophan to a premature stop codon (Aghoram et al., 2006). *KASII* encodes the keto-acyl-ACP synthase that catalyzes the condensation reaction of the FAS complex involved in elongation of 16:0-ACP to 18:0-ACP (Fig. 12.3).

Δ-12 Desaturases

Plant Δ-12 desaturases are plastid membrane-bound or ER membrane-bound enzymes. Arabidopsis plastidal Δ-12 desaturases were isolated using degenerate oligonucleotides, based on amino acid sequences conserved between plant and cyanobacterial desaturases, to screen cDNA libraries (Falcone et al., 1994). The Arabidopsis Δ-12 desaturase was also used to screen rape and soybean cDNA libraries and the homologous sequences isolated (Falcone et al., 1994). These plant chloroplast Δ-12 desaturases all show a high degree of similarity (around 50%) with cyanobacterial Δ-12 desaturases, but less with cyanobacterial and plant ω-3 desaturases.

The Δ-12 desaturase is particularly active in microsomal preparations from developing seed cotyledons of some oilseed species where it is associated with the biosynthesis of triacylglycerols (Stymne and Stobart, 1986; Griffiths et al., 1988). The microsomal Δ-12 desaturase requires NAD(P)H as reductant and molecular oxygen and is inhibited by cyanide but not carbon monoxide, suggesting that cytochrome P450 is not involved in the electron transport chain (Griffiths et al., 1985).

More than 10 plant microsomal and a similar number of plastid Δ-12 cDNAs and genes were isolated and reported to date. Arabidopsis mutants lacking both microsomal Δ-12 (*fad*2) and ω-3 desaturases (*fad*3) were isolated (Browse et al., 1986, 1993). Mutants at the *Fad*2 locus of Arabidopsis that are deficient in the major and, perhaps, only Δ-12 desaturase of the eukaryotic pathway were isolated and characterized. It was shown that the Arabidopsis *fad*2 mutants had similar growth characteristics to wild type at 22 °C but at 12 °C, their growth was greatly impaired and, at 6 °C, the mutants died (Miquel et al., 1993). This experiment showed that Arabidopsis requires polyunsaturated fatty acids for low temperature survival (Tocher et al., 1998). Subsequently, (Okuley et al., 1994) isolated the entire Arabidopsis *fad*2 cDNA sequence with T-DNA tagged line with higher 18:1 content in seeds, roots and leaves than the wild-type line.

After screening soybean libraries with the Arabidopsis *fad*2 cDNA, two different Δ-12 desaturase cDNAs, *FAD*2-1 and *FAD*2-2 were isolated (Heppard et al., 1996). *FAD*2-1 was expressed in developing seeds, whereas *FAD*2-2 was expressed in several tissues (leaves, roots, and stems) in addition to developing seeds.

Comparison of available sequence information reveals that there is a high degree of similarity between the same class of membrane-bound desaturases in different plant species, but much less similarity between different classes of desaturases, even in the same species (Murphy and Piffanelli, 1998). Membrane-bound enzymes most likely contain similar di-iron complexes (Fox et al., 1993). The most strictly conserved feature is the presence of eight histidines in three separate clusters. These clusters are held in position by a different ligation sphere, which may involve the three histidine boxes (Shanklin et al., 1994), which are characteristic for this group of enzymes [HX3-4H, HX2-3HH, (H/Q)X2HH]. This motif was also found in the Δ-12 oleate hydroxylases from castor bean and *Lesquerella fendleri* (van de Loo et al., 1995; Broun et al., 1997), epoxygenase from *Vernonia galamensis* (Hitz, 1998), acetylenase and epoxygenase from *Crepis* spp. (Lee et al., 1998) and the Δ-6 linoleate desaturase from borage (Beremand et al., 1997).

Another major goal of plant breeding has been to develop oils with high oxidative stability without the need for hydrogenation that are liquid at room temperature. Oils high in 18:1 are one way to achieve this. Mutant alleles affecting oleate levels in soybean are given the *ol* designation. An oleate content of >70% has been achieved by conventional breeding/mutagenesis (Alt et al., 2005b). The high oleate mutant, M23, has a deletion in *FAD2-1a* (Alt et al. 2005a). Using sense-mediated PTGS (co-suppression) targeting the Δ 12-desaturase (that converts oleic acid to linoleic acid), Toni Kinney and colleagues at DuPont Co. (Heppard et al., 1996; Kinney, 1998a; Kinney, 1998b) succeeded in producing a soybean with an oxidatively stable oil with a total polyunsaturated content of less than 5% and oleic acid content of 85% by suppressing the *Fad* 2-1 gene, whereas normal soybeans have about 20% oleic acid (18:1). This increase in oleate levels was accompanied by reduced levels of 18:2 from 55% to less than 1% and saturated fatty acids down to 10% (Beisson et al., 2003; Heppard et al., 1996). An oleate content of >90% by seed-specific suppression of *FAD*2-1 was reported by Buhr et al. (2002).

Fatty acid desaturases in all organisms are subject to several different types of regulation, depending on their localization and function. Those desaturases involved in membrane lipid biosynthesis have important 'housekeeping' functions and are therefore constitutively regulated (Murphy and Piffanelli, 1998). A cold-inducible plastidial ω-3 desaturase gene was isolated from Arabidopsis (Gibson et al., 1994) and there are several other reports that are consistent with the presence of cold-inducible ω-3 and Δ-12 desaturase genes in soybeans (Kinney, 1994); (Rennie and Tanner, 1989). However, there are other reports of the isolation of Arabidopsis and soybean Δ-12 desaturase genes that are not regulated by low temperature (Okuley et al., 1994); (Heppard et al., 1996). Since multigene families encode many desaturases, it is possible that some plant species may have both cold-inducible and non-cold-inducible forms of the same class of desaturase enzyme and/or gene (Murphy and Piffanelli, 1998).

ω-3 Desaturases

The Δ-12 and ω-3 desaturases introduce the second and the third double bonds in the biosynthesis of 18:2 and 18:3 fatty acids (which are important constituents of plant membranes). In most species, the fatty acids present in the galactolipids of the chloroplast membrane are ~70–80% trienoic fatty acids. In leaf tissue, there are two distinct pathways for polyunsaturated fatty acid biosynthesis, one located in the microsomes and the other located in the plastid membranes. In non-green tissues and developing seeds, the microsomal pathway predominates. Cytosolic and plastid ω-3 desaturation that result in the production of triene fatty acids are controlled by the *FAD*3, *FAD*7 and *FAD*8 loci in Arabidopsis (Lemieux et al., 1990; Arondel et al., 1992; Yadav et al., 1993; Browse et al., 1986; Lemieux et al., 1990; McConn et al., 1994).

Microsomal ω-3 desaturases are responsible for the production of extraplastidal 18:3. This enzyme accounts for over 80% of the 18:3 in Arabidopsis root tissues.

Arabidopsis *FAD*3 mutants are characterized by reduced levels of 18:3 and correspondingly increased 18:2 levels. However studies with the Arabidopsis *FAD*3 mutants revealed that exchange of lipid between chloroplast and ER allows the chloroplast desaturase to provide highly unsaturated lipid to the extrachloroplast membranes of leaf cells (Browse et al., 1993). Changing 18:3 levels of soybean seed oil has long been a goal of plant breeders and a number of low 18:3 mutants have been generated. Soybean genotypes A5 and A23 have reduced linolenic acid contents when compared with current cultivators. Byrum et al. (1997) reported that the reduced linolenic acid concentration in A5 was at least partially the result of partial or full deletion of a microsomal ω-3 desaturase gene. Alleles for reduced 18:3 in soybeans are designated *fan* alleles with the allele for reduced linolenate in A5 controlled by the *fan1* allele (Byrum et al., 1997). The soybean genome has at least three *FAD*3 ω-3 desaturase genes designated *GmFAD3A, GmFAD3B* and *GmFAD3C* (Bilyeu et al., 2005). Combining mutations *GmFAD3A, GmFAD3B* and *GmFAD3C* into single soybean lines (e.g. A29) can result in linolenate levels ~1% (Anai et al., 2005; Bilyeu et al., 2006; Sarmiento et al., 1997). Experimental soybean lines with >50% 18:3 were reported by (Cahoon, 2003) by increased expression of a *FAD*3 gene in transgenic soybeans.

Membrane and TAG Synthesis

Both membrane and TAG synthesis begins with the acylation of *sn*-glycerol 3 phosphate producing lysophosphatidic acid, catalyzed by glycerol-3-phosphate acyltransferase (GPAT). A second acylation of lysophosphatidic acid catalyzed by lysophatidic acid acyl transferase (LPAT) produces phosphatidic acid (PA) (Fig. 12.3). The PA formed can be subsequently de-phosphorylated to diacylglycerol (DAG). The DAG then serves as a precursor for TAG. The third acylation step is catalyzed by DAG acyltransferase (DGAT). In oil seeds, phosphatidyl choline (PC) was identified as an intermediate in oil biosynthesis and plays a central role in the production of polyunsaturated fatty acids by serving as a substrate for Δ-6, Δ-9, Δ-12, and Δ-15 desaturases (Jackson et al., 1998). In order to induce large changes in oil composition, LPAT was considered an important target enzyme because of its selective discrimination ability (Śanchez et al., 1998). Rapeseed (*Brassica napus*) and meadowfoam (*Limnanthes*) have 60% and 90% erucic acid in their TAGs. In rapeseed erucic acid is excluded from the *sn*2 position, whereas in meadowfoam it is present in the *sn*2 position of TAGs. This difference was attributed to the substrate specificity of LPAT enzyme in both species (Cao et al., 1990). To alter the stereochemical composition of rapeseed oil, a cDNA encoding *Limnanthes* seed-specific LPAT was expressed in *Brassica napus* plants using a napin expression cassette. In the transgenic plants 22.3% erucic acid was present at the *sn*2 position leading to the production of trierucin. However, alteration of erucic acid at the *sn*2 position did not affect the total erucic acid content. It may be that the meadowfoam LPAT may not increase the erucic acid content of rapeseed (Lassner et al., 1995) because of the limited pool size of the 22:1 coenzyme A in the maturing embryos of *B. napus*. The metabolism of laurate was found to be different in transgenic *Brassica napus* lines (transformed with a California bay lauroyl-acyl carrier protein thiosesterase cDNA driven by napin promoter) and the natural laurate accumulators coconut, oil palm and *Cuphea wrightii*. When tested at the mid-stage of embryo development, the PC had up to 26 mol % of laurate in the transgenic rapeseed high laurate line, whereas other species it ranged between 1 and 4 mol %. The laurate in the *Brassica* TAG was nearly totally confined to the outer *sn*1 and *sn*3 positions whereas the laurate in coconut and *Cuphea* was highest in the *sn2* position. Very low amounts of laurate were found in the *sn*2 position in DAG and PC of the rapeseed lipids indicating no arrangement of laurate between the outer and *sn*2 positions occurred in any of the lipids. There was an enhanced activity of lauroyl-PC metabolizing enzymes in the laurate producing rapeseed when the embryos were fed with radiolabeled ^{14}C-lauroyl-PC and ^{14}C-palmitoyl-PC. The data indicated that DAG was preferentially utilized from natural laurate accumulators like oil palm, coconut and *Cuphea* (Wiberg et al., 1997). Transgenic rapeseed oil expressing California bay thioesterase produced 60% saturated FA with laurate alone comprising 48%. In these plants laurate was presented at *sn*1 and *sn*3 positions only. When these plants were crossed with transgenic lines expressing coconut LPAT in the resulting hybrids, laurate was present at the *sn*2 position along with *sn*1 and *sn*3 positions. An overall increase in the oil content was also observed.

When the yeast LPAT genes SLC1 and SLC1-1 (mutant form of yeast LPAT) were expressed in *Brassica napus* and Arabidopsis under the CaMV35S promoter both TAG and VLCFA contents were increased by 56% and 80% (Zou et al., 1997). In the transgenic plants seed weight increased indicating at least a partial contribution from enhanced oil content. In the total oil content 60–75% consisted of VLCFAs and 40% that of non-VLCFAs such as palmitate, oleate, linoleate and linolenate. No increase in total oil content was reported in coconut or meadowfoam LPAT transformed rapeseed. This could be due to different regulatory properties of the plant and yeast LPAT enzymes. The plant LPAT genes have 62% amino acid identity among themselves, whereas the yeast gene had 24% homology. In transgenic plants, the high expression of the SLC1-1 gene did not correlate with high oil content indicating that even small levels of expression were sufficient to overcome the PA limitations during TAG biosynthesis. Although SLC1-1 levels were stronger in leaves than in seeds, no significant changes were observed in the fatty acid composition in leaves indicating the pools of available LPA and/or acyl-CoAs may be more tightly regulated in leaves (source) than in seeds (sink). We have preliminary results indicating a 1–2% increase in oil content of soybeans expressing the yeast SLC1 gene.

A DGAT was purified to apparent homogeneity from lipid body fractions of an oleaginous fungi, *Mortierella ramanniana* (Kamisaka et al., 1997). The purified DGAT utilized a broad range of molecular species of both DAG and acyl-coenzyme-A as substrates (Gavilano et al., 2006) and higher plants (Vogel and Browse, 1996). The first plant DGAT was cloned recently from Arabidopsis (Hobbs et al., 1999). The amino acid sequence shared 38% identity and 59% similarity with the mouse DGAT. Analysis revealed 9 membrane spanning helices and also 14 kD hydrophilic domain at the N-terminus. It had no significant sequence homology with plant GPAT and LPAT genes. Studies on expression of the homolog in *Brassica napus*, showed that the DGAT mRNA was present in the highest concentrations in developing embryos, petals of flowers, and developing flower buds but in very low amounts in leaf and stem tissues.

Another reaction that appears to be involved in TAG accumulation is the reversible conversion of PC into DAG in presence of CDP choline transferase. Slack et al. (1985) gave indirect evidence for the reversibility of PC by labeling studies in vivo with linseed cotyledons and in vitro with safflower cotyledons. When sunflower microsomes were incubated with radiolabeled PC, the radioactivity was progressively incorporated into DAG. When the concentration of the microsomal protein was increased, the activity also increased indicating the reversible reaction of choline transferase in sunflower (Triki et al., 1998). A soybean cDNA encoding an aminoalcoholphosphotransferase (AAPTase) that demonstrates high levels of CDP-choline:*sn*-1,2-diacylglycerol cholinephosphotransferase activity was isolated by Dewey et al. (1994) by complementation of a yeast strain deficient in this function. AAPTases utilize diacylglycerols and cytidine diphosphate (CDP)-aminoalcohols as substrates in the synthesis of the main membrane lipids phosphatidylcholine and phosphatidylethanolamine and can possibly affect DAG pools for TAG synthesis.

Acyl-CoA: diacylglycerol (DAG) acyltransferase (DGAT; EC 2.3.1.20) activity has long been detected in various animal and plant tissues active in TAG synthesis. DGAT catalyzes the reaction:

$$\text{Acyl-CoA} + \text{DAG} \xrightarrow{\textbf{DGAT}} \text{TAG} + \text{CoASH}$$

As expected this enzyme is membrane bound or associated and difficult to work with biochemically. As such, the first DGAT gene was not cloned until 1998. Cases et al. (1998) reported the cloning and functional expression of a DGAT from mice. Hobbs et al. (1999) and Zou et al. (1999) reported the cloning of a DGAT from Arabidopsis. Lardizabal et al. (2001) reported the cloning of a second class of DGAT, DGAT2, from the oleagineous fungus *Mortierella ramanniana*, which had no homology to the earlier identified DGAT sequences now known as DGAT1s. Cases et al. (2001) also cloned a mammalian DGAT2 and its now known that humans have seven DGAT2s (Turkish et al., 2005). Only a single DGAT1 gene (At2g19450) and a single DGAT2 gene (At3g51520) are present in the Arabidopsis genome (Beisson et al., 2003; Mhaske et al., 2005). Soybeans have at least two DGAT1s (see below).

A second mechanism for biosynthesis of TAG in yeast and plants was discovered and reported in 2000 (Dahlqvist et al., 2000; Oelkers et al., 2000) that has homology to lecithin cholesterol acyltransferases (LCATs). This is catalyzed by an enzyme known as phospholipid: diacylglycerol acyltransferase (PDAT, EC 2.3.1.158) that transfers an acyl group (fatty acid) from a phospholipid (PL) to DAG forming TAG and a lysophospholipid (LPL):

$$\text{PL} + \text{DAG} \xrightarrow{\textbf{PDAT}} \text{TAG} + \text{LPL}$$

Arabidopsis has six PDAT/LCAT homologs (Stahl et al., 2004) of which (At5g 13640) is most closely related to the PDAT identified in yeast. Stahl et al. (2004)

demonstrated that this gene is expressed widely in different Arabidopsis tissues and has PDAT activity. In humans, most all TAG is synthesized by DGAT1 and DGAT2 and the only human gene similar to PDAT has phospholipaseA_2 and phospholipid:ceramide transacylase activities (Hiraoka et al., 2002). Mhaske et al. (2005) generated a knockout for At5g13640 and their studies plus those of Stahl et al. (2004) rule out a role for this gene in TAG synthesis in Arabidopsis seeds. A second Arabidopsis PDAT/LCAT homolog most related to t5g13640 (57% identical) is At5g44830. This PDAT/LCAT-like gene was found mainly expressed in developing seeds (Stahl et al., 2004). The authors speculated that it might have a role in seed oil biosynthesis. However this role has not been directly addressed nor has its activity been assessed.

A third TAG biosynthetic activity involving a DAG/DAG transacylase (DGTA) was reported in animals (Lehner and Kuksis, 1993) and plants (Stobart et al., 1997), including Arabidopsis (Stahl et al., 2004). DGTA catalyzes the reaction:

$$\text{DAG} + \text{DAG} \overset{\text{DGTA}}{\longrightarrow} \text{TAG} + \text{MAG (monoacylglycerol)}$$

To date, no DGTA enzyme was biochemically characterized nor the corresponding gene cloned.

A number of mutants with reduced seed oil contents were reported in Arabidopsis and found to be due to defects in DGAT1 (Focks and Benning, 1998; Katavic et al., 1995; Lu and Hills, 2002; Routaboul et al., 1999; Zou et al., 1999) or were impaired in transfer of carbon from sucrose and glucose to TAG, possibly due to impaired hexokinase and pyrophosphate-dependent phosphofructokinase (Focks and Benning, 1998). Arabidopsis DGAT1 mutants have a $\sim$25–50% reduction in seed oil content. DGAT1 is reported to be maximally expressed in developing seeds at a stage of high oil synthesis (Lu et al., 2003). Silencing of DGAT1 in tobacco was also reported to reduce seed oil content (Zhang et al., 2005). Our preliminary data (see below) indicates a role for DGAT1(s) in soybean oil synthesis but this has not been directly addressed. The role of DGAT2 in oil accumulation in Arabidopsis and common oilseeds such as soybeans has not been investigated.

Several reports indicated a role for DGAT in oil accumulation in developing soybean seeds (Kwanyuen and Wilson, 1986; Kwanyuen and Wilson, 1990; Kwanyuen et al., 1988; Settlage et al., 1998). No soybean mutants with large changes in oil levels or defects in DGAT have been reported. It is not yet clear what roles DGAT1, DGAT2 or possible other DGAT play in soybean oil biosynthesis. We detected transcripts for DGAT1, DGAT2 and PDAT in soybean tissues including developing seeds (our unpublished results). Developing soybean seeds accumulate TAG after most cell division has ceased and cotyledons have been formed and cell expansion initiated (Fig. 12.4) (Dahmer et al., 1991). Like most green tissues, linolenate (18:3) is the most abundant fatty acid of soybean oil early in seed development. The 18:3 levels of soybean oil continues to decline throughout seed development with linoleate (18:2) and oleate (18:1) becoming the predominate fatty acids of soybean oil as seeds mature (Dahmer et al., 1991) (Fig. 12.4). DGAT levels correlate with oil accumulation.

Oilseeds including soybeans accumulate TAG in special organelles known as oil bodies. There is strong evidence that oil bodies form with the accumulation of TAG inside the phospholipid bilayer in specialized regions of the ER ballooning out from the accumulating TAG and the remaining phospholipid forming a monolayer surrounding the growing lipid body. Concurrent with this, the oil body-specific protein, oleosin, is co-translationally inserted into the phospholipid monolayer of the oil bodies (Kalinski et al., 1991; Loer and Herman, 1993; Sarmiento et al., 1997; Siloto et al., 2006; Tzen et al., 1990).

Two full-length DGAT1s were cloned from developing soybean cDNA, designated GmDGAT1a (GenBank # AB257589) and GmDGAT1b (GenBank # AB257590). Soybean DGAT1a looks to be the same as GenBank entry # AY496439 (submitted 08 Dec. 2003) from the Institute of Genetics and Developmental Biology, Beijing, China (Wang et al., 2006). They have 99% identity with only two amino acid differences with our clone having a glycine, instead of aspartate, and a histidine, instead of glutamine, both toward the amino terminus and underlined below. This may be due to allelic differences in the genotypes used (GmDGAT1a cDNA is from the Group II cultivar 'Jack' and Wang et al. used cv. '8904'). GmDGAT1b does not match anything previously reported. A comparison of these with another partial soybean DGAT1 reported in GenBank and the Arabidopsis DGAT1 (CLUSTAL W, 1.83) is as follows:

```
soyDGAT1a       -------MAISDEPETVATALNHSSLRRRP---TAAGLFNSPETTTDSSGDDLAKDSGSD
AY496439        -------MAISDEPETVATALNHSSLRRRP---TAAGLFNSPETTTDSSGDDLAKDSGSD
soyDGAT1b       -------MAISDEPESVATALNHSSLRRRPSATSTAGLFNSPETTTDSSGDDLAKDSGSD
AY652765        ------------------------------------------------------------
AtDGAT1         MAILDSAGVTTVTENGGGEFVDLDRLRRRKSRSDSSNGLLLSGSDNNSPSDDVGAPADVR

soyDGAT1a       DSISSDAANSQPQQ---KQDTDFS-----------------VLKFAYRPSVPAHRKVKES
AY496439        DSISSDAANSQPQQ---KQDTDFS-----------------VLKFAYRPSVPAHRKVKES
soyDGAT1b       DSINSDDAAVNSQQQNEKQDTDFS-----------------VLKFAYRPSVPAHRKVKES
AY652765        ------------------------------------------------------------
AtDGAT1         DRIDSVVNDDAQGTANLAGDNNGGGDNNGGGRGGGEGRGNADATFTYRPSVPAHRRARES


soyDGAT1a       PLSSDTIFRQSHAGLFNLCIVVLVAVNSRLIIENLMKYGWLIKSGFWFSSKSLRDWPLFM
AY496439        PLSSDTIFRQSHAGLFNLCIVVLVAVNSRLIIENLMKYGWLIKSGFWFSSKSLRDWPLFM
soyDGAT1b       PLSSDTIFRQSHAGLFNLCIVVLVAVNSRLIIENLMKYGWLIKSGFWFSSKSLRDWPLFM
AY652765        --SSDAIFKQSHAGLFNLCIVVLVAVNSQLIIENLMKYGWSIKYGFWFSSKSLRDWPLFM
AtDGAT1         PLSSDAIFKQSHAGLFNLCVVVLIAVNSRLIIENLMKYGWLIRTDFWFSSRSLRDWPLFM
                  ***:**:**********:***:****:*********** *: .*****:*********

soyDGAT1a       CCLSLVVFPFAAFIVEKLAQQKCIPEPVVVVLHIIITSASLFYPVLVILRCDSAFLSGVT
AY496439        CCLSLVVFPFAAFIVEKLAQQKCIPEPVVVVLHIIITSASLFYPVLVILRCDSAFLSGVT
soyDGAT1b       CCLSLVVFPFAAFIVEKLAQRKCIPEPVVVVLHIIITSTSLFYPVLVILRCDSAFVSGVT
AY652765        CCLSLAIFPLAAFVVERLAQQKCISEPVVVLLHLIISTVELCYPVLVILRCDSAFVSGVT
AtDGAT1         CCISLSIFPLAAFTVEKLVLQKYISEPVVIFLHIIITMTEVLYPVYVTLRCDSAFLSGVT
                **:** :**:*** **:*. :* *.****:.**:**: ..: *** * *******:****

soyDGAT1a       LMLFACVVWLKLVSYAHTNYDMRALTKSVEKGEALPDTLNMDYPYNVSFKSLAYFLVAPT
AY496439        LMLFACVVWLKLVSYAHTNYDMRALTKSVEKGEALPDTLNMDYPYNVSFKSLAYFLVAPT
soyDGAT1b       LMLFSCVVWLKLVSYAHTNYDMRALTKLVEKGEALLDTLNMDYPYNVSFKSLAYFLVAPT
AY652765        LMLLTCIVWLKLVSYAHTNYDMRALTVSNEKGETLPNTLIMEYPYTVTFRSLAYFMVAPT
AtDGAT1         LMLLTCIVWLKLVSYAHTSYDIRSLAN-------AADKANPEVSYYVSLKSLAYFMVAPT
                ***::*:***********.**:*:*:          :.   : .* *:::*****:****
```

```
soyDGAT1a       LCYQPSYPRTPYIRKGWLFRQLVKLIIFTGVMGFIIEQYINPIVQNSQHPLKGNLLYAIE
AY496439        LCYQPSYPRTPYIRKGWLFRQLVKLIIFTGVMGFIIEQYINPIVQNSQHPLKGNLLYAIE
soyDGAT1b       LCYQPSYPRTPYIRKGWLFRQLVKLIIFTGVMGFIIEQYINPIVQNSQHPLKGNLLYATE
AY652765        LCYQTSYPRTPSVRKGWVFRQLVKLIIFTGVMGFIIEQYMNPIVRNSTHPLKGNLLYAIE
AtDGAT1         LCYQPSYPRSACIRKGWVARQFAKLVIFTGFMGFIIEQYINPIVRNSKHPLKGDLLYAIE
                ****.****:. :****: **:.**:****.********:****:** *****:**** *

soyDGAT1a       RVLKLSVPNLYVWLCMFYCFFHLWLNILAELLRFGGREFYQDWWNAKTVEDYWRMWNMPV
AY496439        RVLKLSVPNLYVWLCMFYCFFHLWLNILAELLRFGDREFYQDWWNAKTVEDYWRMWNMPV
soyDGAT1b       RVLKLSVPNLYVWLCMFYCFFHLWLNILAELLRFGDREFYKDWWNAKTVEDYWRMWNMPV
AY652765        RILKLSVPNVYVWLCMFYCFFHLWLNILAELVRFGDREFYKDWWNAKTVEEYWRMWNMPV
AtDGAT1         RVLKLSVPNLYVWLCMFYCFFHLWLNILAELLCFGDREFYKDWWNAKSVGDYWRMWNMPV
                *:*******:*********************: **.****:******:* :*********

soyDGAT1a       HKWMIRHLYFPCLRHGIPKAVALLIAFLVSALFHELCIAVPCHIFKLWAFGGIMFQVPLV
AY496439        HKWMIRHLYFPCLRHGIPKAVALLIAFLVSALFHELCIAVPCHIFKLWAFGGIMFQVPLV
soyDGAT1b       HKWMIRHLYYPCLRHGLPKAAALLIAFLVSALFHELCIAVPCHIFKLWAFGGIMFQVPLV
AY652765        HKWMVRHIYSPCLRRGIPKGAASLIAFLVSAVFHELCIAVPCHMFKLWAFIGIMFQVPLV
AtDGAT1         HKWMVRHIYFPCLRSKIPKTLAIIIAFLVSAVFHELCIAVPCRLFKLWAFLGIMFQVPLV
                ****:**:* ****  :**  * :*******:**********::****** *********

soyDGAT1a       FITNYLQNKFRNSMVGNMIFWFIFSILGHPMCVLLYYHDLMNRKGKLD
AY496439        FITNYLQNKFRNSMVGNMIFWFIFSILGQPMCVLLYYHDLMNRKGKLD
soyDGAT1b       LITNYLQNKFRNSMVGNMIFWFIFSILGQPMCVLLYYHDLMNRKGKLD
AY652765        LITNYLQNKYRNSMVGNMIFWFIFCILGQPMSVLLYYHDLMNRKGEVD
AtDGAT1         FITNYLQERFG-STVGNMIFWFIFCIFGQPMCVLLYYHDLMNRKGSMS
                :******:::  * **********.*:*:**.*************.:.
```

AA sequence comparison of soyDGAT1a and soyDGAT1b

```
soyDGAT1a       MAISDEPETVATALNHSSLRRRP---TAAGLFNSPETTTDSSGDDLAKDSGSDDSISSDA
soyDGAT1b       MAISDEPESVATALNHSSLRRRPSATSTAGLFNSPETTTDSSGDDLAKDSGSDDSINSDD
                        *                 **                            *  *

soyDGAT1a       ANSQPQQ---KQDTDFSVLKFAYRPSVPAHRKVKESPLSSDTIFRQSHAGLFNLCIVVLV
soyDGAT1b       AAVNSQQQNEKQDTDFSVLKFAYRPSVPAHRKVKESPLSSDTIFRQSHAGLFNLCIVVLV

                 ****
soyDGAT1a       AVNSRLIIENLMKYGWLIKSGFWFSSKSLRDWPLFMCCLSLVVFPFAAFIVEKLAQQKCI
soyDGAT1b       AVNSRLIIENLMKYGWLIKSGFWFSSKSLRDWPLFMCCLSLVVFPFAAFIVEKLAQRKCI
                                                                        *

soyDGAT1a       PEPVVVVLHIIITSASLFYPVLVILRCDSAFLSGVTLMLFACVVWLKLVSYAHTNYDMRA
soyDGAT1b       PEPVVVVLHIIITSTSLFYPVLVILRCDSAFVSGVTLMLFSCVVWLKLVSYAHTNYDMRA
                              *                *        *

soyDGAT1a       LTKSVEKGEALPDTLNMDYPYNVSFKSLAYFLVAPTLCYQPSYPRTPYIRKGWLFRQLVK
soyDGAT1b       LTKLVEKGEALLDTLNMDYPYNVSFKSLAYFLVAPTLCYQPSYPRTPYIRKGWLFRQLVK
                   *       *

soyDGAT1a       LIIFTGVMGFIIEQYINPIVQNSQHPLKGNLLYAIERVLKLSVPNLYVWLCMFYCFFHLW
soyDGAT1b       LIIFTGVMGFIIEQYINPIVQNSQHPLKGNLLYATERVLKLSVPNLYVWLCMFYCFFHLW
                                                  *

soyDGAT1a       LNILAELLRFGGREFYQDWWNAKTVEDYWRMWNMPVHKWMIRHLYFPCLRHGIPKAVALL
soyDGAT1b       LNILAELLRFGDREFYKDWWNAKTVEDYWRMWNMPVHKWMIRHLYYPCLRHGLPKAAALL
                           *    *                            *      *   *

soyDGAT1a       IAFLVSALFHELCIAVPCHIFKLWAFGGIMFQVPLVFITNYLQNKFRNSMVGNMIFWFIF
soyDGAT1b       IAFLVSALFHELCIAVPCHIFKLWAFGGIMFQVPLVLITNYLQNKFRNSMVGNMIFWFIF
                                                    *

soyDGAT1a       SILGHPMCVLLYYHDLMNRKGKLD
soyDGAT1b       SILGQPMCVLLYYHDLMNRKGKLD
                    *
```

Alignment of soyDGAT1a and soyDGAT1b sequences by using CLUSTALW.

Different residues in two sequences are highlighted with bold letters and indicated with a star (22 aa different; ~ 4%)

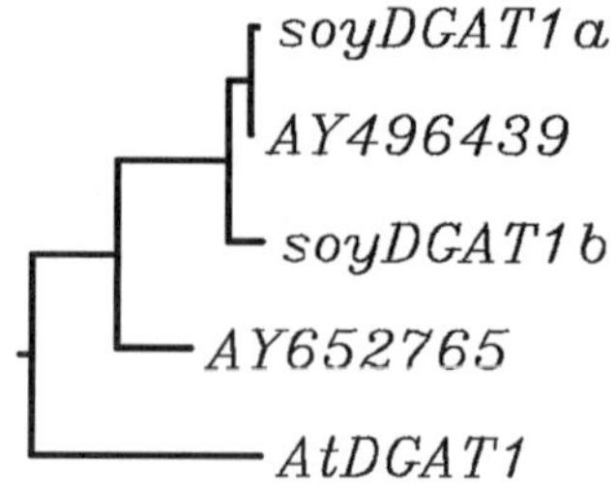

Fig. 12.5 Phylogenetic tree of five DGAT1 cDNA sequences using the CLUSTAL W program (version 1.82) Soy *DGAT1a* (1888 bp) (AB257589); *Soybean DGAT1b* (1960 bp) (AB257590) AY 496439 (accession #): soybean *DGAT1* full sequence (1880 bp) AY 652765 (accession #): soybean DGAT1 partial sequence (1413 bp) At DGAT1: Arabidopsis DGAT1 (1988 bp)

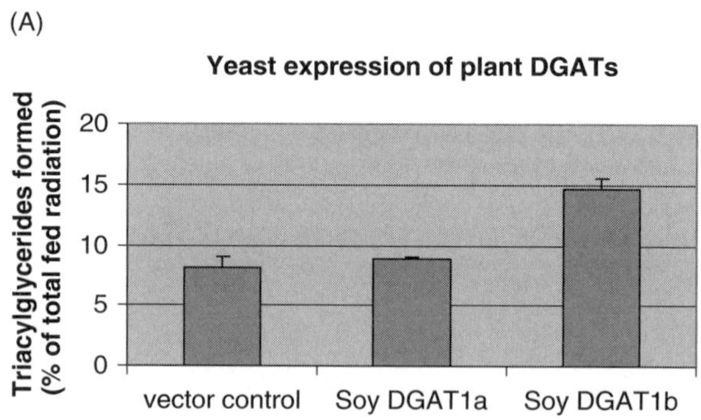

Fig. 12.6A Separately prepared yeast microsomes (50 μg determined by modified Lowry method) fed with 5 μM 18:2-CoA and 200 μM dioleoyl-DAG and incubated at 30 °C for 30 min

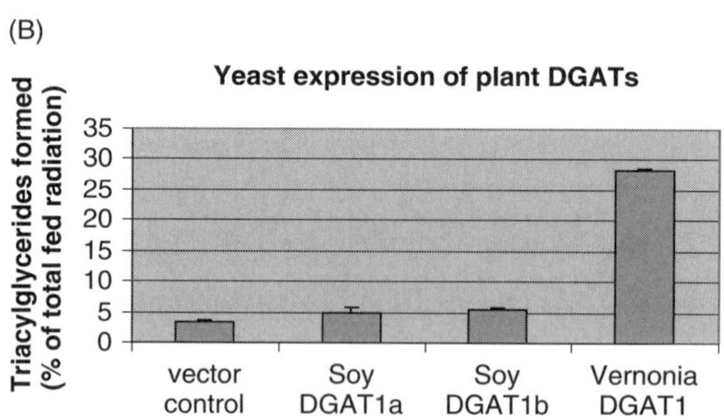

Fig. 12.6B Yeast microsomes assayed as for Fig. 12.6A. The Vernonia galamensis DGAT1 is included for comparison to a high oil accumulator

Interestingly GmDGAT1a and GmDGAT1b show somewhat different expression patterns although both show maximum transcript levels at stages of high TAG biosynthesis. The activity of these soybean DGAT1 cDNAs was analyzed in a yeast expression system. As illustrated in the following figure, we see little increase in DGAT activity with GmDGAT1a compared to the vector control and much greater activity with GmDGAT1b (Fig. 12.6A). Interestingly, we see much higher TAG biosynthetic activity with a DGAT from a much higher oil accumulating plant, *Vernonia galamensis*, than DGAT1s from soybeans (Fig. 12.6B).

The genomic sequence of soybean DGAT1a was recently reported (Wang et al., 2006) and we sequenced the full genomic sequence of DGAT1b. DGAT1a is 7575 bp and DGAT1b is 8164 bp. Both have 14 introns and 15 exons. The 2nd, 6th and 13th introns have the same length with small to large length differences seen with the other introns with most being longer in DGAT1b. Exons 1, 2, 3, 5 and 10 also show length differences between DGAT1a and DGAT1b.

Acknowledgments We thank Keshun Yu for the yeast activity data.

References

Abraham, R., R. Riemersma, D. Wood, R. Elton, and M. Oliver. 1989. Adipose fatty acid composition and the risk of serious arrhythmias in acute myocardial infarction. Am J. Cardiol 63:269–72.

Aghoram, K., R.F. Wilson, J.W. Burton, and R.E. Dewey. 2006. A Mutation in a 3-Keto-Acyl-ACP Synthase II Gene is Associated with Elevated Palmitic Acid Levels in Soybean Seeds. Crop Sci 46:2453–2459.

Ako, H., and D. Okuda. 1995. Healthful new oil from macadamia nuts. Nutr. 11:286–8.

Alt, J.L., W.R. Fehr, G.A. Welke, and D. Sandhu. 2005a. Phenotypic and molecular analysis of oleate content in the mutant soybean line M23. Crop-science; 2005 Sept–Oct; 45(5): 1997–2000 45:1997–2000.

Alt, J.L., W.R. Fehr, G.A. Welke, and J.G. Shannon. 2005b. Transgressive Segregation for Oleate Content in Three Soybean Populations. Crop Sci 45:2005–2007.

Anai, T., T. Yamada, T. Kinoshita, S.M. Rahman, and Y. Takagi. 2005. Identification of corresponding genes for three low-[alpha]-linolenic acid mutants and elucidation of their contribution to fatty acid biosynthesis in soybean seed. Plant Science 168:1615–1623.

Arondel, V., B. Lemieux, I. Hwang, S. Gibson, H. Goodman, and C. Somerville. 1992. Map-based cloning of a gene controlling omega-3 fatty acid desaturation in Arabidopsis. Science 258:1353–1355.

Beisson, F., A.J. Koo, S. Ruuska, J. Schwender, M. Pollard, J.J. Thelen, T. Paddock, J.J. Salas, L. Savage, A. Milcamps, V.B. Mhaske, Y. Cho, and J.B. Ohlrogge. 2003. Arabidopsis genes involved in acyl lipid metabolism. A 2003 census of the candidates, a study of the distribution of expressed sequence tags in organs, and a web-based database. Plant Physiol 132:681–97.

Beremand, P.D., A.N. Nunberg, A.S. AS Reddy, and T. Thomas. 1997. p. 351–353, *In* J. P. Williams, et al., eds. Physiology, Biochemistry and Molecular Biology of Plant Lipids. Kluwer Academic Publishers, Dordrecht, Boston, London.

Bilyeu, K., L. Palavalli, D. Sleper, and P. Beuselinck. 2005. Mutations in Soybean Microsomal Omega-3 Fatty Acid Desaturase Genes Reduce Linolenic Acid Concentration in Soybean Seeds. Crop Sci 45:1830–1836.

Bilyeu, K., L. Palavalli, D.A. Sleper, and P. Beuselinck. 2006. Molecular Genetic Resources for Development of 1% Linolenic Acid Soybeans. Crop Sci 46:1913–1918.

Bloomfield, D.K., and K. Bloch. 1960. Formation of D9-unsaturated fatty acids. J Biol Chem 235:337–345.

Broun, P., N. Hauker, and C.R. Somerville. 1997. p. 342–344, *In* J. P. Williams, et al., eds. Physiology, Biochemistry and Molecular Biology of Plant Lipids. Kluwer, Dordrecht, Boston, London.

Browse, J., N. Warwick, C. Somerville, and C. Slack. 1986. Fluxes through the prokaryotic and eukaryotic pathways of lipid synthesis in the '16:3' plant Arabidopsis thaliana. Biochem J 235:25–31.

Browse, J., M. McConn, D.J. James, and M. Miquel. 1993. Mutants of Arabidopsis deficient in the synthesis of alpha-linolenate. Biochemical and genetic characterization of the endoplasmic reticulum linoleoyl desaturase. J Biol Chem 268:16345–16351.

Budziszewski, G., K.P.C. Croft, and D.F. Hildebrand. 1996. Uses of biotechnology in modifying plant lipids. Lipids 31:557–569.

Buhr, T., S. Sato, F. Ebrahim, A. Xing, Y. Zhou, M. Mathiesen, B. Schweiger, A. Kinney, P. Staswick, and T. Clemente. 2002. Ribozyme termination of RNA transcripts down-regulate seed fatty acid genes in transgenic soybean. Plant Journal 30:155–163.

Byrum, J.R., A.J. Kinney, K.L. Stecca, D.J. Grace, and B.W. Diers. 1997. Alteration of the omega-3 fatty acid desaturase gene is associated with reduced linolenic acid in the A5 soybean genotype. Theor Appl Genet 94:356–359.

Cahoon, E., C. Becker, J. Shanklin, and J. Ohlrogge. 1994. cDNAs for isoforms of the delta 9-stearoyl-acyl carrier protein desaturase from Thunbergia alata endosperm. Plant Physiol 106:807–808.

Cahoon, E., Y. Lindqvist, G. Schneider, and J. Shanklin. 1997. Redesign of soluble fatty acid desaturases from plants for altered substrate specificity and double bond position. Proc Natl Acad Sci U S A 94:4872–4877.

Cahoon, E.B. 2003. Genetic enhancement of soybean oil for industrial uses: Prospects and challenges. AgBioForm 6:11–13.

Cao, Y., K.C. Oo, and A.H.C. Huang. 1990. Lysophosphatidate acyltransferase in the microsomal fraction from maturing seeds of meadowfoam (*Limnanthes alba*). Plant Physiol 9:1199–1206.

Cases, S., S.J. Stone, P. Zhou, E. Yen, B. Tow, K.D. Lardizabal, T. Voelker, and R.V. Farese, Jr. 2001. Cloning of DGAT2, a second mammalian diacylglycerol acyltransferase, and related family members. J Biol Chem 276:38870–6.

Cases, S., S.J. Smith, Y.W. Zheng, H.M. Myers, S.R. Lear, E. Sande, S. Novak, C. Collins, C.B. Welch, and A.J. Lusis. 1998. Identification of a gene encoding an acyl CoA:diacylglycerol acyltransferase, a key enzyme in triacylglycerol synthesis. Proc Natl Acad Sci U S A 95: 13018–13023.

Curb, J., G. Wergowske, J. Dobbs, R. Abbott, and B. Huang. 2000. Serum lipid effects of a high-monounsaturated fat diet based on macadamia nuts. Arch. Intern. Med. 160:1154–1158.

Dahlqvist, A., U. Stahl, M. Lenman, A. Banas, M. Lee, L. Sandager, H. Ronne, and S. Stymne. 2000. Phospholipid:diacylglycerol acyltransferase: an enzyme that catalyzes the acyl-CoA-independent formation of triacylglycerol in yeast and plants. Proc Natl Acad Sci U S A 97:6487–92.

Dahmer, M.L., G.B. Collins, and D.F. Hildebrand. 1991. Lipid concentration and composition of soybean zygotic embryos maturing in vitro and in planta. Crop Sci. 31:735–740.

Dewey, R.E., R.F. Wilson, W.P. Novitzky, and J.H. Goode. 1994. The AAPT1 Gene of Soybean Complements a Cholinephosphotransferase-Deficient Mutant of Yeast. Plant Cell 6: 1495–1507.

Downey, R.K., and D.I. McGregor. 1975. Breeding for modified fatty acid composition. Curr. Adv. Plant Sci. 12:151–167.

Facciotti, M.T., P.B. Bertain, and L. Yuan. 1999. Improved stearate phenotype in transgenic canola expressing a modified acyl-acyl carrier protein thioesterase. Nature Biotechnology 17: 593–97.

Falcone, D., S. Gibson, B. Lemieux, and C. Somerville. 1994. Identification of a gene that complements an Arabidopsis mutant deficient in chloroplast omega 6 desaturase activity. Plant Physiol 106:1453–1459.

Fawcett, T., W.J. Simon, R. Swinhoe, J. Shanklin, I. Nishida, W.W. Christie, and A.R. Slabas. 1994. Expression of mRNA and steady-state levels of protein isoforms of enoyl-ACP reductase from *Brassica napus*. Plant Molec Biol 26:155–163.

Focks, N., and C. Benning. 1998. wrinkled1: A Novel, Low-Seed-Oil Mutant of Arabidopsis with a Deficiency in the Seed-Specific Regulation of Carbohydrate Metabolism. Plant Physiol. 118:91–101.

Fox, B., J. Shanklin, C. Somerville, and E. Munck. 1993. Stearoyl-acyl carrier protein delta 9 desaturase from Ricinus communis is a diiron-oxo protein. Proc Natl Acad Sci USA 90: 2486–2490.

J. Sánchez, et al. (ed.) 1998. Advances in Plant Lipid Research, Sevilla, Spain. Universidad De Sevilla. Secretariado de Publicaciones.

Gavilano, L.B., N.P. Coleman, L.E. Burnley, M.L. Bowman, N.E. Kalengamaliro, A. Hayes, L. Bush, and B. Siminszky. 2006. Genetic Engineering of *Nicotiana tabacum* for Reduced Nornicotine Content. J. Agric. Food Chem. 54:9071–9078.

Gibson, S., V. Arondel, K. Iba, and C. Somerville. 1994. Cloning of a temperature-regulated gene encoding a chloroplast omega-3 desaturase from Arabidopsis thaliana. Plant Physiol 106: 1615–1621.

Gonzalez, C.I., and C.E. Martin. 1996. Fatty Acid-responsive Control of mRNA Stability. Unsaturated fatty acid-induced degradation of the saccharomyces ole1 transcript. J Biol Chem 271:25801–25809.

Graef, G.L., W.R. Fehr, and E.G. Hammond. 1985a. Inheritance of Three stearic Acid Mutants of Soybean. Crop Sci. 25:1076–1079.

Graef, G.L., L.A. Miller, W.R. Fehr, and E.G. Hammond. 1985b. Fatty Acid Development in a Soybean Mutant with High Stearic Acid. JAOCS. 62:773–775.

Grayburn, W.S., and D.F. Hildebrand. 1995. Progeny analysis of tobacco that express a mammalian D9 desaturase. J. Am. Oil Chem. Soc. 72:317–321.

Griffiths, G., A. Stobart, and S. Stymne. 1985. The acylation of sn-glycerol 3-phosphate and the metabolism of phosphatidate in microsomal preparations from the developing cotyledons of safflower (*Carthamus tinctorius* L.) seed. Biochem J 230:379–388.

Griffiths, G., A.K. Stobart, and S. Stymne. 1988. Biochem J 252:641–647.

Hammond, E.G., and W.R. Fehr. 1983. Registration of A5 Germplasm line of Soybean. Crop Sci. 23:192–3.

Harwood, J. 1998. What's so special about plant lipids?, p. 1–26., *In* e. J.L.Harwood, ed. Plant Lipid Biosynthesis Fundamentals and Agricultural Applications. Cambridge University Press, New York.

Harwood, J.L., and R.A. Page. 1994. Biochemistry of oil synthesis., p. 165–194, *In* D. J. Murphy, ed. Designer Oil Crops. VCH, Weinheim.

Hayatsu, H., S. Arimoto, and T. Negishi. 1988. Dietary inhibitors of mutagenesis and carcinogenesis. Mutat Res 202:429–46.

Hegsted, D.M., A.L. M, J.J. A, and D.G. E. 1993. Dietary fat and serum lipids: an evaluation of experimental data. Am J Clin Nutr. 57:875–883.

Heppard, E.P., A.J. Kinney, K.L. Stecca, and G.H. Miao. 1996. Developmental and growth temperature regulation of two different microsomal omega-6 desaturase genes in soybeans. Plantphysiology; Jan 1996; 110(1): 311–319 110:311–319.

Hiraoka, M., A. Abe, and J.A. Shayman. 2002. Cloning and Characterization of a Lysosomal Phospholipase A2, 1-O-Acylceramide Synthase. J. Biol. Chem. 277:10090–10099.

Hitz, W.D. 1998. United States Patent 5,846,784 1998.

Hobbs, D.H., C. Lu, and M.J. Hills. 1999. Cloning of a cDNA encoding diacylglycerol acyltransferase from Arabidopsis thaliana and its functional expression. FEBS Lett 452:145–9.

Hymowitz, T., R.G. Palmer, and H.H. Hadley. 1972. Seed weight, protein, oil, and fatty acid relationships within the genus *Glycine*. Trop Agr. 49:245–250.

Jackson, F.M., L. Michaelson, T.C.M. Fraser, A.K. Stobart, and G. Griffiths. 1998. Biosynthesis of triacylglycerol in the filamentous fungus Mucor circinelloides. Microbiology 144:2639–2645.

Kalinski, A., D.S. Loer, J.M. Weisemann, B.F. Matthews, and E.M. Herman. 1991. Isoforms of soybean seed oil body membrane protein 24 kDa oleosin are encoded by closely related cDNAs. Plant-molecular-biology 17:1095–8.

Kamisaka, Y., S. Mishra, and T. Nakahara. 1997. Purification and characterization of diacylglycerol acyltransferase from the lipid body fraction of an oleaginous fungus. J Biochem 121:1107–14.

Katan, M.B., P.L. Zook, and R.P. Mensik. 1995. Trans-fatty acids and their effects on lipoproteins in humans. Ann. Rev. Nutr. 15:473–493.

Katavic, V., D.W. Reed, D.C. Taylor, E.M. Giblin, D.L. Barton, J. Zou, S.L. Mackenzie, P.S. Covello, and L. Kunst. 1995. Alteration of seed fatty acid composition by an ethyl methanesulfonate-induced mutation in Arabidopsis thaliana affecting diacylglycerol acyltransferase activity. Plant Physiol 108:399–409.

Kaup, M.T., C.D. Froese, and J.E. Thompson. 2002. A role for diacylglycerol acyltransferase during leaf senescence. Plant physiol 129:1616–1626.

Kinney, A.J. 1994. Current Opinion in Biotechnology 5:144–151.

Kinney, A.J. 1998a. Plants as industrial chemical factories-new oils from genetically engineered soybeans. Fett/Lipid 100:173–176.

Kinney, A.J. 1998b. Production of specialised oils for industry, p. 273–285, *In* J. L. Harwood, ed. Plant Lipid Biosynthesis; Fundamentals and Agricultural Applications, Vol. 12. Cambridge University Press, Cambridge, UK.

Klaus, D., J.B. Ohlrogge, H.E. Neuhaus, and P. Dormann. 2004. Increased fatty acid production in potato by engineering of acetyl-CoA carboxylase. Planta 219:389–96.

Knutzon, D.S., G.A. Thompson, S.E. Radke, W.B. Johnson, V.C. Knauf, and K.J. C. 1992. Modification of Brassica seed oil by antisense expression of a stearoyl-acyl carrier protein desaturase gene. Proceedings of the National Academy of Sciences of the United States of America 89:2624–2628.

Kridl, J. 2002. Method for increasing stearate content in soybean oil. U.S. Patent.

Kris-Etherton, P.M., D. Krummel, M.E. Russell, D. Dreon, S. Mackey, J. Borchers, and P.D. Wood. 1988. The effect of diet on plasma lipids, lipoprteins, and coronary heart disease. Journal of the American Dietetic Association. 88:1373–1400.

Kwanyuen, P., and R.F. Wilson. 1986. Isolation and purification of diacylglycerol acyltransferase from germinating soybean cotyledons. Biochim Biophys Acta 877:238–245.

Kwanyuen, P., and R.F. Wilson. 1990. Subunit and amino acid composition of diacylglycerol acyltransferase from germinating soybean cotyledons. Biochim Biophys Acta 1039:67–72.

Kwanyuen, P., R.F. Wilson, and J.W. Burton. 1988. Substrate specificity of diacylglycerol acyltransferase purified from soybean. Proceedings : World Conference on Biotechnology for the Fats and Oils Industry/edited by Thomas H. Applewhite. Champaign, Ill. American Oil Chemists' Society, c1988. p.:294–297.

Ladd, S.L., and P.F. Knowles. 1970. Inheritance of Stearic Acid in the Seed Oil of Safflower (Carthamus tinctorius L.). Crop Sci. 10:525–527.

Lardizabal, K.D., J.T. Mai, N.W. Wagner, A. Wyrick, T. Voelker, and D.J. Hawkins. 2001. DGAT2 is a new diacylglycerol acyltransferase gene family: purification, cloning, and expression in insect cells of two polypeptides from Mortierella ramanniana with diacylglycerol acyltransferase activity. J Biol Chem 276:38862–9.

Lassner, M.W., C.K. Levering, H.M. Davies, and D.S. Knutzon. 1995. Lysophosphatidic acid acyltransferase from meadowfoam mediates insertion of erucic acid at the sn-2 position of triacylglycerol in transgenic rapeseed oil. Plant Physiol 109:1389–1394.

Lee, M., M. Lenman, A. Banas, M. Bafor, S. Singh, M. Schweizer, R. Nilsson, C. Liljenberg, A. Dahlqvist, P.-O. Gummeson, S. Sjodahl, A. Green, and S. Stymne. 1998. Science 280:915–918.

Lehner, R., and A. Kuksis. 1993. Triacylglycerol synthesis by an sn-1,2(2,3)-diacylglycerol transacylase from rat intestinal microsomes. J Biol Chem 268:8781–6.

Lemieux, B.M.M., C. Somerville, and J. Browse. 1990. Mutants of Arabidopsis with alterations in seed lipid fatty acid composition. Theor Appl Genet 80:234–240.

Lin, J.T., J.M. Chen, L.P. Liao, and T.A. McKeon. 2002. Molecular Species of Acylglycerols Incorporating Radiolabeled Fatty Acids from Castor (*Ricinus communis* L.) Microsomal Incubations. J. Agric. Food Chem. 50:5077–5081.

Lindqvist, Y., W. Huang, G. Schneider, and J. Shanklin. 1996. Crystal structure of delta9 stearoyl-acyl carrier protein desaturase from castor seed and its relationship to other di-iron proteins. EMBO J 15:4081–4092.

Liu, W., R.S. Torisky, K.P. McAllister, S. Avdiushko, D. Hildebrand, and G.B. Collins. 1996. A mammalian desaturase gene lowers saturated fatty acid levels in transgenic soybean embryos. Plant Cell, Tissue & Organ Culture 47:33–42.

Loer, D.S., and E.M. Herman. 1993. Cotranslational integration of soybean (*Glycine max*) oil body membrane protein oleosin into microsomal membranes. Plant-physiology; Mar 1993; 101(3): 993–998 101:993–998.

Lu, C., and M.J. Hills. 2002. Arabidopsis mutants deficient in diacylglycerol acyltransferase display increased sensitivity to abscisic acid, sugars, and osmotic stress during germination and seedling development. Plant physiol 129:1352–1358.

Lu, C.L., S.B. de Noyer, D.H. Hobbs, J. Kang, Y. Wen, D. Krachtus, and M.J. Hills. 2003. Expression pattern of diacylglycerol acyltransferase-1, an enzyme involved in triacylglycerol biosynthesis, in Arabidopsis thaliana. Plant Mol Biol 52:31–41.

Martinez-Force, E., V. Fernandez-Moya, and R. Garces. 2002. Plant, seeds and oil with saturated triacylglycerol content and oil having a high stearic content. Patent 6,410,831 & 6,486,336 2002.

McConn, M., S. Hugly, J. Browse, and C. Somerville. 1994. A mutation at the *fad*8 locus of *Arabidopsis* identifies a second chloroplast w-3 desaturase. Plant Physiol 106:1609–1614.

McKeon, T., S.-T. Kang, C. Turner, X. He, G. Chen, and J.-T. Lin. 2006. Enzymatic synthesis of intermediates in castor oil biosynthesis. 97th AOCS Annual Meeting & Expo:15.

Mhaske, V., K. Beldjilali, J. Ohlrogge, and M. Pollard. 2005. Isolation and characterization of an Arabidopsis thaliana knockout line for phospholipid: diacylglycerol transacylase gene (At5g13640). Plant Physiol Biochem 43:413–7.

Miquel, M., D. James, H. Dooner, and J. Browse. 1993. Proc Natl Acad Sci USA 90:6208–6212.

Mudd, J.B., and P.K. Stumpf. 1961. Fat metabolism in plants. IXV. Factors affecting the synthesis of oleic acid by particulate preparations from avocado mesocarp. J Biol Chem 236:2602–2609.

Murphy, and Piffanelli. 1998. Fatty acid desaturases:structure,mechanism and regulation, p. 1–26., *In* e. J.L.Harwood, ed. Plant Lipid Biosynthesis Fundamentals and Agricultural Applications. Cambridge University Press, New York.

Nelson, G.J. 1998. Dietary fat, Trans-fatty acids, and risk of coronary heart disease. Nutr. Rev. 56:250–252.

O'Neill, C.M., S. Gill, D. Hobbs, C. Morgan, and I. Bancroft. 2003. Natural variation for seed oil composition in Arabidopsis thaliana. Phytochemistry 64:1077–1090.

Oelkers, P., A. Tinkelenberg, N. Erdeniz, D. Cromley, J.T. Billheimer, and S.L. Sturley. 2000. A Lecithin Cholesterol Acyltransferase-like Gene Mediates Diacylglycerol Esterification in Yeast. J. Biol. Chem. 275:15609–15612.

Ohlrogge, J., and J. Browse. 1995. Lipid biosynthesis. Plant Cell 7:957–970.

Ohlrogge, J.B., and J.G. Jaworski. 1997. Regulation of fatty acid synthesis. Annu. Rev. Plant Physiol. Plant Molec. Biol. 48:109–136.

Okuley, J., J. Lightner, K. Feldmann, N. Yadav, E. Lark, and J. Browse. 1994. Plant Cell 6:146–158.

Quittnat, F., Y. Nishikawa, T.T. Stedman, D.R. Voelker, J.Y. Choi, M.M. Zahn, R.C. Murphy, R.M. Barkley, M. Pypaert, K.A. Joiner, and I. Coppens. 2004. On the biogenesis of lipid bodies in ancient eukaryotes: synthesis of triacylglycerols by a Toxoplasma DGAT1-related enzyme. Mol Biochem Parasitol 138:107–22.

Rahman, S., T. Anai, T. Kinoshita, and Y. Takagi. 2003. A Novel Soybean Germplasm with Elevated Saturated Fatty Acids. Crop Sci. 43:527–531.

Rahman, S.M., Y. Takagi, and T. Kinshita. 1997. Genetic control of high stearic content in seed oil of two soybean mutants. Ther Appl Genet. 95:772–776.

Rennie, B.D., and J.W. Tanner. 1989. J Amer Oil Chem Soc 86:1622–1624.

Routaboul, J.-M., C. Benning, N. Bechtold, M. Caboche, and L. Lepiniec. 1999. The TAG1 locus of Arabidopsis encodes for a diacylglycerol acyltransferase. Plant Physiology and Biochemistry 37:831–840.

Ruuska, S.A., J. Schwender, and J.B. Ohlrogge. 2004. The capacity of green oilseeds to utilize photosynthesis to drive biosynthetic processes. Plant Physiol 136:2700–9.

Sarmiento, C., J.H. Ross, E. Herman, and D.J. Murphy. 1997. Expression and subcellular targeting of a soybean oleosin in transgenic rapeseed. Implications for the mechanism of oil-body formation in seeds. Plant Journal 11:783–796.

Schnebly, S.R., W.R. Fehr, G.A. Welke, E.G. Hammond, and D.N. Duvick. 1994. Inheritance of reduced and elevated palmitate in mutant lines of soybean. Crop Sci. 34:829–833.

Schnurr, J.A., J.M. Shockey, G.-J. de Boer, and J.A. Browse. 2002. Fatty Acid Export from the Chloroplast. Molecular Characterization of a Major Plastidial Acyl-Coenzyme A Synthetase from Arabidopsis. Plant Physiol. 129:1700–1709.

Settlage, S.B., P. Kwanyuen, and R.F. Wilson. 1998. Relation between diacylglycerol acyltransferase activity and oil concentration in soybean. J Am Oil Chem Soc 75:775–781.

Shanklin, J., and C. Somerville. 1991. Stearoyl-acyl-carrier-protein desaturase from higher plants is structurally unrelated to the animal and fungal homologs. Proc. Natl. Acad. Sci. 88:2510–2514.

Shanklin, J., and B.E. Cahoon. 1998. Desaturation and related modifications of fatty acids. Mol Biol 49:109–136.

Shanklin, J., E. Whittle, and B.G. Fox. 1994. Biochemistry 33:12787–12794.

Shockey, J.M., M.S. Fulda, and J.A. Browse. 2002. Arabidopsis contains nine long-chain acyl-coenzyme A synthetase genes that participate in fatty acid and glycerolipid metabolism. Plant Physiology 129:1710–1722.

Siloto, R.M.P., K. Findlay, A. Lopez-Villalobos, E.C. Yeung, C.L. Nykiforuk, and M.M. Moloney. 2006. The Accumulation of Oleosins Determines the Size of Seed Oilbodies in Arabidopsis. Plant Cell 18:1961–1974.

Simonsen, N.R., J.F.-C. Navajas, J.M. Martin-Moreno, J.J. Strain, J.K. Huttunen, B.C. Martin, M. Thamm, A.F. Kardinaal, P.v.t. Veer, F.J. Kok, and L. Kohlmeier. 1998. Tissue Stores of individual monounsaturated fatty acids and breast cancer: the EURAMIC study. Am. J Clin. Nutr 68:134–41.

Slack, C.R., P.G. Roughan, J.A. Browse, and S.E. Gardiner. 1985. Some properties of choline phosphotransferase from developing safflower cotyledons. Biochim Biophys Acta 833: 438–448.

Slocombe, S.P., P. Piffanelli, D. Fairbairn, S. Bowra, P. Hatzopoulos, M. Tsiantis, and D.J. Murphy. 1994. Temporal and tissue-specific regulation of a Brassica napus stearoyl-acyl carrier protein desaturase gene. Plant Physiol 104:1167–1176.

Spiller, G.A., J.D. J, C.L. N, G.J. E, B. O, B. K, R. C, S. M, and S. R. 1992. Effects of a diet high in monounsaturated fat from almonds on plasma cholesterol and lipoproteins. Am J Clin Nutr. 11:126–130.

Stahl, U., A.S. Carlsson, M. Lenman, A. Dahlqvist, B. Huang, W. Banas, A. Banas, and S. Stymne. 2004. Cloning and functional characterization of a phospholipid:diacylglycerol acyltransferase from Arabidopsis. Plant Physiol 135:1324–35.

Stobart, K., M. Mancha, M. Lenman, A. Dahlqvist, and S. Stymne. 1997. Triacylglycerols are synthesised and utilized by transacylation reactions in microsomal preparations of developing safflower (*Carthamus tinctorius* L.) seeds. Planta New York:Springer-Verlag, 1925-. Sept 1997. v. 203 (1) p. 58–66.

Stoltzfus, D.L., W.R. Fehr, and G.A. Welke. 2000. Relationship of Elevated Palmitate to Soybean Seed Traits. Crop Sci 40:52–54.

Stymne, S., and A.K. Stobart. 1986. Biochem J 240:385.

Swarts, H., S.S. FM, and D.P. JJ. 1990. Binding of unsaturated fatty acids to N+, K(+)-ATPase leading to inhibition and inactivation. Biochim Biophys Acta 1024:32–40.

Theret, N., J. Bard, M. Nuttens, J. Lecerf, C. Delbart, M. Romon, J. Salomez, and J. Fruchart. 1993. The relationship between the phospholipids fatty acid composition of red blood cells, [lasma lipids, and apolipoproteins. Metabolism 42:526–8.

Tocher, D.R., M.J. Leaver, and P.A. Hodgson. 1998. Prog Lipid Res 37:73–117.

Triki, S., J. Ben Hamida, and P. Mazliak. 1998. About the reversibility of the cholinephosphotransferase in developing sunflower seed microsomes, p. 236–239, *In* J. Sanchez, et al., eds. Advances in Plant Lipid Research. Univ. Sevilla, Sevilla.

Turkish, A.R., A.L. Henneberry, D. Cromley, M. Padamsee, P. Oelkers, H. Bazzi, A.M. Christiano, J.T. Billheimer, and S.L. Sturley. 2005. Identification of Two Novel Human Acyl-CoA Wax Alcohol Acyltransferases: MEMBERS OF THE DIACYLGLYCEROL ACYLTRANSFERASE 2 (DGAT2) GENE SUPERFAMILY. Journal of biological chemistry 280: 14755–14764.

Tzen, J.T.C., Y.K. Lai, K.L. Chan, and A.H.C. Huang. 1990. Oleosin isoforms of high and low molecular weights are present in the oil bodies of diverse seed species. Plant-physiology 94:1282–1289.

van de Loo, F.J., P. Broun, S. Turner, and C. Somerville. 1995. An oleate 12-hydroxylase from *Ricinus communis* L. is a fatty acyl desaturase homolog. Proc Natl Acad Sci USA 92: 6743–6747.

Voelker, T., and A. Kinney. 2001. Variations in the biosynthesis of seed-storage lipids. Annual Review of Plant Physiology and Plant Molecular Biology 52:335–361.

Voelker, T.A., A.C. Worell, L. Anderson, J. Bleiaum, C. Fan, D.J. Hawkins, S.E. Radke, and H.M. Davies. 1992. Fatty acid biosynthesis redirected to medium chains in transgenic oilseed plants. Science 257:72–74.

Vogel, G., and J. Browse. 1996. Cholinephosphotransferase and diacylglycerol acyltransferase: Substrate specificities at a key branch point in seed lipid metabolism. Plant physiol 110: 923–931.

Wang, C., C.-K. Chin, C.-T. Ho, C.-F. Hwang, J.J. Polashoc, and C.E. Martin. 1996. Changes of Fatty Acids and Fatty Acid-Derived Flavor Compounds by Expressing the Yeast D-9 Desaturase Gene in Tomato. J. Agri. Food Chem 44:3399–3402.

Wang, H.-W., J.-S. Zhang, J.-Y. Gai, and S.-Y. Chen. 2006. Cloning and comparative analysis of the gene encoding diacylglycerol acyltransferase from wild type and cultivated soybean. Theoretical and Applied Genetics 112:1086–1097.

Wiberg, E., A. Banas, and S. Stymne. 1997. Fatty acid distribution and lipid metabolism in developing seeds of laurate-producing rape (Brassica napus L.). Planta 203:341–348.

Wiberg, E., P. Edwards, J. Byrne, S. Stymne, and K. Dehesh. 2000. The distribution of caprylate, caprate and laurate in lipids from developing and mature seeds of transgenic Brassica napus L. Planta 212:33–40.

Yadav, N.S., A. Wierzbicki, M. Aegerter, C.S. Caster, L. Perez-Grau, A.J. Kinney, W.D. Hitz, J.R.J. Booth, B. Schweiger, and K.L. Stecca. 1993. Cloning of higher plant omega-3 fatty acid desaturases. Plant Physiol 103:467–476.

Yamori, Y., N. Y, T. T, S. Y, I. K, and H. R. 1986. Dietary prevention of stroke and its mechanism in stoke-prone spontaneously hypertensive rats-Peventive effect of dietary fibre and palmitoleic acid. J Hypertens Suppl 4:S449–52.

Yasumoto, K., K. Fujimoto, O. Igarashi, R. Sasaki, R. Hayashi, and Y.I. Matsumoto. 1993. p. 49–73., *In* E. S. K. Kenkyukai, ed. Food Chemistry, 3rd ed. Doubunshoin, Tokyo.

Zhang, F.Y., M.F. Yang, and Y.N. Xu. 2005. Silencing of DGAT1 in tobacco causes a reduction in seed oil content. Plant science 169:689–694.

Zou, J., Y. Wei, C. Jako, A. Kumar, G. Selvaraj, and D.C. Taylor. 1999. The Arabidopsis thaliana TAG1 mutant has a mutation in a diacylglycerol acyltransferase gene. Plant J. 19:645–653.

Zou, J.T., V. Katavic, E.M. Giblin, D.L. Barton, S.L. MacKenzie, W.A. Keller, X. Hu, and D.C. Taylor. 1997. Modification of seed oil content and acyl composition in the *Brassicaceae* by expression of a yeast *sn*-2 acyltransferase gene. Plant Cell 9:909–923.

Chapter 13
Genomics of Secondary Metabolism in Soybean

Terry Graham, Madge Graham, and Oliver Yu

An Overall Perspective on Plant Secondary Metabolites

Secondary product pathways in plants are extraordinarily diverse. Although classical chemical and biochemical analyses have gone a long way in delineating the major pathways and metabolites in many plant species, no plant species has been completely characterized for all of the secondary products that it produces *or is capable of producing*. While examinations of plants at the metabolic level (e.g., through metabolic profiling/metabolomics) is powerful in that it provides a picture of the ultimate net accumulation of endproducts in various cells, tissues and organs, unexpected or new metabolites are often found in a given species when the plant is examined under previously unexamined conditions. A very simple example is the production of large quantities of formononetin by soybean following treatment by the disease resistance-inducing herbicide lactofen (Landini et al. 2002). While this metabolite is found at high levels in some other legumes, such as chickpea, it is normally undetectable or at trace levels in soybean. Thus, it is very difficult by analyses of metabolites per se to know all of the potential metabolites that can be produced by a plant. This inherent limitation is an aspect where genomic analyses will have particular impact in that the presence of a gene for a particular enzyme may be suggested even though the product of its action is not.

Quantification of the secondary metabolites in classical biochemical or metabolomics approaches is also a challenge. While quantitative data is readily obtained from various chromatographic and spectroscopic techniques, the actual tissue concentration of individual metabolites requires the use of standards for each metabolite, something that is extremely difficult to achieve across all metabolites and tissues. Moreover, many important biological processes involve changes in metabolites in individual cells or groups of cells, which is beyond the limits of detection of most

Terry Graham
Department of Plant Pathology, Ohio State University, Columbus, OH 43210, USA
e-mail: graham.1@osu.edu

G. Stacey (ed.), *Genetics and Genomics of Soybean*,

biochemical analyses. New techniques such as laser capture microdissection (LCM, see e.g., Klink et al. 2005) promise to provide genome wide mRNA expression information at a cellular level. Such analyses may provide additional critical and complementary insights.

Although secondary product metabolism across the plant kingdom is diverse, extensive characterization of the natural products made by plants has suggested that the major metabolites normally produced by a given plant family, genus or species are somewhat limited. While fundamental natural products such as some of the simple flavonoids or anthocyanins are widespread in plants, beyond these common metabolites, plant families tend to specialize in a few major pathways. As just one example, the phenylpropanoid derived isoflavones are predominant metabolites in the Leguminosae. Likewise the major defense related phytoalexins in the legumes are coumestans and pterocarpans further derived from the isoflavonoids. In contrast, in the Solanaceae, phenylpropanoid-derived metabolites are relatively simple and alkaloids and sesquiterpenoid-derived phytoalexins predominate. Further specialization is found even within a family. Within the family Leguminosae, to which soybeans belong, the specific major isoflavones and pterocarpans that predominate are different at a genus level. Thus, the specific pterocarpan phytoalexins that predominate in soybean (the glyceollins) are different from those that predominate in alfalfa (the medicarpins), etc. At the ultimate level of specialization, even different cultivars of a specific species, for instance soybeans (*Glycine max*), can produce very different mixtures of specific endproducts (e.g., different flavonols and flavonol glycosides, Buttery and Buzzell 1973, 1975).

In this short review, we will focus on (1) the information on secondary product enzymes, pathways and their expression that can be mined from the extensive soybean EST and derived databases and (2) the information that is just starting to emerge from a few interesting genome wide microarray studies on secondary products produced in *Glycine max*. Since the phenylpropanoid pathway in soybean has been particularly well characterized at a biological and biochemical level, we focus primarily on this pathway and its many branch pathways (e.g., for the flavonoids, isoflavonoids, anthocyanins, etc). It is beyond the scope of this current review to cover other major secondary metabolites, such as the saponins, waxes, fatty acids, and alkaloids.

Secondary Product Metabolism in Soybean

Figures 13.1A and 13.1B outline the major secondary metabolic pathways in soybean that will be the focus of this review. Much of the work on these pathways in soybean was the result of their importance to various phenotypes, such as flower color or seed coat pigmentation (anthocyanins), protection against UV light (flavonols), and interaction with various pests or symbionts (isoflavones). In Fig. 13.1A, we show an abbreviated schematic for lignin, flavonol and anthocyanin

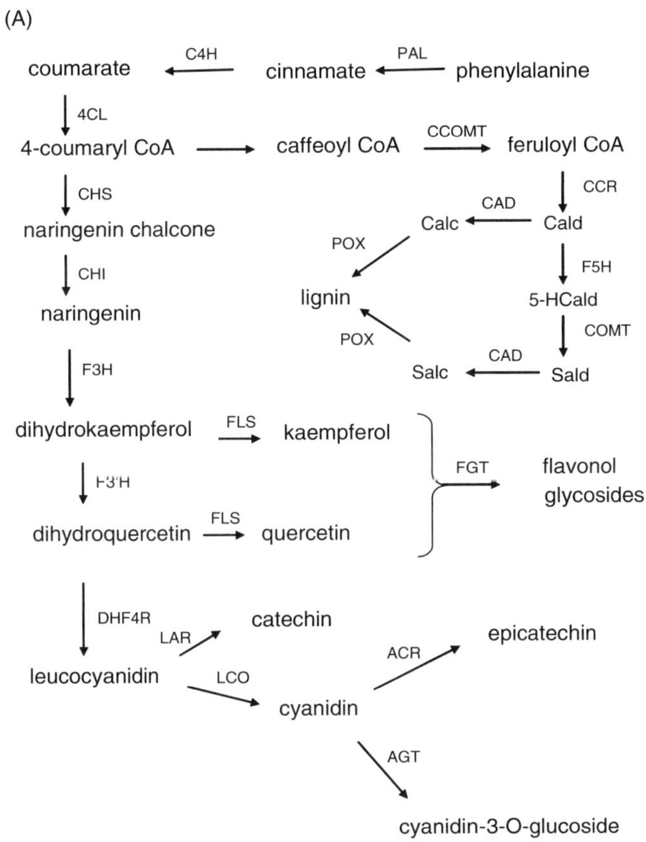

Fig. 13.1A Lignin, flavonol and anthocyanin biosynthesis Abbreviations: 4CL, 4-coumaryl-CoA ligase; ACR, anthocyanidin reductase; AGT, anthocyanidin-3-O-glucosyltransferase; C4H, cinnamic acid 4-hydroxylase; CAD, cinnamyl alcohol dehydrogenase; CCOMT, caffeoyl-CoA O-methyltransferase; CCR, cinnamoyl-CoA reductase; CHI, chalcone isomerase; CHS, chalcone synthase; COMT, caffeic acid O-methyltransferase; DHF4R, dihydroflavonol 4-reductase; F3′H, flavonoid 3′-hydroylase; F3H, flavanone 3-hydroxylase; F5H, ferulate 5-hydroxylase; FLS, flavonol synthase; LAR, leucoanthocyanidin reductase; LCO, leucocyanidin oxygenase; POX, lignin peroxidase; PAL, phenylalanine ammonia lyase

biosynthesis and in Fig. 13.1B we show a simplified pathway for the formation of the isoflavones and the pterocarpans, including the phytoalexin glyceollin. Although we tried to illustrate most of the key enzymes, each of these pathways is considerably more complex than shown. In the sections below, we first describe each branch of these pathways in more detail, including some important aspects of their genetics, biology and biochemistry. We then address them from a genomics perspective, including information that can be mined from the various EST libraries and recent microarray work.

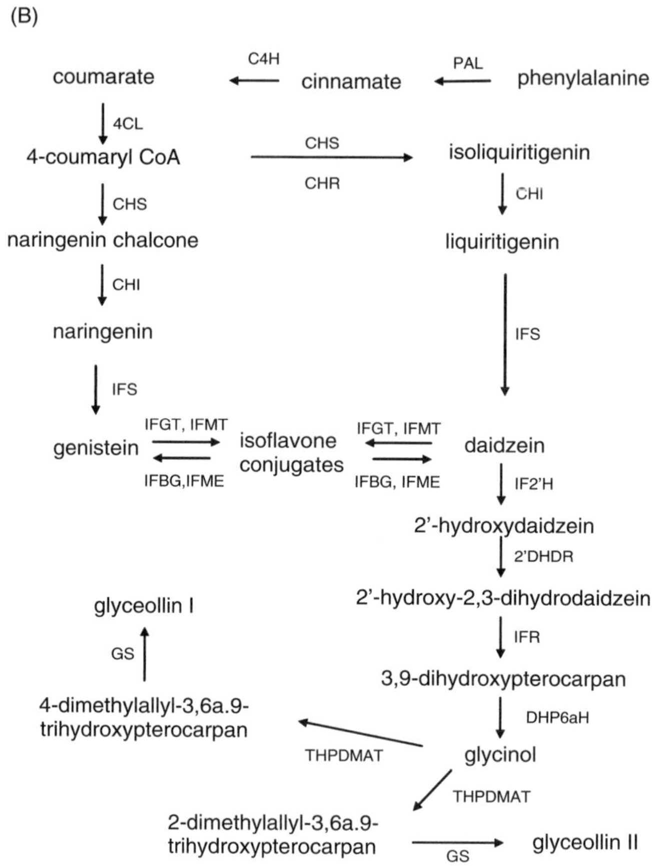

Fig. 13.1B Isofavone and glyceollin biosynthesis Abbreviations: 2′DHDR, 2′-hydroxydihydrodaidzein reductase; CHR, chalcone reductase; DFR, dihydroflavonol-4-reductase, DHP6aH, 3,9-dihydroxypterocarpan 6a-hydroxylase; GS, glyceollin synthase; IF2′H, isoflavone 2′-hydroxylase; IFBG, isoflavone 7-O-beta glucosidase; IFGT, isoflavone 7-0-glucosyl transferase; IFME, isoflavone malonyl esterase; IFMT, isoflavone malonyl transferase; IFR, isoflavone reductase; IFS, isoflavone synthase; THPDMAT, trihydroxypterocarpan dimethylallyltransferase

Early Phenylpropanoids

The early phenylpropanoids include those metabolites from phenylalanine to the simple phenylpropanoic acids. Early metabolic profiling of various soybean organs (Graham 1991a, b) suggested that these metabolites do not normally accumulate to high levels, suggesting that they are efficiently used as substrates for the more complex metabolites shown in Fig. 13.1. Phenylalanine ammonia lyase (PAL) is considered the entry point for phenylpropanoid metabolism and a major point of regulation for loading the entire metabolic grid deriving from the simple phenylpropanoids.

Cinnamate-4-hydroxylase (C4H) adds a hydroxyl group to the 4 position, giving rise to *p*-coumaric acid. Following this, additional hydroxylations and/or methylations of the aromatic ring occur, which at least partially determine if these simple phenylpropanoids are destined to be diverted into lignin/suberin, flavonoids/anthocyanins or simple esters (Hahlbrock 1981). Formation of the CoA derivatives through the action of 4-coumarate:CoA ligase (4CL) is the entry point for the simple phenylpropanoids into the various alternative pathways leading to more complex metabolites.

Phenolic Polymers: Lignin and Suberin

Lignin is synthesized mostly as a component of secondary cell walls, which provide additional structural rigidity to support the plants. Although the biosynthesis of lignin involves relatively few enzymes, there are some unresolved complexities in the actual enzymatic pathways (Humphreys and Chapple 2002; Boerjan et al. 2003). Basically, the process involves a series of reductions of CoA derivatives of phenylpropanoic acids first to an aldehyde, through the action of cinnamoyl-CoA reductase (CCR), and then to an alcohol by cinnamyl alcohol dehydrogenase (CAD). This is followed by reoxidation and polymerization by ligninperoxidases. Much of the complexity of the pathway involves how and where in the grid of enzymes the hydroxylation and methylation reactions occur that are needed for the formation of the corresponding caffeic, ferulic, 5-hydroxyferulic and sinapic acids derivatives. Although ferulate 5-hydroxylase (F5H), caffeic acid O-methyl transferase (COMT) and caffeoyl-CoA O-methyltransferase (CCOMT) are thought to be involved, other enzymes are possible (Humphreys and Chapple 2002; Boerjan et al. 2003). Laccases (diphenol oxidases) are also thought to play a potential role in lignin formation (Gavnholt and Larsen 2002).

Unlike lignin which has a polyaromatic structure, suberin consists of both polyaromatic and polyaliphatic domains. The polyaliphatic domain makes suberin a very hydrophobic polymer, which participates primarily in exclusion of water, for instance in cell walls of the casparian strip in roots. The composition of the monomeric constituents of suberin varies in different species. Common aliphatic monomers include α-hydroxyacids and α, ω-diacids. Like lignin, the monomeric constituents of the polyaromatic domain are hydroxycinnamic acids and derivatives (for a recent review, see Ma and Peterson 2003).

As a part of general defense mechanisms, lignin content increases markedly during biotic (e.g., infection) and abiotic (e.g., wounding) stress. Formation of lignin and suberin-like polymers in plants is developmentally important for the formation of the cell walls of many cells, but particularly in secondary cell walls (Humphreys and Chapple 2002) and the endodermis (Ma and Peterson 2003), respectively. In soybean, the formation of these phenolic polymers is also a relatively early event in infected or pathogen elicitor treated tissues (Graham and Graham 1991a; Mohr and Cahill 2001; Lozovaya et al. 2004). Some genomic aspects of the induction of enzymes for lignin formation in wounded tissues and pathogen resistance are discussed in detail below.

Flavonols

The precise roles of the flavonols are not fully understood, although they are very widespread in the plant kingdom. In soybean, the flavonols are present in all mature aerial tissues, but not in roots or seedling organs (Graham 1991b). A possible function of flavonols in protection against UV-B irradiation was suggested (see e.g., Landry et al. 1995; Logemann et al. 2000). The flavonols are formed through the action of chalcone synthase (CHS) and chalcone isomerase (CHI) which lead to the flavanone naringenin. Flavanone 3-hyroxylase (F3H) then leads to dihydrokaempferol, which can be further hydroxylated by flavonoid 3′-hydroxylase (F3′H) to form dihydroquercetin. Kaempferol and quercetin are then formed through the action of flavonol synthase (FLS). Not all soybean cultivars have both kaempferol and quercetin. The presence of quercetin requires the presence of the *T* gene (Buttery and Buzzell 1973; Buzzell et al. 1987), which encodes F3′H (Zabala and Vodkin 2003). Soybean TC216289, which encodes a F3′H, is a close homolog to the *TT7* gene in *Arabidopsis* which helps impart testa (seed coat) color. The flavonol, isorhamnetin, was also identified in certain soybean lines (Le- Van, N. and Graham, T. L., unpublished).

Like many end products of secondary product metabolism, the flavonols are usually present as conjugates (e.g., as glycosides). The patterns of glycosylation can be very complex, perhaps suggesting roles that we still do not fully understand. Several of the genes involved in the accumulation of flavonol 3-O-glycosides in soybean leaves were identified through a combination of biochemistry and classical genetics. Buttery and Buzzell (1975) identified nine different 3-O-glycosides for each of the flavonols, kaempferol or quercetin. Genes for glycosylation of kaempferol or quercetin are the same. The kaempferol monoglucoside (designated as K5) is found in all soybean cultivars, and both K5 and Q5 are found in all cultivars carrying the *T* gene (for F3′H). At least four additional enzymes give rise to the various diglycosides and triglycosides by simple additions to these monoglucosides. These include glucosyl transferases that transfer glucose in β linkages to the 6 position *(FG1)* or the 2 position *(FG3)* of the monoglucoside and rhamnosyl transferases that transfer rhamnose in α linkages to either the 6 position *(FG2)* or the 2 position *(FG4)* of the monoglucoside. Taken together, the *T* and various *FG* genes can give rise to over 32 flavonoid combinations for kaempferol and quercetin alone.

Flavonols can also serve as co-pigments in flowers. A *wm* locus that is associated with magenta flower color dramatically reduces flavonol glycoside levels in leaves and flowers, but does not affect their presence in pod pubescence (Buttery and Buzzell 1987). The magenta flower color is in fact thought to be due to the lack of flavonol glycosides in the otherwise purple flower background (caused by anthocyanins, Buzzell et al. 1977). Thus the *WM* locus seems to regulate the organ specific formation or glycosylation of the flavonols. In summary, the genetics of flavonol glycoside formation in soybean generally support the notion that addition of O-glycosidic bonds to flavonoids occurs after synthesis of the aglycone, and that specific sugar linkages are added in single steps, each conferred by a different gene.

While the biochemistry and genetics of flavonol glycoside metabolism is fairly well established, the accompanying genomic aspects are very limited due to the fact that the cultivars used in the various current EST libraries are quite limited. Although we will not treat the leaf flavonols further in this chapter, this is an important area for further molecular genetic analyses.

Another type of common co-pigments of flowers are the flavones. However, unlike other legumes such as *Medicago truncatula* where various flavones are dominant flavonoid compounds in leaves and roots (Farag et al. 2007), flavones have not been reported in soybean. Consistent with this, the key enzyme for flavone synthesis, flavone synthase (FNS, CYP93B), was not found in the soybean EST genomic database. It is possible that soybean does not have a flavone biosynthesis pathway.

Anthocyanins and Proanthocyanidins

The anthocyanins and proanthocyanidins (condensed tannins) are downstream from the flavonols (Fig. 13.1) and contribute to flower and seed coat color in soybean. In addition, proanthocyanidins are important to protect ruminant animals from pasture bloat (Waghorn and McNabb 2003). There are interesting parallels between the genes controlling the synthesis of anthocyanins and proanthocyanidins in soybeans and those responsible for the pigmentation of the testa in Arabidopsis. The genes for testa pigmentation in Arabidopsis were identified through analysis of a series of transparent testa (*tt*) mutants. As with Arabidopsis, several major genes control the accumulation of flavonols in the seed coat of soybean. These include the *T* gene (Buttery and Buzzel 1973), which as noted already is responsible for the F3′H-mediated conversion of dihydrokaempferol to dihydroquercetin (Zabala and Vodkin 2003). Both of these flavonoids are also precursors for the anthocyanins, although we only show the pathway for cyanidin formation from dihydroquercetin in Fig. 13.1. The *TT7* gene, which has an analogous function in Arabidopsis, was cloned and shown to be a cytochrome P450 that indeed functions as a flavonoid 3′-hydroylase (Schoenbohm et al. 2000). It and a related, though unpublished, Arabidopsis sequence (AF155171) have by far the highest homology (tblastx, 1.4e-166) to a single TC (TC216289) in the soybean EST database and somewhat lower homology to a soybean singleton derived from AY117551. Thus, it is likely, by homology to the Arabidopsis gene that TC216289 or AY117551 may function in the conversion of dihydrokaempferol to dihydroquercetin in the seed coat.

For biosynthesis of proanthocyanidin, the key enzymes are the *Banyuls* homologs that function as anthocyanidin reductase (ACR or BAN, TC220896 and TC226859), and the leucoanthocyanidin reductase (LAR, TC232038). The expression of the homologs of these two genes was recently investigated in the legume *Lotus corniculatus* (Paolocci et al. 2007), but not in soybean.

The function of another gene that controls anthocyanin accumulation in soybean, the soybean *R* gene, remains unknown although it is thought to function in the anthocyanin pathway before accumulation of the anthocyanins per se (Todd and Vodkin 1993). Finally, in soybean, an extra copy of the chalcone synthase *CHS1* gene, *ICHS1*, at the I (inhibitor) locus (Senda et al. 2002) actually causes suppression of pigment accumulation through homology-dependent gene silencing of CHS (Tuteja et al. 2004; Clough et al. 2004). The presence of this duplicated *CHS1* gene in most commercial soybean lines is responsible for the yellow seed coat of these lines. In homozygous recessive (*i*) lines, the entire seed coat is pigmented, whereas additional *i* alleles restrict pigmentation to specific areas of the seed. CHS is also required for anthocyanin/proanthocyanidin based pigmentation of the testa in Arabidopsis and TT4 encodes a CHS.

Isoflavones and Isoflavone-Derived Metabolites

Due to their pharmacological activities, there has been much interest in recent years in the biosynthesis of isoflavones in soybean seed. However, very early work on the isoflavone metabolic pathways in soybean was greatly stimulated by an interest in accumulation of defense-related metabolites, including the phytoalexins. The soybean-*Phytophthora sojae* association was one of the classical interactions in which early host-pathogen biochemistry focused. This was in part due to the identification of the cell wall glucan elicitor (Ps-WGE) from this oomycetic pathogen (Ayers et al. 1976), an event that not only provided one of the earliest characterized pathogen-derived elicitors, but greatly facilitated biochemical characterization of the pathways stimulated by this elicitor. Pioneering work in several laboratories (see especially work from the Albersheim, Grisebach, Keen, Yoshikawa and Ward labs, reviewed in Ebel 1986) was instrumental in making this system one of the best understood in terms of the metabolic pathways involved and their regulation. More recent work led to more detailed insight into the importance of these pathways in *P. sojae* resistance (for instance, recent work from the labs of Ebel, Hahn, Graham and Yoshikawa, partially reviewed in Hammerschmidt 1999) and their roles in soybean's interactions with the symbiont *Bradyrhizobium japonicum* (reviewed in Stacey et al. 2006).

The major pathways involved in isoflavone and isoflavone-derived metabolism are outlined in Fig. 13.1B. While the presence of isoflavones in soybean seed was known for some time, the finding of isoflavone conjugates as major metabolites in all seedling organs (Graham 1991b) highlighted the importance of the flux of these metabolites into and out of the multiple conjugated forms. In particular, daidzein is a precursor for a variety of other metabolites, including the coumestans (coumestrol, not shown in Fig. 13.1B) and the pterocarpans, including the soybean phytoalexins, the glyceollins (Ebel 1986). For the sake of simplicity, enzymes for the formation of glycitein, an additional isoflavone, are not shown in Fig. 13.1B. Glycitein contributes 5–15 % of total isoflavone contents in soybean seeds. The enzymes responsible for isoflavone-6-hydroxylation and related methyltransferase have not been identified.

The enzymes for the glyceollin branch pathway starting from the isoflavone daidzein are also shown in Fig. 13.1B. In simple terms, these reactions include a cyclization to form the pterocarpan structure, the addition of an isopentenyl side chain, followed by an additional cyclization to form glyceollin I or II. Of the various enzymes involved, TCs for IF2′H, 2′DHDR, IFR, and DHP6aH can be identified based on homologies to cloned sequences. The genes for THPDMAT and GS have not yet been identified so it is not possible to assign a TC to them.

The cytochrome P450 proteins are important to many of the enzymatic steps outlined in Fig. 13.1B. This is a very complex family of proteins (Ralston and Yu 2006) with many subfamilies based on sequence homology. The soybean P450's have been classified (Kim et al. 2004) and their expression examined in response to various stimuli (Kim et al. 2004) including Ps-WGE (Schopfer and Ebel 1998a). The cytochrome P450's are particularly difficult to work with, due to the very high degree of homology among family members even at the DNA level. Most P450 enzymes are associated with ER-membranes, which also makes them difficult to study. Thus expression and enzymatic analyses are usually needed to rigorously establish the function and identity of family members. As an example, although 9 cytochrome P450's were identified by differential display following treatment with Ps-WGE (Schopfer and Ebel 1998a), only the enzymatic function of two of them, cinnamate 4-hydroxylase (C4H, Schopfer and Ebel 1998a) and dihydroxypterocarpan 6a-hydroxylase (DHP6aH, Schopfer et al. 1998b), were verified by functional expression. So, while it is possible to broadly classify these genes and their products, this is an area where a combination of genomics with in depth biochemistry and techniques such as gene silencing (see, e.g., Subramanian et al. 2005) will be necessary for further functional analysis.

Transcription Factors Regulating the Phenylpropanoid Pathways

Several classes of transcription factors were shown to coordinately regulate biosynthesis in the flavonoid pathways. Among them, Myb-like transcription factors are the most extensively studied (Jin and Martin 1999). The *PAP1* gene of Arabidopsis and *C1* gene of maize are both Myb-like transcription factors that regulate anthocyanin biosynthesis by recognizing consensus *cis*-elements that exist in the promoters of almost all phenylpropanoid structural genes (Borevitz et al. 2000; Martin and Paz-Ares 1997). Ectopic expression of these transcription factors led to specific activation of the pathway and accumulation of anthocyanins in the tissues where the transcription factors are activated. However, Myb-like transcription factors are one of the biggest transcription factor families in plants. It will be difficult to predict the function of these regulators simply based on their sequence homology (Stracke et al. 2001). There are at least 92 Myb-like transcription factors in soybean EST database. The functions of these Myb genes have not been reported. Furthermore, soybean isoflavonoids and flavonoids clearly have different functions during growth and development. The defense related isoflavonoid compounds may be regulated by a different set of transcription factors than the anthocyanin specific Myb genes. None of these regulators have been identified.

Genomics of Phenylpropanoid Pathways

In addition to GenBank, there are several useful web-based soybean genomic browsers, including one maintained at Southern Illinois University (http://soybean-genome.siu.edu/) and SoyBase at Iowa State University (http://soybase.agron.iastate.edu/). Finally, the public soybean EST and related databases discussed immediately below can be accessed at http://compbio.dfci.harvard.edu/tgi/cgi-bin/tgi/gimain.pl?gudb=soybean.

Mining the Soybean EST Database

The public soybean EST database is comprised of over 80 cDNA libraries with over 300,000 expressed sequence tags (ESTs) derived from mRNA expressed in a variety of tissues from various cultivars under a diverse conditions. Private EST databases (DuPont and Monsanto) are also currently available for analysis under a signed agreement. The ESTs within the public database were assembled into tentative contigs (TCs) many of which were assigned possible coding sequences for individual genes (Shoemaker et al. 2003). Information on the ESTs comprising these TCs and many tools for their analysis can be found at the DFCI Soybean Gene Index, formerly the TIGR Soybean Gene Index (http://compbio.dfci.harvard.edu/tgi/cgi-bin/tgi/gimain.pl?gudb=soybean). Information available includes descriptions of the EST libraries, putative functional annotations for the TCs, etc. A useful tool is provided for mining the EST database to estimate the relative level of "expression" of any particular TC in a given library. This is accomplished electronically by determining the number of ESTs for that TC in the library of interest. For instance, for TC203618 (Chalcone synthase 7), there are 4 ESTs in the Phytophthora infection library, E5T. Since this library has a total of 3524 ESTs, this TC accounts for 0.11 % of the ESTs sequenced in that library. Importantly, there are several considerations that limit the usefulness of such analyses to more global or qualitative interpretation. However, they can be useful to gain an initial overall perspective on the relative importance of different pathways in different libraries and can in some cases lead to the identification of specific family members of a gene for further verification by functional analysis.

Of the genomic resources currently available for soybean, the EST and related databases have been very useful. We used it in this review to "mine" global information on the secondary product enzymes present and then some qualitative information on their expression under certain conditions. We began by examining the entire public soybean EST database (all libraries) for TCs that putatively encode enzymes in the pathways shown in Fig. 13.1. This was done by keyword searching of the TC annotations as well as independently BLASTing the database (tblastx) with known homologs of key enzymes from GenBank. Generally, we excluded from our final list TCs with less than a 40–50 % stretch of similarity to known genes and/or with a tblastx e-value greater than e −30 to a known gene. The results are grouped by branch pathway and then by specific enzyme in Table 13.1. A preliminary grouping of secondary product TCs from soybean and a number of other species was previously compiled for comparison (Dixon et al. 2002), but the TC numbers for soybean have changed completely from this previous tabulation due to more detailed and accurate sequence alignments.

Nearly all of the enzymes in Fig. 13.1 are represented in Table 13.1. Several specific branch enzymes not included in the simplified pathways in Fig. 13.1 are also listed. While the biochemical function of a few cytochrome P450's were verified (e.g., C4H, Schopfer and Ebel 1998a, and dihydroxypterocarpan 6a-hydroxylase, Schopfer et al. 1998b), due to their complexity and relatively high degree of homology to one another, we simply listed most of the unidentified cytochrome P450's as a separate group. Likewise, the functions of the very diverse peroxidases were not verified, so we again simply listed them a separate class. We also included a class of enzymes in a "Miscellaneous" category. These are enzymes for which the precise fit in the pathways (Fig. 13.1) is unknown. In a few cases (e.g., with the isoflavone synthase family members) we listed specific family members corresponding to a particular TC. This kind of analysis can provide a preliminary road map toward understanding the biosynthetic pathways.

Global Perspectives on Phenylpropanoid Pathways Associated with Various Selected Phenotypes

We next undertook a comparative analysis of the expression of the various genes in Fig. 13.1 (and Table 13.1) by examination of selected key libraries for a few of the phenotypes or events most often associated with various branches of these pathways. Specifically, we focused on libraries corresponding to seed coats, flowers and disease. Due to its potential interest to disease resistance, we also included a salicylic acid response library. Because of the lack of definitive libraries for differentiation of enzymes for the leaf flavonols, we did not examine these. Details of the libraries that we examined in depth are shown in Table 13.2. Although other potentially relevant libraries exist, since library size can limit the usefulness of expression evaluation by EST counts, we rejected any library with fewer than 1,000 ESTs. After deriving complete lists of the TCs for secondary product metabolic enzymes from each of the libraries of interest, we examined EST expression levels for the same TCs in target and "control" (or reference) libraries whenever relevant (e.g., untreated hypocotyls for the *P. sojae* hypocotyl infection libraries, see Table 13.2). After subtraction of

Table 13.1 TCs for soybean phenylpropanoid/miscellaneous secondary product enzymes

Enzyme class	Specific enzyme	Abbreviation	EC number	TCs
Early phenylpropanoid	Phenylalanine ammonia lyase	PAL	4.3.1.5	TC216852, TC225162, TC225165, TC225168
	Cinnamate-4-hydroxylase	C4H	1.14.13.11	TC230826, TC204341
	4-coumarate-CoA ligase	4CL	6.2.1.12	TC205641, TC216083, TC216190, TC219164, TC226636, TC226637, TC227531
Lignin	Caffeoyl-CoA O-methyltransferase	CCOMT	2.1.1.104	TC204613, TC204614, TC221351, TC225339, TC225342, TC225343, TC225345, TC228561, TC228993
	Cinnamoyl-CoA reductase	CCR	1.2.1.44	TC204999, TC208377, TC210404, TC219427, TC225292, TC225293, TC226045, TC226046, TC226919
	Caffeate O-methyltransferase	COMT	2.1.1.68	TC226262, TC226264, TC226266
	Cinnamyl-alcohol dehydrogenase	CAD	1.1.1.195	TC204437, TC204440, TC206457, TC208568, TC215695, TC225589, TC225590, TC225591, TC227396, TC229375, TC229704
	Laccase (Diphenol oxidase)	LAC	1.10.3.2	TC206546, TC209344, TC210109, TC217098, TC217105, TC219245, TC222550, TC229326
Flavonol	Chalcone synthase	CHS	2.3.1.74	TC203259, TC203618 (CHS7), TC210490, TC214207, TC214443 (CHS1), TC216931, TC215169, TC225256

Table 13.1 (continued)

Enzyme class	Specific enzyme	Abbreviation	EC number	TCs
	Chalcone isomerase	CHI	5.5.1.6	TC205496 (CHI4), TC205536 (CHI1a), TC219615, TC216784, TC216785, TC216488, TC216489
	Flavanone 3-hydroxylase	F3H	1.14.11.9	TC205989, TC208325, TC209692, TC226020
	Flavonoid 3′-hydroxylase	F3′H	1.14.13.21	TC211750, TC216289
	Flavonol synthase	FS	1.14.11.23	TC208467
	Flavonol 3-O-glucosyltransferase	F3GT	2.4.1.91	TC208101, TC203723, TC208339, TC229034, TC231164, TC234035
	Chalcone O-methyltransferase	ChOMT	2.1.1.154	TC208485, TC208486
Anthocyanin	Dihydroflavonol-4-reductase	DHF4R	1.1.1.219	TC217923, TC228195
	Leucoanthocyanidin reductase	LAR	1.17.1.3	TC232038
	Leucoanthocyanidin dioxygenase	LCO	1.14.11.19	TC208819, TC217747, TC217748, TC218608, TC220111, TC220906, TC221541, TC232276
	Anthocyanidin reductase	ACR (BAN)		TC220896, TC226559
	Anthocyanin glucosyl transferase	AGT	2.4.1.91	TC234097
Isoflavone	Chalcone reductase	CHR	1.1.1.–	TC203399, TC227002, TC229251
	Isoflavone synthase	IFS	1.14.14.–	TC215321 (IFS1), TC204612 (IFS2)
	Isoflavone-7-O-methytransferase		2.1.1.150	TC206764, TC206765, TC210019, TC215839
	Trihydroxyisoflavone O-methyltransferase	THIOMT	2.1.1.46	TC216576
	Isoflavone 7-O-glucosyl transferase	IFGT	2.4.1.–	TC211462, TC220510, TC231296
Glyceollin	Isoflavone 2′-hydroxylase	I2′H	1.14.13.89	TC204662, TC204663, TC204664, TC204665
	2′-hydroxydihydrodaidzein reductase	2′DHDR	1.1.1.–	TC224892, TC226859, TC213000
	Isoflavone reductase	IFR	1.3.1.45	BE611038, TC204692, TC206041, TC207832, TC210161, TC215001, TC215002, TC215003, TC215004, TC215005, TC215632, TC217830, TC226270
	Dihydroxypterocarpan 6a-hydroxylase	DHP6aH		TC227063

Table 13.1 (continued)

Enzyme class	Specific enzyme	Abbreviation	EC number	TCs
Cytochrome P450s	Cytochrome P450	P450	1.14.-.-	TC204664, TC205134, TC205337, TC205410, TC205411, TC206973, TC207603, TC207959, TC209682, TC211282, TC212296, TC216063, TC216663, TC217041, TC219004, TC219972, TC220823, TC221496, TC227063, TC228930, TC229792, TC230633, TC231530, TC231675
Peroxidases	Peroxidase	POX	1.11.1.7	TC203271, TC203397, TC203489, TC204153, TC204257, TC204493, TC204494, TC204496, TC204669, TC204671, TC205158, TC205794, TC205795, TC205894, TC206016, TC206181, TC206280, TC206281, TC206384, TC206605, TC206606, TC207185, TC207747, TC209015, TC209413, TC212050, TC213104, TC214084, TC214085, TC214302, TC214651, TC214657, TC214752, TC214928, TC215014, TC215015, TC215016, TC215017, TC216323, TC216324, TC216834, TC216945, TC217066, TC217976, TC218583, TC218639, TC218640, TC219609, TC221519, TC221869, TC225212, TC225213, TC226619, TC226717, TC226843, TC226844, TC227439, TC227916, TC228199, TC228613, TC229562, TC229710, TC229834, TC230675, TC230907, TC231488, TC232832, TC233091, TC234043, TC221178, TC230884
Miscellaneous	Beta-glucosidase		3.2.1.21	TC205953, TC206214, TC208892
	Hydroxycinnamoyl transferase		2.3.1.110	TC215706, TC215707, TC215709
	Carboxy-lyase		4.1.1.–	TC208106
	cis-trans-Isomerase		5.2.1.–	TC209912
	Oxidoreductase acting on the CH-CH group of donors		1.3.–.–	TC208996

Table 13.2 EST libraries used for comparisons of phenylpropanoid pathways

Library	Total ESTs	Cultivar	Tissue	Age	Treatment	Lab
	Berberine bridge enzyme-like protein			TC222911		
81E (Gm-c1074)	6953	Will 82	Leaves	9–11 day	9–11 day old seedlings that were induced for HR by vacuum infiltrating plant tissue with *Pseudomonas syringae* pv. *glycinea* carrying the *avrB* gene (Genetics 141:1597–1604). Plant tissue (expanded unifoliate leaves) was collected at 2, 4, 8, 12, 24, 36, and 53 hrs after inoculation and their mRNA pooled equally.	S. Clough/L. Vodkin
8J2 (Gm-c1076)	4354	Will 82	Cotyledon	11 day	Wounded cotyledons were treated with 2 ugs/ml of a crude glucan elicitor preparation isolated from the mycelial walls of *Phytophthora sojae*. Treated surface tissue was harvested after 24 hr in the dark.	M. Hahn
9IS (Gm-c1084)	4484	Will 82	Hypocotyl	NA	Tissue was inoculated with *Phytophthora sojae* (race 1, incompatible) and tissues were harvested 2 and 4 hrs following infection. The library is the pool of these two time points.	M. Bhattacharyya
E5T	3524	Harosoy	Hypocotyl	NA	Hypocotyls were inoculated with *Phytophthora sojae* culture P6497 (race 2, compatible) and harvested 48 hr following infection.	B. Tyler

Table 13.2 (continued)

Library	Total ESTs	Cultivar	Tissue	Age	Treatment	Lab
2567 (Gm-c1023)	3556	T157	Seed Coats	NA	mRNA was isolated from seed coats (100–200 mgs) of greenhouse grown plants	L. Vodkin, A. Khanna
5649 (GmaxSC)	2051	Harosoy 63	Seed Coats	NA	Library was constructed from polyA+ enriched mRNA from green seed coats in mid to late developmental stage, average fresh weight 250 mg per seed. Traces of pod and embryo tissue also present.	M. Gijzen
GM18 (Gm-c1019)	5385	Williams	Seed Coats	NA	cDNA library was constructed from mRNA isolated from immature seed coats (200–300 mgs) of greenhouse grown plants	L. Vodkin, A. Khanna
6112 (Gm-c1061)	4045	Raiden	Flowers	NA	mRNA isolated from mature flowers of field grown plants	R. Shoemaker
GM16 (Gm-c1016)	8532	Williams 82	Flowers	NA	mRNA isolated from immature flowers of field grown plants	R. Shoemaker, J. Erpelding
GM20 (Gm-c1015)	5334	Williams 82	Flowers	NA	mRNA isolated from mature flowers of field grown plants	R. Shoemaker, J. Erpelding
DE8 (Soybean induced by Salicylic Acid)	28520	Kefeng 1		2 weeks	mRNA isolated from two-week seedlings (cultivar Kefeng 1) treated by spraying 2.0 mM salicylic acid for 24, 36, 48 and 72 hr	S.-Y. Chen
Some reference libraries used						
Gm-c1013	2922	Williams	Whole seedlings	2–3 weeks	mRNA isolated from whole seedlings of 2–3 week old greenhouse grown plants	R. Shoemaker, J. Erpelding

Table 13.2 (continued)

Library	Total ESTs	Cultivar	Tissue	Age	Treatment	Lab
5567 (Gm-c1045)	5384	Will 82	Hypocotyl	9–10 day	mRNA isolated from etiolated hypocotyl tissue of 9–10 day old seedlings of the cultivar Williams 82	R. Shoemaker
GM26 (Gm-c1027)	7571	Williams	Cotyledon	3 and 7 day	mRNA isolated from cotyledons of 3- and 7-day-old Williams seedlings which were propagated on paper towels with distilled water. The cotyledons were flash-frozen in liquid nitrogen, then lyophilized for 72 hrs. Unequal amounts of mRNA was used for cDNA synthesis	P. Keim, V. Coryell
6109 (Gm-c1050)	3393	Clark	Leaf	Various develop stages	mRNA isolated from leaf tissue at various developmental stages of 3 week old greenhouse grown plants	R. Shoemaker, J. Specht, P. Keim

the reference EST expression levels, we finally rejected from these TC lists any TC for which the number of ESTs matching that TC did not comprise at least 0.05 % of the library. By using this percent, we normalized "expression" to library size. With a very few exceptions, we found that enzymes of the phenylpropanoid pathways are highly expressed in the libraries examined, with comparatively very low constitutive expression in control or reference libraries. In fact, expression of a given TC in the libraries chosen for our examinations often accounted for 50–100 % of the total expression for that TC in all soybean EST libraries.

Employing these various limitations, Fig. 13.2A shows a Venn diagram for the presence of secondary metabolic TCs in various global categories of libraries. For this global analysis, we combined the TC lists for each category (e.g. all flower libraries were combined, etc). From this global perspective, only 2 TCs were shared by all libraries, which represents only 4 % of those in *P. sojae* libraries, 7 % of Flower libraries and 17 % of the Seed Coat libraries. While the numbers of TCs involved in these correlations may seem relatively small compared to other genome-wide comparisons, this is a result of limiting the comparisons to enzymes of phenylpropanoid metabolism and to setting relatively stringent limitations on inclusion of a specific TC.

Using the same limitations for the data, Fig. 13.2B shows a Venn diagram for comparison of three separate *P. sojae* related libraries (combined for the analysis in Fig. 13.2A). As described in Table 13.2, these include two hypocotyl infection libraries (an incompatible infection at 2–4 hr and a compatible infection at 48 hr) and a cotyledon *P. sojae* cell wall glucan elicitor (Ps-WGE) response library. Based on expression analysis of responses to infection or Ps-WGE by many years of HPLC metabolic profiling or mRNA analyses, it was concluded that Ps-WGE is of primary importance in soybean – *P. sojae* interactions (see, e.g., Graham et al. 1990; Graham

(A)

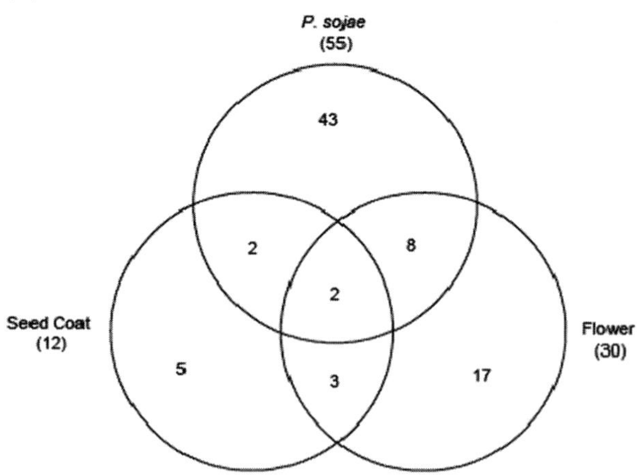

Fig. 13.2A Relationships of secondary product metabolic genes between various broad categories of EST libraries

(B)

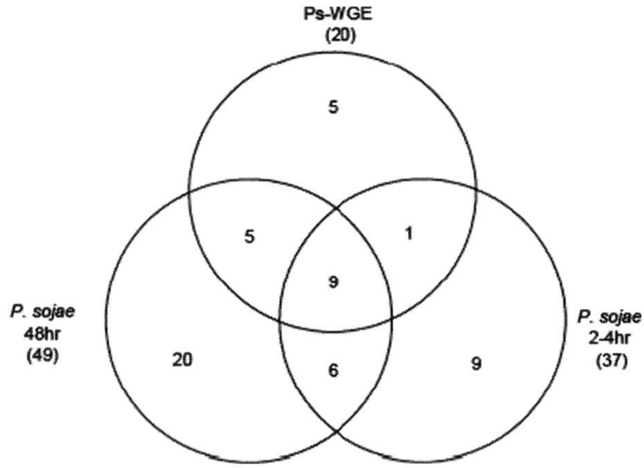

Fig. 13.2B Relationships of secondary product metabolic genes between defense related EST libraries

1991a,Graham and Graham1991b; Graham et al. 2003). Indeed, the majority (15/20) of TCs expressed in the Ps-WGE library are common to either one or the other of the infection libraries. There are 9 TCs common to all libraries, which represents 45 % of the TCs expressed in the Ps-WGE library, 24 % of those in the 2–4 hr *P. sojae* infection library and 18 % of those in the 48 hr *P. sojae* infection library. Thus, there is considerable overlap between the *P. sojae* related libraries, particularly that of the elicitor with the two infections. On the other hand, the two infection libraries share only 15 TCs, amounting to ca. 30–40 % of these two libraries. This is not surprising given the different nature and timing of the two libraries (compatible at 48 hr or incompatible at 2–4 hr). In this same context, however, it is interesting that the compatible library shows as much expression of phenylpropanoid metabolism genes as it does. This may reflect the common observation, in many studies of the phenylpropanoid responses in soybean – *P. sojae* interactions, that the differences in incompatible and compatible infections is more a matter of timing, with the latter expressing key phenylpropanoid responses later, after the infection front has passed through the tissue. It is also consistent with the fact that with more pathogen biomass in the compatible infection, it is likely that there will also be much more Ps-WGE present. These issues cannot be easily addressed from the study of whole organs and is an excellent example of where use of laser capture microdissection techniques will be useful. Recently, we completed microarrays on Ps-WGE treated cotyledon tissues (4 hr post treatment). The resulting data are discussed below in comparison to the TC expression shown in Fig. 13.2B.

Finally, another analysis of global interest is the relationship of infection by *P. sojae* to that by *Pseudomonas syringae* pv. *glycinea* (Psg) and the relationships of both to salicylic acid treatment. Figure 13.2C shows a Venn diagram comparing the TCs expressed in library Gm-C1084 (*P. sojae*, incompatible, 2–4 hr), library

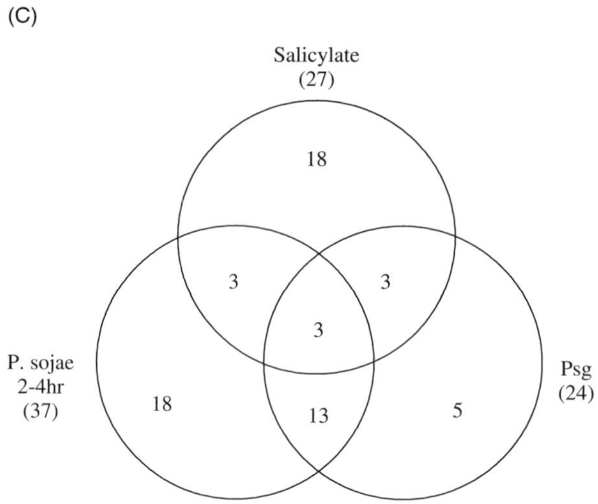

Fig. 13.2C Relationships of secondary product metabolic genes between pathogen and salicylic acid EST libraries

8IE (Psg, incompatible, mixed time points, 2–53 hr) and library DE8 (SA induced). There is little apparent overlap in EST expression in the SA library compared to the two incompatible infection libraries. There are only 3 TCs in common to all three libraries, representing 8 %, 11 %, and 12 % of selected TCs for the *P. sojae*, SA and Psg libraries, respectively. On the other hand the Psg and *P. sojae* infection libraries have 16 TCs in common, accounting for 55 % and 42 % of the Psg and *P. sojae* libraries respectively. Thus, at least in terms of secondary product response genes, there seem to be some similarities between resistance responses induced by infection with either oomycetic or bacterial pathogens. The lack of a clear correlation of SA responses to either infection library is consistent with our long term negative results with SA in terms of activation of phenylpropanoid responses or local or systemic resistance in soybean (T.L. Graham, unpublished). In contrast, ACC oxidase (involved in ethylene generation) and 12-oxophytodienoic acid 10,11-reductase (OPR, involved in jasmonic acid accumulation) are both strongly activated in the various disease resistance libraries, consistent with published conclusions (Creelman et al. 1992; Graham et al. 2003) that these wound signals participate in up-regulation of phenylpropanoid pathway enzymes. These results are further discussed in the comparisons of EST library expression and microarray results described below.

While it is tempting to examine the specific enzymes that contribute to the differences and similarities suggested by the Venn diagrams, due to the limited nature of the data, this is really not feasible. However, we found that examination of differences in the branch pathways for which TCs were present gave some additional and interesting global insights. In Fig. 13.3A, we first compare the classes of metabolic

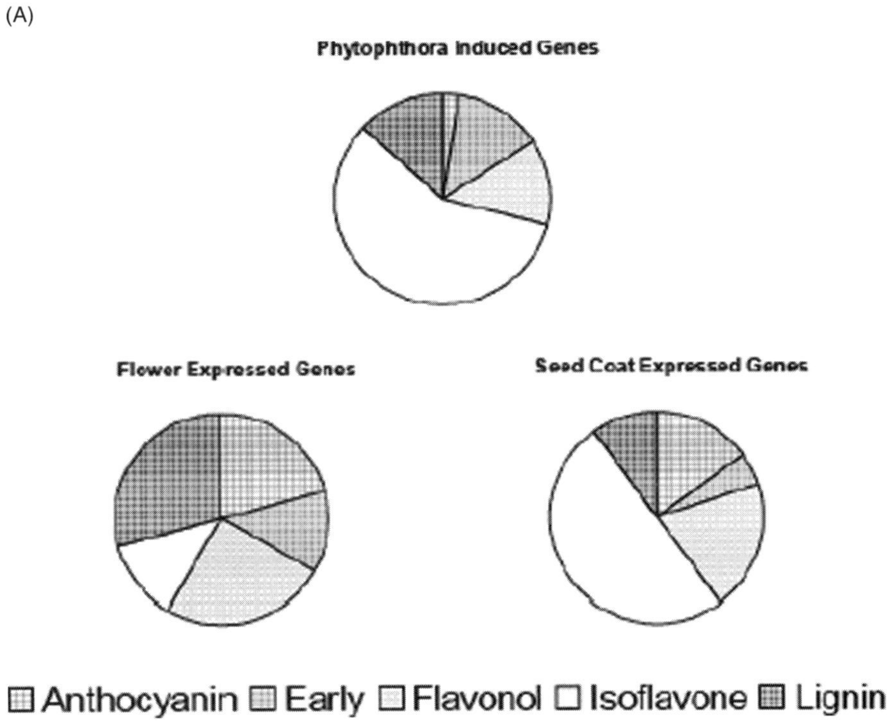

Fig. 13.3A Comparison of metabolic genes expressed in widely different libraries

genes expressed in the various *P. sojae* infection libraries to those in the seed coat and flower libraries. The enzymes for isoflavone metabolism (including the pterocarpans) predominate in the *P. sojae* infection libraries. In the flower libraries the expression of various genes is much more balanced, with a clear and much stronger contribution from the anthocyanin pathway. Although anthocyanin genes also contribute to the seed coat libraries, an interesting finding was that isoflavone enzymes also contribute strongly to these libraries. It is difficult to know if this is actually true or if it is due to slight contamination of this library with tissues from the underlying endosperm, which is known to be producing high levels of isoflavones (Dhaubhadel et al. 2003). Indeed, the description of one of the seed coat libraries (Table 13.2) points out this possibility.

In Fig. 13.3B, the libraries for various defense responses are compared. Isoflavone enzymes predominate for both *P. sojae* and Psg, consistent with the overlap in the Venn diagrams. This is also true for the Ps-WGE library. On the other hand, the SA library has a more balanced distribution, with much lower expression of isoflavone related genes and significant contributions from the lignin and anthocyanin pathways. This mirrors the discussion above based on Venn diagrams and the conclusion that the SA and disease resistance libraries are quite different.

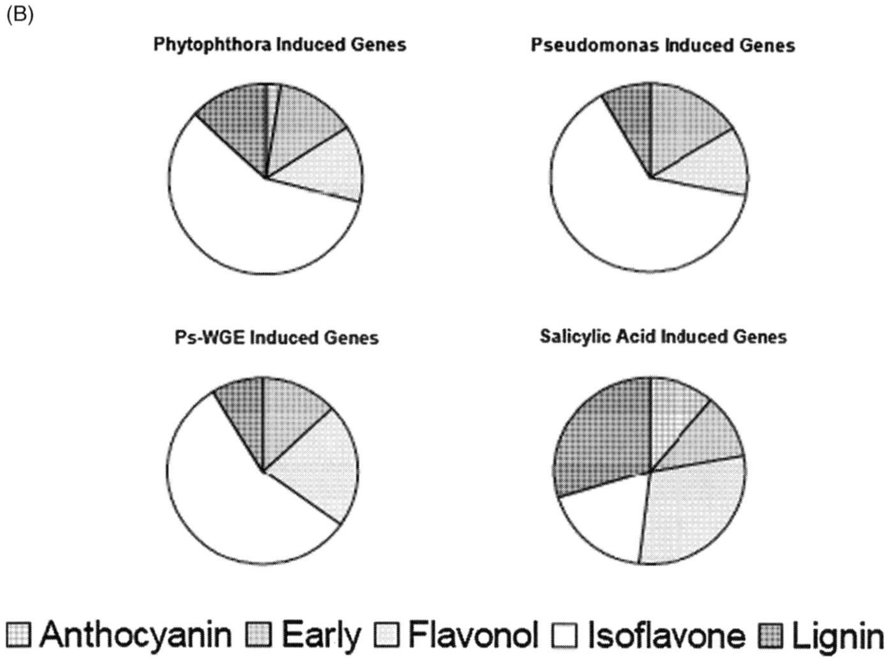

Fig. 13.3B Comparison of metabolic genes induced in defense libraries

Phenylpropanoid Pathways in EST Libraries and Microarray Experiments Associated with Pathogen Resistance

To compare expression in the various *P. sojae* and Psg EST libraries to that in three microarray experiments, we first tabulated the TCs that had 0.1 % EST expression or greater in at least one of these defense EST libraries. We then examined expression for each of these TCs in three microarray experiments: (1) a *P. sojae* hypocotyl infection experiment (Moy et al. 2004) involving a custom array of a smaller number of soybean genes previously seen to be induced in an earlier cDNA library (E5S); (2) an Affymetrix microarray experiment (MY Graham, unpublished) involving a 4 hr treatment of cotyledons by either wounding alone or wounding and WGE elicitor treatment; (3) a microarray experiment involving the unigene re-racked cDNA microarrays and infection by Psg (Zou et al. 2005; Zabala et al. 2006). The TCs chosen for these analyses and information on them is presented in Table 13.3. The results of the comparisons of their expression across the selected libraries and arrays are shown in Table 13.4.

For comparative purposes, Array 1 is somewhat related to both *P. sojae* infection EST libraries. All are hypocotyl infections. The soybean – *P. sojae* EST library E5T is a compatible infection at 48 hr, whereas library 91S is an incompatible infection at 2–4 hr. Array 1 is an incompatible infection over a time frame of 3–48 hr. The data shown in Table 13.3 was derived from the 48 hr data for better comparison to E5T.

Table 13.3 TCs examined in depth in various defense libraries

TC number	Group	Enzyme	Specific information	GB # of Closest Homolog	Tblastx e-value	Genus
TC204341	Early	C4H	elicitor induced cinnamate 4-hydroxylase	X92437	0	Glycine
TC216852		PAL	Phenylalanine ammonia-lyase 1	X52953	0	Glycine
TC225162		PAL	Phenylalanine ammonia-lyase (class II)	AB236800	0	Trifolium
TC225165		PAL	Phenylalanine ammonia-lyase (class II)	AB236800	0	Trifolium
TC215695	Lignin	CAD	Cinnamyl-alcohol dehydrogenase.	AF053084	0	Malus
TC225589		CAD	Cinnamyl-alcohol dehydrogenase.	AF053084	8e-169	Malus
TC204999		CCR	Cinnamoyl-CoA reductase	NM_127952	3e-50	Arabidopsis
TC204153		POX	Peroxidase.	AJ011939	5e-174	Trifolium
TC205496	Flavonoid	CHI	Chalcone isomerase 4	AY595417	0	Glycine
TC205536		CHI	Chalcone isomerase 1A	AY595413	0	Glycine
TC203618		CHS	Chalcone synthase 7	M98871	5e-163	Glycine
TC214443		CHS	Chalcone synthase 1	DQ239918	0	Glycine
TC215169		CHS	6′-deoxychalcone synthase	X55730	0	Glycine
TC225256		CHS	6′-deoxychalcone synthase	X55730	2e-145	Glycine
TC224552		SAMS	S-adenosylmethionine synthetase	Z71271	0	Catharanthus
TC215321	Isoflavone	IFS	Isoflavone synthase 1	AF195798	0	Glycine
TC204612		IFS	Isoflavone synthase 2	AF135484	0	Glycine
TC226859	Glyceollin	2′DHDR	2′-hydroxydihydrodaidzein reductase	AJ003246	0	Glycine
TC204663		I2′H	Isoflavone 2′-hydroxylase	AY278227	0	Medicago
TC204664		I2′H	Isoflavone 2′-hydroxylase	DQ340238	0	Glycine
TC204692		IFR	Isoflavone reductase-like protein	AB265592	3e-176	Lotus
TC215001		IFR	Isoflavone reductase homolog 2	AF202183	0	Glycine
TC215002		IFR	Isoflavone reductase	AJ003245	1e-178	Glycine
TC215632		IFR	Isoflavone reductase homolog 1	AF202184	0	Glycine
TC227063		DHP6aH	Dihydropterocarpan 6a-hydroxylase	D83968/Y10492	0	Glycine
TC206973	P450	P450	Elicitor induced P450	Y10493	0	Glycine
TC216063		P450	Elicitor induced P450	Y10491	0	Glycine
TC225031	Signaling	ACCO	1-aminocyclopropane-1-carboxylate oxidase	AB062359	0	Phaseolus
TC225757		OPR	12-oxophytodienoate reductase (OPR2)	NM_106319	0	Arabidopsis

Table 13.4 Expression of selected genes in various defense libraries and arrays*

			Expression in *P. sojae* – related EST libraries			Expression in *P. sojae* microarray (Array 1)	Expression in WGE microarray (Array 2)	Expression in wound microarray (Array 2)	Expression in Psg EST library	Expression in Psg microarray (Array 3)	
TC number	Group	Enzyme	Library E5T	Library 91S	Library 8J2	48 hr time point, 2X or greater	4 hr time point, 3X or greater	4 hr time point, 3X or greater	Library 8IE	I, 24 hr	C, 24 hr
TC204341	Early	C4H	0.23	0.18	0.05		X	X	0.03		
TC216852		PAL	0.06	0.09		X		X			
TC225162		PAL	0.60	0.02	0.07				0.03	2.3	0.0
TC225165		PAL	0.14		0.05				0.01		
TC215695	Lignin	CAD	0.11								
TC225589		CAD	0.11						0.01	2.0	0.0
TC204999		CCR	0.06	0.18				X			
TC204153		POX	0.23	0.22	0.34			X	0.04		
TC205496	Flavonoid	CHI 4	0.03	0.09				X	0.03	7.0	0.6
TC205536		CHI 1a	0.11	0.09	0.05			X	0.01	13.0	6.0
TC203618		CHS 7	0.11	0.02	0.05			X	0.03		
TC214443		CHS 1	0.34	0.42	0.02			X	0.03	17.0	6.0
TC215169		CHS	0.11	0.18	0.02	X			0.04	9.0	4.0
TC225256		CHS	0.20	0.09	0.11				0.09	31.0	3.0
TC224552		SAMS	0.06	0.13	0.09			X	0.09		

Table 13.4 (continued)

			Expression in *P. sojae* – related EST libraries			Expression in *P. sojae* microarray (Array 1)	Expression in WGE microarray (Array 2)	Expression in wound microarray (Array 2)	Expression in Psg EST library	Expression in Psg microarray (Array 3)	
TC number	Group	Enzyme	Library E5T	Library 91S	Library 8J2	48 hr time point, 2X or greater	4 hr time point, 3X or greater	4 hr time point, 3X or greater	Library 8IE	I, 24 hr	C, 24 hr
TC215321	Isoflavone	IFS 1	0.26	0.13	0.07	X		X	0.01	6.0	3.0
TC204612		IFS 2	0.31	0.18	0.14			X	0.09	17	7
TC226859	Glyceollin	2′HDDR	0.11	0.04	0.11	X	X		0.04		
TC204663		I2′H	0.51		0.09		X		0.01		
TC204664		I2′H	0.11		0.02	X	X		0.01		
TC204692		IFR	0.06	0.13							
TC215001		IFR	0.11		0.05		X		0.10		
TC215002		IFR	0.09	0.36	0.05			X	0.06	16.0	0.0
TC215632		IFR	0.14		0.02				0.04		
TC227063		DHP6aH	0.20	0.02	0.02	X	X		0.04		
TC206973	P450	P450	0.14	0.02	0.02		X		0.01		
TC216063		P450	0.17	0.04	0.07	X			0.01		
TC225031	Signaling	ACCO	0.06	0.09	0.18			X	0.24		
TC225757		OPR	0.17	0.2	0.07			X	0.10		

*Values for EST expression libraries represent the percent of the total ESTs for that library which correspond to the specific TC. Values for the Psg microarrays are the fold difference for the treatment versus a mock inoculation (Zabala et al. 2006).

Array 2 is similar in treatment to the Ps-WGE EST library 8J2, except that the time point is much earlier in Array 2 (4 hr compared to 24 hr). In addition, in library 8J2 wounded cotyledon effects were not separated out as they were in Array 2. Array 3 is similar to the Psg infection EST library 81E, except that in the array work both incompatible and compatible infections were examined (Zabala et al. 2006). Array 3 involved sampling over the period 2–53 hr. For the data shown in Table 13.3 for Array 3, we averaged expression in these arrays for each TC at 24 hr.

Although expression of individual genes varies across the various libraries and arrays, an interesting result from the analyses is that nearly all (26/30) of the major genes up-regulated in the various *P. sojae* EST libraries were also up-regulated in at least one of the microarray experiments. Moreover, all of the major pathways (early phenylpropanoid, lignin, flavonoid, isoflavonoid, glyceollin) showed enhanced expression, consistent with these pathways representing the major commitments of phenylpropanoid metabolism in infected tissues (see Venn diagrams and pie charts above). Other enzymes that are up-regulated in at least one library or microarray (data not shown) include homologs of other key enzymes in the early phenylpropanoid (4CL) or lignin pathways (CCR and COMT). That these enzymes were more library/array specific might reflect differences in timing, pathogen and/or the organ examined. The up-regulation of the putative I2′H TCs in library E5T and the Ps-WGE microarray was among the strongest of all genes. Up-regulation of TC 204663 was 0.51 % of E5T and over 500X for the Ps-WGE array. Up-regulation of TC 204664 was somewhat lower, being 0.11 % and over 300X, respectively. This is quite interesting since this is the first committed enzyme for the mobilization of daidzein into the glyceollin branch pathway. It suggests that this critical step is under very strong induction control. Again, it is interesting that there is such strong induction of this enzyme in the compatible infection library, E5T. As discussed previously, this may reflect the possibility that quite a bit of Ps-WGE may be released in the compatible infection (in which more pathogen biomass is present) and the late time point for this library. These observations once again emphasize the need for analysis of more specific tissues than whole organs, for instance those at the infection front.

Confirming the analyses discussed above, salicylic acid was a weak inducer of all of the enzymes shown in Table 13.3. On the other hand, TCs related to two wound induced defense signal molecules, ethylene and jasmonic acid were dominantly induced in many of the libraries or microarrays (Table 13.3, under "Signaling" genes). These included ACCO and OPR, involved in the synthesis of ethylene and jasmonic acid respectively. Interestingly, while there are 22 TCs for ACCO in the soybean EST database, the specific TC for this enzyme induced in the various libraries shown in Table 13.3 appears to be strongly correlated to defense.

Other Relevant Macroarray or Microarray Studies

There are several other studies involving macroarrays or microarrays that describe some general aspects of phenylpropanoid related responses. For the most part these studies do not go into specifics of the phenylpropanoid enzymes up-regulated, but discuss them as part of the general categories of genes that are expressed.

A study of root responses to *Fusarium solani* f. sp. *glycines*, which causes sudden death syndrome (SDS) examined the transcript abundance of 191 ESTs on macroarrays at five time points post inoculation (Iqbal et al. 2005). Root responses of both susceptible (Essex) and partially resistant line RIL23 were compared. The ESTs examined were chosen based on earlier studies from the same lab (Iqbal et al. 2002). Expression of some of the enzymes of early phenylpropanoid metabolism (PAL, C4H, 4CL) and isoflavone metabolism (CHS, IOMT) was noted as being stronger in the partially resistant line, although no quantitative data is presented. However, some interesting specific quantitative analyses of the induction of phenylalanine ammonia lyase (PAL) are described. Much as we described above for the *P. sojae* and Psg libraries, we also undertook an electronic examination of expression of ESTs in the EST libraries for sudden death syndrome [catalog 81C (Gm-c1072) and 81D (Gm-c1073)] corresponding to cultivar PI567374 (partially resistant) and Williams 82 (susceptible) to the disease, respectively. These libraries were constructed from pooled RNA isolated at 4 time points after inoculation. Consistent with the study by Iqbal et al. (2002), expression of phenylpropanoid enzymes (including those of early, lignin, isoflavone and glyceollin pathways) was much higher in the partially resistant line. Interestingly, the pie chart for expression of genes by category (see Fig. 13.3) was quite similar to that for *P. sojae* and Psg (not shown), again suggesting that phenylpropanoid defense responses to a range of pathogens (bacterial, fungal and oomycetic) may involve a similar basic grid of phenylpropanoid enzymes.

Another interesting set of experiments involve microarray analyses of responses of the susceptible cultivar Kent to soybean cyst nematode (Khan et al. 2004; Alkharouf et al. 2005). Although the discussions in these papers are again largely limited to the broad categories of genes, an online resource is available for examination of the microarray data for all ca. 6,000 genes examined. Moreover, microarray data for both resistant and susceptible cultivars are available at the website: http://psi081.ba.ars.usda.gov/SGMD/Publications/OLAP/. There are also soybean EST libraries, though relatively small, for soybean cyst nematode interactions that can be explored (Catalog #'s 9NO, C1G, C1H, C1I).

There are also a series of EST libraries for soybean interactions with *Bradyrhizobium japonicum* that could be mined for interesting data on phenylpropanoid metabolic enzymes. These include libraries for both the cultivar Bragg and its supernodulating mutant (Catalog #s 9DO, 9DP, 9F9, 9G3, GM28).

Recently, a transcriptome analysis following soybean embryo development was carried out using a set of soybean cDNA microarrays (Dhaubhadel et al. 2007). By selecting two cultivars different in isoflavone content, it was discovered that the expression patterns of key isoflavonoid biosynthesis enzymes, such as PAL and IFS were higher at 70 day after pollination in both the cultivars. Therefore, expression of these genes coincides with the onset of isoflavone accumulation. The most interesting discovery was that only CHS7 and CHS8 are highly expressed in the high isoflavone containing lines, suggesting that these two genes may represent the rate limiting enzyme of the pathway.

Finally, an interesting observation from a microarray analysis of soybean leaves grown under elevated carbon dioxide was that two enzymes required for lignin formation, caffeic acid methyl transferase and cinnamoyl-CoA reductase are both moderately up-regulated under high carbon dioxide concentrations (Ainsworth et al. 2006).

Quantitative Trait Loci Associated with Isoflavones

The total isoflavone levels in soybean are clearly controlled by multiple factors. Several groups have identified soybean quantitative trait loci (QTL) for isoflavone levels. Kassem et al. (2004) established more than 100 recombinant inbred lines using the high isoflavone cultivar Essex with the low isoflavone cultivar Forrest. In two separate reports with a total of 390 markers, the group identified 8 QTL markers that relate to the levels of isoflavones (Meksem et al. 2001). Unfortunately, earlier soybean genetic maps and physical maps were notoriously difficult to match. Later, QTL mapping of isoflavone content using a different soybean population (AC756 X RCAT Angora) was reported (Primomo et al. 2005). These QTL markers will be useful for molecular breeding, but there are still major challenges to derive the underlying genes that control isoflavone levels.

Conclusions and Future Perspectives

Our understanding of the genomics of secondary product metabolism in soybean is still in its infancy. Here, we provide a few interesting analyses and perspectives on the data that can currently be mined from the EST and related databases and a few microarray experiments. Much of the promise of the genomic analysis of secondary product pathways in soybean lies in future work and will in many cases (e.g., particularly with the cytochrome P450's and peroxidases) require the careful integration of complementary information from microarrays, laser capture microdissection, biochemical verification of enzymatic function, and the complementary use of metabolomics with emerging tools such as gene silencing (see, e.g., Subramanian et al. 2005).

References

Ainsworth, E. A., Rogers A., Vodkin, L.O., Walter, A., and Schurr, U. (2006) The effects of elevated CO2 concentration on soybean gene expression. An analysis of growing and mature leaves. Plant Physiology 142, 135–147.

Alkharouf, N.W., Jamison, D. C. and Matthews B.F. (2005) Online analytical processing (OLAP): a fast and effective data mining tool for gene expression databases. J Biomed. Biotechnol. 2005(2): 181–188.

Ayers A.R., Ebel J., Valent B.S., Albersheim P. (1976) Host-pathogen interactions. X. Fractionation and biological activity of an elicitor isolated from the mycelial walls of *Phytophthora megasperma* var. *sojae*. Plant Physiol. 57, 760–765.

Boerjan, W., Ralph, J., Baucher, M. (2003) Lignin biosynthesis. Annual Review of Plant Biology 54, 519–546.

Borevitz JO, Xia Y, Blount J, Dixon RA, Lamb C. Activation tagging identifies a conserved MYB regulator of phenylpropanoid biosynthesis. Plant Cell. 2000 Dec; 12, 2383–2394.

Buttery, B.R. and Buzzell, R. I. (1973). Varietal differences in leaf flavonoids of soybeans. Crop Sci. 13, 103–106.

Buttery, B.R. and Buzzell, R. I. (1975) Soybean flavonol glylcosides: identification and biochemical genetics. Can. J. Bot. 53, 219–224.

Buttery, B. R. and Buzzell, R. I. (1987) Leaf traits associated with flavonol glycoside genes in soybean. Plant Physiol. 85, 20–21.

Buzzell, R. I., Buttery, B. R. and Bernard, R. L. (1977) Inheritance and linkage of a magenta flower gene in soybean. Can. J. Genet. Cytol. 19, 749–751.

Buzzell R. I., Buttery B. R., Mactavish D.C. (1987) biochemical genetics of black pigmentation of soybean seed. Journal of Heredity 78, 53–54.

Clough, S. J., Tuteja, J. H., Li, M., Marek, L. F., Shoemaker, R. C., Vodkin, L. O. (2004) Features of a 103-kb gene-rich region in soybean include an inverted perfect repeat cluster of CHS genes comprising the I locus. Genome 47, 819–831

Creelman R.A., Tierney M.L., Mullet J.E. (1992) Jasmonic Acid/Methyl Jasmonate Accumulate in Wounded Soybean Hypocotyls and Modulate Wound Gene Expression. Proc. Natl. Acad. Sci. 89, 4938–4941.

Dhaubhadel, S., McGarvey, B. D., Williams, R., Gijzen, M. (2003) Isoflavonoid biosynthesis and accumulation in developing soybean seeds. Plant Molecular Biology 53,733–743.

Dhaubhadel S, Gijzen M, Moy P, Farhangkhoee M. (2007) Transcriptome Analysis Reveals a Critical Role of CHS7 and CHS8 Genes for Isoflavonoid Synthesis in Soybean Seeds. Plant Physiology 143, 326–338.

Dixon, R. A., Achnine L., Kota, P., Liu, C-J., Reddy, M. S. S., Wang, L. (2002) The phenylpropanoid pathway and plant defence: A genomics perspective. Molecular Plant Pathology 3, 371–390.

Ebel J. (1986) Phytoalexin synthesis: the biochemical analysis of the induction process. Ann. Rev. Phytopathology 24, 235–264.

Farag, M. A., Huhman, D., V., Lei, Z., Sumner, L. W. (2007) Metabolic profiling and systematic identification of flavonoids and isoflavonoids in roots and cell suspension cultures of Medicago truncatula using HPLC-UV-ESI-MS and GC-MS. Phytochemistry 68, 342–354.

Gavnholt, B., Larsen, K. (2002) Molecular biology of plant laccases in relation to lignin formation. Physiologia Plantarum 116, 273–280.

Graham M.Y., Graham T.L. (1991a) Rapid accumulation of anionic peroxidases and phenolic polymers in soybean cotyledon tissues following treatment with *Phytophthora megasperma* f. sp. *glycinea* wall glucan. Plant Physiol. 97, 1445–1455.

Graham M.Y., Weidner J., Wheeler K., Pelow M.L., Graham T.L. (2003) Induced expression of pathogenesis-related protein genes in soybean by wounding and the *Phytophthora sojae* cell wall glucan elicitor Physiol. Molec. Plant Pathol. 63,141–149.

Graham T.L. (1991a) A rapid, high resolution HPLC profiling procedure for plant and microbial aromatic secondary metabolites. Plant Physiol. 95, 584–593.

Graham T.L. (1991b) Flavonoid and isoflavonoid distribution in developing soybean seedling tissues and in seed and root exudates. Plant Physiology 95, 594–604.

Graham T.L., Graham M.Y. (1991b) Glyceollin elicitors induce major but distinctly different shifts in isoflavonoid metabolism in proximal and distal soybean cell populations. Mol. Plant Microbe Interact. 4, 60–68.

Graham T.L., Kim J.E., Graham M.Y. (1990) Role of constitutive isoflavone conjugates in the accumulation of glyceollin in soybean infected with *Phytophthora megasperma*. Molec. Plant Microbe Interact. 3, 157–166.

Hahlbrock, K. (1981) Flavonoids In: P. K. Stumpf and E. E. Conn (eds.) *The Biochemistry of Plants,* Vol 7, *Secondary Plant Products*, Academic Press, NY, pp. 425–456.

Hammerschmidt, R. (1999) Phytoalexins: What Have We Learned After 60 Years? Annual Review of Phytopathology. 37, 285–306.

Humphreys, J.M., Chapple, C. (2002) Rewriting the lignin roadmap. Current Opinion in Plant Biology 5, 224–9.

Iqbal, M.J., Yaegashi, S., Njiti, V. N., Ahsan, R., Cryder, K. L. and Lightfoot, D. A. (2002). Resistance locus pyramids alter transcript abundance in soybean roots inoculated with *Fusarium solani* f.sp. *glycines*. Molecular Genetics and Genomics 268, 407–417.

Iqbal, M. J., Yaegashi, S., Ahsan, R., Shopinski, K. L., Lightfoot, D. A. (2005). Root response to Fusarium solani f. sp. glycines: temporal accumulation of transcripts in partially resistant and susceptible soybean. Theoretical and Applied Genetics 110, 1429–1438.

Jin H, Martin C. Multifunctionality and diversity within the plant MYB-gene family. Plant Mol Biol. 1999; 41, 577–585.

Kassem MA, Meksem K, Iqbal MJ, Njiti VN, Banz WJ, Winters TA, Wood A, Lightfoot DA. (2004) Definition of Soybean Genomic Regions That Control Seed Phytoestrogen Amounts. Journal of Biomedicine & Biotechnology 2004(1), 52–60.

Khan, R., Alkharouf N., Beard H., MacDonald, M., Chouikha, I., Meyer, S., Grefenstette, J., Knap, H. and Matthews, B. (2004) Microarray analysis of gene expression in soybean roots susceptible to the soybean cyst nematode two days post invasion. Journal of Nematology 36, 241–248.

Kim, B.-G., Ko J.H., Hur, H-G. Ahn, J.-H. (2004) Classification and expression analysis of cytochrome P450 genes from soybean. Agricultural Chemistry and Biotechnology 47, 173–177.

Klink, V.P., Alkharouf, N., MacDonald, M., Matthews, B. (2005) Laser capture microdissection (LCM) and expression analyses of *Glycine max* (soybean) syncytium containing root regions formed by the plant pathogen Heterodera glycines (soybean cyst nematode). Plant Molecular Biology 59, 965–79.

Landini S., Graham M.Y., Graham T.L. (2002) Lactofen induces isoflavone accumulation and glyceollin elicitation competency in soybean. Phytochemistry 62, 865–874.

Landry, L.G., Chapple, C.C., Last, R.L. (1995) Arabidopsis mutants lacking phenolic sunscreens exhibit enhanced ultraviolet-B injury and oxidative damage. Plant Physiology, 109, 1159–1166.

Logemann, E., Tavernaro, A., Schulz, W., Somssich, I. E., and Hahlbrock, K. (2000) UV light selectively coinduces supply pathways from primary metabolism and flavonoid secondary product formation in parsley. Proc. Natl. Acad. Sci. (USA) 97, 1903–1907.

Lozovaya, V. V., Lygin, A. V., Li, S., Hartman, G. L., Widholm, J. M. (2004) Biochemical response of soybean roots to *Fusarium solani* f. sp. *glycines* infection. Crop Science 44, 819–826.

Ma, F., Peterson, C. A. (2003) Current insights into the development, structure, and chemistry of the endodermis and exodermis of roots. Canadian Journal of Botany 81, 405–421.

Martin C, Paz-Ares J. MYB transcription factors in plants. Trends Genet. 1997 Feb; 13, 67–73.

Meksem K, Njiti VN, Banz WJ, Iqbal MJ, Kassem MM, Hyten DL, Yuang J, Winters TA, Lightfoot DA. (2001) Genomic Regions That Underlie Soybean Seed Isoflavone Content. Journal of Biomedicine & Biotechnology 2001 (1), 38–44.

Mohr, P. G. and Cahill, D. M. (2001) Relative roles of glyceollin, lignin and the hypersensitive response and the influence of ABA in compatible and incompatible interactions of soybeans with Phytophthora sojae. Physiological and Molecular Plant Pathology 58, 31–41.

Moy, P., Qutob D., Chapman, B. P., Atkinson, I. and Gijzen, M. (2004) Patterns of gene expression upon infection of soybean plants by Phytophthora sojae. Molecular Plant-Microbe Interactions 17, 1051–1062.

Paolocci F, Robbins MP, Madeo L, Arcioni S, Martens S, Damiani F. (2007) Ectopic Expression of a Basic Helix-Loop-Helix Gene Transactivates Parallel Pathways of Proanthocyanidin Biosynthesis. Structure, Expression Analysis, and Genetic Control of Leucoanthocyanidin 4-Reductase and Anthocyanidin Reductase Genes in Lotus corniculatus. Plant Physiology 143, 504–516.

Primomo, V.S., Poysa, V., Ablett, G.R., Jackson, C.J., Gijzen, M. and Rajcan, I. (2005) Mapping QTL for individual and total isoflavone content in soybean seeds. Crop science, 45, 2454–2464.

Ralston L., Yu O. (2006) Metabolons involving plant cytochrome P450s. Phytochemistry Review 5, 459–472.

Schoenbohm, C., Martens, S., Eder, C., Forkmann, G. and Weisshaar,B. (2000) Identification of the Arabidopsis thaliana flavonoid 3′-hydroxylase gene and functional expression of the encoded P450 enzyme. Biol. Chem. 381, 749–753.

Schopfer, C. R. and Ebel J. (1998a) Identification of elicitor-induced cytochrome P450s of soybean (Glycine max L.) using differential display of mRNA. Molecular and General Genetics 258, 315–322.

Schopfer, C.R., Kochs, G., Lottspeich, F., Ebel, J. (1998b) Molecular characterization and functional expression of dihydroxypterocarpan 6a-hydroxylase, an enzyme specific for pterocarpanoid phytoalexin biosynthesis in soybean (Glycine max L.) FEBS Lett. 432, 182–186.

Senda, M., Jumonji, A., Yumoto, S., Ishikawa, R., Harada, T., Niizeki, M., Akada, S. (2002) Analysis of the duplicated CHS1 gene related to the suppression of the seed coat pigmentation in yellow soybeans. Theoretical and Applied Genetics 104, 1086–1091.

Shoemaker, R. C., Schlueter J. A., Cregan, P., Vodkin, L. (2003) The status of soybean genomics and its role in the development of soybean biotechnologies. AgBioForum 6, 4–7.

Stacey, G., Libault M., Brechenmacher L., Wan, J. and May, G.D. (2006) Genetics and functional genomics of legume nodulation. Current Opinion in Plant Biology 9, 110–121.

Stracke R, Werber M, Weisshaar B. (2001) The R2R3-MYB gene family in Arabidopsis thaliana. Curr Opin Plant Biol. 2001 Oct; 4(5), 447–56.

Subramanian S., Graham M.Y., Yu O., Graham T.L. (2005) RNA interference of soybean isoflavone synthase genes leads to silencing in tissues distal to the transformation site and to enhanced susceptibility to *Phytophthora sojae.* Plant Physiol. 137, 1345–1353.

Todd, J. J. and Vodkin, L.O. (1993) Pigmented soybean (Glycine max) seed coats accumulate proanthocyanidins during development. Plant Physiol. 102, 663–670.

Tuteja, J. H., Clough S. J., Chan, W.-C., and Vodkin, L. O. (2004) Tissue-specific gene silencing mediated by a naturally occurring chalcone synthase gene cluster in *Glycine max*. Plant Cell 16, 819–835.

Waghorn G.C. and McNabb, W.C. (2003) Consequences of plant phenolic compounds for productivity and health of ruminants. Proceedings of the Nutrition Society 62, 383–392.

Zabala, G. and Vodkin L. (2003) Cloning of the pleiotropic T locus in soybean and two recessive alleles that differentially affect structure and expression of the encoded flavonoid 3' hydroxylase. Genetics 163, 295–309.

Zabala G., Zhou, J. Jigyasa, T., Gonzalez, D.O., Clough, S.J. and Vodkin, L.O. (2006) Transcriptome changes in the phenylpropanoid pathway of Glycine max in response to Pseudomonas syringae infection. BMC Plant Biology 6:26.

Zou, J., Rodriguez-Zas S., Aldea, M., Li, M. Zhu, J., Gonzalez, D.O., Vodkin, L.O., DeLucia, E. and Clough, S.J. (2005) Expression profiling soybean response to *Pseudomonas syringae* reveals new defense-related genes and rapid HR-specific downregulation of photosynthesis. Molecular Plant-Microbe Interactions 18, 1161–1174.

Chapter 14
Genomics of Fungal- and Oomycete-Soybean Interactions

Brett M. Tyler

Introduction

Plant pathogens historically called fungi include organisms belonging to the kingdom Mycota (true fungi) and those belonging to the kingdom Stramenopila (oomycetes). As a result of convergent evolution, true fungi and oomycetes evolved similar structures and physiological strategies for overcoming plant defense responses and causing disease. Therefore, the two groups of pathogens will be discussed together in this chapter. Diseases caused by fungal and oomycete pathogens account for approximately 50 % of all soybean disease losses in the US and around the world (Wrather et al. 2001a, 2003; Wrather and Koenning 2006). Of these, the economically most important are root and stem rot caused by the oomycete *Phytophthora sojae*, sudden death syndrome caused by *Fusarium virguliforme* (syn. *Fusarium solani* f.sp. *glycines*), and charcoal rot caused by *Macrophomina phaseolina* (Wrather et al. 2001a, 2003; Wrather and Koenning 2006). Soybean rust caused by *Phakopsora pachyrhizi* causes severe soybean yield losses in Asia, and is also a problem in Africa, South America and Hawaii (Miles et al. 2003). The pathogen recently reached the USA mainland, but has not yet caused significant losses, although projections suggest that losses in the USA could exceed 50 % in southern states and 10 % elsewhere if the pathogen becomes well established (Kuchler et al. 1984). *Diaporthe phaseolorum* var *sojae* (Pod and stem blight; *Phomopsis* seed rot), *Phialophora gregata* (brown stem rot) and *Fusarium oxysporum* (*Fusarium* root rot) also cause significant yield reduction in soybean (Wrather et al. 2001a, 2003; Wrather and Koenning 2006).

Most of these fungal and oomycete diseases are refractory to fungicide control, because they are soil-borne. Therefore, control relies heavily on cultural practices

B.M. Tyler
Virginia Bioinformatics Institute, Virginia Polytechnic Institute and State University, Blacksburg, VA 24061-0477, USA
e-mail: bmtyler@vt.edu

G. Stacey (ed.), *Genetics and Genomics of Soybean*,

and on genetic resistance. Major gene resistance (R genes) has proven of value against *Phytophthora sojae* (Burnham et al. 2003a; Buzzell et al. 1987; Schmitthenner Schmitthenner), and has recently been deployed against *Phialophora gregata* (Eathington et al. 1995). However, major gene resistance is vulnerable to genetic changes in pathogen populations. For example, genetic changes in *Phytophthora sojae* populations, and especially sexual reproduction by the pathogen, have steadily eroded the protection provided by R genes against this pathogen (*Rps* genes) (Dorrance et al. 2003b; Förster et al. 1994; Schmitthenner 1985; Schmitthenner et al. 1994). Similarly all four known *Rpp* genes conferring resistance against *Phakopsora pachyrhizi* have been defeated (Monteros et al. 2007). As a result, there is increased interest in breeding for quantitative or multigenic resistance, as well as introducing resistance genes from other *Glycine* species. However, mobilizing both these sources of resistance presents significant challenges for conventional breeding. In the case of quantitative resistance, the incremental nature of the resistance contributed by each gene, and the lack of rapid and reliable seedling assays in many cases make scoring for the presence of the resistance genes difficult and time consuming. As a result, there is strong interest in the use of molecular genetic markers to tag quantitative resistance genes. Mobilizing sources of resistance from other *Glycine* species is impeded by the barriers against inter-species crosses and would benefit from improved resources and technologies for transferring the genes to soybean.

The advent of genomics revolutionized the study of plant-pathogen interactions, from both a practical point of view and from a more fundamental point of view. The ability of genomic approaches to speed the identification of molecular markers for marker-assisted breeding, or the isolation of resistance genes themselves, are some of the most important and immediate practical benefits of genomics. Studies aimed at understanding the molecular mechanisms used by pathogens to attack their hosts and at understanding the molecular mechanisms of disease resistance have also been greatly advanced by the technologies and resources of genomics. For practical reasons, the first decades of molecular plant-microbe interactions research were focused on searches for single genes that had dominating contributions to the success of infection, epitomized by the concept of gene-for-gene interactions (e.g. Day 1974; Yoder and Turgeon 1996; Yoder and Turgeon 2001). A number of such genes were found, but the individual contributions of many others resisted characterization due to high levels of functional redundancy (e.g. Van Etten et al. 1995). Today, however, high throughput tools, such as global gene expression profiling, enable plant-pathogen interactions to be investigated from a far more realistic perspective of “many-genes-to-many-genes” interactions; that is, from the perspective that a complex network of interactions among many genes in both plant and pathogen governs the outcome of an infection. Global genomics-enabled approaches are especially appropriate for dissecting quantitative or multigenic resistance. A deeper understanding of the genetic interactions between soybean and its pathogens holds the potential for developing novel and more durable forms of genetic resistance.

Soybean Genomics Resources

This section will briefly summarize a variety of genomics resources that have been developed for the study of the interaction of soybean with its fungal and oomycete pathogens. Other chapters in this book describe many of these resources in depth.

Genetic Maps

Genetic maps constructed with molecular markers such as restriction fragment length polymorphisms (RFLPs), simple sequence repeats (SSRs), single nucleotide polymorphisms (SNPs) or single feature polymorphisms (SFPs) are fundamental resources required for marker assisted breeding (Cregan 2007). In combination with physical maps such as those constructed using bacterial artificial chromosomes (BACs) (Shoemaker 2007) they can enable map-based cloning of resistance genes or other genes of agronomic interest.

A large diversity of soybean populations were developed for the purpose of mapping fungal and oomycete disease resistance genes and QTL in soybean. These include resistance genes and QTL against *Phytophthora sojae* (Bhattacharyya et al. 2005; Burnham et al. 2003b; Gordon et al. 2006; Lohnes and Schmitthenner 1997; Polzin et al. 1994; Sandhu et al. 2005); *Fusarium virguliforme* (de Farias Neto et al. 2006; Iqbal et al. 2001), *Phakopsora pachyrhizi* (Hyten et al. 2007), *Phialophora gregata* (Klos et al. 2000; Bachman et al. 2001) and *Sclerotinia sclerotiorum* (Arahana et al. 2001) and *Cercospora sojina* (Mian et al. 1999). The mapping of *P. sojae* resistance genes also led to the cloning of resistance genes *Rps*1k (Gao et al. 2005), and of regions containing *Rps*2 (Graham et al. 2002), and *Rps*4 and *Rps*6 (Sandhu et al. 2004). In addition to the genetic mapping of resistance genes, molecular methods were used to identify genes with sequence similarity to known disease resistance genes from other species (resistance gene analogs; RGAs) (Kanazin et al. 1996; Yu et al. 1996). A highlight of these studies was the identification of a number of regions of the linkage map that contain multiple disease resistance genes and QTL, together with clusters of RGAs (Fig. 14.1). As the soybean genome sequence is completed these regions will be of major interest for cloning R genes and QTL, and for studies of the evolution of resistance genes.

Since most resistance genes were mapped in separate populations, there is a strong need to integrate the genetic maps developed from each of these populations into a common resource. Efforts in this direction include the integration of five maps using SSR markers (Song et al. 2004) and more recently 1,500 SNPs (Cregan 2007). Most recent efforts are focused on using physical maps of the soybean genome (Cregan 2007; Shoemaker 2007; Shultz et al. 2006, 2007) and the soybean genome sequence (Rokhsar 2007) as a common reference for integrating diverse genetic maps. A genetic map based on expressed sequence tags (ESTs) was also created by mapping SNPs in each of 1,141 genes identified from the EST collection (Choi et al.

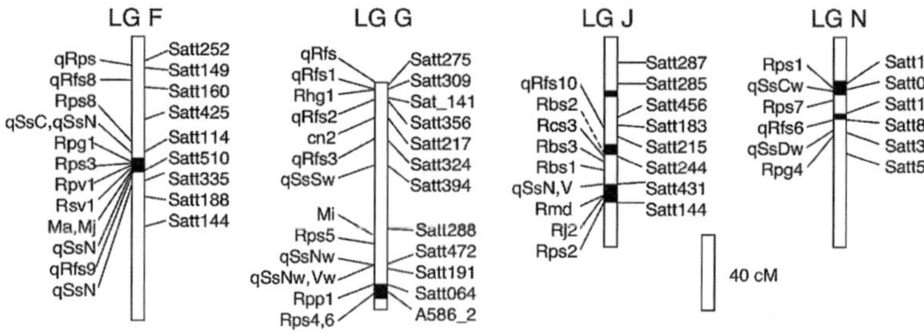

Fig. 14.1 Examples of clustering of disease resistance genes and QTL in soybean. Four linkage groups containing large clusters of resistance genes and QTL are shown. The positions of the molecular markers (shown to the right of each LG) were taken from the consensus map (Song et al. 2004). Gene symbols (left of each LG) are: *cn2* and *Rhg*, QTL against *Heterodera glycines* (Abdelmajid et al. 2007); *Ma*, *Mi* and *Mj*, QTL against *Meloidogyne arenaria*; *M. incognita* and *M. javanica* respectively (Tamulonis et al. 1997), *qRfs*, QTL against *Fusarium virguliforme* (Abdelmajid et al. 2007; Iqbal et al. 2001); *qRps*, QTL against *Phytophthora sojae* (Burnham et al. 2003b); *qSs*, QTL against *Sclerotinia sclerotiorum* (Arahana et al. 2001); *Rbs*, major QTL against *Phialophora gregata* (Bachman et al. 2001); *Rj*, "R gene" causing ineffective nodulation by *Bradyrhizobium japonicum* (Polzin et al. 1994); *Rpg*, R genes against *Pseudomonas syringae* (Ashfield et al. 2003); *Rpp*, R gene against *Phakopsora pachyrhizi* (Hyten et al. 2007); *Rps*, R genes against *Phytophthora sojae* (Bhattacharyya et al. 2005; Gordon et al. 2006; Lohnes and Schmitthenner 1997; Polzin et al. 1994; Sandhu et al. 2005); *Rcs*, R gene against *Cercospora sojima* (Mian et al. 1999); *Rmd*, R gene against *Microsphaera diffusa* (Polzin et al. 1994); *Rpv* and *Rsv*, R genes against peanut mottle virus and soybean mosaic virus respectively (Gore et al. 2002). Resistance genes and QTL were positioned relative to the nearest molecular markers as described in the referenced publications. The location of *Rbs2* is known only within a 20 cM interval. Black bars indicate the locations of some clusters of resistance gene analogs (RGAs) that have been genetically mapped (Ashfield et al. 2003; Bhattacharyya et al. 2005; Graham et al. 2002; Kanazin et al. 1996; Song et al. 2004; Yu et al. 1996)

2007).These efforts will greatly speed the ability to identify resistance genes, QTL and genetic markers that can be used to aid marker assisted selection.

Expressed Sequence Tags (ESTs) and Genome Sequence

A very large collection of soybean mRNA sequences was created through the public soybean EST project (Shoemaker et al. 2002; Tian et al. 2004; Vodkin et al. 2004). The NCBI dbEST database currently contains 371,817 soybean ESTs. The Gene Index Project clustered 330,436 of these ESTs into 31,928 contigs and 31,636 singletons (Quackenbush 2007), while Tian et al. (2004) resolved 314,254 ESTs into 32,278 contigs and 23,869 singletons. These sequences provide not only a comprehensive survey of the soybean transcriptome, but also enable preliminary insights into patterns of gene expression in the tissues from which EST libraries were constructed. The public soybean EST project contains sequences obtained from tissue infected with *Phytophthora sojae* or treated with culture filtrates from *Fusarium virguliforme.* The soybean genome sequence that is currently being developed (Rokhsar 2007) will provide an even more comprehensive resource for investigating all aspects of soybean biology including its interaction with fungal and oomycete pathogens.

Microarrays

Microarrays provide a highly parallel technology for simultaneously estimating the mRNA levels of a large number of genes in an organism, for example in soybean tissue during pathogen infection. The transcriptional profiles obtained in this way can provide a global view of the physiology of the organisms during the infection process and can serve as the basis for inferring gene regulatory networks that may control the interaction. Several microarray platforms were developed for the transcriptional profiling of soybean tissue.

Vodkin and collaborators (see Chapter 11 in this volume) produced an amplified cDNA array containing a low redundancy set of 27,513 sequences derived from the public EST project (Vodkin et al. 2004). The sequences were derived from a wide variety of developmental stages and soybean organs, and from a variety of physiological challenges, including *P. sojae* infection. The clones selected for the array were sequenced from the 3′ end in addition to the original 5′ end sequencing used to generate the ESTs. The resource was arrayed as three sets of 9,216 clones, each singly spotted. One of these sets represents clones from libraries of specific developmental tissues, namely immature cotyledons, flowers, pods, and seed coats. A second set that is especially useful for the study of root infection contains sequences from 8-day old seedling roots, seedling roots inoculated with *Bradyrhizobium japonicum*, 2-month old roots, and whole seedlings. The arrays were used to profile the soybean response to infection by the bacterial pathogens *Pseudomonas syringae* (Zou et al. 2005) and are currently being used by S. Clough (Univ. of Illinois) to study the response to infection by *Sclerotinia sclerotiorum* (Calla et al. 2007) and *Fusarium virguliforme* (Li et al. 2004a). A smaller cDNA array containing 3,937 soybean ESTs and 969 *P. sojae* ESTs was used by Moy et al. (2004) to study the expression of both host and pathogen genes using *P. sojae* infection of soybean hypocotyls.

A more comprehensive microarray platform was created by Affymetrix™ Inc. (Santa Clara, California) using their GeneChip™ technology. This array contains probe sets for over 37,500 soybean unigenes. It also carries probe sets for around 15,800 *P. sojae* genes and for 7,500 EST unigenes from the soybean cyst nematode, *Heterodera glycines*. The presence of the pathogen probe sets makes the GeneChip very useful for profiling both host and pathogen gene expression during infection by these two serious pathogens. Each probe set contains 11 oligonucleotides of length 25 nucleotides that have a perfect match to the target gene sequence, and a control set of 11 oligonucleotides that have a single nucleotide mismatch. The GeneChip has been used to profile soybean gene expression during infection by *Phakopsora pachyrhizi* (van de Mortel et al. 2007; Panthee et al. 2007) and is being used to profile both host and pathogen gene expression during infection by *Heterodera glycines* (Ithal et al. 2007), and during *P. sojae* infection of soybean cultivars varying in quantitative resistance (Tyler et al. 2007).

Pathogen Genomics Resources

Soybean pathogens evolved the ability to evade, suppress or tolerate the defense responses of soybean. Pathogen genomic resources provide the means to dissect the molecular mechanisms that these pathogens use for this purpose. Ultimately, comparison of the pathogenicity mechanisms used by a variety of soybean pathogens should shed light on the principal mechanisms that soybean uses for defense against infection.

Phytophthora sojae (Root Rot Pathogen)

An extensive set of genomics resources were developed for *P. sojae*. An EST collection consisting of 26,943 high quality sequences obtained from mycelia grown under a variety of conditions, swimming zoospores, germinating cysts and from soybean tissues heavily infected with *P. sojae* (Qutob et al. 2000; Torto-Alalibo et al. 2007). The ESTs clustered into 7,863 unigenes comprised of 2,845 contigs and 5,018 singletons. The collection included 9,244 pathogen sequences obtained from infected soybean tissue, providing an invaluable window into the pathogen's physiology inside infected tissue.

A more comprehensive resource is represented by the *P. sojae* draft genome sequence (Govers and Gijzen 2006; Tyler et al. 2006). *P. sojae* has a 95 Mb genome containing an estimated 19,000 genes. Approximately 55 % of these genes are well conserved by comparison to the genome sequence of the sudden oak death pathogen *Phytophthora ramorum*, which was determined at the same time as that of *P. sojae* (Tyler et al. 2006). A high degree of gene colinearity was observed in the order of the well-conserved genes in the two genomes. In contrast, approximately 10 % of the predicted genes in *P. sojae* had diverged so much from *P. ramorum* that they appeared to be unique to *P. sojae*. The remaining 35 % of genes belonged to similar gene families in *P. sojae* and *P. ramorum*, but the degree of divergence was such that orthologs could not be identified. These rapidly diverging gene families included many families encoding hydrolytic enzymes and protein toxins that are candidates for involvement in the pathogen infection process (Jiang et al. 2006; Tyler et al. 2006). The most striking gene family was a large, very diverse family of more than 350 genes with sequence similarity to several oomycete avirulence genes that encode products that interact with host R gene products. All members of this large protein family contain an N-terminal motif, RXLR, which was shown to be required to transport the pathogen proteins into the cytoplasm of soybean cells. Thus this family likely encodes a large array of effector proteins that, in the case of *P. sojae*, can infiltrate soybean cells to render the tissue more susceptible to infection.

The *P. sojae* EST and genome sequences were used to create microarrays for transcriptional profiling of the pathogen. As described above, both a small amplified cDNA microarray (Moy et al. 2004) and a comprehensive Affymetrix GeneChip (Tyler et al. 2007) were created that contain not only probe sets for the pathogen but also for soybean.

Phakopsora pachyrhizi (Soybean Rust Pathogen)

A small collection of ESTs was generated from *Phakopsora pachyrhizi* urediniospores (Posada-Buitrago and Frederick 2005). This collection was recently expanded to 34,394 by the Department of Energy Joint Genome Institute (DOE JGI) and placed in NCBI's dbEST database. The collection includes 29,601 ESTs from germinating urediniospores, 2,295 from resting urediniospores, and 641 from infected tissue. Genome sequencing of *Phakopsora pachyrhizi* was begun at the DOE JGI, and 840,789 reads were released into the NCBI trace archives (July 2007). A smaller number of reads was generated from the "new world" rust pathogen *Phakopsora meibomiae*. The *Phakopsora pachyrhizi* genome has proven to be very large and repetitive (>500 Mb), making assembly difficult (J.L. Boore, personal communication). Therefore, genome sequencing is presently focusing on finished sequencing of selected 35 kb fosmids. So far, 200 fosmid sequences were released into GenBank (July 2006). The mitochondrial genomes of both *P. pachyrhizi* and *P. meibomiae* were assembled from the genome sequence reads.

Fusarium virguliforme (SDS Pathogen)

Genomic resources for the sudden death syndrome pathogen are still at an early stage. A small set of 37 ESTs was deposited into GenBank by H. Iqbal and D. Lightfoot. Furthermore a BAC library was constructed and 4,152 BAC end sequences were generated (Jayaraman et al. 2007). These were mapped onto the completed genome sequence of *Fusarium graminearum* to produce a synteny-based physical map of *Fusarium virgulliforme*. The map consists of 350 *Fusarium virgulliforme* contigs encompassing 1,063 of the BAC clones, providing 3X coverage of the estimated 35 Mbp genome (Jayaraman et al. 2007).

Sclerotinia sclerotiorum (White Mold Pathogen)

Genomic resources are also well advanced in *Sclerotinia sclerotiorum,* which is a broad host range pathogen that infects many plant species in addition to soybean. A draft genome sequence of the 38 Mb genome was produced by the Broad Institute

(Cambridge Massachussetts) (Rollins et al. 2007). The genome was sequenced to a depth of 8 genome equivalents (8X) and assembled into 36 scaffolds. The genome contains 14,522 predicted genes, and community annotation of these genes is in progress. As part of the same project, more than 70,000 ESTs were also generated, representing six different developmental stages and growth conditions, adding to a small collection produced from mycelia, appressoria and infected *Brassica napus* tissue (Sexton et al. 2006). The EST collection included 4,797 ESTs that did not overlap with a predicted gene, indicating that a number of genes were likely missed during the annotation and so the transcriptome may be somewhat larger than 14,522.

The genome sequence and EST resource are being used to design a microarray based on the Agilent eArray platform, which utilizes 60 nucleotide oligonucleotides (Rollins et al. 2007). The Array will contain two oligonucleotides for each for the 14,522 predicted genes plus two for each of the 4,797 ESTs that did not overlap with any predicted genes. The entire set of oligonucleotides will be printed four times on each array slide.

Other Pathogens

Important fungal pathogens of soybean for which genome resources are currently lacking include *Macrophomina phaseolina, Diaporthe phaseolorum* var *sojae, Phialophora gregata* and *Fusarium oxysporum.* The genome sequence of the tomato pathogen *Fusarium oxysporum* f.sp. *lycopersici*, currently being developed by the Broad Institute, will likely prove useful for understanding the soybean *Fusarium* root rot pathogen, but genes specific to soybean infection are likely to be missing.

Genomics Studies of Individual Soybean-Pathogen Interactions

In this section, detailed structural and functional genomics studies of several soybean–pathogen interactions are summarized. An overall summary of these studies is presented in Table 14.1.

Phytophthora sojae-Soybean Interaction

P. sojae can infect all parts of the soybean plant, from germinating seedlings to mature plants (Schmitthenner 1989). The economically most important diseases caused by *P. sojae* are damping of seedlings and root and stem rot of mature plants. Damping off is most problematic during warm wet springs, while root rot is most common when the soil is wet for extended periods of time and the soil is poorly drained (Schmitthenner 1989). *P. sojae* causes $1–2 billion in losses worldwide annually and $100–300 million annually in the US (Wrather et al. 2001a, 2003,

Table 14.1 Summary of genetic and genomic studies of four soybean-pathogen interactions

Interaction	Organism	Genome	cDNAs	Functional Genomics
Phytopthora sojae	Host	14 R genes[a] 5 QTL[b]	ESTs[c]	cDNA arrays[d] Affymetrix arrays[e] Proteomics[f]
	Pathogen	Genome sequence[g]	ESTs[h]	cDNA arrays[j]
			SSH[i]	Affymetrix arrays[k]
Phakopsora pachyrhizi	Host	5 R genes[l]	ESTs[m]	Affymetrix arrays[n]
	Pathogen	genome sequence in progress[o]	ESTs[p]	
Fusarium virguliforme	Host	14 QTL[q]	ESTs[r]	Macroarrays[t]
			SSH[s]	cDNA arrays in progress[u]
	Pathogen	Physical map in progress[v]	ESTs[v]	
Sclerotinia sclerotiorum	Host	10+ QTL[w]		cDNA arrays in progress[x]
	Pathogen	genome sequence in progress[y]	ESTs[y]	Oligonucleotide array planned[y]

[a] Bhattacharyya et al. 2005; Gordon et al. 2006; Lohnes and Schmitthenner 1997; Polzin et al. 1994; Sandhu et al. 2005;

[b] Burnham 2007; Tyler et al. 2007;

[c] Narayanan et al. 2004; Qutob et al. 2000; Shoemaker et al. 2002; Tyler et al. 2007; Bhattacharyya 2007;

[d] Moy et al. 2004;

[e] Tyler et al. 2007;

[f] Mithofer et al. 2002; Bhattacharyya and Saravanan 2007;

[g] Tyler et al. 2006;

[h] Qutob et al. 2000; Torto-Alalibo et al. 2007;

[i] Chen et al. 2007; Wang et al. 2006;

[j] Moy et al. 2004;

[k] Tyler et al. 2007;

[l] Hartwig 1986; Hartwig and Bromfield 1983; Monteros et al. 2007;

[m] Shoemaker et al. 2002; Vodkin et al. 2004;

[n] van de Mortel et al. 2007;

[o] www.jgi.doe.gov, www.ncbi.nlm.nih.gov;

[p] Posada-Buitrago and Frederick 2005, www.jgi.doe.gov, www.ncbi.nlm.nih.gov;

[q] reviewed by Kassem 2007;

[r] Iqbal et al. 2002;

[s] Iqbal et al. 2005;

[t] Li et al. 2004a;

[u] Jayaraman et al. 2007;

[v] Li et al. 2004b, www.ncbi.nlm.nih.gov;

[w] Arahana et al. 2001; Kim and Diers 2000;

[x] Calla et al. 2007;

[y] Rollins et al. 2007, www.broad.mit.edu.

2001b; Wrather and Koenning 2006). Under field conditions, infection generally occurs via swimming zoospores that are released from submerged hyphae. The zoospores transform into a sticky cyst on the root surface, then germinate to produce a hyphal tube that penetrates the root surface. The hyphae proliferate biotrophically for the first 12–18 hr forming haustoria (Enkerli et al. 1997). In susceptible tissue, the pathogen transitions to necrotrophic proliferation once the vascular tissue of the root or stem has been colonized (Enkerli et al. 1997).

Both major (*Rps*) genes and QTL for resistance against *P. sojae* were identified. Fifteen *Rps* genes were genetically mapped to eight loci (Bhattacharyya et al. 2005; Gordon et al. 2006; Lohnes and Schmitthenner 1997; Polzin et al. 1994b; Sandhu et al. 2005). Six resistance alleles were mapped to the *Rps*1 locus, and *Rps*7 is linked to *Rps*1. Three resistance alleles map to the *Rps*3 locus, and *Rps*8 is linked to *Rps*3. *Rps*4 and *Rps*6 map to the same locus and *Rps*5 is linked to these two genes. *Rps*1k was cloned by map-based cloning and encodes an intracellular protein of the coiled-coil class of NBS-LRR resistance proteins (Gao et al. 2005). The soybean genome contains approximately 38 genes with strong similarity to *Rps*1k (Bhattacharyya et al. 2005), and 10 are located in the vicinity of the *Rps*1 locus. At least two genes at this locus can confer resistance to *P. sojae* strains expressing *Avr1k* (Gao et al. 2005). From 3 to 10 copies of this sequence were polymorphic in lines containing other *Rps*1 alleles, implying that the other *Rps*1 resistance alleles also encode NBS-LRR proteins (Gao et al. 2005). Coiled-coil NBS-LRR genes similar to those at the *Rps*1 locus are also polymorphic at a locus to which both *Rps*4 and *Rps*6 map (Sandhu et al. 2004), suggesting that these two genes also encode this class of resistance protein. A locus spanning or near *Rps*2 has alsobeen cloned and sequenced (Graham et al. 2002). *Rps*2 confers a form of resistance that is expressed in the roots and more closely resembles quantitative resistance (Mideros et al. 2007). The region near *Rps*2 contains several other resistance genes (Polzin et al. 1994), and also at least nine resistance gene analogs similar to the Toll-Interleukin Receptor (TIR) class of NBS-LRR resistance genes (Graham et al. 2002). Thus, *Rps*2 may be a TIR-NBS-LRR resistance gene.

Although major gene resistance against *P. sojae* was effective in the past, increased emphasis has been placed on quantitative resistance (also called partial resistance, multigenic resistance, field resistance, rate-reducing resistance, general resistance, or tolerance) (Dorrance et al. 2003a; Dorrance and Schmitthenner 2000; Tooley and Grau 1982, 1984) due to the appearance of new strains of *P. sojae* able to overcome commonly used *Rps* genes. The principal mechanism of quantitative resistance in the soybean lines characterized to date is the ability of the plant to reduce the rate of lesion expansion following infection (Mideros et al. 2007; Tooley and Grau 1982, 1984). Some molecular characterization of quantitative resistance suggests that resistance may involve both expression of defense genes (Tyler et al. 2007; Vega-Sánchez et al. 2005) and also pre-formed suberin levels (Thomas et al. 2007). However, the full basis for resistance and the nature of the genes underlying the QTL have yet to be determined.

In the pathogen, eleven avirulence genes involved in gene-for-gene interactions with *Rps* genes were defined by genetic analysis (May et al. 2002; Whisson et al.

1995). There is some clustering of these genes also: *Avr1b* and *Avr1k* are tightly linked (Shan et al. 2004). *Avr4* and *Avr6* are tightly linked (Gijzen et al. 1996; Whisson et al. 1995), and in fact may be allelic since *Rps4* and *Rps6* are allelic (Sandhu et al. 2004). *Avr5* and *Avr3a* are also linked (Whisson et al. 1995). One of these genes, *Avr1b*-1, was cloned by map-based cloning (Shan et al. 2004), and cloning of *Avr1a* (MacGregor et al. 2002) and *Avr4/6* (Whisson et al. 2004) is well advanced. As described above, the *P. sojae* genome contains 350 or more genes with similarity to *Avr1b*-1 and to several other oomycete avirulence genes (Tyler et al. 2006), and it is presumed that at least some of these genes contribute in some way to the infection process. However, it remains an open question which of the genes contribute to infection and by what mechanisms. As many as 9,000 predicted genes in the *P. sojae* genome show evidence of rapid evolution, as orthologs cannot unambiguously be identified in the *P. ramorum* genome (Tyler et al. 2006). If much of this rapid evolution was driven by co-evolution with the plant host, then many of these genes could potentially contribute to infection. Other than the avirulence-like genes, the most interesting of the rapidly evolving genes are several large families of protein toxin genes, including those of the NPP, PcF and crinkler families (Tyler et al. 2006). *P. sojae* contains 29, 19 and 40 copies of these toxin genes, respectively (Tyler et al. 2006). Since these protein toxins can kill host tissue, it was hypothesized that they contribute to the necrotrophic stages of the infection process (Qutob et al. 2002).

Transcriptional profiling was used to evaluate the potential role of both host and pathogen genes in the infection process. Profiling of the host can potentially reveal not only genes involved in resistance, but also physiological changes induced by the pathogen to promote susceptibility. Similarly, profiling of the pathogen may not only identify genes that might promote infection, but may also indicate genes that reveal stresses on the pathogen triggered by host defense mechanisms. Profiling approaches include EST profiling by conventional sequencing (Narayanan et al. 2004; Qutob et al. 2000; Torto-Alalibo et al. 2007; Tyler et al. 2007) or by pyrosequencing (Bhattacharyya 2007), suppressive subtractive hybridization (Chen et al. 2007; Wang et al. 2006) and microarray analysis (Moy et al. 2004; Tyler et al. 2007).

EST libraries were constructed from both resistant and susceptible soybean tissues infected with *P. sojae*. The resistant tissue, etiolated hypocotyl tissues from a line containing resistance gene *Rps1k*, was sampled 2 and 4 hr following inoculation, and 0.2 % of the ESTs were from *P. sojae* (Narayanan et al. 2004). The susceptible tissue, also etiolated hypocotyl tissues, was sampled 48 hr after infection and 70 % of the ESTs were from *P. sojae* (Qutob et al. 2000; Torto-Alalibo et al. 2007). Thus these two libraries represent extremes of incompatible and compatible interactions. L. Zhou and B. Tyler (Personal Communication) compared the distribution of soybean gene sequences in the two libraries. As expected, ESTs representing many pathogenesis-related (PR) proteins, and phytoalexin biosynthesis proteins were well-represented in both libraries. There were several interesting sequences specific to or over-represented in the incompatible interaction library, namely a copper amine oxidase, an ascorbate oxidase, a superoxide dismutase, an oxalate oxidase (germin)-like protein, a glucan endo-1,3-beta-D-glucosidase and a xyloglucan endo-1,4-beta-D-glucanase. Several peroxidases were also elevated compared to the compatible interaction. A proteomic study of soybean cell wall proteins also found elevated copper amine oxidase and several peroxidases in an incompatible interaction with *P. sojae* (Mithofer et al. 2002). The elevation of a number of enzymes involved in the generation of reactive oxygen species (ROS) is congruent with the known importance of this process in plant defense responses (Levine et al. 1994). Copper amine oxidase in particular is a secreted protein that oxidizes extracellular polyamines, generating hydrogen peroxide and ammonia (Tipping and McPherson 1995). It has been implicated in cross-linking and suberization of cell walls (Rea et al. 2002). In chickpea, the copper amine oxidase is strongly induced by jasmonate and is required for resistance to fungal infection (by *Ascochyta rabei*) (Rea et al. 2002). The soybean copper amine oxidase may play a similar key role in defense against *P. sojae* infection. The elevation of an oxalate oxidase is curious as oxalate was reported as a pathogenicity factor in *Sclerotinia* infection (Bolton et al. 2006) but not in *Phytophthora* infection. An alternative explanation may be that the oxalate oxidase, which produces carbon dioxide and hydrogen peroxide, is part of the ROS generating machinery. Consistent with this hypothesis, constitutive expression of oxalate oxidase in potato conferred increased resistance to *Phytophthora infestans* (Chatot et al. 2002). The elevation of 1,3-beta-D-glucosidase and a xyloglucan endo-1,4-beta-D-glucanase suggests that these cell-wall-degrading enzymes are also important to protection against *P. sojae*.

In the compatible interaction library, some proteins considered to be involved in resistance responses, such as PR1a, phenylalanine ammonia lyase and chitinase were much more highly represented than in the incompatible library. PR1a was reported to contribute to resistance against *Phytophthora* infection (Alexander et al. 1993), and phenylalanine ammonia lyase is required for phytoalexin biosynthesis and lignification (Graham 1995). It is possible that these enzymes appear to be elevated simply because of the much higher pathogen load. Chitinase on the other hand is expected to be ineffective against oomycetes, since oomycetes do not have chitinous cell walls, raising the possibility that the pathogen may be actively directing the defense responses into pathways that are ineffective.

Qutob et al. (2000) and Torto-Alalibo et al. (2007) analyzed the transcription profile of *P. sojae* during infection as represented by the pathogen ESTs present in the compatible interaction library. Approximately 70 % of the ESTs in the library were from *P. sojae* and could be distinguished from the soybean ESTs on the basis of base composition and matches to the *P. sojae* genome sequence and the soybean non-infection ESTs (Qutob et al. 2000; Torto-Alalibo et al. 2007). Comparison of the EST profile from infected tissue compared to that from axenic cultures revealed that the physiology of the pathogen during infection most resembled that during growth on a defined sugar-salts medium rather than growth on complex medium derived from plant materials such as soybeans, lima beans or vegetable juice (Torto-Alalibo et al. 2007). The profile from the infectious zoospores also showed similarity to the infection profile. Genes expressed during infection included many putative pathogenicity factors such as ABC transporters, oxidative stress response genes, hydrolytic enzymes, proteins toxins and secreted proteins with an RXLR motif (Qutob et al. 2000; Torto-Alalibo et al. 2007). The genes most specific to infected tissue included alcohol, aldehyde and formate dehydrogenases suggesting that mixed alcohol/formic acid fermentation was an important complement to glycolysis for substrate catabolism and energy generation during infection (Qutob et al. 2000).

Suppressive subtractive hybridization was used to identify transcripts that are distinct to *P. sojae* mycelia grown on leaf surfaces (Wang et al. 2006) and transcripts specific to a strain selected for increased virulence (Chen et al. 2007). Many of the transcripts identified as specific to leaf-surface growth corresponded to those identified in the EST analysis as infection-related. The strain with increased virulence exhibited increased expression of a number of growth related genes, and also an infection-expressed myb-domain transcription factor gene and a polygalacturonase gene that are both interesting candidates for having roles in the infection process.

Microarrays carrying both soybean and *P. sojae* sequences were used to profile the transcriptomes of both organisms simultaneously during infection (Moy et al. 2004; Tyler et al. 2007). Moy et al. (2004) used a small cDNA array containing 3,937 soybean ESTs and 969 *P. sojae* ESTs to profile gene expression in both organisms 3, 6, 12, 24, and 48 hr after infection of a susceptible soybean cultivar. Over this time period, *P. sojae* initiates infection in a biotrophic mode, then switches to necrotrophy between 12 and 24 hr. Congruent with the EST profiles, sequences encoding enzymes of phenylpropanoid phytoalexin biosynthesis and defense and pathogenesis-related proteins were strongly up-regulated during infection relative to a mock-inoculated control, while peroxidases and lipoxygenases were down-regulated. In particular, pathogenesis-related protein PR-1a and a flavin adenine dinucleotide (FAD)-linked oxidoreductase encoding a berberine bridge-like enzyme (BBE) were up-regulated from the earliest time point (Moy et al. 2004). Genes encoding proteins with similarity to ABC transporters, β-1,4 glucanases, proteinases, and the crinkling and necrosis-inducing toxin CRN2 from *P. infestans* (Torto et al. 2003), were among the earliest *P. sojae* genes detectable (at 6 hr). However, since the percentage of pathogen RNA at that point was very low, only the most abundant pathogen transcripts were detectable (Moy et al. 2004). From 12 to 24 hr there was a major increase in the proportion of pathogen RNA (to around 40 %) and a concomitant large increase in the number of differentially regulated pathogen and host genes. Transcripts for the pathogen necrosis-inducing protein, NPP1, that was hypothesized to be a major driver of the switch to necrotrophy, appeared at 24 hr and increased in level at 48 hr (Moy et al. 2004). Host lipoxygenase transcripts were strongly down-regulated (60 fold by 48 hr) during pathogen infection. Lipoxygenase is required for tobacco resistance to *Phytophthora parasitica*, leading Moy et al. (2004) to speculate that *P. sojae* suppresses lipoxygenase expression as part of its pathogenic strategy. The host defense genes induced during early infection, such as PR1a, are typical of salicylate-mediated defense pathways that respond to biotroph infection, whereas down-regulated genes such lipoxygenase are typically induced via jasmonate mediated pathways that are activated during nectrotroph infection. Moy et al. (2004) speculate that the switch from biotrophy to nectrotrophy is a strategy to take advantage of the down-regulation of the jasmonate-mediated necrotroph-defense pathway that occurred by 24 hr.

Tyler et al. (2007) initiated a microarray study of the soybean-*P. sojae* interaction in cultivars with different levels of quantitative resistance. The goals of the study are to investigate possible mechanisms of quantitative resistance. A particular goal is to determine if the resistant cultivars utilize different mechanisms of resistance and whether different QTL for resistance within a given cultivar also act via different mechanisms. The study utilizes the Affymetrix^{TM} GeneChip^{TM} that carries probes representing over 37,500 soybean genes and approximately 15,800 *P. sojae* genes. In the initial step in the study, four cultivars with strong quantitative resistance, two with moderate resistance and two with low resistance were profiled (Tyler et al. 2007). Utilizing a growth chamber assay that reliably reflects field resistance, RNA was extracted from leading edge of a *P. sojae* lesion that was progressing up the root, 3 and 5 days after inoculation. The experimental design included the pooling of samples from 60 plants for each measurement and four overall replications of the experiment, providing the power to detect transcriptional variation as low as 20 % with statistical significance. Linear mixed model analysis was used to identify genes that varied in expression among cultivars, according to infection or to time, or to combinations of those factors. Very large numbers of genes showed statistically significant variation in each category. For example, 97 % of all the genes with detectable mRNAs (29,102 genes out of 29,953) showed significant variation among the eight cultivars, irrespective of the infection status or time of sampling. In the category of most interest for this study, 78 % of detectable genes (23,446 genes) showed significant variation as a result of cultivar x infection interaction, i.e. showed infection responses that were significantly affected by which cultivar was assayed, with little difference between day 3 and day 5. Of the changes, 70 % were less than two-fold in magnitude. Tyler et al. (2007) speculated that the large numbers of small changes reflect a large number of small adjustments that occur in the transcriptional program of the sampled tissues as result of the treatments, in concert with the major response to treatment enacted by a much smaller number of genes. Contrast analysis was used to identify soybean genes that were significantly up-regulated in all four of the most resistant cultivars and also genes that were significantly up-regulated in particular resistance cultivars. In the four resistant cultivars, 213 genes were up-regulated in common in response to infection. Individual resistant cultivars 53 to 144 up-regulated genes unique to the cultivar, suggesting that there may be mechanisms of resistance unique to each cultivar. In one cultivar, the uniquely up-regulated genes contained many genes associated with phytoalexin biosynthesis, while in another cultivar, many of the genes were related to the generation of reactive oxygen species. The project is currently assaying transcriptional changes in a set of 300 recombinant inbred lines that is segregating for quantitative resistance in order to map the soybean genetic loci responsible for both resistance and any associated transcriptional changes.

Phakopsora pachyrhizi-Soybean (Soybean Rust) Interaction

Soybean rust is caused by two closely related species, *P. pachyrhizi* and *P. meibomiae*. Of the two species, *P. pachyrhizi* is the more aggressive pathogen on soybeans and causes considerably more yield losses (Miles et al. 2003). Unlike most other rust pathogens, both *Phakopsora* species have broad host ranges. *P. pachyrhizi* naturally infects 31 species in 17 genera of Leguminosae, and the ability of this pathogen to establish reservoirs in weedy legumes such as kudzu is one of the factors that makes it so difficult to control (Bromfield 1984; Miles et al. 2003). *P. pachyrhizi* spreads predominantly via the production of urediniospores which are produced in large quantities, easily wind disseminated, and produced in multiple spore cycles throughout the season. Telia, teliospores and basidiospores have been observed on infected plants (Bromfield 1984; Saksirirat and Hoppe 1991; Yeh et al. 1981). No alternate hosts have been identified. Following adhesion and germination of a urediniospore, the pathogen invades the host plant tissue by piercing the cuticle via an appressorium, penetrating an epidermal cell or guard cell directly (Bromfield 1984).

Five major genes for resistance against *P. pachyrhizi* were identified, *Rpp*1 through *Rpp*4 (Hartwig 1986; Hartwig and Bromfield 1983), plus an unidentified gene in the accession Hyuuga (Monteros et al. 2007). In general, major gene resistance has not proved to be durable. *Rpp*4 has not yet been defeated in the field, but isolates that overcome it were identified in greenhouse tests. *Rpp*1 was mapped to linkage group G (Hyten et al. 2007). A resistance gene in the accession Hyuuga was mapped to LG C2 (Monteros et al. 2007) and may represent a new *Rpp* gene. Pathogen isolates varying in ability to overcome the *Rpp* genes were identified (Miles et al. 2003), implying the existence of specific avirulence genes. Variation in ability to infect a panel of legume species was also identified. Partial or rate-reducing resistance and also tolerance were identified in some accessions, but the genetic basis of these traits was not well defined due to the lack of rapid assays for these traits (Miles et al. 2003).

The soybean response to rust infection was analyzed by transcriptional profiling using the Affymetrix GeneChip (van de Mortel et al. 2007; Panthee et al. 2007). Infection of a susceptible line and a line containing *Rpp*2 was profiled over ten time points from 6 hr to 168 hr post inoculation, spanning disease development up to the formation of uredinia. Infected tissue was compared to mock-inoculated tissue at each time point. The level of pathogen colonization was followed by measuring the level of pathogen alpha-tubulin mRNA by quantitative RT-PCR. Up until 96 hr, the level of pathogen accumulation was similar in both genotypes. However, after 96 hr, the pathogen proliferated aggressively in the susceptible cultivar but only moderately in the resistance cultivar (van de Mortel et al. 2007).

The most striking finding of the study was a bi-phasic response to the pathogen. From 6 to 36 hr (corresponding to spore germination and penetration), there was an initial increase in differential gene expression between pathogen to mock inoculated, with a peak at 12 hr (van de Mortel et al. 2007). During this period, 879 genes in the susceptible line and 240 in the resistant line showed differential expression. From 36 to 72 hr (haustoria formation and initial growth of intercellular, secondary hyphae), only 16 and 5 genes, respectively, showed differential expression in the susceptible and resistant lines compared to the mock inoculated tissue. Panthee et al. (2007) also observed only a small number of differentially expressed genes at 72 hr post inoculation. From 72 to 168 hr (colonization, lesion/uredinia formation), 180 and 238 genes were differentially expressed in the susceptible and resistant genotypes, respectively. Overall, 470 genes were identified as having significant expression changes in both genotypes, whereas 1046 and 424 genes showed changes that were unique to either the susceptible or the resistant genotypes, respectively. Comparison of the expression profiles of the 470 common genes revealed that the program of expression of the genes was similar from 6 to 36 hr, but during the second phase of the response, the increase in gene expression was more rapid in the resistance line, about one day earlier in each case. These genes included those encoding enzymes of phenylpropanoid phytoalexin biosynthesis, as well as a number of WRKY transcription factors. The pattern of gene expression changes during early infection (6–36 hr) was suggestive of a non-specific recognition of the pathogen and activation of basal defense responses. The authors speculate that the biphasic response occurs because the pathogen secretes effector proteins that suppress the initial defense response as haustoria are formed; recognition of an effector protein by the *Rpp2* resistance gene product then triggers the rapid resistance response in the resistance line (van de Mortel et al. 2007).

The pathogen has a very complex (500 Mb) genome, which has hampered genomic studies (J.L. Boore, personal communication). Furthermore, the fungus is an obligate parasite, making in vitro culture impossible, and confining functional genomics studies to the spore stages. Posada-Buitrago and Frederick (2005) profiled 908 EST sequences from germinating urediniospores, representing 488 unique sequences. As expected, significant percentages of the ESTs represented genes involved in growth, metabolism and signaling. Intriguingly, an extraordinary 20 % of all the ESTs (183/908) in the EST library encoded a small secreted, cysteine-rich protein with a high degree of similarity to proteins expressed by *Blumeria graminis* f. sp. *hordei* and *Magnaporthe grisea* during spore germination and appressorium formation (Posada-Buitrago and Frederick 2005), suggesting that the protein has a very important role in these early stages of infection. A similar protein is also found in many other fungi including animal pathogens and saprotrophs, but this does not preclude its having developed a specialized usage in the appressoria of plant pathogens. There are currently (July 2007) 34,394 *P. pachyrhizi* ESTs in NCBI's dbEST database, deposited by the DOE JGI, but these have not yet been characterized.

Fusarium virguliforme-soybean (SDS) Interaction

Fusarium virguliforme causes two associated diseases of soybean. Infection of the roots causes root rot symptoms, while release of toxins during infection results in leaf scorch symptoms (Roy et al. 1997). Like *P. sojae,* the pathogen is hemi-biotrophic, initially colonizing the root tissue with formation of haustoria, then progressing to necrotrophic destruction of the root tissue. As the root infection progresses, the fungus produces toxins that translocate to the upper parts of the plant, causing the characteristic symptoms of interveinal chlorosis, leaf necrosis, premature defoliation and pod abortion (Roy et al. 1997). Losses to *Fusarium virguliforme* result from both the root rot and the leaf scorch symptoms. A related species, *F. tucumaniae*, causes a similar disease in South America (Aoki et al. 2003).

Soybean lines with increased resistance to *F. virguliforme* were identified. These lines show a reduced frequency of root infections and the latent period before appearance of lead scorch symptoms is longer in the resistant lines (reviewed by de Farias Neto et al. 2006; Abdelmajid et al. 2007). QTL were identified in several crosses. In a cross of the resistant line Forrest with the susceptible line Essex, six QTL were mapped (Iqbal et al. 2001). Four QTL from Forrest were located on linkage group (LG) G, where they clustered with the *Rhg*1 resistance gene against cyst nematode (Iqbal et al. 2001). An additional two QTL originating from Essex were located on LGs I and C2 (Iqbal et al. 2001). Additional QTL were mapped to LGs C1, C2, D2, G, I, J and N (summarized by de Farias Neto et al. 2006; Abdelmajid et al. 2007) and progress was made in identifying BAC clones in the regions of several of the QTL (Shultz et al. 2007).

Iqbal et al. (2005) carried out transcriptional profiling of resistant and susceptible lines using a selected set of 191 soybean genes. Of the genes, 28 were previously identified using suppressive subtractive hybridization (SSH) to select transcripts of increased abundance in Forrest following inoculation with *F. virguliforme* (Iqbal et al. 2002). The remaining 163 sequences were selected from a soybean root EST collection on the basis of their involvement in plant defense, stress responses and related pathways (Iqbal et al. 2005). Amplified DNA from each cDNA was spotted onto a nylon macroarray. The two soybean lines profiled were Essex and a recombinant inbred line from the cross of Forrest x Essex, RIL23, which contained all 6 QTL from that population. The plant roots were dip-inoculated with spores or with a spore-free mock solution. Roots were collected from inoculated and mock-inoculated plants at 1, 2, 3, 7 and 10 days after inoculation. In RIL23 there were significant increases following inoculation in mRNA levels of genes associated with phenylpropanoid biosynthesis over days 3–10 (Iqbal et al. 2005), consistent with the reported increase in glyceollin biosynthesis following inoculation of resistant soybean lines by *F. virguliforme* (Lozovaya et al. 2004). Transcripts characteristic of ethylene and jasmonate mediated responses to necrotrophs (Glazebrook 2005) also increased. In Essex, most of these genes remained unchanged or even decreased in abundance, suggesting some suppression of defense responses by the pathogen (Iqbal et al. 2005). More comprehensive profiling of the soybean response to *F. virguliforme* and to its toxin are currently underway (Li et al. 2004a).

Sclerotinia sclerotiorum-Soybean (White Mold) Interaction

Sclerotinia sclerotiorum is a necrotrophic pathogen that causes stem rot, not only in soybean, but in a very wide range of crop plants (reviewed by Bolton et al.

2006). The disease is initiated by ascospores that are released from apothecia that develop from sclerotia when the leaf canopy of the growing soybean plants closes or is near to closing. Infection of healthy tissue requires an external nutrient supply, so infection is often initiated after senescing blossoms have been colonized. Initial stem lesions may be dark or water-soaked. Lesions subsequently necrotize and develop patches of fluffy white mycelium, which gave the disease its name (Bolton et al. 2006). As the fungus progresses into the main stems, wilting typically occurs. There has been extensive molecular genetic characterization of the pathogen, which demonstrated that oxalic acid production and cell wall-degrading enzymes are key agents of pathogenicity (Bolton et al. 2006).

No sources of complete genetic resistance, such as R genes, against *S. sclerotiorum* have been identified. However, a number of QTL sources for resistance were identified (Arahana et al. 2001; Bolton et al. 2006; Kim and Diers 2000). Kim and Diers (2000) reported three QTL on Linkage Groups C2, K, and M from a cross of Williams x S19–90. Analysis of the phenotypes conditioned by the QTL suggested that the QTL on LG K conferred physiological resistance, whereas the QTL on LG's C2 and M conferred agronomic traits that reduced disease incidence (Kim and Diers 2000). The QTL on LG C2 was associated with shorter plants and reduced lodging while the QTL on LG M was associated with earlier flowering; both traits are known to be associated with reduced disease incidence (Kim and Diers 2000). Arahana et al. (2001) used a detached leaf assay to measure physiological resistance directly in five populations segregating for quantitative resistance. Twenty-eight QTL were identified on 15 different linkage groups in the five RIL populations. Of these, seven QTL on different linkage groups were identified in more than one population (Arahana et al. 2001). Four of the QTL mapped to regions of linkage groups F, G, J and N that contain clusters of resistance genes and QTL against other pathogens including fungi, oomycetes, viruses and nematodes (Arahana et al. 2001). Numerous resistance gene analogs were also mapped to these clusters (Fig. 14.1), suggesting that the molecular basis of these QTL resembles that of major R genes. Resistance against *S. sclerotiorum* was also achieved by introducing the wheat oxalate oxidase gene into transgenic soybean lines (Donaldson et al. 2001), underlining the importance of oxalate as a pathogenicity factor for the fungus (Bolton et al. 2006).

Functional genomics analysis of this disease interaction is still at an early stage. Calla et al. (2007) reported using soybean microarrays to study host gene expression following oxalate exposure and during infection, in susceptible lines, in more resistant lines, and in transgenic lines expressing oxalate oxidase. Preliminary results revealed strong expression of genes associated with defense, oxidative stress and secondary metabolism in proportion to the progress of the disease, i.e. most strongly in the most susceptible plants. This suggests that these responses are actually ineffective against the pathogen and that the mechanisms of effective resistance lie elsewhere, such as resistance to oxalate (Calla et al. 2007).

The sequencing of the *S. sclerotiorum* genome and of 70,000 pathogen ESTs has enabled design of a pathogen microarray for studying infection (Rollins et al. 2007). As this microarray is deployed, further information will be gathered regarding the mechanisms of pathogenicity of this necrotroph.

Summary and Conclusions

The availability of genomics tools for soybean and for several of its oomycete and fungal pathogens is paving the way for rapid advancements in our knowledge of the molecular mechanisms governing these interactions. Several themes are beginning to emerge. One is that interactions with oomycete or fungal pathogens cause a massive reprogramming of the soybean transcriptome. Very large numbers of defense-related genes are induced, particularly those involved in the generation of reactive oxygen species and in secondary metabolism, including the biosynthesis of phenylpropanoid compounds, such as phytoalexins and lignin precursors. However, in several cases, the induction of defense-related genes is more extensive in compatible interactions than in incompatible interactions, suggesting that the pathogens can tolerate many of the defense activities of the plant, and in fact act to promote ineffective responses. The plant responses which *are* effective against the pathogens are still not fully understood. Tools are still needed to probe the physiology of the pathogen during an unsuccessful infection, to determine how it is being stressed by the plant. Extremely high throughput transcript tagging, using technology like that of Solexa (now Illumina) holds promise in this area. The genome sequence of soybean that will shortly be available will enable the identities of clustered disease resistance genes to be quickly determined. Genetical genomics, which combines microarray analysis, QTL mapping and genome sequences (Jansen and Nap 2001), will enable genes underlying quantitative resistance, and the mechanisms they control, to be deciphered. Identifying genes for major resistance and quantitative resistance, or markers very close to them, will greatly facilitate the pyramiding of disease resistance genes and QTL both through marker-assisted selection and through transgenic approaches. Genome sequences of more pathogens are needed, particularly for damaging pathogens such as *Fusarium virguliforme*, *Macrophomina phaseolina* and *Phakopsora pachyrhizi*.

References

Abdelmajid, K.M., Meksem, K., Wood, A.J., and Lightfoot, D.A. (2007) Loci Underlying SDS and SCN Resistance Mapped in the 'Essex' by 'Forrest' Soybean Recombinant Inbred Lines. Reviews in Biology and Biotechnology 6, 2–10.

Alexander, D., Goodman, R. M., Gut-Rella, M., Glascock, C., Weymann, K., Friedrich, L., Maddox, D., Ahl-Goy, P., Luntz, T., Ward, E., and Ryals, J. (1993) Increased tolerance to two oomycete pathogens in transgenic tobacco expressing pathogenesis-related protein 1a. Proc. Natl. Acad. Sci. USA 90, 7327–7331.

Aoki, T., O'Donnell, K., Homma, Y., and Lattanzi, A.R. (2003) Sudden-death syndrome of soybean is caused by two morphologically and phylogenetically distinct species within the *Fusarium solani* species complex—*F. virguliforme* in North America and *F. tucumaniae* in South America. Mycologia 95, 660–684.

Arahana, V., Graef, G.L., Specht, J.E., Steadman, J.R., and Eskridge, K.M. (2001) Identification of QTL for Resistance to *Sclerotinia sclerotiorum* in Soybean. Crop Science 41, 180–188.

Ashfield, T., Bocian, A., Held, D., Henk, A. D., Marek, L. F., Danesh, D., Penuela, S., Meksem, K., Lightfoot, D. A., Young, N. D., Shoemaker, R. C., and Innes, R. W. (2003) Genetic and physical localization of the soybean *Rpg*1-b disease resistance gene reveals a complex locus containing several tightly linked families of NBS-LRR genes. Mol Plant Microbe Interact 16, 817–826.

Bachman, M.S., Tamulonis, J.P., Nickell, C.D., and Bent, A.F. (2001) Molecular Markers Linked to Brown Stem Rot Resistance Genes, *Rbs*1 and *Rbs*2, in Soybean. Crop Science 41, 527–535.

Bhattacharyya, M. K., Narayanan, N. N., Gao, H., Santra, D. K., Salimath, S. S., Kasuga, T., Liu, Y., Espinosa, B., Ellison, L., Marek, L., Shoemaker, R., Gijzen, M., and Buzzell, R. I. (2005) Identification of a large cluster of coiled coil-nucleotide binding site–leucine rich repeat-type genes from the *Rps*1 region containing *Phytophthora* resistance genes in soybean. Theor Appl Genet 111, 75–86.

Bhattacharyya, M.K., Sandhu, D., Gao, H., MacMil, S., Wiley, G., Roe, B., and Cannon, S. (2007) Application of 454 Technology (Pyrosequencing) in Transcript Analyses of the Soybean-*Phytophthora sojae* Interaction. Plant and Animal Genomes XV Conference. San Diego, California. http://www.intl-pag.org/15/abstracts/PAG15_W22_160.html

Bhattacharyya, M.K. and Saravanan, R.S. (2007) Phosphoproteomics Of The Soybean-*Phytophthora sojae* Interaction. Plant and Animal Genomes XV Conference. San Diego, California. http://www.intl-pag.org/15/abstracts/PAG15_W55_353.html

Bolton, M. D., Thomma, B., and Nelson, B. D. (2006) *Sclerotinia sclerotiorum* (Lib.) de Bary: biology and molecular traits of a cosmopolitan pathogen. Molecular Plant Pathology 7, 1–16.

Bromfield, K. R. (1984) *Soybean Rust.* American Phytopathological Society, St. Paul, MN.

Burnham, K. D., Dorrance, A. E., Francis, D. M., Fioritto, R. J., and St-Martin, S. K. (2003a) *Rps*8, A New Locus in Soybean for Resistance to *Phytophthora sojae*. Crop Sci. 43, 101–105.

Burnham, K. D., Dorrance, A. E., VanToai, T. T., and St. Martin, S. K. (2003b) Quantitative Trait Loci for Partial Resistance to *Phytophthora sojae* in Soybean. Crop Science 43, 1610–1617

Buzzell, R.I., Anderson, T.R., and Rennie, B.D. (1987) Harosoy *Rps* isolines. Soybean Genet. Newsl. 14, 79–81.

Calla, B., Zhang, Y., Simmonds, D., and Clough, S.J. (Year) Genomic analysis of soybean resistance to *Sclerotinia sclerotiorm* [abstract]. ARS Sclerotinia Initiative Annual Meeting. Minneapolis (Bloomington), MN January 17–19, 2007. http://www.whitemoldresearch.com/files/012507.pdf

Chatot, C., Malno, P., Droz, E., Schneider, M., Mtraux, J. P., and Bonnel, E. (2002) Transgenic potato plants expressing oxalate oxidase have increased resistance to oomycete and bacterial pathogens. Potato Research 45, 177–185,

Chen, X.R., Shen, G., Wang, Y.L., Zheng, X.B., and Wang, Y.C. (2007) Identifcation of *Phytophthora sojae* genes upregulated during the early stage of soybean infection. Federation of European Microbiological Societies 269, 280–288.

Choi, I. Y., Hyten, D. L., Matukumalli, L. K., Song, Q., Chaky, J. M., Quigley, C. V., Chase, K., Lark, K. G., Reiter, R. S., Yoon, M. S., Hwang, E. Y., Yi, S. I., Young, N. D., Shoemaker, R. C., van Tassell, C. P., Specht, J. E., and Cregan, P. B. (2007) A Soybean Transcript Map: Gene Distribution, Haplotype and SNP Analysis. Genetics 176, 685–696.

Cregan, P. (2007) Soybean Genetic Map. In: Stacey,G. (Ed.), *Soybean Genomics.* Springer, New York, pp. xxx–xxx.

Day, P. (1974) *Genetics of Host-Parasite Interaction.* W.H. Freeman and Co., San Francisco.

de Farias Neto, A. L., Hashmi, R., Schmidt, M., Carlson, S. R., Hartman, G. L., Li, S.X., Nelson, R. L., and Diers, B. W. (2006) Mapping and confirmation of a new sudden death syndrome resistance QTL on linkage group D2 from the soybean genotypes PI 567374 and 'Ripley'. Molecular Breeding 20, 53–62.

Donaldson, P.A., Anderson, T., Lane, B.G., Davidson, A.L., and Simmonds, D.H. (2001) Soybean plants expressing an active oligomeric oxalate oxidase from the wheat gf-2.8 (germin) gene are resistant to the oxalate-secreting pathogen *Sclerotina sclerotiorum.* Physiol. Molec.Plant Pathol. 59, 297–307.

Dorrance, A. E., McClure, S. A., and St. Martin, S. K. (2003a) Effect of partial resistance on *Phytophthora* stem rot incidence and yield of soybean in Ohio. Plant Dis. 87, 308–312.

Dorrance, A. E. and Schmitthenner, A. F. (2000) New sources of resistance to *Phytophthora sojae* in the soybean plant introductions. Plant Disease 84, 1303–1308.

Dorrance, A.E., McClure, S.A., and DeSilva, A. (2003b) Pathogenic diversity of *Phytophthora sojae* in Ohio soybean fields. Plant Disease 87, 139–146.

Eathington, S. R., Nickell, C. D., and Gray, L. E. (1995) Inheritance of Brown Stem Rot Resistance in Soybean Cultivar BSR 101. J. Heredity 86, 55–60.

Enkerli, K., Hahn, M. G., and Mims, C. W. (1997) Ultrastructure of compatible and incompatible interactions of soybean roots infected with the plant pathogenic oomycete *Phytophthora sojae*. Can. J. Bot. 75, 1494–1508.

Förster, H., Tyler, B.M., and Coffey, M.D. (1994) *Phytophthora sojae* races have arisen by clonal evolution and by rare outcrosses. Mol. Plant-Microbe Interact. 7, 780–791.

Gao, H., Narayanan, N. N., Ellison, L., and Bhattacharyya, M. K. (2005) Two classes of highly similar coiled coil-nucleotide binding-leucine rich repeat genes isolated from the *Rps*1-k locus encode *Phytophthora* resistance in soybean. Mol Plant Microbe Interact 18, 1035–1045.

Gijzen, M., Forster, H., Coffey, M. D., and Tyler, B.M. (1996) Cosegregation of *Avr4* and *Avr6* in *Phytophthora sojae*. Can. J. Bot. 74, 800–802.

Glazebrook, J. (2005) Contrasting mechanisms of defense against biotrophic and necrotrophic pathogens. Annu Rev Phytopathol 43, 205–227.

Gordon, Stuart G., St. Martin, Steven K., and Dorrance, Anne E. (2006) Rps8 Maps to a Resistance Gene Rich Region on Soybean Molecular Linkage Group F. Crop Science 46, 168–173.

Gore, M. A., Hayes, A. J., Jeong, S. C., Yue, Y. G., Buss, G. R., and Maroof, M.A.S. (2002) Mapping tightly linked genes controlling potyvirus infection at the *Rsv*1 and *Rpv*1 region in soybean. Genome 45, 592–599.

Govers, F. and Gijzen, M. (2006) *Phytophthora* genomics: the plant destroyers' genome decoded. Mol Plant Microbe Interact 19, 1295–1301.

Graham, M.A., Marek, L.F., and Shoemaker, R.C. (2002) Organization, Expression and Evolution of a Disease Resistance Gene Cluster in Soybean. Genetics 162, 1961–1977.

Graham, T. L. (1995) Cellular biochemistry of phenylpropanoid responses of soybean to infection by *Phytophthora sojae*. In: Daniel and Purkayastha (Eds.), *Handbook of Phytoalexin Metabolism and Action*. Marcel Dekker, Inc., New York, pp. 85–116.

Hartwig, E.E. (1986) Identification of a fourth major gene conferring resistance to soybean rust. Crop Sci. 26, 1135–1136.

Hartwig, E.E. and Bromfield, K.R. (1983) Relationships among three genes conferring specific resistance to rust in soybeans. Crop Sci. 23, 237–239.

Hyten, D. L., Hartman, G. L., Nelson, R. L., Frederick, R. D., Concibido, V. C., Narvel, J. M., and Cregan, P. B. (2007) Map Location of the *Rpp*1 Locus That Confers Resistance to Soybean Rust in Soybean. Crop Science 47, 837–840.

Iqbal, M. J., Yaegashi, S., Ahsan, R., Shopinski, K. L., and Lightfoot, D. A. (2005) Root response to *Fusarium solani* f. sp. *glycines*: temporal accumulation of transcripts in partially resistant and susceptible soybean. Theor Appl Genet 110, 1429–1438.

Iqbal, M. J., Yaegashi, S., Njiti, V. N., Ahsan, R., Cryder, K. L., and Lightfoot, D. A. (2002) Resistance locus pyramids alter transcript abundance in soybean roots inoculated with *Fusarium solani* f.sp. *glycines*. Mol Genet Genomics 268, 407–417.

Iqbal, M.J., Meksem, K., Njiti, V.N., Kassem, M. A., and Lightfoot, D.A. (2001) Microsatellite markers identify three additional quantitative trait loci for resistance to soybean sudden-death syndrome (SDS) in Essex × Forrest RILs. Theor Appl Genet 102, 187–192.

Ithal, N., Recknor, J., Nettleton, D., Hearne, L., Maier, T., Baum, T. J., and Mitchum, M. G. (2007) Parallel genome-wide expression profiling of host and pathogen during soybean cyst nematode infection of soybean. Mol Plant Microbe Interact 20, 293–305.

Jansen, R. C. and Nap, J.-P. (2001) Genetical genomics: The added value from segregation. Trends Genet. 17, 388–391.

Jayaraman, D., Shultz, J., Ali, S., Yuan, J., and Lightfoot, D.A. (2007) The *Fusarium virguliforme* Genome Database (FVGD): A Tool For Integrated Pathogen Biology. Plant & Animal Genomes XV Conference. San Diego, California. http://www.intl-pag.org/15/abstracts/PAG15_C01_944.html

Jiang, R. H., Tyler, B. M., and Govers, F. (2006) Comparative analysis of *Phytophthora* genes encoding secreted proteins reveals conserved synteny and lineage-specific gene duplications and deletions. Mol Plant Microbe Interact 19, 1311–1321.

Kanazin, V., Marek, L. F., and Shoemaker, R. C. (1996) Resistance gene analogs are conserved and clustered in soybean. Proc. Nat. Acad. Sci. USA 93, 11746–11750.

Kim, H. S. and Diers, B. W. (2000) Inheritance of Partial Resistance to *Sclerotinia* Stem Rot in Soybean. Crop Science 40, 55–61.

Klos, K.L.E., Paz, M.M., Marek, L.Fredrick, Cregan, P.B., and Shoemaker, R.C. (2000) Mo-lecular Markers Useful for Detecting Resistance to Brown Stem Rot in Soybean. Crop Science 40, 1445–1452.

Kuchler, F., Duffy, M., Shrum, R. D., and Dowler, W. M. (1984) Potential economic consequences of the entry of an exotic fungal pest: the case of soybean rust. Phytopathology 74, 916–920.

Levine, A., Tenhaken, R., Dixon, R., and Lamb, C. (1994) H2O2 from the oxidative burst orchestrates the plant hypersensitive disease resistance response. Cell 79, 583–593.

Li, M., Zou, J.J., Li, S.X., Vodkin, L., and Clough, S.J. (2004a) Microarray analysis of sudden death syndrome in soybeans. Plant & Animal Genomes XII Conference. San Diego, California. http://www.intl-pag.org/12/abstracts/P7a_PAG12_809.html

Li, S., Liu, L., Hernandez, A.G., Hartman, G.L., Domier, L.L., and Schweitzer, P.A. (2004b) Generation and Analysis of Expressed Sequence Tags of *Fusarium solani* f. sp. *glycines* during Infection of Soybean. Plant & Animal Genomes XII Conference. San Diego, California. http://www.intl-pag.org/12/abstracts/P01_PAG12_72.html

Lohnes, D. G. and Schmitthenner, A. F. (1997) Position of the *Phytophthora* resistance gene *Rps*7 on the soybean molecular map. Crop Science 37, 555–556.

Lozovaya, V. V., Lygin, A. V., Li, S., Hartman, G. L., and Widholm, J. M. (2004) Biochemical Response of Soybean Roots to *Fusarium solani* f. sp. *glycines* Infection. Crop Science 44, 819–826.

MacGregor, T., Bhattacharyya, M., Tyler, B. M., Bhat, R., Schmitthenner, A. F., and Gijzen, M. (2002) Genetic and physical mapping of *Avr1a* in *Phytophthora sojae*. Genetics 949–959.

May, K.J., Whisson, S.C., Zwart, R.S., Searle, I.R., Irwin, J.A.G., Maclean, D.J., Carroll, B.J., and Drenth, A. (2002) Inheritance and mapping of eleven avirulence genes in *Phytophthora sojae*. Fungal Genet. Biol. 37, 1–12.

Mian, M.A.R., Wang, T., Phillips, D.V., Alvernaz, J., and Boerma, H.R. (1999) Molecular Mapping of the *Rcs*3 Gene for Resistance to Frogeye Leaf Spot in Soybean. Crop Science 39, 1687–1691.

Mideros, S., Nita, M., and Dorrance, A. E. (2007) Characterization of components of partial resistance, *Rps*2, and root resistance to *Phytophthora sojae* in soybean. Phytopathology in press.

Miles, M.R., Frederick, R.D., and Hartman, G.L. (2003) Soybean Rust: Is the U.S. Soybean Crop At Risk? APSNet feature. American Phytopathological Society. http://www.apsnet.org/online/feature/rust/.

Mithofer, Axel, Muller, Bernd, Wanner, Gerhard, and Eichacker, Lutz A. (2002) Identification of defence-related cell wall proteins in Phytophthora sojae-infected soybean roots by ESI-MS/MS. Molecular Plant Pathology 3, 163–166.

Monteros, Maria J., Missaoui, Ali M., Phillips, Daniel V., Walker, David R., and Boerma, H. Roger (2007) Mapping and Confirmation of the 'Hyuuga' Red-Brown Lesion Resistance Gene for Asian Soybean Rust. Crop Science 47, 829–836.

Moy, P., Qutob, D., Chapman, B. P., Atkinson, I., and Gijzen, M. (2004) Patterns of gene expression upon infection of soybean plants by *Phytophthora sojae*. Mol Plant Microbe Interact 17, 1051–1062.

Narayanan, N.N., Tasma, M., Grant, D., Shoemaker, R., and Bhattacharyya, M. K. (2004) Early Gene Expression in Soybean Hypocotyls Following Infection With *Phytopthora sojae*. Plant and Animal Genomes XII Conference. San Diego, California. http://www.intl-pag.org/pag/12/abstracts/P7a_PAG12_811.html

Panthee, D. R., Yuan, J. S., Wright, D. L., Marois, J. J., Mailhot, D., and Jr., C. N. Stewart (2007) Gene expression analysis in soybean in response to the causal agent of Asian soybean rust (*Phakopsora pachyrhizi* Sydow) in an early growth stage. Funct Integr Genomics DOI 10.1007/s10142-10007-10045-10148.

Polzin, K.M., Lohnes, D. G., Nickell, C. D., and Shoemaker, R. C. (1994) Integration of *Rps*2, *Rmd*, and *Rj*2 Into Linkage Group J of the Soybean Molecular Map. J. Heredity 85, 300–303.

Posada-Buitrago, M. L. and Frederick, R. D. (2005) Expressed sequence tag analysis of the soybean rust pathogen *Phakopsora pachyrhizi*. Fungal Genet Biol 42, 949–962.

Quackenbush, J. (2007) Soybean Gene Index. http://compbio.dfci.harvard. edu/tgi/cgi-bin/tgi/gimain.pl?gudb=soybean

Qutob, D., Hraber, P., Sobral, B., and Gijzen, M. (2000) Comparative Analysis of Expressed Sequences in *Phytophthora Sojae*. Plant Phys. 123, 243–253.

Qutob, D., Kamoun, S., and Gijzen, M. (2002) Expression of a *Phytophthora sojae* necrosis-inducing protein occurs during transition from biotrophy to necrotrophy. Plant J 32, 361–373.

Rea, G., Metoui, O., Infantino, A., Federico, R., and Angelini, R. (2002) Copper amine oxidase expression in defense responses to wounding and *Ascochyta rabiei* invasion. Plant Physiol 128, 865–875.

Rokhsar, D. (2007) Sequencing the soybean genome. In: Stacey (Eds.), *Soybean Genomics*. Springer, New York, pp. xxx–xxx.

Rollins, J., Cuomo, C., Kodira, C., Mauceli, E., DeCaprio, D., Galagan, J., Birren, B., Dickman, M., and Kohn, L. (2007) Moving Beyond An Assembled Genome With *Sclerotinia sclerotiorum*. Plant & Animal Genomes XV Conference. San Diego, California. http://www.intl-pag.org/15/abstracts/PAG15_W23_165.html

Roy, K.W., Rupe, J. C., Hershman, D. E., and Abney, T. S. (1997) Sudden Death Syndrome of Soybean. Plant Disease 81, 1100–1111.

Saksirirat, W. and Hoppe, H. H. (1991) Teliospore germination of soybean rust fungus (*Phakopsora pachyrhizi* Syd.). J. Phytopathol. 132, 339–342.

Sandhu, D., Gao, H.Y., Cianzio, S., and Bhattacharyya, M.K. (2004) Deletion of a Disease Resistance Nucleotide-Binding-Site Leucine-Rich-Repeat-like Sequence Is Associated With the Loss of the *Phytophthora* Resistance Gene *Rps*4 in Soybean. Genetics 168, 2157–2167.

Sandhu, D., Schallock, K. G., Rivera-Velez, N., Lundeen, P., Cianzio, S., and Bhattacharyya, M. K. (2005) Soybean *Phytophthora* resistance gene *Rps*8 maps closely to the *Rps*3 region. J Heredity 96, 536–541.

Schmitthenner, A.F. (1985) Problems and progress in controlling *Phytophthora* root rot of soybean. Plant Dis. 69, 362–368.

Schmitthenner, A.F. (1989) *Phytophthora* rot. In: Sinclair and Backman (Eds.), *Compendium of Soybean Diseases, 3rd Ed.* APS Press, St. Paul, MN, pp. 35–38.

Schmitthenner, A.F., Hobe, M., and Bhat, R.G. (1994) *Phytophthora sojae* races in Ohio over a 10-year interval. Plant Disease 78, 269–276.

Sexton, A. C., Cozijnsen, A. J., Keniry, A., Jewell, E., Love, C. G., Batley, J., Edwards, D., and Howlett, B. J. (2006) Comparison of transcription of multiple genes at three developmental stages of the plant pathogen *Sclerotinia sclerotiorum*. FEMS Microbiol Lett 258, 150–160.

Shan, W., Cao, M., Leung, D., and Tyler, B.M. (2004) The *Avr1b* Locus of *Phytophthora sojae* Encodes an Elicitor and A Regulator Required for Avirulence on Soybean Plants Carrying Resistance Gene *Rps1b*. Mol. Plant Microbe Interact 17, 394–403.

Shoemaker, R. (2007) Elucidating soybean genome structure. In: Stacey, G. (Ed.), *Soybean Genomics*. Springer, New York, pp. xxx–xxx.

Shoemaker, R., Keim, P., Vodkin, L., Retzel, E., Clifton, S. W., Waterston, R., Smoller, D., Coryell, V., Khanna, A., Erpelding, J., Gai, X., Brendel, V., Raph-Schmidt, C., Shoop, E. G.,

Vielweber, C. J., Schmatz, M., Pape, D., Bowers, Y., Theising, B., Martin, J., Dante, M., Wylie, T., and Granger, C. (2002) A compilation of soybean ESTs: generation and analysis. Genome 45, 329–338.

Shultz, J. L., Kazi, S., Bashir, R., Afzal, J. A., and Lightfoot, D. A. (2007) The development of BAC-end sequence-based microsatellite markers and placement in the physical and genetic maps of soybean. Theor Appl Genet 114, 1081–1090.

Shultz, J. L., Kurunam, D., Shopinski, K., Iqbal, M. J., Kazi, S., Zobrist, K., Bashir, R., Yaegashi, S., Lavu, N., Afzal, A. J., Yesudas, C. R., Kassem, M. A., Wu, C., Zhang, H. B., Town, C. D., Meksem, K., and Lightfoot, D. A. (2006) The Soybean Genome Database (SoyGD): a browser for display of duplicated, polyploid, regions and sequence tagged sites on the integrated physical and genetic maps of Glycine max. Nucleic Acids Res 34, D758–765.

Song, Q. J., Marek, L. F., Shoemaker, R. C., Lark, K. G., Concibido, V. C., Delannay, X., Specht, J. E., and Cregan, P. B. (2004) A new integrated genetic linkage map of the soybean. Theor Appl Genet 109, 122–128.

Tamulonis, J. P., Luzzi, B. M., Hussey, R. S., Parrott, W. A., and Boerma, H. R. (1997) RFLP Mapping of Resistance to Southern Root-Knot Nematode in Soybean. Crop Science 37, 1903–1909.

Thomas, R., Fang, X., Ranathunge, K., Anderson, T. R., Peterson, C. A., and Bernards, M. A. (2007) Soybean Root Suberin: Anatomical Distribution, Chemical Composition, and Relationship to Partial Resistance to *Phytophthora sojae*. Plant Physiol 144, 299–311.

Tian, A. G., Wang, J., Cui, P., Han, Y. J., Xu, H., Cong, L. J., Huang, X. G., Wang, X. L., Jiao, Y. Z., Wang, B. J., Wang, Y. J., Zhang, J. S., and Chen, S. Y. (2004) Characterization of soybean genomic features by analysis of its expressed sequence tags. Theor Appl Genet 108, 903–913.

Tipping, A. J. and McPherson, M. J. (1995) Cloning and molecular analysis of the pea seedling copper amine oxidase. J Biol Chem 270, 16939–16946.

Tooley, P.W. and Grau, C.R. (1982) Identification and Quantitative Characterization of Rate-Reducing Resistance to *Phytophthora megasperma* f.sp. *glycinea* in Soybean Seedlings. Phytopathology 72, 727–733.

Tooley, P.W. and Grau, C.R. (1984) Field characterization of rate-reducing resistance to *Phytophthora megasperma* f.sp. *glycinea* in soybean. Phytopathology 74, 1201–1208.

Torto, T. A., Li, S., Styer, A., Huitema, E., Testa, A., Gow, N. A., van West, P., and Kamoun, S. (2003) EST mining and functional expression assays identify extracellular effector proteins from the plant pathogen *Phytophthora*. Genome Res 13, 1675–1685.

Torto-Alalibo, T., Tripathy, S., Smith, B. M., Arredondo, F., Zhou, L., Li, H., Qutob, D., Gijzen, M., Mao, C., Sobral, B.W.S., Waugh, M.E., Mitchell, T. K., Dean, R.A., and Tyler, B.M. (2007) Expressed sequence tags from *Phytophthora sojae* reveal genes specific to development and infection. Mol. Plant-Microbe Interact. in press.

Tyler, B. M., Tripathy, S., Zhang, X., Dehal, P., Jiang, R.H.Y., Aerts, A., Arredondo, F., Baxter, L., Bensasson, D., Beynon, J. L., Damasceno, C. M. B., Dickerman, A., Dorrance, A. E., Dou, D., Dubchak, I., Garbelotto, M., Gijzen, M., Gordon, S., Govers, F., Grunwald, N., Huang, W., Ivors, K., Jones, R. W., Kamoun, S., Krampis, K., Lamour, K., Lee, M.-K., McDonald, W. H., Medina, M., Meijer, H.J.G., Nordberg, E., Maclean, D. J., Ospina-Giraldo, M. D., Morris, P., Phuntumart, V., Putnam, N., Rash, S., Rose, J.K.C., Sakihama, Y., Salamov, A., A.Savidor, Scheuring, C., Smith, B., Sobral, B.W. S., Terry, A., Torto-Alalibo, T., Win, J., Xu, Z., Zhang, H., Grigoriev, I., Rokhsar, D., and Boore, J.L. (2006) *Phytophthora* Genome Sequences Uncover Evolutionary Origins And Mechanisms Of Pathogenesis. Science 313, 1261–1266.

Tyler, B.M., Jiang, R. H.Y., Zhou, L., Tripathy, S., Dou, D., Torto-Alalibo, T., Li, H., Mao, Y., Liu, B., Vega-Sanchez, M., Mideros, S.X., Hanlon, R., Smith, B.M., Krampis, K., Ye, K., Martin, S. St, Dorrance, A.E., Hoeschele, I., and Maroof, M.A.S. (2007) Functional genomics and bioinformatics of the *Phytophthora sojae*-soybean interaction. In: Gustafson, Stacey, and Taylor (Eds.), *The Genomics of Disease.* Kluwer Academic/Plenum Publisher, New York, pp. in press.

van de Mortel, M., Recknor, J.C., Graham, M.A., Nettleton, D., Dittman, J.D., Nelson, R.T., Godoy, C.V., Abdelnoor, R.V., Almeida, M.R., Baum, T.J., and Whitham, S.A. (2007) Distinct biphasic mRNA changes in response to Asian soybean rust infection. Molec Plant Microbe Interact in press.

Van Etten, H. D., Sandrock, R. W., Wasman, C. C., Soby, S. D., McCluskey, K., and Wang, P. (1995) Detoxification of phytoanticipins and phytoalexins by phytopathogenic fungi. Can. J. Bot. 73, S518–S525.

Vega-Sánchez, M.E., Redinbaugh, M.G., Costanzo, S., and Dorrance, A.E. (2005) Spatial and temporal expression analysis of defense-related genes in soybean cultivars with different levels of partial resistance to *Phytophthora sojae*. Physiological and Molecular Plant Pathology 66, 175–182.

Vodkin, L. O., Khanna, A., Shealy, R., Clough, S. J., Gonzalez, D. O., Philip, R., Zabala, G., Thibaud-Nissen, F., Sidarous, M., Stromvik, M. V., Shoop, E., Schmidt, C., Retzel, E., Erpelding, J., Shoemaker, R. C., Rodriguez-Huete, A. M., Polacco, J. C., Coryell, V., Keim, P., Gong, G., Liu, L., Pardinas, J., and Schweitzer, P. (2004) Microarrays for global expression constructed with a low redundancy set of 27,500 sequenced cDNAs representing an array of developmental stages and physiological conditions of the soybean plant. BMC Genomics 5, 73.

Wang, Z., Wang, Y., Chen, X., Shen, G., Zhang, Z., and Zheng, X. (2006) Differential screening reveals genes differentially expressed in low- and high-virulence near-isogenic *Phytophthora sojae* lines. Fungal Genet Biol 43, 826–839.

Whisson, S. C., Basnayake, S., Maclean, D. J., Irwin, J. A., and Drenth, A. (2004) *Phytophthora sojae* avirulence genes *Avr*4 and *Avr*6 are located in a 24kb, recombination-rich region of genomic DNA. Fungal Genet Biol 41, 62–74.

Whisson, S.C., Drenth, A., Maclean, D.J., and Irwin, J.A.G. (1995) *Phytophthora sojae* avirulence genes, RAPD and RFLP markers used to construct a detailed genetic linkage map. Mol. Plant-Microbe Interact. 8, 988–995.

Wrather, J.A., Anderson, T.R., Arsyad, D.M., Tan, Y., Ploper, L.D., Porta-Puglia, A., Ram, H.H., and Yorinori, J.T. (2001a) Soybean disease loss estimates for the top ten soybean-producing countries in 1998. Canada J. Plant Pathol. 23, 115–121.

Wrather, J.A., Koenning, S.R., and Anderson, T.R. (2003) Effect of Diseases on Soybean Yields in the United States and Ontario (1999 to 2002). Online. Plant Health Progress March 25, doi:10.1094/PHP-2003-0325-1001-RV.

Wrather, J.A., and Koenning, S.R. (2006). Estimates of disease effects on soybean yields in the United States 2003 to 2005. J. Nematology 38, 173–180.

Wrather, J.A., Stienstra, W.C., and Koenning, S.R. (2001b) Soybean disease loss estimates for the United States from 1996 to 1998. Can. J. Plant Pathol. 23, 122–131.

Yeh, C. C., Tschanz, A. T., and Sinclair, J. B. (1981) Induced teliospore formation by *Phakopsora pachyrhizi* on soybeans and other hosts. Phytopathology 71, 1111–1112.

Yoder, O. C. and Turgeon, B. G. (1996) Molecular-genetic evaluation of fungal molecules for roles in pathogenesis in plants. Journal of Genetics 75, 425–440.

Yoder, O.C. and Turgeon, B. Gillian (2001) Fungal genomics and pathogenicity. Current Opinion in Plant Biology 4, 315–321.

Yu, Y. G., Buss, G. R., and Maroof, S. M. A. (1996) Isolation of a superfamily of candidate disease-resistance genes in soybean based on a conserved nucleotide-binding site. Proc. Nat. Aca. Sci. USA 93, 11751–11756.

Zou, J., Rodriguez-Zas, S., Aldea, M., Li, M., Zhu, J., Gonzalez, D. O., Vodkin, L. O., DeLucia, E., and Clough, S. J. (2005) Expression profiling soybean response to *Pseudomonas syringae* reveals new defense-related genes and rapid HR-specific down-regulation of photosynthesis. Mol Plant Microbe Interact 18, 1161–1174.

Chapter 15
Genomics of Insect-Soybean Interactions[1]

Wayne Parrott, David Walker, Shuquan Zhu, H. Roger Boerma, and John All

Introduction

On-going efforts to lower costs of production, along with increased concerns over insecticide residues in the food chain and in the environment are incentives to conduct research on crop resistance to insect pests. While multiple genes for resistance to insects have been identified in many plants (Yencho et al. 2000), understanding their molecular basis remains far behind the understanding achieved for disease resistance genes. Given that plants and insects have co-evolved for millions of years, it is not surprising that a wide range of plant-insect relationships exist and that many of these have a very different genetic basis. The following review is an attempt to categorize the various plant-host relationships that exist within soybean and provide a summary of what is known about the genetics of resistance and its characterization at the genomic level.

Damage Caused by Soybean Insect Pests

Insect pests of soybean are well recognized, though their economic impact is hard to quantify, as data on insect damage are seldom collected. Georgia is perhaps the only state that historically has kept data on insect damage to the various crops grown in the state. In 2004, the last year for which data are available, there were 72,800 ha planted in soybean, for a crop valued at $43.1 million. During 2004,

W. Parrott
Department of Crop and Soil Sciences, University of Georgia, Athens, GA 30602-6810, USA
e-mail: wparrott@uga.edu

[1] **Abbreviations of commonly cited insect names**
BAW = beet armyworm, *Spodoptera exigua* Hübner; BLB = bean leaf beetle, *Cerotoma trifurcata* Forster; CEW = corn earworm, *Helicoverpa zea* Boddie; LCB = Lesser cornstalk borer, *Elasmopalpus lignosellus* Zeller; MBB = Mexican bean beetle, *Epilachna varivestis* Mulsant; PLH = potato leafhopper, *Empoasca fabae* Harris; SBA = soybean aphid, *Aphis glycines* Matsumura; SBL = soybean looper *Pseudoplusia includens* Walker; TBW = tobacco budworm, *Heliothis virescens* Fabricius; VBC = velvetbean caterpillar, *Anticarsia gemmatalis* Hübner

G. Stacey (ed.), *Genetics and Genomics of Soybean*,

Georgia soybean farmers spent $3.3 million on insect control and lost an additional $4.25 million due to insect-induced yield reduction. Thus the combined cost of insect control and yield loss was equal to 17.5 % of total crop value in 2005 (McPherson 2004).

Insect-caused injury to soybean depends on the feeding behavior and biology of the pest, the potential of the pest to vector pathogens and, in some cases, on the development stage of the plant. The majority of the principal soybean insect pests can be classified on the basis of their feeding technique into piercing-sucking, chewing, or tunneling pest species. Chewing insects can damage foliage, reproductive organs, and/or stems.

Piercing-Sucking Insects

The major pests in this category are in the order Hemiptera, which now includes the true bugs, such as stink bugs, and the aphids (suborder Sternorrhyncha) and leafhoppers (suborder Auchenorrhyncha). The green [*Acrosternum hilare* Say], southern green [*Nezara viridula* L.] and brown [*Euschistus servus* Say] stink bugs comprise the bulk of the soybean stink bug complex in North America. The primary economic damage in soybean results when nymphs and adults pierce pods and developing seeds to feed on plant juices. Damage to young pods can cause shriveled seed and pod abortion, while feeding on developing seed can result in abnormal development, wrinkling, and a stained seed coat (Turnipseed and Kogan 1987). Stink bugs are polyphagous, so they can easily move into soybean fields from alternative hosts such as cotton. Economic damage from stink bugs can outweigh damage from all other insects in the South (McPherson 2004)

The soybean aphid [*Aphis glycines* Matsumura] was first reported in the United States in 2000, and has rapidly become a major insect pest in the Midwest (Hill et al. 2004). Aphid feeding can result in yield loss, severe stunting, leaf distortion, reduced pod set, lower seed weight, and nutrient deficiencies (Mensah et al. 2005). Indirect damage is caused by transmission of viruses like soybean mosaic virus, alfalfa mosaic virus, soybean dwarf virus, and bean yellow mosaic virus (Hill et al. 2001). Growth of sooty mold on 'honeydew,' a sticky substance excreted by feeding aphids, can further reduce photosynthesis and yield (Mensah et al. 2005). Feeding by other hemipteran pests such as the potato leafhopper [*Empoasca fabae* (Harris)] and the three-cornered alfalfa hopper [*Spissistilus festinus* (Say)] can cause damage to leaves or make weakened stems more prone to lodging (Turnipseed and Kogan 1987).

Chewing Insects

These include lepidopteran and coleopteran larvae and adults that feed on soybean leaves and seed pods. Major lepidopteran pests include the green cloverworm

[*Plathypena scabra* (F.)] in the Midwest, and the corn earworm [*Helicoverpa zea* (Boddie)], soybean looper [*Pseudoplusia includens* (Walker)], and velvetbean caterpillar [*Anticarsia gemmatalis* (Hübner)] in the southern states (Turnipseed and Kogan 1987). These are primarily defoliators, with the exception of CEW, which frequently feeds on flowers, pods, and developing seed as well as leaves. Because of its feeding behavior, CEW is generally assumed to be the lepidopteran pest responsible for the greatest yield losses in the USA, even though supporting data are lacking. In addition, the ravenous appetites of VBC larvae make this species one of the most destructive defoliators in the southeastern USA, especially since they may also damage meristems after foliage becomes limited.

The Mexican bean beetle [*Epilachna varivestis* Mulsant] and the bean leaf beetle [*Cerotoma trifurcata* (Forster)] are among the most destructive coleopteran pests of soybean (Turnipseed and Kogan 1987). Both larvae and adults of the MBB and adult BLB feed on foliage, sometimes causing economic losses from defoliation, especially when populations develop during the plants' reproductive growth stage. In addition, BLB is capable of transmitting bean pod mottle virus and of injuring pods on mature plants and facilitating secondary invasion of fungi (Turnipseed and Kogan 1987). BLB larvae feed on the roots and nodules of soybean plants. Several other coleopteran species may cause infrequent yield losses.

Tunneling Insects

Lesser cornstalk borer (*Elasmopalpus lignosellus* Zeller) is the main pest that causes damage by tunneling into the stems of seedlings at the soil surface (Turnipseed and Kogan 1987). This pest tends to attack soybean growing on sandy soils in hot, dry weather, which may compound the effects of the injury caused by the larvae. Small plants are often killed outright, while surviving plants are more prone to lodging as a result of having weakened stems. The Dectes stem borer (*Dectes texanus texanus* LeConte, Cerambycidae) produces sporadic but sometimes severe damage to early planted soybean.

Seed Storage Insects

Stored soybean seed is subject to insect feeding as well. Such pests include the almond moth (*Ephestia cautella* Walker), tobacco borer beetle (*Lasioderma serricone* Fab.), red grain beetles (*Tribolium castenum* Herbst and *T. confusum* Kackuelinduval), khapra beetle (*Trogoderma granarium* Everts) and the pulse beetles (*Callosobruchus analis* Fab. and *C. chinensis* L.; Islas-Rubio and Higuera-Ciapara 2003).

Types of Resistance

Based on Effect on Insects

Painter (1941) suggested that host plant resistance to insects could be classified as non-preference, antibiosis, or tolerance. The term '*antixenosis*' was later suggested by Kogan and Ortman (1978) as a substitute for 'non-preference.' Antixenosis refers to a host plant effect on pest behavior which discourages feeding and/or oviposition. The underlying traits can be morphological (e.g., dense pubescence) or biochemical (presence of a deterrent compound or absence of an attractant). Figure 15.1 shows near isogenic soybean lines, whereby the one on the right has antixenotic insect resistance.

Antibiosis refers to a type of resistance in which a host plant has a detrimental effect on the physiology and life history of an insect pest (Painter 1951). This can be manifested as decreased growth rate, lower pupal weights, decreased fitness, or other factors that interfere with the pest's ability to survive, mature and reproduce. CEW larvae fed near isogenic lines of soybean, one of which has antibiotic resistance, are shown in Fig. 15.2.

Experimentally, a distinction between these two types of resistance is based on whether a choice (antixenosis) or no-choice (antibiosis) assay was used to quantify the resistance. In choice assays, the insect has an assortment of genotypes to feed on. In a no-choice assay, the insect is only provided one genotype as a food source. These two types of resistance are not mutually exclusive: a trait that deters a larva from feeding may slow its development and/or reduce the adult's ability to reproduce. For example, an allele at the *Pb* locus that conditions sharp pubescence tips was detected using both antixenosis and antibiosis feeding assays (Hulburt et al. 2004).

Fig. 15.1 Soybean (*right*) showing antixenotic resistance towards SBL (*See* also Color Insert)

Fig. 15.2 Corn earworms (*right*) showing an antibiotic effect on their growth (*See* also Color Insert)

Tolerance is technically not a type of resistance, but rather a greater ability of certain plant genotypes to maintain yields after sustaining levels of insect damage that would cause yield losses in other genotypes of the species (Haile et al. 1998). For example, narrow-leafed isolines of soybean were found that can sustain greater levels of defoliation than conventional soybeans can before yield is affected, perhaps because of the greater light-interception capacity of narrow leaves (Haile et al. 1998).

Based on Gene Expression Patterns

Insect resistance factors can also be classified as constitutive (e.g., trichome density) or induced, which involve some change in gene expression levels in response to injury caused by an insect pest. Constitutive resistance factors may or may not be uniformly effective in a plant. For example, SBL larvae can grow faster when fed newly expanded leaves from soybean accession PI 227687, than when fed older leaves from the same plant (Reynolds and Smith 1985).

Induced resistance refers to a reactive response by plants to specific or non-specific assault by predators (e.g., insects), pathogens or environmental disturbance. Induced resistance typically involves a cascade of biochemical defenses that are otherwise not present, or are only present at very low levels, within plant tissues, and innate resistance targeted towards specific pest species (Cooper et al. 2004). Innate resistance often depends on the presence of single, dominant resistance genes (R-genes) that are involved in direct or indirect recognition of the presence of a pest or pathogen. Recognition may involve detection of elicitors produced by the pest or detection of biochemical changes in host tissue or cells in response to injury caused by the pest. Salicylic acid and jasmonic acid have been implicated in the signaling that activates or suppresses resistance responses to insects and pathogens (Cooper et al. 2004).

The degree of resistance that a plant has can be determined in part by the interaction of its genotype, its environment, and the age of the plant or plant tissue. Water availability is one of the environmental factors that can alter resistance (Hammond et al. 1995; Lambert and Heatherly 1991). Resistance of soybean plants to both coleopteran and lepidopteran pests was observed to decline as the plants reach the late flowering and pod-filling stages of maturity (Hammond et al. 1995; Nault et al. 1992; Rowan et al. 1993).

Based on Traits that Confer Resistance

Biochemical: Compounds involved in resistance and susceptibility of soybean genotypes to various insect pests likely include leaf volatiles (Liu et al. 1989), variations in nutrient concentrations, feeding excitants and deterrents, and antibiotic substances (Fischer et al. 1990). Allelochemicals produced by plants can discourage colonization, feeding, and/or oviposition by one or more pest species, or interfere with metabolic processes in pests that are feeding on the host plant. Proteins such as trypsin inhibitors and secondary metabolites such as alkaloids were associated with resistance to insect pests. Mutations that alter the structure or reduce the quantity of volatile compounds that attract insect pests may also result in a higher level of perceived resistance.

Phytoalexins are antibiotic metabolites that undergo enhanced or *de novo* synthesis and accumulation in response to physiological stress (Hart et al. 1983). Flavonoids are important phenylpropanoids in *Glycine* spp., and some of them may be phytoalexins with an antiherbivory role (Burden and Norris 1992). For example, the isoflavonoid, glyceolin, was shown to function as an antifeedant against MBB and some other coleopterans, but did not have a significant effect on SBL (Hart et al. 1983). Coumestrol, another isoflavonoid, appears to contribute to antixenosis resistance in 'Davis' soybean plants (Burden and Norris 1992). Hart et al. (1983) suggested that pest deterrence in some soybean genotypes may involve the combined actions of multiple flavonoids, including glyceollin, glycinol, coumestrol, sojagol, daidzein, and genistein. There has been a notable lack of supportive research into the role of phytoalexins as deterrents to herbivory since the original research was conducted.

Physical: Physical and morphological traits that discourage or impede feeding or oviposition also contribute to host resistance. Studies of isolines that differed in the length and orientation of trichomes showed that the long, erect trichomes inherited from the resistant soybean PI 229358 contributed to antixenosis towards PLH (Turnipseed 1977). Elevated lignin content in soybean stems may reduce the damage caused by tunneling insects like the LCB.

History of Breeding Efforts

Efforts to develop soybean cultivars with insect resistance can be divided into three eras, the first two beginning with discoveries that had a major influence on subsequent breeding work, and a contemporary period that is characterized by (i) the use of biotechnological tools to address limitations of a conventional breeding approach, and (ii) the recent emergence of resistance to the soybean aphid (*Aphis glycines* Matsumura) as an important breeding objective. In the 1930s, Hollowell and Johnson found that pubescence provided resistance to the PLH, in comparison with glabrous plants (Hollowell and Johnson 1934; Johnson and Hollowell 1935). Subsequent selection for pubescence eliminated PLH as a soybean pest, and this resistance has proven durable for 70 years.

Breeding for Resistance to Defoliating Insects

Insect resistance became a major objective in several breeding programs during the 1970s after the identification of three germplasm accessions with resistance to the MBB (Van Duyn et al. 1971, 1972). PI 171451 ('Kosamame'), PI 227687 ('Mikayo White'), and PI 229358 ('Sodendaizu') also were found to be resistant to several major lepidopteran soybean pests (Clark et al. 1972; Lambert and Kilen 1984a,b). By the mid 1980s, soybean breeding programs in 10 states were using one or more of these accessions as a source of insect resistance genes (Turnipseed and Kogan 1987). Unfortunately, it proved difficult to capture the full resistance levels of these accessions in progenies derived from crosses with adapted high-yielding germplasm. For example, when five PI 229358-derived breeding lines selected for resistance to MBB and SBL were evaluated for their resistance to these two insect pests plus CEW and TBW, the results showed that, while all five lines were as resistant as PI 229358 to SBL and CEW larvae, only two of the lines equaled PI 229358 resistance to all four pests (Hatchett et al. 1979). This led the authors to suggest that the genes conditioning resistance to CEW and SBL are likely to be the same, or at least tightly linked. Smith and Brim (1979a,b) similarly found that PI 229358- and PI 227687-derived lines selected for resistance to MBB were not very effective in deterring feeding by CEW.

The presence of F_3 lines exhibiting transgressive segregation for both susceptibility and resistance suggests that these three PIs might possess at least some different genes for resistance (Kilen and Lambert 1986). Furthermore, there is some evidence that the resistance of each PI differs to different insects. PI 229358 and PI 227687 were reported to be resistant to VBC, SBL, CEW, TBW, and BAW, with the resistance of PI 227687 being greater to CEW and BAW than that of PI 229358. PI 171451 was reported resistant to CEW and TBW, but not to VBC, SBL, or BAW (Lambert and Kilen 1984b). However, similar differences were not found in a subsequent study (Gray et al. 1985). The inference is that screening segregating lines for resistance to a single insect pest may prevent selection of lines with resistance genes to other major insects. Furthermore, results may be confounded by the different soybean genotypes, and affected by the type of assay used, and perhaps even by possible geographical differences in the genetic makeup of the insect pests. The recent availability of near-isolines of soybean for different insect resistance genes may finally make it possible to start evaluating insect-specific effects of the various soybean insect-resistance genes.

It has been equally difficult to recapture the yield potential and agronomic qualities of the elite parent in crosses between these insect-resistant breeding lines and elite parents (Kilen and Lambert 1986). Although numerous breeding lines with insect resistance were released from breeding programs that used the Japanese accessions as resistance sources, only three resistant cultivars were released: 'Lamar', 'Crockett', and 'Lyon', and none could compete agronomically with the best contemporary cultivars (Boethel 1999; Bowers 1990; Hartwig et al. 1990, 1994; Lambert and Tyler 1999). 'Shore', another cultivar with both antibiosis and antixenosis towards MBB, was not derived from any of the three resistant Japanese accessions (Smith et al. 1975).

Classical genetic studies were hampered by the inability to properly identify plants segregating for genes that provide the same phenotype. Nevertheless, research on the inheritance of insect resistance from PI 171451, PI 227687, and PI 229358 consistently pointed to quantitative inheritance involving a few major genes (Sisson et al. 1976; Kenty et al. 1996).

Kenty et al. (1996) estimated the broad-sense heritability of PI 229358-derived antibiosis towards SBL to be 21–47 %, with two to six genes being involved. A similar heritability (33–48 %) and number of resistance genes was estimated for MBB antibiosis derived from PI 171451 and 229358 (Rufener et al. 1989). Sisson et al. (1976) investigated resistance to MBB in F_3 populations derived from PI 227687, PI 229358, and PI 229321, a somewhat less resistant accession. They concluded that the resistance was quantitative, probably involving only two or three major genes expressing primarily additive gene action. Studies of other progenies derived from crosses to PI 229358 also indicated that resistance to SBL depended on a few major genes, but with partial dominance towards susceptibility (Kilen et al. 1977; Kenty et al. 1996). Investigations of VBC resistance of F_3 progenies from intercrosses among PI 171451, PI 227687, and PI 229358 also supported partial dominance of resistance (Kilen and Lambert 1986). In crosses between 'Davis' and PI 171451, PI 227686, and PI 229358, the average size of lepidopteran larvae from five pest species fed tissue from F_1 intercross progeny was similar to that of larvae from the resistant parent, suggesting that the genes conditioning antibiosis resistance were at least partially dominant (Lambert and Kilen 1984b). Taken collectively, these studies suggest there is a gene-dosage effect.

The limitations that hindered classical genetics studies were largely overcome by using DNA markers, which made it possible to identify the number, locations, and relative contributions of quantitative trait loci (QTLs) associated with soybean resistance.

Rector et al. (1998, 1999 and 2000) used restriction fragment length polymorphisms (RFLPs) to identify and map quantitative trail loci (QTL) associated with antibiosis and antixenosis resistance to CEW in populations derived from PI 171451, PI 227687, and PI 229358. QTL for both types of resistance were mapped in all of the PIs, and each of the PIs appeared to possess both unique and shared QTLs, as originally suggested by the classical genetic studies. This approach made it possible for the first time to quantify the contributions of the individual QTL to the resistance phenotype. The subsequent availability of abundant simple sequence repeat (SSR) markers allowed greater precision in mapping the locations of the QTL and facilitated the use of marker-assisted selection to transfer the resistance alleles into elite genetic backgrounds (Narvel et al. 2001). DNA markers made it possible to confirm the locations and effects of resistance QTL in different genetic backgrounds, and to determine which QTL contribute to resistance against a range of insect pests (Terry et al. 2000; Komatsu et al. 2005).

Of the various QTL from the three PIs, the QTL on linkage group M (QTL-M) is the major QTL conditioning resistance to defoliating insects, accounting for 37 % of the antixenotic effect and 22 % of the antibiotic effect (Rector et al. 1998, 1999 and 2000; Komatsu et al. 2005). Markers flanking the QTL are also useful to identify plants in a breeding population that have the smallest introgressed regions of donor parent DNA; therefore, mitigating the risk of linkage drag, the major historical obstacle in utilizing resistance genes from these three PIs. SSR markers were used to develop near-isogenic lines of 'Jack' and 'Benning' that carry various combinations of resistance alleles at mapped QTL, allowing the main effects and interactions of these genes on insect resistance to be studied with greater precision than previously possible (Walker et al. 2004; Zhu et al. 2006). Mean defoliation of lines with QTL-M by both SBL (16.8 vs 11.1 %) and CEW (15.4 vs 9.5 %) was about one-third lower than that of lines without QTL-M.

QTL-M appears to have a synergistic effect with the *cry1Ac* Bt gene, as determined by evaluation of Jack near-isogenic lines (Walker et al. 2004). The near-isolines with both the Bt and QTL-M resistance genes averaged 2.0 % defoliation by CEW and 5.5 % defoliation by SBL, compared with 7.0 and 10.7 % defoliation, respectively, for near-isolines with only Bt.

Another intriguing aspect of QTL-M is that a TBW strain selected for resistance to *Cry1Ac* Bt (Gould et al. 1995) is more sensitive to QTL-M than is a closely related wild-type strain of TBW (Walker et al. 2004). Wild-type TBW larvae feeding on near-isolines of Jack containing an introgressed genomic segment containing QTL-M weighed about 180 mg, while the Bt-resistant larvae raised on the same near-isoline only weighed about 80 mg, a highly significant difference. Such fitness costs associated with resistance to Bt were hypothesized to be a reason why insect resistance in transgenic crops has held up better than initially anticipated (Tabashnik et al. 2003).

Although several other QTL were identified that also confer resistance to insects (Rector et al. 1998, 1999 and 2000; Terry et al. 2000; Komatsu et al. 2005), the most relevant may be ones on linkage groups G (QTL-G) and H (QTL-H; Zhu et al. 2006). QTL-G conditions antibiosis, while QTL-H conditions antixenosis. However, the effect of these QTL is only evident in the presence of QTL-M. Near-isolines having no resistance genes or only QTL-H suffered about 30 % defoliation with CEW, those with QTL-M suffered about 24 % defoliation, but those with both QTLs only suffered 19 % defoliation, a significantly lower value.

A similar pattern is found with QTL-G. CEW larvae fed near isolines having no resistance genes or only QTL-G grew to weigh about 130 mg each. Those feeding on lines with QTL-M only grew to 80 mg each, but those feeding on lines with both QTLs only weighed 50 mg each (Zhu et al. 2006).

A retrospective genetic analysis of soybean breeding lines and cultivars phenotypically selected for insect resistance further emphasized the importance of QTL-M. (Narvel et al. 2001). Most of the 15 genotypes analyzed had PI 229358 as the donor parent, though PI 171451 was reported as the donor for three lines/cultivars. All lines were selected phenotypically for resistance to pests such as MBB, SBL, and/or VBC. Thirteen of the soybean genotypes examined had the PI 229358-derived allele at QTL-M. In contrast, PI 229358-derived alleles at QTL-G and QTL-H only were introgressed into two genotypes each. None of the derived genotypes had the PI 229358-derived alleles at all three QTLs. The inability of traditional breeding programs to consistently introgress QTL-G and QTL-H along with QTL-M helps explain why breeders failed in their attempts to recover levels of insect resistance as high as that originally found in PI 229358.

The high introgression rate of QTL-M into various conventionally bred lines and cultivars developed by several independent breeding programs (Narvel et al. 2001) indicates its action is not limited by environment or genetic background. Furthermore, it confers resistance to multiple lepidopteran pests (at least CEW, SBL, VBC, common cutworm, and LCB) and to a coleopteran pest (MBB). QTL effective against different taxonomic orders or genera appear to be rare in crop plants. In a comprehensive review on mapped insect resistance loci, Yencho et al. (2000) listed 233 QTL in six crop species. None of these were reported to confer resistance against insects from different orders or genera. Of the 29 major single-genes reviewed, 20 were reported to provide resistance to a single insect species or a closely related species within the same genus. Furthermore, QTL-M does not map to any known resistance gene cluster, suggesting that its mode of action is probably different from that of the tomato *Mi-1.2* gene, which is the first cloned gene for insect resistance and described below.

QTL-M from PI 229358 is also of special interest because of its contribution to both antixenosis and antibiosis resistance, its ability to enhance the effectiveness of Bt, its detrimental effect on Bt-resistant insects, its ability to 'activate' the effectiveness of QTL G and H, and its broad range of effectiveness against several lepidopteran and at least one coleopteran species. Furthermore, the major antibiosis/antixenosis QTL from PI 171451 maps to the same position on LG-M. Accordingly, QTL-M was the subject of intensive fine-mapping efforts, and by using soybean Willams 82 genomic sequence released by Department of Energy Joint Genomic Institute (DOE-JGi), we were able to develop a new simple sequence repeat (SSR) marker Satt729, and map it to the 0.52-cM interval. QTL-M was recently narrowed down to about 0.25-cM interval (Fig. 15.3). A more recent study

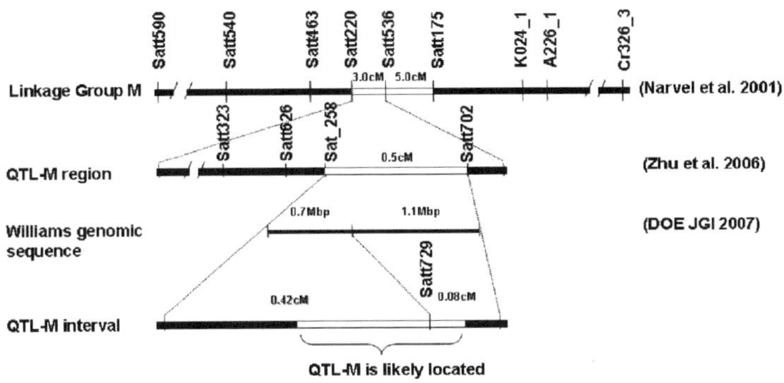

Fig. 15.3 Current map of the Linkage Group M section around QTL-M, showing the linked markers. Satt626 and Sat_258 are on the same BAC clone

by Komatsu et al. (2005) indicates that QTL-M (= *CCW-1*) is also found in the Japanese forage soybean 'Himeshirazu.' It is expected that cloning QTL-M will provide a key insight into the molecular mechanism involved in soybean resistance to defoliating insects, providing an effective tool for insect resistance management strategies.

As a final note, the use of molecular markers has made it possible to finally introgress QTL-M and QTL-H into elite germplasm without the loss of agronomic quality that plagued earlier attempts. QTL-G still has some linkage drag (or possible pleiotropy) associated with it, so additional breeding and selection is needed to disassociate this QTL from undesirable linked alleles (Zhu et al. 2007).

Breeding for Resistance to Aphids

The appearance of the SBA in the USA in 2000 prompted efforts to identify resistance to this new arrival. This pest rapidly spread throughout the Midwest and has now moved into other regions of North America. Hill et al. (2004) found that 'Dowling', 'Jackson', and PI 71506 have antixenosis resistance, and that Dowling and Jackson are also antibiotic, having a negative effect on aphid fecundity. This resistance was expressed at all plant stages. Palmetto and CNS, which are ancestors of Dowling and Jackson, were also resistant. Four additional germplasm accessions, 'Sugao Zarai', 'Sato', 'T260H', and PI 230977 were also found to exhibit antixenosis at a level equivalent to that of Dowling, Jackson, and PI 71506. Although Dowling, Jackson, and PI 71506 are ancestors of modern North American cultivars, no resistance was found among 1,425 genotypes tested (Hill et al. 2004).

Four Maturity Group III accessions from Shandong Province, China, also showed resistance to the SBA (Mensah et al. 2005). PI 567541 B and PI 567598 B have antibiosis resistance, while PI 567543 C and PI 567597 C have both antibiosis and antixenosis resistance. In another evaluation of 240 accessions, 11 were found to exhibit resistance. Of these, nine were moderately antibiotic, and two (K1639 and Pioneer 95B97) were antixenotic and strongly antibiotic (Diaz-Montano et al. 2006).

The antibiosis resistance of Dowling and Jackson to SBA was shown to be conferred by a single dominant gene in each genotype (Hill et al. 2006a,b). *Rag1*, identified in Dowling, and the resistance gene from Jackson both map to a locus that is tightly linked to Satt435 on LG-M, though it is not yet known whether these genes represent the same or different alleles (Li et al. 2007). With the help of SSR markers linked to *Rag1*, the resistance gene was successfully backcrossed into elite material adapted to the Midwest. There are concerns, however, that this single-gene resistance will not be durable (Brian Diers, personal communication).

A summary of all the known QTLs and loci for insect resistance in soybean is in Table 15.1. Both *Rag1* and QTL-M are on LG-M, approximately 10 cM apart. Besides having different flanking markers, the QTL-M-containing PI 229358 is SBA-susceptible, further reinforcing the fact that *Rag1* and QTL-M are not the same locus.

Molecular Bases for Resistance

Given the multiple types of resistance, it is reasonable to predict that several types of genetic mechanisms are involved, each with a different molecular basis. Furthermore, no insect resistance genes have been cloned from soybean, and only a limited number from other plants. To date, the *Mi-1.2* gene from tomato (*Lycopersicon esculentum*) is the only insect resistance gene cloned, though it was initially identified and cloned as a nematode-resistance gene (Kaloshian et al. 2000). Thus, this gene appears to be a general resistance gene against piercing–sucking pests.

Two conceivable functions for a gene conferring biochemically based insect resistance are: (1) involvement in the synthesis of a compound or compounds with antibiotic or antixenotic properties, or (2) a role in direct or indirect recognition of attack by an insect pest leading to an increased defense response in the plant. Examples of the first group would be genes that encode proteinase inhibitors or enzymes that enhance the biosynthesis of secondary metabolites that function as anti-feedants. The second group includes genes involved in triggering a local or systemic defense response following recognition of foreign compounds introduced into the plant tissue during insect feeding (i.e., elicitors in a classical gene-for-gene interaction), or recognition of changes in one or more of the plant's own proteins that are structurally modified as a result of injury caused by the pest (i.e., analogous to the 'guard hypothesis' of van der Biezen and Jones 1998).

Resistance to Piercing-Sucking Insects

A brief summary of what is known about *Mi-1.2* may provide some clues about the function of related genes in soybean. This gene confers resistance to diverse phloem-feeding pests such as potato aphid [*Macrosiphum euphorbiae* (Thomas)], sweet-potato whitefly (*Bemisia tabaci*), and root-knot nematodes (*Meloidogyne* spp.), from which the name of the gene is derived (Kaloshian et al. 2000). This gene deters potato aphid feeding and reduces population growth (Cooper et al. 2004), and is a member of the major class of resistance (R) genes that encode proteins characterized by nucleotide binding sites (NBS) and leucine-rich-repeats (LRRs, Milligan et al. 1998). The class of genes encoding these proteins includes many R genes involved in classic gene-for-gene interactions with pathogens. Pest or pathogen recognition is usually highly specific, and this is the case with the *Mi-1.2* gene, which provides greater resistance to European isolates of potato aphid than to North American isolates (Goggin et al. 2001).

Kaloshian et al. (2000) found that while potato aphids began probing host tissue equally rapidly and were able to locate sieve elements equally well on resistant and susceptible lines, the amount of time spent feeding once contact with a sieve element was made was 7- to 10-fold longer on the susceptible line. Insects on the resistant line made briefer and more frequent probes, indicating that the resistance mechanism involves shorter duration of contact with sieve elements, resulting in decreased salivation and ingestion of phloem fluids. Application of a salicylic acid analog, benzothiodiazole, to induce salicylic acid-dependent defenses enhanced aphid control on a tomato cultivar carrying *Mi-1.2*, as well as on a near-isogenic cultivar that lacked the gene (Cooper et al. 2004). In contrast, application of jasmonic acid enhanced resistance of the susceptible cultivar, but did not significantly increase the resistance of the cultivar with *Mi-1.2*.

A gene from *Medicago truncatula* conditions both antixenosis resistance and phloem-specific antibiosis to the blue alfalfa aphid (*Acyrthosiphon kondoi* Shinji; Klinger et al. 2005). This gene maps to a region of the genome containing resistance gene analogs that are predicted to encode resistance proteins of the coiled coil (CC)-NBS-LRR subfamily. The antibiosis resistance is inducible, since aphids on previously infested plants spent significantly less time ingesting phloem sap, and is also systemic (Klinger et al. 2005). The resistance requires an intact plant, since aphids feeding on excised shoots had enhanced survival and growth compared to those feeding on intact plants of the same resistant genotype. The resistance conditioned by this gene differs from that mediated by *Mi-1.2* in that aphid reproduction on resistant *M. truncatulata* plants was possible, albeit reduced.

Resistance to Defoliating/Tunneling Insects

Much less is known about genes which condition resistance to defoliating insects. Mechanical wounding such as that induced by caterpillars is known to activate jasmonic-acid-regulated gene expression, and some caterpillars regurgitate compounds that are thought to induce a novel defense pathway in plants (Reymond et al. 2000; Baldwin and Preston 1999; Walling 2000). Insect herbivory in lima bean can induce the upregulation of isoprenoid production (Bartram et al. 2006). However, it

Table 15.1 QTLs and genes reported to confer insect resistance in soybean

QTL or gene	Source	Insect target	Linkage group	Flanking markers	Segment size (cM)	% Variation explained	Confirmed?	References
CCW-2	Hime-shirazu	Defoliators	M	Sat_567 Satt463	17	16	ND[a]	Komatsu et al. 2005
QTL A1	Cobb	Defoliators	A1	A0831 Satt382	11	16	ND	Hulburt 2001
QTL B2#1	PI 227687	Defoliators	B2	A343b Satt126	25	12	ND	Rector et al. 2002; Hulburt 2001
QTL B2#2	PI 227687	Defoliators	B2	Sat_230 Satt126	4	17	ND	Hulburt 2001
QTL C1[b]	PI 227687	Defoliators	C1	A132T-1 A670T	–	12	ND	Rector et al. 1999; Hulburt 2001
QTL D1b	PI 229358	Defoliators	D1b	Satt141 Satt290	4	10	No	Rector et al. 1998, 1999; Narvel et al. 2001
QTL E, *Pb*, QTL U2	PI 227687 Minsoy	Defoliators	E	Satt124 Satt411	3	20–26	Yes	Terry et al. 2000; Hulburt 2001; Hulburt et al. 2004
QTL F#1	Cobb	Defoliators	F	B212V-1 A757V-2	–	20	ND	Rector et al. 1999
QTL F#2[c]	PI 227687	Defoliators	F	Sat_090 Sat_074	24	12	ND	Rector et al. 2000; Hulburt 2001
QTL G	PI 229358	Defoliators	G	Satt472 Satt191	4	14	Yes	Rector et al. 2000; Narvel et al. 2001; Zhu et al. 2006
QTL H	PI 171451 PI 227687 PI 229358	Defoliators	H	Sat_122 Satt541	0.5[d]	15	Yes	Rector et al. 1998, 1999; Narvel et al. 2001; Zhu et al. 2006
QTL J	Cobb	Defoliators	J	A064V K411T-1	41	19	ND	Rector et al. 2000

Table 15.1 Continued

QTL or gene	Source	Insect target	Linkage group	Flanking markers	Segment size (cM)	% Variation explained	Confirmed?	References
QTL M *CCW-1*	PI 229358	Defoliators Borers	M	Sat_258 Satt702	0.5	21–37	Yes	Rector et al. 1998, 1999, 2000; Narvel et al. 2001; Walker et al. 2004; Komatsu et al. 2005; Zhu et al. 2006
QTL O	Cobb	Defoliatiors	O	Satt358	–	19	ND	Hulburt 2001
QTL U10[e]	Minsoy	Defoliators	H	Satt192 Satt302	42	6–9	ND	Terry et al. 2000
Rag1	Dowling	Suckers	M	Satt435 Satt463	12	–	Yes	Hill et al. 2006a; Li et al. 2007
Rag-Jackson	Jackson	Suckers	M	Satt435 Satt463	10	–	Yes	Hill et al. 2006b; Li et al. 2007

[a] ND- Confirmation study either not done or not reported
[b] Originally identified as QTL C2 in LG C2 by Rector et al. 1999
[c] Positive allele was incorrectly identified by Rector et al. 2000
[d] Distance between markers = 8 cM in Song et al. 2004
[e] Probably the same as QTL-H. The flanking markers for QTL-H are nested within those for QTL-U10.

is important to note that these studies were conducted on plant genotypes without any special resistance to defoliating insects, so these responses probably represent a general response to herbivory, rather than a response that limits the resulting herbivory to a greater extent than takes place in susceptible genotypes.

Smith and Fischer (1983) investigated the nature of the chemical basis of SBL resistance in PI 227687 using artificial diets containing leaf powder or extracts. They reported the most pronounced allelochemic effects on larval weight and mortality resulted when larvae were fed diets containing compounds extracted using methanol, though chloroform extracts also affected larvae. Later Caballero et al. (1986) found two fractions from methanol leaf extracts that were active against SBL, and isolated coumestrol, phaseol, and afrormosin from these fractions. Subsequent experiments showed that antibiotic effects of PI 227687 on SBL larvae were enhanced by mechanical wounding of foliage 24 hr prior to allowing larvae to consume the tissue (Smith 1985). On the basis of these experiments, Smith (1985) speculated that antibiosis was due to combined effects of a feeding deterrent and a growth inhibitor. Rutin, a glycosyl flavone, was isolated from leaves of PI 227687. Incorporation of rutin into the diet of VBC larvae inhibited their growth (Hoffman-Campo et al. 2006).

In this regard, resistance to defoliating insects in soybean may be similar to CEW resistance in maize. CEW is inhibited by several C-glycosyl flavones, predominantly maysin, found in maize silks. Maysin and related compounds are synthesized via a complex metabolic pathway that results in a variety of end products, only a few of which have an antibiotic effect towards CEW. In a loose sense, any gene along the pathway that eventually leads to maysin production could be considered an insect resistance gene. For example, the *p1* locus is a Myb-homologous transcription activator that up-regulates the entire pathway. However, the genes near the ends of the metabolic branches that actually result in the production of maysin and related compounds have not been identified (Cortés-Cruz et al. 2003; Zhang et al. 2003).

Resistance to Seed Storage Insects

Many of the seed storage proteins found in soybean and other legumes double as protease or amylase inhibitors. Protease inhibitors are particularly relevant, as midgut proteases are essential for insect larval survival, and their inactivation will have detrimental effects on insect growth. For example, the Kunitz trypsin inhibitor from soybean was shown to provide effective resistance to the Angoumois moth (*Sitotroga cerealella* Oliver; Shukle and Wu 2003). It also gives resistance to brown planthopper (*Nilaparvata lugens* Stal) when engineered and expressed in rice (Lee et al. 1999). Soybean vicilins were shown to inhibit development of cowpea weevil (*Callosobruchus maculatus* Fabricius; Yunes et al. 1998). The mode of action appears to be the ability of the vicilins to bind to chitin in larval midgut microvilli (Sales et al. 2001).

Physical Resistance

Sharp pubescence tip (sharp as opposed to blunt, Fig. 15.4) is associated with antixenosis to CEW, BAW, and SBL, and with antibiosis to CEW and BAW, as determined in bioassays with sharp-tipped and blunt-tipped near-isolines of 'Clark' and 'Harosoy' (Hulburt et al. 2004). The effect of this type of resistance is as large as that of QTL-M. According to the latest consensus soybean linkage map of LG-E (Song et al. 2004), *Pb* is 0.7 cM away from Satt411 on one side, and 2.3 cM from Sat_124 on the other. The *Pb* locus determines whether pubescence tips are sharp (*Pb*) or blunt (*pb*).

A QTL mapped to linkage group E (QTL-E) is associated with both antixenosis ($R^2 = 20\,\%$) and antibiosis ($R^2 = 26\,\%$) resistance to CEW in a Cobb × PI 227687

Fig. 15.4 Sharp (*top*) and blunt (*bottom*) pubescence

mapping population (Hulburt et al. 2004; Boerma and Walker 2005). This QTL maps to an interval spanning the *Pb* locus, which determines whether pubescence tips are sharp (*Pb*) or blunt (*pb*). The pubescence of PI 227687 has sharp tips, suggesting that *Pb* and QTL-E are the same locus. Other lines of evidence also suggest QTL-E and *Pb* are the same. Insect resistance and *Pb* remain associated in Harosoy and Clark near-isolines for *Pb*. A major QTL for insect resistance was found on LG-E (previously U2) in a cross between Minsoy and Noir 1 (Terry et al. 2000), which has since been found to be also segregating for pubescence tip.

Sharp-tipped pubescence should be very useful for improving insect resistance in soybean, since it contributes to both antixenosis and antibiosis, and provides some level of resistance to several lepidopteran pests. This mode of resistance is not related to that conditioned by QTL-M. Sharp pubescence is rare among soybean accessions, but is found in most wild soybean (*G. soja*) accessions (Broich and Palmer 1981). Because PI 227687 has very poor agronomic qualities, there is a high probability of linkage drag. Therefore, it has been used less frequently than PI 229358 as a donor parent in breeding programs aimed at achieving resistance to insects.

Conclusions and Future Prospects

It is clear that a variety of insect resistance mechanisms are operational in soybean. Initial genomic efforts were geared towards mapping major genes for resistance, and to use molecular markers to dissect the interactions among the various QTL for insect resistance. Future research will determine the extent to which genetic resistance will relate to increased soybean productivity. Syntenic relationships among plant species and their resistance genes allow information and knowledge about resistance genes from one plant to be applied to others. A better understanding of insect resistance in soybean should facilitate development of insect-resistant cultivars of important food legumes like the common bean (*Phaseolus lunatus* L.), for which funding of genomics research has historically been more limited than for soybean.

For insect-resistant soybean to be adopted by growers, cultivars must have superior yield and other agronomic characteristics. Genomic approaches are enabling the development of elite genotypes with specific insect resistance traits. The availability of such tagged genes and the understanding of how the QTLs interact with each other should facilitate breeding for insect resistance, and achieve the decades-old goal of obtaining modern, high-yielding cultivars that are as resistant as the original Plant Introductions from Japan.

Furthermore, pyramiding these QTLs with different modes of resistance, such as sharp pubescence, promises to achieve economically viable levels of resistance to a broad spectrum of defoliating insects. Enough progress has been made towards cloning QTL-M, that this QTL may be the first locus for resistance to a defoliating insect ever cloned, thus providing an understanding of the biological basis to resistance to defoliating insects.

References

Baldwin, I.T. and Preston, C.A. (1999) The eco-physiological complexity of plant responses to insect herbivores. Planta 208, 137–145.

Bartram, S., Jux, A, Gleixner, G. and Boland, W. (2006) Dynamic pathway allocation in early terpenoid biosynthesis of stress-induced lima bean leaves. Phytochem. 67:1661–1672.

Boerma, H.R. and Walker, D.R. (2005) Discovery and utilization of QTLs for insect resistance in soybean. Genetica 123, 181–189.

Boethel, D.J. (1999) Assessment of soybean germplasm for multiple insect resistance. In S.L. Clement and S.S. Quisenbury (Eds.) Global Plant Genetic Resources for Insect-Resistant Crops. CRC Press, Boca Raton, FL, pp. 101–129.

Broich, S.L. and Palmer, R.G. (1981) Evolutionary studies of the soybean - the frequency and distribution of alleles among collections of *Glycine max* and *Glycine soja* of various origin. Euphytica 30: 55–64.

Burden, B.J. and Norris, D.M. (1992) Role of the isoflavenoid coumestrol in the constitutive antixenosic properties of "Davis" soybeans against and oligophagous insect, the Mexican bean beetle. J. Chem. Ecol. 18, 1069–1081.

Bowers, G.R. (1990) Registration of 'Crockett' soybean. Crop Sci. 30, 427.

Caballero, P., Smith, C.M., Fronczek, F.R. and Fischer, N.H. (1986) Isoflavones from an insect-resistant variety of soybean and the molecular structure of afrormosin. J. Nat. Prod. 49, 1126–1129.

Clark, W.J., Harris, F.A., Maxwell, F.G. and Hartwig, E.E. (1972) Resistance of certain soybean cultivars to bean leaf beetle, striped blister beetle and bollworm. J. Econ. Entomol. 65, 1669–1671.

Cooper, W.C., Jia, L. and Goggin, F.L. (2004) Acquired and R-gene mediated resistance against the potato aphid in tomato. J. Chem. Ecol. 30, 2527–2541.

Cortés-Cruz, M., Snook, M. and McMullen, M.D. (2003) The genetic basis of C-glycosyl flavone B-ring modification in maize (*Zea mays* L.) silks. Genome 46, 182–194.

Desclaux, D., Roumet, P. (1996) Impact of drought stress on the phenology of two soybean cultivars. Field Crops Res. 46, 61–70.

Diaz-Montano, J., Reese, J.C., Schapaugh, W.T. and Campbell, L.R. (2006). Characterization of antibiosis and antixenosis to the soybean aphid, (Hemiptera: Aphidae) in several soybean genotypes. J. Econ. Entomol. 99. 1884–1889.

Fischer, D.C., Kogan, M. and Paxton, J. (1990) Effect of glyceollin, a soybean phytoalexin, on feeding by three phytophagous beetles (Coleoptera: Coccinellidae and Chrysomelidae): dose versus response. Environ. Entomol. 9, 1278–1282.

Goggin, F.L., Williamson, V.M. and Ullman, D.E. (2001) Variability in the response of *Macrosiphum euphorbiae* and *Myzus persicae* (Hemiptera : Aphididae) to the tomato resistance gene *Mi*. Environ. Entomol. 30, 101–106.

Goldberg, R.B., Bake, S.J., Perez-Grace, L. (1989) Regulation of gene expression during plant embryogenesis. Cell 56, 149–160.

Gould, F., Anderson, A. Reynolds, A., Bumbarner, L. and W. Moar, W. (1995) Selection and genetic analysis of a *Heliothis virescens* (Lepidoptera: Noctuidae) strain with high levels of resistance to *Bacillus thuringiensis* toxins. J. Econ. Entomol. 88, 1545–1559.

Gray, D.J., Lambert, L. and Ouzts, J.D. (1985). Evaluation of soybean plant introductions for resistance to foliar feeding insects. J. Miss. Acad. Sci. 30, 67–82.

Haile, F.J., Higley, L.G. and Specht, J.E. (1998) Soybean cultivars and insect defoliation: Yield loss and economic injury levels. Agron. J. 90, 344–352.

Haile, F.J., Higley, L.G., Specht, J.E. and Spomer, S.M. (1998) Soybean leaf morphology and defoliation tolerance. Agron. J. 90, 353–362.

Hammond, R.B., Bledsoe, L.W. and Anwar, M.N. (1995) Maturity and environmental effects on soybean resistant to Mexican bean beetle (Coleoptera: Coccinellidae). J. Econ. Entomol. 88, 175–181.

Hart, S.V., Kogan, M. and Paxton, J.D. (1983) Effect of soybean phytoalexins on the herbivorous insects Mexican bean beetle and soybean looper. J. Chem. Ecol. 9, 657–673.

Hartwig, E.E., Kilen, T.C. and Young, L.D (1994) Registration of 'Lyon' soybean. Crop Sci. 34, 1412.

Hartwig, E.E., Lambert, L. and Kilen, T.C. (1990) Registration of 'Lamar' soybean. Crop Sci. 30, 231.

Hatchett, J.H., Beland, G.L. and Kilen, T.C. (1979) Identification of multiple insect resistant soybean lines. Crop Sci. 19, 557–559.

Hill, C.B., Li, Y. and Hartman, G.L. (2004) Resistance to the soybean aphid in soybean germplasm. Crop Sci. 44, 98–106.

Hill, C.B., Li, Y. and Hartman, G.L. (2006a) A single dominant gene for resistance to the soybean aphid in the soybean cultivar Dowling. Crop Sci. 46, 1601–1605.

Hill, C.B., Li, Y. and Hartman, G.L. (2006b) Soybean aphid resistance in soybean Jackson is controlled by a single dominant gene. Crop Sci. 46, 1606–1608.

Hill, J.H., Alleman, R., Hogg, D.B. and Grau, C.R. (2001) First report of transmission of Soybean mosaic virus and Alfalfa mosaic virus by *Aphis glycines* in the New World. Plant Dis. 85, 561.

Hoffman-Campo, C.L., Ramos Neto, J.A., Neves de Oliveira, M.C. and Jacob Oliveira, L. (2006). Detrimental effect of rutin on *Anticarsia gemmatalis*. Pesq. agropec. bras. 41, 1453–1459.

Hollowell, E.A. and Johnson, T.W. (1934) Correlation between rough-hairy pubescence in soybean and freedom from injury by *Empoasca fabae*. Phytopath. 24, 12.

Hubick, K.T., Farquhar, G.D., Shorter, R. (1986) Correlation between water-use efficiency and carbon isotope discrimination in diverse pea nut (Arachis) germplasms. Aust. J. Plant Physiol. 13, 803–816.

Hulburt, D.J. (2001) Identifying additional insect resistance quantitative trait loci in soybean using simple sequence repeats. M.S. Thesis, University of Georgia, Athens, GA.

Hulburt, D.J., Boerma, H.R. and All, J.N. (2004) Effect of pubescence tip on soybean resistance to lepidopteran insects. J. Econ. Entomol. 97, 621–627.

Islas-Rubio, A.R. and Higuera-Ciapara, I. (2003) XXI: Soybean: Post-harvest operations. In: Mejía, D. And Lewis, B. (Eds). Compendium on Post-harvest Operations. Information Network on Post-harvest Operations (INPhO). FAO, www.fao.org/inpho.

Johnson, T.W. and Hollowell, E.A. (1935) Pubescent and glabrous character of soybeans as related to injury by the potato leafhopper, J. Agric. Res. 51, 371–381.

Kaloshian, I., Kinsey, M.G., Williamson, V.M. and Ullman, D.E. (2000) Mi-mediated resistance against the potato aphid *Macrosiphum euphorbiae* (Hemiptera: Aphidae) limits sieve element ingestion. Environ. Entomol. 29, 690–695.

Kenty, M.M., Hinson, K., Quesenberry, K.H. and Wofford, D. (1996) Inheritance of resistance to the soybean looper in soybean. Crop Sci. 36, 1532–1537.

Kilen, T.C., Hatchett, J.H. and Hartwig, E.E. (1977) Evaluation of early generation soybean for resistance to soybean looper. Crop Sci. 17, 397–398.

Kilen, T.C. and Lambert, L. (1986) Evidence for different genes controlling insect resistance in three soybean genotypes. Crop Sci. 26, 869–871.

Klinger, J., Creasy, R., Gao, L., Nair, R.M., Calix, A.S., Jacob, H.S., Edwards, O.R. and Singh, K.B. (2005) Aphid resistance in *Medicago truncatulata* involves antixenosis and phloem-specific, inducible antibiosis and maps to a single locus flanked by NBS-LRR resistance gene analogs. Plant Physiol. 137, 1445–1455.

Kogan, M. and Ortman, E.E. (1978) Antixenosis - a new term proposed to define Painter's "nonpreference" modality of resistance. Bull. Entomol. Soc. Am. 24, 175–176.

Komatsu, K., Okuda, S., Takahashi, M., Matsunaga, R. and Nakazawa, Y. (2005) QTL mapping of antibiosis to common cutworm (*Spodoptera litura* Fabricius) in soybean. Crop Sci. 45, 2044–2048.

Lambert, L. and Heatherly, L.G. (1991) Soil water potential: effects on soybean looper feeding on soybean leaves. Crop Sci. 31, 125–1628.

Lambert, L. and Kilen, T.C. (1984a) Multiple insect resistance in several soybean genotypes. Crop Sci. 24, 887–890.

Lambert, L. and Kilen, T.C. (1984b) Influences of three soybean genotypes and their F1 intercrosses on the development of five insect species. J. Econ. Entomol. 77, 622–625.

Lambert, L. and Tyler, J. (1999) An appraisal of insect resistant soybeans. In: J.A. Webster and B.R. Wiseman (Eds.), Economic, Environmental, and Social Benefits of Insect Resistance in Field Crops. Thomas Say, Lanham, MD. pp. 131–148.

Lee, S.I., Le,e S.H., Koo, J.C., Chun, H.J., Lim, C.O., Mun, J.H., Song. Y.H. and Cho, M.J. (1999) Soybean Kunitz trypsin inhibitor (SKTI) confers resistance to the brown planthopper (*Nilaparvata lugens* Stal) in transgenic rice. Mol. Breed. 5, 1–9.

Li, Y., Hill, C.B., Carlson, S.B., Diers, D.W., and Hartman, G.L. (2007) Soybean aphid resistance genes in the soybean cultivars Dowling and Jackson map to linkage group M. Mol. Breed. 19, 25–34.

Liu, S.H., Norris, D.M. and Lyne, P. (1989) Volatiles from the foliage of soybean, *Glycine max*, and lima bean, *Phaseolus lunatus*: their behavioral effects on the insects *Trichoplusia ni* and *Epilachna varivestis*. J. Agric. Food Chem. 37, 496–501.

Liu, F., VanToai, T.T., Moy, L., Bock, G., Linford, L. D., Quackenbush, J. (2005) Global transcription profiling reveals novel insights into hypoxic response in Arabidopsis. Plant Physiol. 137, 1115–1129.

Mensah, C., DiFonzo, C., Nelson, R.L. and Wang, D. (2005) Resistance to soybean aphid in early maturing soybean germplasm. Crop Sci. 45, 2228–2233.

McPherson, R.M. (2004) XVIII. Soybean insects. In: Gillebeau, P., Hinkle, N., and Roberts, P. (Eds). Summary of Losses from Insect Damage and Cost of Control in Georgia 2004. Georgia Cooperative Extension Service. Pp 37–38.

Milligan, S.B., Bodeau, J., Yaghoobi, J., Kaloshian, I., Zabel, P. and Williamson, V.M. (1998) The root-knot nematode resistance gene *Mi* from tomato is a member of leucine zipper, nucleotide binding, leucine-rich repeat family of plant genes. Plant Cell 10, 1307–1319.

Miranda, M., Borisjuk, L., Tewes, A., Dietrich, D., Rentsch, D., Weber, H., Wobus, U. (2003) Peptide and amino acid transporters are differentially regulated during seed development and germination in faba bean. Plant Physiol. 132, 1950–1960.

Momen, N.N., Carlson, R.E., Shaw, R.H., Arjmand, O. (1979) Moisture stress effects on the yield components of two soybean cultivars. Agron. J. 71, 86–90.

Narvel, J.M., Walker, D.R., Rector, B.G., All, J.N., Parrott, W.A. and Boerma, H.R. (2001) A retrospective DNA marker assessment to the development of insect resistant soybean. Crop Sci. 41, 1931–1939.

Nault, B.A., All, J.N. and Boerma, H.R. (1992) Resistance in vegetative and reproductive stages of a soybean breeding line to three defoliating pests (Lepidoptera: Noctuidae). J. Econ. Entomol. 85, 1507–1515.

Nielsen, N.C. (1996) Soybean seed composition. In: *Soybean: Genetics, Molecular Biology, and Biotechnology.* Biotechnology in Agriculture, D.P.S. Verma, R.C Shoemaker (Eds.) No. 14, CAB International, Wallingford, UK, pp. 127–163.

Painter, R.H. (1941) The economic value and biological significance of insect resistance in plants. J. Econ. Entomol. 34, 358–367.

Painter, R.H. (1951) Insect Resistance in Crop Plants. Macmillan & Sons, New York.

Rector, B.G., All, J.N., Parrott, W.A. and Boerma, H.R. (1998) Identification of molecular markers associated with quantitative trait loci for soybean resistance to corn earworm. Theor. Appl. Genet. 96, 786–790.

Rector, B.G., All, J.N., Parrott, W.A. and Boerma, H.R. (1999) Quantitative trait loci for antixenosis resistance to corn earworm. Crop Sci. 39, 531–538.

Rector, B.G., All, J.N., Parrott, W.A. and Boerma, H.R. (2000) Quantitative trait loci for antibiosis resistance to corn earworm in soybean. Crop Sci. 40, 233–238.

Rennie, B.D., Tanner, J.W. (1989) Fatty acid composition of oil from soybean seeds grown at extreme temperatures. J. Am. Oil Chem. Soc. 66, 1622–1624.

Reymond, P., Weber, H., Damond, M. and Farmer, E.E. (2000) Differential gene expression in response to mechanical wounding and insect feeding in *Arabidopsis*. Plant Cell 12, 707–719.

Reynolds, G.W. and Smith, C.M. (1985) Effects of leaf position, leaf wounding, and plant age of two soybean genotypes on soybean looper (Lepidoptera: Noctuidae) growth. Environ. Entomol. 14, 475–478.

Rowan, G.B., Boerma, H.R., All, J.N. and Todd, J. (1993) Soybean maturity effect on expression of resistance to Lepidopteran insects. Crop Sci. 33, 618–622.

Rufener, G.K., II, St. Martin, S.K., Cooper, R.C. and Hammond, R.B. (1989) Genetics of antibiosis resistance to Mexican bean beetle in soybean. Crop Sci. 29, 618–622.

Sales, M.P., Pimenta, P.P., Paes, N.S., Grossi-de-Sa, M.F. and Xavier, J. (2001) Vicilins (7S storage globulins) of cowpea (*Vigna unguiculata*) seeds bind to chitinous structures of the midgut of *Callosobruchus maculatus* (Coleoptera : Bruchidae) larvae. Braz. J Med. Biol. Res. 34. 27–34.

Shukle, R.H. and Wu, L. (2003) The role of protease inhibitors and parasitoids on the population dynamics of *Sitotroga cerealella* (Lepidoptera: Gelechiidae). Environ. Entomol. 32, 488–498.

Sisson, V.A., Miller, P.A., Campbell, W.V. and Van Duyn, J.W. (1976) Evidence of inheritance of resistance to the Mexican bean beetle in soybeans. Crop Sci. 16:835–837.

Smith, C.M. (1985) Expression, mechanisms and chemistry of resistance in soybean, *Glycine max* L. (Merr.) to the soybean looper, *Pseudoplusia includens* (Walker). Insect Sci. Applic. 6, 243–248.

Smith, C.M. and Brim,C.A. (1979a) Field and laboratory evaluations of soybean lines for resistance to corn earworm feeding. J. Econ. Entomol. 72, 78–80.

Smith, C.M. and Brim,C.A. (1979b) Resistance to the Mexican bean beetle and corn earworm in soybean genotypes derived from PI 227687. Crop Sci. 19, 313–314.

Smith, C.M. and Fischer, N.H. (1983) Chemical factors of an insect resistant soybean genotype affecting growth and survival of the soybean looper. Entomol. Exper. Applic. 33, 343–345.

Smith, T.J., Camper Jr., H.M. and Schillinger, J.A. (1975) Registration of Shore soybean. Crop Sci. 15, 100.

Song, Q.J., Marek, L.F., Shoemaker, R.C, Lark, K.G., Concibido, V.C., Delannay, X., Specht, J.E. and Cregan, P.B. (2004) A new integrated genetic linkage map of the soybean. Theor. Appl. Genet. 109, 122–128.

Tabashnik, B.E., Carriere, Y., Dennehy, T.J., Morin, S., Sisterson, M.S., Roush, R.T., Shelton, A.M. and Zhao, J.Z. (2003) Insect resistance to transgenic Bt crops: Lessons from the laboratory and field. J. Econ. Entomol. 96:1031–1038.

Terry, L.I., Chase, K., Jarvik, T., Orf, J., Mansur, L. and Lark, K.G. (2000) Soybean quantitative trait loci for resistance to insects. Crop Sci. 40, 375–382.

Turner, N.C. (2000) Drought resistance: A comparsion of two frame works. In: *Management of Agricultural Drought: Agronomic and Genetic Options.* N.P. Saxena, C. Johansen, Y.S. Chauhan, R.C.N. Rao (Eds.), Oxford and IBH, New Delhi.

Turnipseed, S.G. (1977) Influence of trichome density on populations of small phytophagous insects on soybean. Environ. Entomol. 6, 815–817.

Turnipseed, S.G. and Kogan, M. (1987) Integrated control of insects. In: J.R. Wilcox (Ed.), Soybeans: Improvement, Production, and Uses. ASA, CSSA and SSSA, Madison, pp. 779–817.

van der Biezen, E.A. and Jones, J.D.G. (1998) Plant disease resistance proteins and the gene-for-gene concept. Trends Biochem. Sci. 23, 454–456.

Van Duyn, J.W., Turnipseed, S.G. and Maxwell, J.D. (1971) Resistance in soybeans to the Mexican bean beetle. I. Sources of resistance. Crop Sci. 11, 572–573.

Van Duyn, J.W., Turnipseed, S.G. and Maxwell, J.D. (1972) Resistance in soybeans to the Mexican bean beetle. II. Reactions of the beetle to resistant plants. Crop Sci. 12, 561–562.

Walker, D.R., Narvel, J.M., Boerma, H.R., All, J.N. and Parrott. W.A. (2004) A QTL that enhances and broadens Bt insect resistance in soybean. Theor. Appl. Genet. 109, 1051–1057.

Walling, L.L. (2000) The myriad plant responses to herbivores. J. Plant Growth Reg. 19, 195–216.

Yencho, G.C., Cohen, M.B. and Byrne, P.F. (2000) Applications of tagging and mapping insect resistance loci in plants. Ann. Rev. Entomol. 45, 393–422.

Yunes, A.N.A., de Andrade, M.T., Sales, M.P., Morais, R.A., Fernandes, K.V.S., Gomes, V.M. and Xavier-Filho, J. (1998) Legume seed vicilins (7S storage proteins) interfere with the development of the cowpea weevil (*Callosobruchus maculatus* (F)). J. Sci. Food Agric. 76, 111–116.

Zhang, P., Wang, Y., Zhang, J., Maddock, S., Snook, M. and Peterson, T. (2003) A maize QTL for silk maysin levels contains duplicated Myb-homologous genes which jointly regulate flavone biosynthesis. Plant Mol. Biol. 52, 1–15.

Zhu, S., Walker, D.R., Boerma, H.R., All, J.N. and Parrott W.A.(2006) Fine mapping of a major insect resistance QTL in soybean and its interaction with minor resistance QTLs. Crop Sci. 46, 1094–1099.

Zhu, S., Walker, D.R, Warrington, C.V., Parrott, W.A., All, J.N., Wood, E.D. and Boerma, H.R. (2007) Registration of four soybean germplasm lines containing defoliating insect resistance QTLs from PI 229358 introgressed into 'Benning'. J. of Plant Registrations 1, 162–163.

Chapter 16
Genomics of Viral–Soybean Interactions

M.A. Saghai Maroof, Dominic M. Tucker, and Sue A. Tolin

Introduction

Virus infected soybean can be found in all soybean growing areas of the world. To date, over 67 viruses have been identified that are capable of replicating in the soybean plant. Twenty-seven of the 67 are currently a concern or have the potential to be a problem in soybean production systems (Tolin and Lacy 2004). The most prevalent viruses causing significant crop losses are made of single-stranded, positive-sense RNA from the *Potyviridae* and *Comoviridae* families, which this chapter will focus on. *Soybean mosaic virus* (SMV), which belongs to the *Potyviridae*, has been known to cause total crop loss (Kwon and Oh 1980) and occurs in all soybean growing areas of the world. *Bean pod mottle virus* (BPMV) is a member of the *Comoviridae* family and has increased its geographical distribution throughout the United States and poses a significant risk to soybean growers, particularly those in the North Central and Northern Great Plains states. Together the two viruses, BPMV and SMV, interact synergistically and drastically reduce yield and seed quality compared to each disease alone. Therefore, priority was placed on improving cultivar resistance often through utilization of molecular and genomic approaches.

The most common method of controlling SMV in soybean is through development of resistant cultivars often carrying a single hypersensitive resistant (R) gene. However, recent reports of resistance-breaking (RB) stains in Japan and Korea (Choi et al. 2005; Koo et al. 2005; Saruta et al. 2005) shifted breeders and molecular biologists to incorporate durable forms of resistance into cultivars through gene pyramiding or transgenic methods. As SMV is seed borne, seed distribution processes such as those used in germplasm exchange programs, serve to spread the virus to different environments and other areas of the world. Disease management of BPMV through genetic resistance is not possible as no soybean cultivars with resistance to BPMV are commercially available. Only a few experimental transgenic lines conferring resistance against BPMV are available.

M.A.S. Maroof
Department of Crop and Soil Environmental Sciences, Virginia Polytechnic Institute and State University, Blacksburg, VA 24061, USA
e-mail: Smaroof@vt.edu

G. Stacey (ed.), *Genetics and Genomics of Soybean*,

Viruses, Genomes and Diversity

Data on genetic or genomic interactions in soybean are available for only a few viruses. These include *Cowpea chlorotic mottle virus* (CCMV), *Bean pod mottle virus* (BPMV), *Tobacco ringspot virus* (TRSV), and the potyviruses *Soybean mosaic virus* (SMV), *Peanut mottle virus* (PMV), and *Peanut stripe virus* (PStV).

As SMV and BPMV will be the focal point of this chapter, details of viral genomes will be given for these viruses. Unique, similar features of these two viruses include structure of the 5′ and 3′ termini, and expression strategy as a polyprotein that is cleaved by viral proteases through a series of steps into 8–10 mature proteins. The total genome size is about the same, but BPMV is in two RNA species packaged in separate icosahedral capsids transmitted by beetles, and SMV is in one piece encapsidated in a filamentous rod transmitted by aphids.

Soybean Mosaic Virus

Like all potyviruses, SMV has filamentous particles approximately 750 nm in length and 11–15 nm in diameter composed of approximately 2,000 copies of a 29.5 kDa coat protein (CP) arranged in a helix, encapsidating one molecule of single-stranded, positive-sense, 9,588 nucleotide RNA with a 3′ polyA and a single, 5′-linked protein (VPg). Analyses of the complete nucleotide sequence of potyviruses identified, in addition to the 5′ untranslated region, a single open reading frame, a 3′ untranslated region, and a polyadenylated 3′ terminus (Reichmann et al. 1992; Shukla et al. 1991). The map of the SMV genome shown in Fig. 16.1 indicates the predicted protease cleavage sites of the polyprotein and the name and location of the resulting mature proteins (Jayaram et al. 1992). Functional genomic analysis of potyvirus sequences showed that several proteins are multifunctional, with no single protein having host specificity or movement functions. One of the important functions of HC-Pro is as a silencing suppressor of post-transcriptional gene silencing. A small polypeptide of 6 kDa lies between the CI, and the NIa. The VPg protein is encoded in the N-terminal part of NIa, from which it is cleaved and attached covalently to the 5′ end of the RNA to function as a primer of translation.

Strains of SMV are defined by their virulence on a series of differential cultivars (Cho and Goodman 1979). The coat protein of several strains and isolates were nearly identical in peptide profiles and differed from closely related potyviruses (Jain et al. 1992). Sequence analysis of the CP region of 25 SMV isolates from

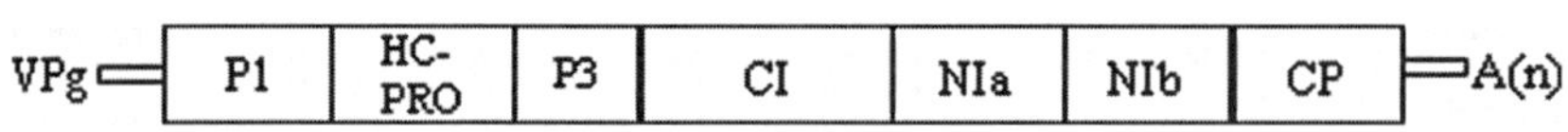

Fig. 16.1 Genome map of a potyvirus showing mature proteins: **P1**: protein 1, a serine protease; **HC-Pro**: helper component-protease; **P3**: protein 3; **CI**: cytoplasmic inclusion, a helicase; **VPg:** genome-linked protein with primer activity; **NIa:** nuclear inclusion a, a protease; **NIb:** nuclear inclusion b – replicase; and **CP:** coat protein

North America and Asia confirmed similarity, even among isolates from different geographic regions (Domier et al. 2003). The total RNA sequence is known for several SMV strains including G2 and G7 (Hajimorad et al. 2003; Jayaram et al. 1992), N, G5 and G7H from Korea (Lim et al. 2003), an attenuated isolate Aa15-M2 from Japan, and a severe strain from China. Partial sequences are available for many more SMV isolates, and are useful in distinguishing between strains (Omunyin et al. 1996) as well as locating domains of the SMV associated with virulence and pathogenicity (Hajimorad and Hill 2001). Domier et al. (2003) showed that the P1 and HC/Pro regions are more variable than CP, and separated 18 SMVs according to geographic origin. Comparisons predicted a phylogenetic relationship among all North American isolates, suggesting a common progenitor more than likely brought in seed-borne during early germplasm exchange.

Bean Pod Mottle Virus

Like other comoviruses, the BPMV genome is in two segments, RNA1 and RNA2, each packaged separately in an icosahedral capsid of 28 nm diameter. Each RNA is a single-stranded and positive-sense, with a 5′-linked VPg and a polyadenylated 3′ terminus. The genomic map of BPMV (Fig. 16.2) is based on sequence data and on its close relationship with the well-described *Cowpea mosaic virus* (CPMV). The approximately 6.0 kb RNA1 encodes all information needed for replication of the both viral RNAs, and the smaller, 3.6 kb RNA2 encodes a cell-to-cell movement protein and two coat proteins. The gene expression strategy of BPMV is similar to that of SMV in that a single polyprotein is translated from each RNA, and is post-translationally cleaved by RNA1-encoded proteases into intermediate and mature proteins. RNA1 also encodes a protease cofactor, a putative helicase, VPg, and a putative RNA-dependent RNA polymerase (Goldbach et al. 1995).

In the last decade, BPMV has re-emerged as an important pathogen of soybean and has increased its geographic distribution from the Southern states into the upper Midwest (Giesler et al. 2002). Close examination of new isolates of the virus demonstrated that they could be classified into two distinct strain subgroups, I and II, on the basis of RNA1 sequence detected by hybridization. Some of the more

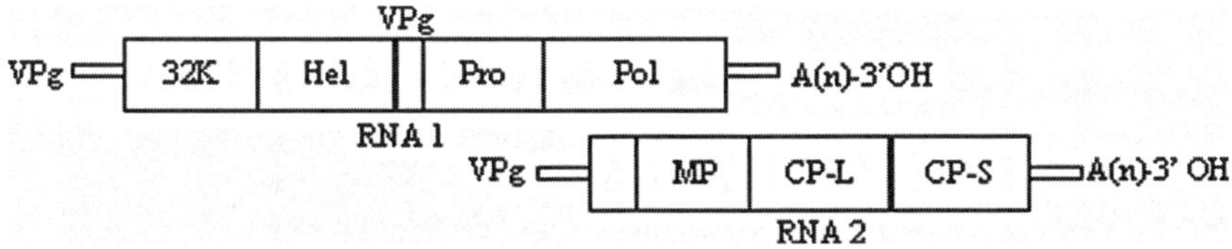

Fig. 16.2 Genome of bipartite BPMV. RNA1 consists of a protease cofactor (**32K**), helicase binding domain (**Hel**), a viral genome linked protein (**VPg**), protease (**Pro**), and RNA-dependent RNA polymerase (**Pol**). RNA2 encodes a movement protein (**MP**) and two viral coat proteins designated large and small (**CP-L** and **CP-S**)

severe isolates, such as Hancock (K-Ha-1), were natural reassortants between the two subgroups (Gu et al. 2002). More recently, naturally occurring severe isolates were found to be partial diploid reassortant strains, which are diploid for RNA1 with two distinct sequences and are haploid for RNA2 (Gu et al. 2006). Although mixed infections of distinct viruses are common in nature, this is the first report of partial diploid populations in a plant virus.

Viral RNAs were extracted from purified virus particles, thus capturing the encapsidated RNA species, in order to examine plants for BPMV genome type. It is known that RNA1 is preferentially degraded or decreased over time (Kartaatmadja and Sehgal 1990). The significance of this was not recognized at the time, but it is now thought to be due to a host anti-viral defense involving RNAi. As soybean genomics projects examined host DNA, including polyadenylated ESTs, a report suggested that BPMV sequences are found in genomic DNA (Sundararaman et al. 2000). Recently, available ESTs were examined *in silico* and found to contain contigs homologous to BPMV (1,097), SMV (132) and CCMV (423), although source plants were not inoculated (Strömvik et al. 2006). This represents the largest collection of sequence data available for these three viruses. Contig consensus sequences of CCMV were all from RNA3, which encodes the CP and M proteins. For SMV, contigs were mostly from the 3′ end of the genome which encodes CP. BPMV RNA2 contigs were also from the region encoding the large and small CPs. For BPMV RNA1, they were mostly from the 3′ end corresponding to polymerase gene. A greater frequency was found in field-grown than in greenhouse-grown plants. Diversity of the over 1,000 BPMV ESTs was greater than the diversity described by Gu et al. (2006, 2002), with most contigs clustering with the severe strain rather than the milder strains.

Resistance Interactions of Soybean and Viruses

Plant viruses have evolved the ability to infect and replicate in soybean, causing yield losses and decreasing seed quality. Dominant and recessive genes were characterized for a variety of plant viruses providing breeders resources for incorporating virus resistance into cultivars. For detailed reviews on virus resistance, see Kang et al. (2005) and Maule et al. (2007). These genes follow Flor's gene-for-gene recognition of pathogen Avr factors often, but not in all cases, leading to complete resistance and are associated with cell death and tissue necrosis.

Recessive genes confer resistance where the virus is able to infect the plant but the movement and accumulation are restricted. This reaction is closely associated with recessive genes for resistance to potyviruses. In lettuce and pepper, eukaryotic translation initiation factors were identified as host factors involved in recessiveresistance to potyviruses. The interaction of potyviral genome-linked portein VPg with the host factor eIF4E was essentialin the initiation of translation of viral RNA and up-regulation of viral genome amplification (Diaz-Pendon et al. 2004). Lack of compatibility between viral VPg and eIF4 is suggested to be the mechanism of resistance in these pathosystems.

Complicating issues in breeding for soybean virus resistance is the emergence of *Aphis glycines*, an aphid that colonizes soybean, identified in the United States in 2000 (Hartman et al. 2001), which subsequently spread throughout North America. *A. glycines* was (See Parrott Chapter 15) reported to transmit at least 10 soybean viruses (Clark and Perry 2002; Hill et al. 2001; Iwaki et al. 1980) including the most common and widespread virus, SMV. Therefore, the new vector places extensive pressure on deployed R genes in commercial cultivars to control virus diseases in soybean, highlighting the need for multiple genes for resistance. This section will briefly summarize the availability of resistance genes to various soybean viruses in nature.

Genes and Alleles Conditioning Resistance to SMV

SMV strain diversity was first discussed in the USA by Ross (1969). Cho and Goodman (1979, 1982) collected many isolates of SMV in the USA from germplasm collections and experimental plots, and inoculated them to a set of cultivars known to have genetic resistance to SMV. The isolates were classified into seven strain groups, based on the set of responses of the soybean differential cultivars 'Ogden', 'Marshall', 'York', 'Kwanggyo', and 'PI96983' (Table 16.1). The isolate designated by them as the type isolate for each strain group became known as SMV strains G1 through G7 (Table 16.1) that were used in studies of the SMV-soybean pathosystem.

A single dominant gene was first identified in soybean plant introduction (PI) 96983 and subsequently designated as *Rsv1* (Kiihl and Hartwig 1979). Resistance alleles at the *Rsv1* locus were identified in each of the differential cultivars: *Rsv1-y* (York, Hutcheson), *Rsv1-m* (Marshall), *Rsv1-t* (Ogden), and *Rsv1-k* (Kwanggyo) (Chen et al. 1991, 2002) (Table 16.1). Additional alleles at the *Rsv1* locus include *Rsv1-s* (PI486355), *Rsv1-r* (Raiden), *Rsv1-sk* (Suweon97) and non-designated allele in OX686. These resistance genes confer differential reactions to SMV strains G1 through G7 (Cho 1979, 1982). The *Rsv1* locus confers a complete resistance reaction to strains G1 through G3, while strains G4 through G7 produce a necrotic or mosaic reaction on plants carrying *Rsv1* (Chen et al. 1994). The exception is *Rsv1-n* in PI507389, which is overcome by all strains (Ma et al. 2003).

Two other resistance genes at different loci, designated as *Rsv3* and *Rsv4*, were identified (Table 16.1). Lines with the *Rsv3* locus are susceptible to lower numbered strains (SMV-G1–G4) but resistant to higher numbered strains (Gunduz et al. 2002). Initially, it was reported that *Rsv4* conferred resistance to all known SMV strains and was a single dominant gene (Ma et al. 1995). However, Ma et al. (2002) and Gunduz et al. (2004) observed that SMV G1 and G7 are capable of limited local movement and long-distance vascular spread, restricting virus in systemically-invaded leaves to cells along veins in heterozygous, *Rsv4*-containing plants. The appearance was unlike a typical susceptible mosaic reaction, and was termed a late susceptible (LS)

Table 16.1 *Soybean mosaic virus* strain groups and differential interaction with resistance genes

		Symptom induced by SMV strain group							
Source	Gene	G1	G2	G3	G4	G5	G6	G7	Reference
PI 96983[a]	Rsv1	R[b]	R	R	R	R	R	Nt	Kiihl and Hartwig 1979
RIL 800-46	3gG2	R	R	R	–	R	R	Nt	Hayes et al. 2004
Ogden[a]	Rsv1-t	R	R	Nt	R	R	R	Nt	Chen et al. 1991
Marshall[a]	Rsv1-m	R	Nv	Nv	R	R	Nv	Nv	Chen et al. 1991
RIL 613-10	G-class	R	Nv	Nv	R	R	R	Nv	Hayes et al. 2004
RIL 1044-98	G-class	R	Nv	Nv	R	R	R	Nv	Hayes et al. 2004
Davis[a]	Rsv1-y	R	R	R	Nt	S	S	S	Cho and Goodman 1979
York	Rsv1-y	R	R	R	Nt	S	S	S	Chen et al. 1991
Kwanggyo[a]	Rsv1-k	R	R	R	R	Nt	N	N	Chen et al. 1991
PI 507389	Rsv1-n	Nt	N	S	S	N	N	S	Ma et al. 2003
LR1[c]	Rsv1-s	R	R	R	R	N	N	R	Ma et al. 1995
OX686[d]	Rsv1 ?	N	–	–	Nt	–	–	R	Buzzell and Tu 1989
Raiden	Rsv1-r	R	R	R	R	N	N	R	Chen et al. 2001
Suweon 97	Rsv1-sk	R	R	R	R	R	R	R	Chen et al. 2002
L29[e]	Rsv3-hd	S	S	S	S	R	R	R	Buss et al. 1999
Harosoy	Rsv3-hs	S	S	S	S	R	R	R	Gunduz et al. 2001
Tousan	Rsv1Rsv3	R	R	R	–	R	R	R	Gunduz et al. 2002
Hourei	Rsv1Rsv3	R	R	R	–	R	R	R	Gunduz et al. 2002
Ox 670	Rsv1Rsv3	R	R	R	R	R	R	R	Gunduz et al. 2001
Zao 18	Rsv1Rsv3	R	R	R	–	R	R	R	Liao et al. 2002
LR2[c]	Rsv4[f]	R	R	R	R	R	R	R	Ma et al. 1995
Peking	Rsv4	R	R	R	R	R	R	R	Gunduz et al. 2004
PI 88788	Rsv4	R	R	R	R	R	R	R	Gunduz et al. 2004
Columbia	Rsv3Rsv4	R	R	R	Nt	R	R	R	Ma et al. 2002

[a] Differential cultivar used by Cho and Goodman 1979
[b] R = resistant symptomless; S = susceptible (systemic mosaic symptoms); N = necrotic, local and systemic; Nv = limited veinal necrosis; Nt = extensive tip necrosis; – = not tested
[c] LR1 and LR2 are derived from cv. Essex x PI 486355
[d] OX 686 is derived from cv. Columbia (*Rsv3Rsv4*) x cv. Harosoy)
[e] L29 is a cv. Williams BC_5 isoline with SMV resistance derived from cv. Hardee
[f] For *Rsv4* lines, no symptoms develop. Heterozygous progeny of crosses with S lines exhibit late susceptibility to all SMV strains (Gunduz et al. 2004)

phenotype (Ma et al. 2002). In their genetic studies, LS phenotypes were classed with R phenotypes to establish that the gene was dominant (Gunduz et al. 2004; Ma et al. 2002).

Resistance to Bean Pod Mottle Virus

Resistance to BPMV has not been reported in soybeans, but a few accessions of the ancestral species *G. tomentella* displaying resistance to systemic infection were observed. However, incorporation of this resistance into *G. max* is difficult in an interspecific cross (Zheng et al. 2005b). Some cultivars vary significantly in their response to BPMV allowing selection for field tolerance. Hill et al. (2007) identified seven accessions that were tolerant to BPMV based on relative level of virus antigen in seed and mottling of seed coats. There have been no reports identifying the nature of this tolerance. As described in the "Interactions of BPMV and Soybean Section" BPMV strains vary in their severity on the cultivar Essex, but no studies were done to examine if this was consistent across cultivars.

Resistance to Tobacco Ringspot Virus

Tobacco ringspot virus is a member of the genus *Nepovirus* in the family *Comoviridae* and infects a wide range of host plants including soybean, in which it often causes the disease known as bud blight – necrosis of the tip of apical meristem. Tolerance to TRSV was reported in some soybean lines where the plant recovered from infection and produced new leaves (Shakiba et al. 2006). However, the genetics of host resistance in these cultivars is unknown. Studies with a TRSV grape strain by Lee et al. (1996) in Arabidopsis found that no lines were completely resistant but many were tolerant to the virus and developed only a mild mosaic symptom producing flowers and seed. Genetic studies with the tolerant resistant lines indicated a single incompletely dominant locus controlling tolerance to the TRSV grape strain. The locus was designated TTR1 for '*t*olerance to *t*obacco *r*ingspot', and mapped to chromosome V. Similar results were reported in soybean with a major QTL identified in 'Young' a well adapted, commercial Maturity Group VI cultivar (Burton et al. 1987). Plants were evaluated for bud blight resistance on the basis of terminal bud death under natural field pressure. One major QTL was identified explaining 82 % of the phenotypic variation (Fasoula et al. 2003). The mapped location of this QTL will be discussed in further sections.

Resistance to Cowpea Chlorotic Mottle

Seven CCMV strains were identified by differential reaction of a set of soybean genotypes (Paguio et al. 1988). The virus is a member of the *Bromoviridae* family with single stranded, tripartite RNA genomes. Cultivars Davis, Jackson, and Hood were reported to be resistant to CCMV, giving necrotic lesions upon inoculation. Genetic studies by Boerma et al. (1975) on these resistant lines concluded that a single dominant gene, designated as *Rcv*, conditioned resistance. Allelism tests were not conducted to examine if the cultivars contain the same or different genes for CCMV resistance. Goodrick et al. (1991), examining the non-necrotic (resistance to systemic movement) type of resistance in PI346304, identified two recessive genes based on segregation ratios. The extent of allelism between the two recessive genes and the *Rcv* locus was not determined.

Resistance to Peanut Mottle Virus and Peanut Stripe Virus

The potyvirus PMV is a significant soybean-infecting virus, particularly in the southeastern United States where soybean and peanuts are grown in close proximity. Peanuts act as the primary source of inoculum in the spread of the virus which is further spread by aphids (Paguio and Kuhn 1974). Symptoms range from a mild to severe mottle and vary by virus isolate and soybean cultivar (Bays et al. 1986; Paguio and Kuhn 1974). Screening by natural and mechanical inoculation identified numerous soybean lines and plant introductions carrying resistance to PMV (Bays et al. 1986; Demski and Kuhn 1975; Shipe et al. 1979). A dominant gene, designated *Rpv1*, was identified in 'Dorman' (Boerma and Kuhn 1976). The same gene is present in 'York', closely linked with *Rsv1* (Roane et al. 1983). In the SMV differential cutlivars, Marshall and Ogden are susceptible, but Kwanggyo was resistant. A second dominant gene was described in 'CNS' but was not assigned a gene symbol (Buss et al. 1985). A recessive gene was identified in 'Peking' and designated as *rpv2* (Shipe et al. 1979). Peking is also resistant to SMV and carries the dominant gene *Rsv4*. No allelism tests were done between *Rsv4* and *rpv2*.

The potyvirus PStV is now classified as a strain of *Bean common mosaic virus*. Like PMV, PStV can infect soybeans growing in close proximity to peanuts (Gillaspie and Hopkins 1991). Warwick and Demski (1988) screened 121 soybean classifying approximately 35 % of the lines as total resistant as ELISA confirmed no virus present in the lines. Genetic studies conducted by Choi et al. (1989) with SMV and PStV resistance found the two loci were linked by approximately 9 cM and both acted in a dominant manner.

Genomics of Resistance to Viruses

Numerous consensus maps of restriction fragment length polymorphisms (RFLPs), simple sequence repeats (SSRs), amplified fragment length polymorphisms (AFLPs) and single nucleotide polymorphisms (SNPs) are now available for soybean (http://soybase.org). Many of these markers were previously linked to disease resistance and other traits in soybean providing opportunities for utilization of MAS. Recently, Choi et al. (2007) created a transcript map of soybean by mapping one SNP in each of 1,141 genes filling many gaps that were previously present in the SSR consensus map. Physical mapping projects provide additional sequence information whereby PCR-based markers can be designed directly from bacterial artificial chromosome (BAC) sequences connecting physical and genetic maps. Within the next couple of years, the soybean genome will be completed providing an unlimited resource to locate and clone additional genes. To date, PCR-based markers have assisted breeders in developing soybean lines with pyramided SMV resistance genes (Maroof et al. unpublished), identifing novel disease resistance genes (Hayes et al. 2000b) and performing high-resolution tagging of a potyvirus resistance genes (Gore et al.

2002). Continuing advancements in soybean genomics will assist in the creation of cultivars with pyramided, durable virus resistance and other diseases.

Mapping of Soybean Virus Resistance Genes

Initial mapping with RFLPs and SSRs by Yu et al. (1994) placed the single dominant gene contained in PI 96983, *Rsv1*, near HSP176 (a soybean heat shock protein gene), and two RFLP markers, pA186 and pK644a. These markers were linked to *Rsv1* at distances of 0.5, 1.5, and 2.1 cM, respectively and were located on MLG-F (Fig. 16.3). Additional SSR markers Satt120 and Satt510, were also mapped in the vicinity of the virus resistance gene (Gore et al. 2002). Fine mapping and cloning of this virus resistance gene using resistant gene candidates is discussed in detail in continuing sections.

The less characterized *Rsv3* and *Rsv4* resistance genes were, respectively, mapped to linkage groups MLG-B2 and MLG-D1b with molecular markers. *Rsv3* was flanked (Jeong et al. 2002) with RFLP marker A519 and SSR marker M3Satt at 0.8 cM away near a minor soybean cyst nematode QTL (Yue et al. 2001). Other SSR markers Satt560 and Satt063 are in close proximity (Fig. 16.3) (Jeong et al. 2002) proving additional markers for MAS. Furthermore, for rapid genotyping, gel-based SNP assays were developed for *Rsv1* and *Rsv3* (Jeong 2004). The *Rsv4* locus is flanked with molecular markers Satt542 at 4.7 cM and Satt558 at 7.8 cM on MLG-D1b (Hayes et al. 2000b) in a disease resistance-gene-poor region (Fig. 16.3). Additional fine mapping schemes for the *Rsv4* gene are discussed in sections below using expressed sequence tags (ESTs) and comparative genomics approaches.

Similar to SMV, genes for resistance to TRSV, PStV and PMV in soybean were also mapped to MLG-F in the same region as *Rsv1*. Fasoula et al. (2003) using RFLP and SSR markers found a major QTL controlling bud blight, disease caused by TRSV, in cultivar ‘Young’. The major QTL was closely linked to Satt510. The PMV resistant gene *Rpv1* was found closely linked to *Rsv1* by 1.1 cM (Gore et al. 2002). The respective area on MLG-F is a highly concentrated disease resistance cluster as other non-virus resistance genes were mapped in the area including: *Rps3* (Demirbas et al. 2001) and *Rps8* (Gordon et al. 2006; Sandhu et al. 2005) (conditioning resistance to *Phytophthora sojae*) and *Rpg1* (resistance to bacterial blight) (Ashfield et al. 1995). Quantitative trait resistance for Javanese root knot nematode (Tamulonis et al. 1997b), peanut rot nematode (Tamulonis et al. 1997a), corn earworm (Rector et al. 1999) and Sclerotinia resistance (Arahana et al. 2001) were also mapped to the region on MLG-F (Fig. 16.3). The *Rsv1-Rpg1* region of MLG-F is the subject of a project, led by Roger Innes of Indiana University, on comparative analysis of legume genome evolution. A large 1- megabasepair region is currently being sequenced from the SMV-resistant PI 96983, susceptible Williams cultivar and several taxa from Leguminosae, providing insights into gene and genome evolution (Ashfield et al. 2007; Cannon et al. 2007). Construction of this large contig identified homoeologous regions of MLG-F in the soybean genome as a result of

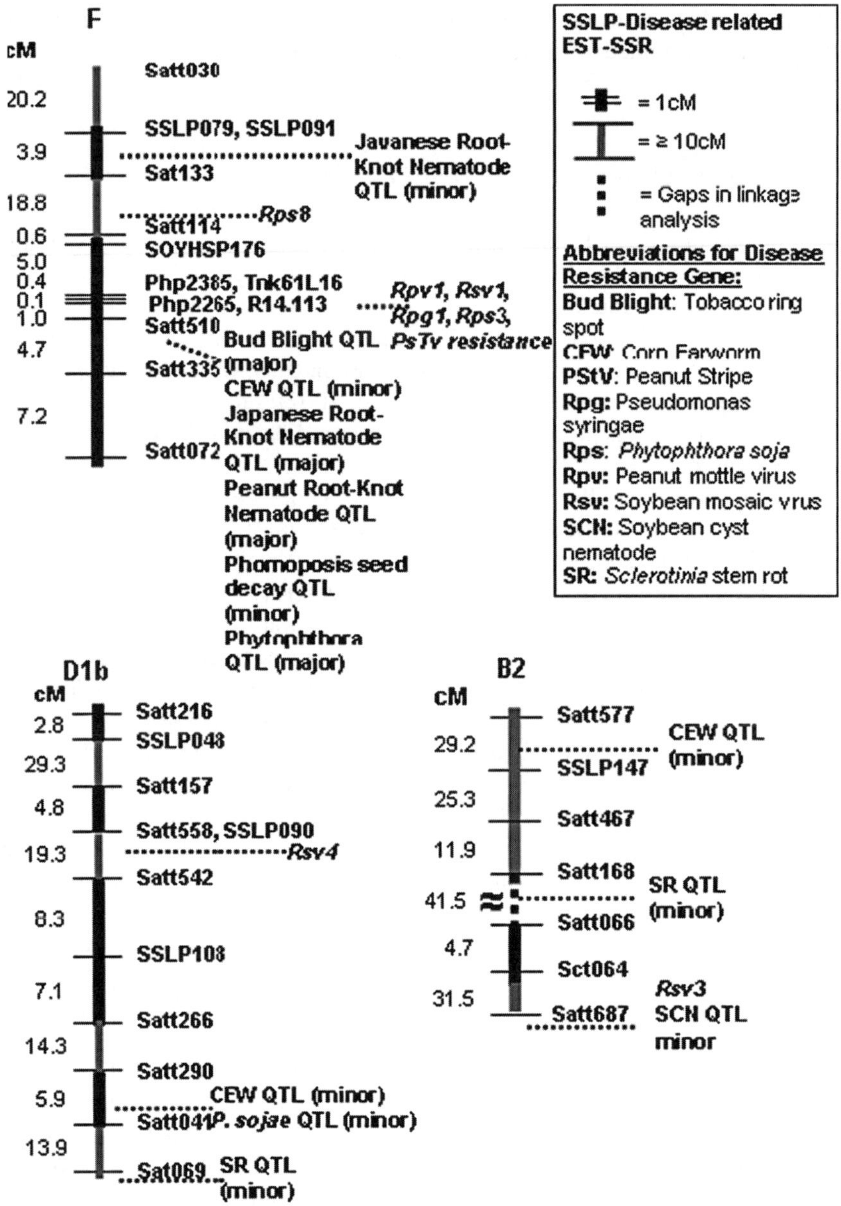

Fig. 16.3 Location of known virus resistance genes (*Rsv1*, *Rsv3*, *Rsv4*, *Rpv*1, PStV, and bud blight QTL) position on molecular linkage groups F, D1b, and B2 with respect to other known bacterial and fungal resistance genes shown on the Soybase website (http://soybase.org) and publicly available markers

an ancient genome duplication. One area on MLG-E located in the current study was first identified by Shoemaker et al. (1996), who reported extensive homology between MLG-E and MLG-F using RFLP markers. Hayes (2000a) using resistance gene candidates (section discussed below) identified two full length resistance gene candidates where the two homologs mapped to MLG-E and MLG-F. However, the candidate gene on MLG-E had several putative stop codons and frame shifts. In addition, both respective areas share a common resistance QTL for corn earworm resistance (Rector et al. 1999, 2000; Terry et al. 2000) and peanut rot nematode (Tamulonis et al. 1997a) providing further evidence of this genome duplication, but for unknown reasons MLG-E is relatively gene poor compared to MLG-F. Information from the above mentioned studies combined with the soybean whole genome sequencing project will be pertinent in understanding virus R gene evolution.

Resistance Gene Analog Utilization in Fine Mapping Rsv1 and Other Resistant Genes

Many cloned defense related genes from plant species encode putative proteins that are components of the integral signal transduction network. These sequences often belong to a class of proteins with a nucleotide binding site (NBS) and a leucine rich repeat (LRR), including the cloned virus resistance genes *N* from tobacco and *Rx* from potato, and the *Pseudomonas syringae* resistance gene *RPS*2 from Arabidopsis. NBS-LRR sequences are quite prevalent throughout plant genomes. Yu et al. (1996) utilized these conserved structural domains to develop resistance gene analogs (RGAs) for mapping and cloning disease resistance genes from soybean. They designed degenerate oligonucleotide primers from the NBS region of *N* and *RPS2*, amplified and cloned NBS sequences from soybean and grouped them into 11 different classes of RGAs based on their distinct, non-overlapping multi-enzyme RFLP patterns on a set of soybean lines. Five of the 11 subfamilies mapped to MLG-J and -N but two, namely NBS5 and NBS61, mapped near *Rpv1* and *Rsv1* on MLG-F. Using the short DNA sequence of NBS5 as a probe, Hayes et al. (2000a) screened a cDNA library and identified a candidate resistance gene, L20a, which belonged to the TIR-NBS-LRR class of R genes but did not cosegregate with *Rsv*1. Further targeted mapping of *Rsv1* and *Rpv1* was conducted by Hayes and Saghai Maroof (2000) using a modified AFLP approach where a degenerate primer corresponding to the p-loop of cloned resistance genes amplified bands in a segregating *Rsv1* population. Bulk segregant analysis revealed four markers that were linked to *Rsv1* with the closest being 0.4 cM away from *Rsv1*. Finally, Jeong et al. (2001) and Gore et al. (2002) using primers specifically targeting the non-TIR-NBS encoding sequences (class j) (Yu et al. 1996) of R gene families, identified several sequences which flanked or co-segregated with *Rpv1* and *Rsv1* loci, placing the two virus resistance genes 1.1 cM apart. Gore et al. (2002) also concluded that one or more tightly linked genes besides *Rsv*1 conditions resistance to SMV, contrary to previous thought that the *Rsv1* was a single dominant gene. Cloning of this cluster of NBS-LRR genes from MLG-F by Hayes et al. (2004) identified a strong candidate, 3gG2, for the major *Rsv1* resistance gene from PI 96983. The 3gG2 ORF sequence encodes a 3,390-bp gene, with a deduced protein product similar to previously cloned non-TIR-NBS-LRR disease resistance genes. Various unique resistant and necrotic reactions were observed coincident with the presence or absence of the other members of this gene cluster. Response of recombinant lines from the cross of PI96983 (resistant) by Lee 68 (susceptible) indicated that more than one gene in the region of PI96983 conditions resistance and/or necrosis to SMV. Unique recombinant lines in the segregating mapping population, RIL 613-10 and RIL 1044-98, lacking specific complement genes produced unique reactions to tested SMV strains similar to Ogden and Marshall carrying alleles *Rsv1*-t and *Rsv1*-m (Table 16.1), respectively (Hayes et al. 2004).

Expressed Sequence Tags (EST) – Simple Sequence Repeats (SSRs)

Expressed sequence tags (ESTs) represent functional genes compared to conventional simple sequence repeat (SSR) markers that are primarily based on amplifying non-coding regions of a genome. Random repetitive or repeat motifs in an organism's genome, the basis of conventional SSRs, were shown to be present in mRNA transcripts (Han et al. 2004; Kantety et al. 2002; Morgante and Olivieri 1993), allowing the development of an additional type of marker known as EST-SSRs. Existence of these common repeat motifs in ESTs enhances their potential for marker development and fine mapping of disease resistance genes, especially in soybean where there are approximately 372,000 ESTs available. (http://www.ncbi.nlm.nih.gov/dbEST/dbEST_summary.html, verified, on 6/28/07)

Godwin et al. (unpublished) examined DuPont's soybean EST database which contained 194,000 ESTs originating from cDNA libraries constructed from roots, stems and leaves of disease-infected soybean tissue. Plants were inoculated with several pathogens including *Fusarium solani* (SDS), *Sclerotinia sclerotiorum,* and soybean cyst nematode (SCN) prior to tissue collection for library construction. Using various bioinformatics tools, Godwin et al. (unpublished) identified a set of 1,218 ESTs that were homologous to known disease-related sequences and constituted the basis of EST-SSR marker development. They designed primers for over 200 of the disease related EST-SSRs in an attempt to integrate them into a soybean map with publicly available SSRs (Cregan et al. 1999). Many of these EST-SSR markers mapped to the chromosomal regions of known R genes/QTL (see SSLP markers in Fig. 16.3). For example, one EST-SSR marker, SSLP090, mapped near Satt542 and Satt558, the known location of *Rsv4* (Fig. 16.3), which may facilitate fine mapping of this important gene.

Hwang et al. (2006) deployed a similar method attempting to fine map *Rsv4* with ESTs but utilized comparative genomics. Genome sequence is now available for the model legume species *Lotus japonicus*. Comparative mapping with soybean and *Lotus* led to the identification of the chromosomal region around the *Rsv4* region in soybean that is microsyntenic with a *Lotus* chromosomal region. Designed EST markers from this chromosomal region from *Lotus* were mapped as close as 2.5 cM flanking the resistant gene and providing new markers more useful for marker-assisted selection and eventual cloning of the *Rsv*4 gene.

Pyramiding Virus Resistance Genes

Assembling genes in a single genotype to obtain resistance to multiple pests or strains of a pathogen has been a priority for plant breeders. Incorporating multiple genes for resistance to a single pathogen (gene pyramiding) was proposed as a method for creating durable resistance where pathogen populations are dynamic and continually evolving. Conventional gene pyramiding requires extensive disease screening with several races of the pathogen due to race specificity of many of these genes after each cycle of crossing. Further complicating issues for pyramiding efforts is the absence of an effective selection method due to a lack of differentiating races. As highly effective resistance genes that are effective to many races of the pathogen are incorporated into a single cultivar, valuable hypostatic genes are easily lost. Availability of PCR-based and tightly linked molecular markers has facilitated pyramiding through marker-assisted selection (MAS).

A few studies reported pyramiding virus resistance genes for the potyviruses *Bean common mosaic virus* (BCMV) in *Phaseolus vulgaris* (Kelly et al. 1995) using molecular markers and *Pepper veinal mottle virus* in *Capsicum annuum* (Caranta et al. 1996) through conventional breeding techniques. Werner et al. (2005) pyramided *rym4*, *rym5*, *rym9* and *rym11*, all with different modes of action, into a single winter barley (*Hordeum vulare*) using double haploid populations and molecular markers to select for F_1 plants containing two, three, and eventually four resistance genes. Selection of lines carrying all four resistant genes without the aid of molecular markers was infeasible due to lack of differentiating virus strains. The resulting cultivar conferred resistance against *Barley mild mosaic virus* (BaMMV) and two different strains of *Barley yellow mosaic virus* (BaYMV and BaYMV-2).

Recently, Maroof et al. (unpublished) pyramided SMV resistance genes *Rsv1*, *Rsv3*, and *Rsv4*. Three different isolines containing one of three virus resistance genes (*Rsv1, Rsv3, or Rsv4*) in the same susceptible Essex background were created using backcrossing techniques followed by MAS. Pair wise crosses followed by MAS of these single gene isolines were utilized to create pyramided lines with two and three gene combinations of the SMV viruses resistance genes. Following that, the Essex isolines containing one, two, or three *Rsv* genes were inoculated with six SMV strains, comparing reactions in the Essex background and the donor source of the Rsv resistance gene. Two gene and three gene isolines of *Rsv1Rsv3*, *Rsv1Rsv4* and *Rsv1Rsv3Rsv4* acted in a complementary manner conferring resistance against all strains of SMV, whereas isolines of *Rsv3Rsv4* displayed a late susceptible reaction to selected SMV strains not displayed in the donor sources. Additionally, unexpected background effects also caused the *Rsv3* isoline to be more susceptible than the donor to selected SMV strains. Therefore, the Essex background or unknown modifier genes lost from the original donor source often resulted in isolines with unpredicted virus reactions. Similar results were reported by Wang et al. (2006) in L92-8580, an isogenic line of Williams. Repeated backcrossing to Williams caused the line to be more susceptible to SMV-G5 whereas the donor parent Suweon 97 was resistant to SMV-G5 leading to the conclusion that a possible rare recombination event at the *Rsv* locus between the two lines may have occurred.

Functional Genomics and Plant-Virus Research

The usefulness of global expression profiling in plant-virus interactions was demonstrated in a number of plant host-virus systems. Cooper (2001) conducted cDNA-AFLP analysis on virus-infected *Chenopodium amaranticlor* and detected up-regulation of defense-related sequences. Interestingly, a number of these sequences were induced after inoculation with taxonomically different viruses. Whitham et al. (2003) inoculated Arabidopsis leaves with cucumber mosaic, turnip vein clearing, potato virus X, and turnip mosaic viruses comparing gene regulation to mock-inoculated lines using GeneChip microarrays (Affymetrix). The experiment revealed virus-induced changes in gene expression associated with defense or stress responses. Espinoza et al. (2007) conducted a gene expression study with a compatible viral-disease interaction in grapevines using the Affymetrix GeneChip. Gene expression in the *Vitis vinifera* red wine cultivars were studied in response to natural infection with GLRaV-3, the leaf roll-associated closterovirus-3. Numerous genes were induced or repressed involving various processes such as metabolism, transport and cell defense. Yang et al. (2007) studied the spatial and temporal relationship between sites of virus accumulation and the resultant host gene expression responses in Arabidopsis infected with TuMV. A number of up- or down-regulated genes were identified and their expression level was primarily correlated with the amount of virus accumulation.

The availability of an extensive soybean EST database, cDNA microarrays (Vodkin et al. 2004) and Affymetrix GeneChip soybean genome array has provided opportunities for global expression profiling studies in soybean-pathogen interactions. Zou et al. (2005) conducted gene expression studies of a susceptible or hypersensitive (resistant) reaction to *Pseudomonas syringe* using cDNA arrays. *P. syringae* strain carrying avirulence gene, *avrB*, provokes a hypersensitive response when paired with *RPG*1, the first resistance gene cloned from soybean (Ashfield et al. 2004). While *P. syringae* lacks *avrB* and produces a susceptible reaction, Zou et al. (2005) observed significant changes in 3,897 genes in response of susceptible and resistant reactions to *P. syringae*. More than 25 % of the genes identified had not been previously reported to be involved in pathogen interaction. Ithal et al. (2007) conducted simultaneous genome-wide analysis of gene expression in soybean and *Heterodera glycines* (soybean cyst nematode) during the course of infection in a compatible interaction. A set of 429 soybean genes showed statistically significant differential expression between uninfected and nemoatode-infected root tissue. A total of 1,850 SCN genes showed significant expression changes during different stages of nematode parasitism and development. A similar simultaneous host-pathogen global gene expression study is being conducted for the *Phytophthora sojae*-soybean interaction system (see Tyler Chapter 14). As for the virus-soybean interaction, the availability of unique genetic materials should facilitate global gene expression studies. The near-isogenic lines that possess one, two, or all three of the identified SMV resistance genes, in the same susceptible background, should help understand the epistatic interactions and genetic mechanisms between these distinct loci from different sources. Identifying additional genes or modifiers for SMV resistance will be necessary to create a durable, broad spectrum resistance in future cultivars.

Pathogen Derived Resistance

Pathogen derived resistance (PDR), where plants are transformed with portions of the pathogen genes, provides additional sources of resistance to viruses where no naturally occurring virus R genes are present in germplasm or resistance was previously overcome by virus. Commonly in plants, PDR is created by inserting functional and nonfunctional virus coat proteins (CPs), replicases, movement proteins, and proteases of the viral genome. The resistance conferred by these virus components was reported to include posttranscriptional gene silencing involving RNAi (Baulcombe 1996). However, the resistance conferred by these virus genes is often variable and unpredictable as varying degrees of resistance were reported ranging from delayed symptom development to complete immunity. Against soybean viruses (*Potyvirus, Comovirus* and *Luteovirus*) the CP portion of the genome was used in creating resistance. This form of PDR does not require protein expression and often confers very high levels of resistance but is more strain-specific causing significant concerns for field applications. We will highlight research using this system as an approach to incorporating additional sources of resistance to soybean viruses.

Wang et al. (2001) were the first to experiment with stable PDR resistance in soybean. They transformed soybean with coat protein and 3′ UTR from SMV strain N, a non-aphid transmissible strain. As SMV-N showed a high degree of similarly to other strains including AL-5, G6, and G7, it was thought it may be effective against a broad spectrum of strains. Two of the transgenic lines were highly resistant to strain AL-5 at high virus load levels while one became infected at moderately low levels of virus inoculum. Additional studies on these PDR lines examined spread of strain AL-5 from a point source in a field environment. Two lines had lower SMV infection rates and significantly higher yield with less seed coat mottling than the control lines (Steinlage et al. 2001). Although the transgenic lines did not have total immunity, their function in reducing virus infection rate likely exerted less pathogen pressure, similar to the pyramiding scheme of *Rsv3Rsv4* and late susceptible reaction in *Rsv4* donor sources described above, possibly increasing durability of resistance genes to dynamic SMV field populations.

PDR for BPMV was also sought since its emergence as a prominent disease in the USA. Di et al. (1996) were first to transform soybean with the CP precursor gene but the integrated transgene was unstable and not present in advanced generations. Further studies to develop PDR to BPMV conducted by Reddy et al. (2001) were more successful in integrating stable, viable resistance utilizing a capsid polyprotein from the BPMV genome using the particle bombardment approach. One of the five transgenic lines showed little or no disease symptom development when rub inoculated with the severe strain, Hopkins, of BPMV. Further feeding assays, confirmed with the highly resistant line, showed no symptom development on non-inoculated leaves and the virus did not go systemic.

Studies conducted with the inverted repeat SbDV coat protein also were able to achieve resistance against the YP strain of SbDV. Three of seven CP expressing plants showed little symptom development and very little virus protein accumulation. These plants contained SbDV CP-specific siRNA suggesting resistance was achieved through an siRNA-mediated silencing pathway (Tougou et al. 2002).

DNA markers were identified for numerous genes/QTL controlling resistance to soybean pests and pathogens. The availability of PDR-based transgenic soybeans provides opportunities to create additional forms of durable resistance by pyramiding transgenes with host resistance genes/QTL through MAS. Such an approach will be especially important for BPMV for which no host resistance gene is currently available. It would be interesting to combine CP-based resistance to BPMV with the 3-gene, *Rsv1Rsv3Rsv4*, pyramids of SMV. A similar strategy was already employed by Walker et al. (2002) who incorporated a QTL conferring resistance to corn earworm with a synthetic *Bacillus thuringiensis crylAc* transgene.

Interactions of Viruses with Soybean

Interactions of SMV with Rsv1

The characteristics of *Rsv1* described above suggest that it mediates recognition of SMV through specific interaction of the soybean *R* gene and a virus avirulence (*Avr*) gene, triggering host defense responses to restrict the virus. Phenotypically, the response varies from no symptoms (R) to varying expressions of necrosis (N) resulting from non-restricted expression of the hypersensitive response (HR) (Table 16.1). All SMV strains/pathotypes are replication competent in *rsv1*-genotype soybeans. On leaves of *Rsv1*-genotypes (PI 96983) mechanically inoculated with SMV-G1 through G6, no symptoms are induced and there is no detectable virus accumulation, typical of extreme resistance (ER). Introduction of SMV-N (G2 strain group) by grafting on PI 96983, however, produced HR-like lesions on stems, petioles and leaf veins and elevated transcription of a pathogenesis-related protein, PR1, associated with defense response (Hajimorad and Hill 2001). When inoculated with SMV-G7,

*Rsv*1-genotype responded with a visible HR that was not restricted to the point of entry but spread systemically to the apical meristem to give a phenotype termed lethal systemic hypersensitive response (LSHR). They concluded that ER and HR were the consequence of the same defense response. This response appears to be equivalent to the phenotype described in inheritance studies as tip necrosis with SMV (Buzzell and Tu 1989) and bud blight with TRSV (Fasoula et al. 2003).

The elicitor of LSHR in *Rsv1*-genotype soybeans was mapped to the P3 region of the SMV genome (Hajimorad et al. 2005, 2006). Analysis of the full length sequences of SMV-N and SMV-G7 indicated differences in a number of regions of the genome (Jayaram et al. 1992). Construction of infectious full-length cDNA clones (Eggenberger and Hill 1997) enabled inoculation with a genetically uniform population of pSMV-G7, passage on resistant soybeans, and experimental evolution of a mutant pSMV-7d capable of evading recognition by *Rsv*1 (Hajimorad et al. 2003). The mutant differed from the parent clone by 7 non-silent substitutions, one each in P1, HC-Pro, CP and 4 in P3, suggesting a major role for P3. It was noted by the authors (Hajimorad et al. 2003) that this is the first demonstration, to their knowledge, of experimental evolution from a cloned RNA plant virus to a mutant that evaded recognition by a resistance gene. Site-specific mutagenesis partially confirmed these conclusions. Chimeras were also constructed that were virulent, inducing LSHR, non-lethal systemic HR (SHR), or avirulent (ER) on *Rsv*1-genotypes. Strain-specific sequences in the N-terminus of the P3 of SMV-N inserted into G7 converted the infectious clone to avirulence, the reciprocal substitution did not result in gain of virulence by SMV-N (Hajimorad et al. 2006). Recent results suggest that sequences in the HC/Pro region are also required (Eggenberger et al. 2007; Hajimorad et al. 2007).

The function of potyviral proteins was analyzed for several different potyviruses (Maule et al. 2007). The role of P3 remains unclear, except that it is associated with the cytoplasmic inclusion bodies on which viral replication occurs and is thus part of the virus replication complex. It apparently has no sequence homologies with proteases or replication-related enzymes, or metabolites. P3, however, was noted as a pathogenicity domain and associated with symptom severity of *Turnip mosaic virus*, together with CI, overcoming resistance genes in *Brassica napus* (Jenner et al. 2002, 2003). Cellular localization studies detected P3 within the arms of the pinwheel inclusions formed by CI, further implicating P3 as part of the replication complex (Rodriguiz-Cerezo et al. 1993).

Functional genomic analysis of the SMV-*Rsv*1 interaction was done with only two strain groups, G2 and G7, and one source of *Rsv*1, PI 96983. It would be of interest to examine additional strains including emerging resistance-breaking strains from Asia to reveal additional virulence or avirulence regions on the viral genome. Durability of resistance to SMV is at risk, although comparative analysis placed SMV in a category in which some resistance factors were durable but others were overcome in less than 25 years (García-Arenal and McDonald 2003). Additionally, the series of alleles at this locus and the complexity of resistance gene candidates in recombinant lines described by Hayes et al. (2004) should be examined. The extent of necrosis in the SHR or LSHR responses is not well-characterized. Studies reported by Hajimorad et al. (2005, 2006) describe LSHR as developing within 3–4 weeks, whereas Hayes et al. (2004) observed similar phenotypes in less than half of that time. The effect of gene dosage in necrotic responses of heterozygotes (Chen et al. 1991) and temperature effects of expression of resistance (Zheng et al. 2005a) should also be examined.

Interactions of SMV and Rsv3 and Rsv4

The *Rsv3*-genotype L-29, Harosoy, and Hardee are resistant to SMV G5, G6 and G7, but are readily infected by SMV G1, G2 and G3, often developing severe symptoms (Buss et al. 1999). Three cultivars (Table 16.1) examined in an inheritance study were reported as the *Rsv*1*Rsv*3 genotype and resistant to all strains (Gunduz et al. 2002). One cultivar, Columbia, contains *Rsv*3 and *Rsv*4 and is resistant to all strains but G4, which provokes LSHR (Ma et al. 2002). Recent work suggests that the N-terminal region of the CI gene is the virulence determinant as well as symptom determinant on *Rsv*3-genotype soybeans (Zhang et al. 2007). It will be of interest to compare their results, the details of which are not yet available, with that of a new G7H strain from Korea which maps virulence to the CI gene (Kim et al. 2003). Other emerging resistance breaking strains infect L29 (*Rsv*3), but also PI 96983 (*Rsv*1) and genotypes with five alleles of *Rsv*1 (Choi et al. 2005), suggesting different viral domains are recognized by *Rsv*3 than by *Rsv*1.

The *Rsv4* genotype was first genetically isolated in LR2 or D26, derived from a cross of a susceptible line Essex X PI 486355, and was resistant to SMV G1–G7. One resistant $F_{3:4}$ line, LR1, had an allele at *Rsv1* (Table 16.1), and the other, LR2, had what became *Rsv4* (Hayes et al. 2000b). However, in the process of scoring progeny of crosses in allelism tests of the two genes in Columbia, a late susceptible (LS) phenotype was observed (Ma et al. 2002). Genetic analyses of crosses demonstrated resistance was inherited as a single dominant gene, when late susceptible phenotypes were scored as resistant (Gunduz et al. 2004; Ma et al. 1995), and mapped to D1b (Hayes et al. 2000b). Marker-assisted selection was used to develop Essex isolines with the same phenotype as the parental resistant lines (Saghai Maroof et al. unpublished). In contrast to ER, virus can be detected in inoculated leaves but infection sites are fewer and expand more slowly on PI 88788 than either susceptible or HR responding lines (Gunduz et al. 2004). The virus moves vascularly from inoculated leaves but does not appear until the third or fourth trifoliolate leaves, 3–4 weeks after inoculation. The mechanism of resistance is thought to involve both reduced accumulation of virus and movement into and from the vascular system.

The impaired susceptibility phenotype of virus resistance was described for several potyvirus pathosystems and in all cases it was associated with recessive virus resistance genes (Maule et al. 2007). Of all recessive resistance genes, over 70% are for resistance to potyviruses (Diaz-Pendon et al. 2004).

Interactions of BPMV with Soybean

Infectious cDNA clones of RNA1 and RNA2 of BPMV enabled extensive analysis of the structure and function of viral-encoded mature proteins and their role in pathogenicity. Reassortants, pseudorecombinants, chimeric constructs, and mutants of three distinct BPMV strains (K-G7 mild; K-Ha1 moderate; and K-Ho1 severe) from Kentucky were observed on cultivar Essex (Gu and Ghabrial 2005). Severe symptoms were induced only with RNA1 from K-Ho1, regardless of RNA2 source, and mapped to coding regions of the protease co-factor (Co-Pro) and the C-terminal half of the Hel, the putative helicase. In CPMV, Co-Pro is a multi-functional protein regulating RNA1 polyprotein and a required co-factor for cleavage of the RNA2 polyprotein (Peters et al. 1994; Pouwels et al. 2002).

Severe symptoms of necrosis on inoculated leaves and systemic severe mottling and blistering resulted in higher viral RNA accumulation, but the mechanism was not that of suppression of RNA silencing (Gu and Ghabrial 2005). Increased symptom severity in soybeans is also attributed to mixed infections of SMV and BPMV, in which BPMV accumulates to a higher level (Anjos et al. 1992). The mechanism of this synergy is believed to be a suppressor of RNA silencing activity of the HC-Pro of SMV which blocks the hosts anti-SMV and anti-BPMV activity (Anandalakshmiki et al. 1998). No comparable suppressor of gene silencing region was found in BPMV. The coding regions of Co-Pro, Hel, Hel+VPg, and the two CPs, were expressed from a potato virus X vector. Following inoculation to *Nicotiana benthamiana*, necrosis was induced on inoculated leaves (Gu and Ghabrial 2005). It was suggested that necrosis was due to toxicity of the mature protein molecules, rather than to induction of an HR resistance or other response.

BPMV as a Vector for Virus-Induced Gene Silencing (VIGS)

Construction of infectious cDNA clones of RNA1 and RNA2 of BPMV provided evidence of the functions of sequences associated with symptom severity, but not with knowledge of precise host-virus interactions. However, recently Zhang and Ghabrial (2006) observed that this binary vector system could be used for protein expression and a sequence-specific virus-induced gene silencing system (VIGS) for soybean. Although VIGS can be performed in several plants with potato virus X (PVX) or tobacco rattle virus (TRV) (Burch-Smith et al. 2004), Zhang and Ghabrial (2006) are the first to design a vector for soybean. A mild strain of BPMV was selected, and an insertion site created at the cleavage site between the movement protein (MP) and the large coat protein (L-CP). Sequences near this site were modified to reduce potential recombination with the virus, and to provide insertion sites. Foreign genes expressed successfully included the *bar* gene for herbicide resistance, the *PDS* gene, and suppressors of gene silencing from *Tomato bushy stunt virus* (p19), *Turnip crinkle virus* (CP), and HC-Pro of SMV and *Tobacco etch virus* (Roth et al. 2004). Interestingly, all silencing suppressors in the VIGS vector strain caused an increase in symptom severity similar to that seen in mixed infections of BPMV and SMV (Zhang and Ghabrial 2006).

The BPMV-based VIGS vector was used to examine the defense-related functionality of soybean sequences. For this purpose, soybean endogenous stearoyl desaturase gene(s) were silenced, which resulted in the accumulation of stearic acid, reduction of oleic acid and induction of cell death and PR gene expression similar to those observed in the Arabidopsis stearoyl-ACP-desaturase mutant (ssi2) (Kachroo and Ghabrial 2007; Sekine et al. 2004) The authors conclude that 'Elucidation of signal transduction pathway(s) downstream of the initial pathogen recognition event would aid in the engineering of novel, enduring, and broad-spectrum disease resistance in the soybean crop.' In addition to applications in resistance gene discovery, the BPMV-based VIGS system should have significant implications for functional genomics research in soybean.

Summary and Conclusion

Soybean genomics will continue to advance and provide additional insight into virus-soybean interaction. Currently, with BPMV and SMV genome sequences, studies have mainly focused on understanding and manipulating the virus genome. These results aided in development of tools of biotechnology such as the VIGS system recently developed from the BPMV genome. This system provides a highly effective gene discovery tool previously unavailable to the soybean research community. Only a few studies examined the actual interaction between the host and virus in a susceptible or resistant reaction in soybean. Recent advancements in global expression profiling similar to studies in soybean conducted with fungi and nematodes will likely identify additional genes needed for resistance or genes that the virus requires or manipulates in the host in a susceptible interaction. Information from these studies will further assist in the development of cultivars with durable virus resistance as these resistance genes will encounter extensive selection pressure due to the emergence of the soybean aphid, *Aphis glycines,* and new reports of resistance breaking strains in all areas of the world. In the case of SMV and BPMV, creating pyramided cultivars with durable resistance incorporating native genes and transgenes to both viruses will be crucial to prevent the synergism that occurs between the two. Stable, highly effective transgenic pathogen-derived resistance for BPMV, with CP or other viral sequences, should be further developed as resistance is currently not available in commercial cultivars and is needed because of the prominence of the disease in the Great Plains states. Additional studies are needed to understand host targets of BPMV in highly susceptible lines to develop this resistance. Within the next couple of years the entire sequence of the soybean genome will be complete, further assisting in cloning of virus resistance genes and providing an understanding of virus R genes evolution.

References

Anandalakshmiki, R., Pruss, G.J., Ge, X., Marathe, R., Mallory, A.C., Smith, T.H., and Vance, V.G. (1998) A viral suppressor of gene silencing in plants. Proc Natl. Acad. Sci. USA. 95, 13079–13084.

Anjos, J.R., Jarlfors, U., and Ghabrial, S.A. (1992) Soybean mosaic potyvirus enhances the titer of two comoviruses in dually infected soybean plants. Phytopathology. 82, 1022–1027.

Arahana, V.S., Graef, G.L., Specht, J.E., Steadman, J.R., and Eskridge, K.M. (2001) Identification of QTLs for resistance to *Sclerotinia sclerotiorum* in soybean. Crop Sci. 41, 180–188.

Ashfield, T., Keen, N.T., Buzzell, R.I., and Innes, R.W. (1995) Soybean resistance genes specific for different *Pseudomonas syringae* avirulence genes are allelic, or closely linked, at the Rpg1 locus. Genetics. 141, 1597–1604.

Ashfield, T., Ong, L.E., Nobuta, K., Schneider, C.M., and Innes, R.W. (2004) Convergent evolution of disease resistance gene specificity in two flowering plant families. Plant Cell. 16, 309–318.

Ashfield, T., Wawrzynski, A., Podicheti, R., Metcalf, M., Pfeil, B., Denny, R., Ameline-Torregrosa, C., Cannon, S., Cannon, E., Ratnaparkhe, M., Mammadov, J., Nguyen, A., Glover, N., Ilut, D., Chacko, B., Seigfreid, M., Deshpande, S., Doyle, J.J., Roe, B., Maroof, S., Young, N., and Innes, R. (2007) Comparative analysis of legume genome evolution in the vicinity of the *Rpg1-b* disease resistance gene. Plant and Animal Genome XV International Conference, San Diego, CA.

Baulcombe, D.C. (1996) Mechanisms of pathogen-derived resistance to viruses in transgenic plants. Plant Cell. 8, 1833–1844.

Bays, D.C., Tolin, S.A., and Roane, C.W. (1986) Interactions of peanut mottle virus strains and soybean germ plasm. Phytopathology. 76, 764–768.

Boerma, H.R., and Kuhn, C.W. (1976) Inheritance of resistance to Peanut mottle virus in soybeans. Crop Sci. 16, 533–534.

Boerma, H.R., Kuhn, C.W., and Harris, H.B. (1975) Inheritance of resistance to cowpea chlorotic mottle virus (soybean strain) in soybeans. Crop Sci. 15, 849–850.

Burch-Smith, T.M., Anderson, J.C., Martin, G.B., and Dinesh-Kumar, S.P. (2004) Applications and advantages of virus-induced gene silencing for gene function studies in plants. Plant J. 39, 734–746.

Burton, J.W., Brim, C.A., and Young, M.F. (1987) Registration of 'Young' soybean. Crop Sci. 27, 1093.

Buss, G.R., Roane, C.W., Tolin, S.A., and Vinardi, T.A. (1985) A second dominant gene for resistance to peanut mottle virus in soybeans. Crop Sci. 2, 314–316.

Buss, G.R., Ma, G., Kristipati, S., Chen, P., and Tolin, S.A. (1999) A new allele at the *Rsv3* locus for resistance to soybean mosaic virus. In: H. E. Kauffman (Ed.) *Proc. World Soybean Res. Conf. VI.* Superior Printing, Champaign, IL., pp.

Buzzell, R.I., and Tu, J.C. (1989) Inheritance of a soybean stem-tip necrosis reaction to soybean mosaic virus. J. Hered. 80, 400–401.

Cannon, S., Ameline-Torregrosa, C., Ashfield, T., Pfeil, B., Mammadov, J., Ratnaparkhe, M., Podiceti, R., O'Bleness, M., Deshpande, S., Cannon, E., Chacko, B., Wawrzynski, A., Glover, N., Nguyen, A., Ilut, D., Denny, R.R., B., Doyle, J., Maroof, S., Young, N., and Innes, R. (2007) Comparative Analysis Of Legume Genome Evolution Around The Rpg1 Locus: Evolution Of Genes And Genomic Regions From Soybean And Medicago truncatula. Plant and Animal Genome XV International Conference, San Diego, CA.

Caranta, C., Palloix, A., Gebre-Selassie, K., Lefebvre, V., Moury, B., and Daubeze, A.M. (1996) A complementation of two genes originating from susceptible *Capsicum annuum* lines confers a new and complete resistance to pepper veinal mottle virus. Phytopathology. 86, 739–743

Chen, P., Buss, G.R., Roane, C.W., and Tolin, S.A. (1991) Allelism among genes for resistance to soybean mosaic virus in strain-differential soybean cultivars. Crop Sci. 31, 305–309.

Chen, P., Buss, G.R., Roane, C.W., and Tolin, S.A. (1994) Inheritance in soybean of resistant and necrotic reactions to soybean mosaic virus strains. Crop Sci. 34, 414–422.

Chen, P., Buss, G.R., Tolin, S.A., Gunduz, I., and Cicek, M. (2002) A valuable gene in Suweon 97 soybean for resistance to soybean mosaic virus. Crop Sci. 42, 333–337.

Chen, P., Ma, G., Buss, G.R., Gunduz, I., Roane, C.W., and Tolin, S.A. (2001) Inheritance and allelism tests of Raiden soybean for resistance to soybean mosaic virus. J. Heredity. 92, 51–55.

Cho, E.K., and Goodman, R.M. (1979) Strains of soybean mosaic virus: classification based on virulence in resistant soybean cultivars. Phytopathology. 69, 467–470.

Cho, E.K., and Goodman, R.M. (1982) Evaluation of resistance in soybeans to soybean mosaic virus strains. Crop Sci. 22, 1133–1136.

Choi, B.K., Koo, J.M., Ahn, H.J., Yum, H.J., Choi, C.W., Ryu, K.H., Chen, P., and Tolin, S.A. (2005) Emergence of *Rsv*-resistance breaking Soybean mosaic virus isolates from Korean soybean cultivars. Virus Res. 112, 42–51.

Choi, I.-Y., Hyten, D.L., Matukumalli, L.K., Song, Q., Chaky, J.M., Quigley, C.V., Chase, K., Lark, K.G., Reiter, R.S., Yoon, M.-S., Hwang, E.-Y., Yi, S.-I., Young, N.D., Shoemaker, R.C., van Tassell, C.P., Specht, J.E., and Cregan, P.B. (2007) A soybean transcript map: gene distribution, haplotype and single-nucleotide polymorphism analysis. Genetics. 176, 685–696.

Choi, S.H., Green, S.K., and Lee, D.R. (1989) Linkage relationship between two genes conferring resistance to peanut stripe virus and soybean mosaic. Euphytica. 44, 163–166.

Clark, A.J., and Perry, K.L. (2002) Transmission of field isolates of soybean viruses by *Aphis glycines*. Plant Dis. 86, 1219–1222.

Cooper, B. (2001) Collateral gene expression changes induced by distinct plant viruses during the hypersensitive resistance reaction in *Chenopodium amaranticolor*. The Plant J. 26, 339–349.

Cregan, P.B., Jarvik, T., Bush, A.L., Shoemaker, R.C., Lark, K.G., Kahler, A.L., Kaya, N., VanToai, T.T., Lohnes, D.G., and Chung, J. (1999) An integrated genetic linkage map of the soybean genome. Crop Sci. 39, 1464–1490.

Demirbas, A., Rector, B.G., Lohnes, D.G., Fioritto, R.J., Graef, G.L., Cregan, P.B., Shoemaker, R.C., and Specht, J.E. (2001) Simple sequence repeat markers linked to the soybean *Rps* genes for phytophthora resistance. Crop Sci. 41, 1220–1227.

Demski, J.W., and Kuhn, C.W. (1975) Resistant and susceptible reaction of soybeans to peanut mottle virus. Phytopathology. 65, 95–99.

Di, R., Purcell, V., Collins, G.B., and Ghabiral, S.A. (1996) Production of transgenic soybeans lines expressing the bean pod mottle virus coat protein precursor gene. Plant Cell Rep. 15, 746–750.

Diaz-Pendon, J.A., Truniger, V., Nieto, C., Garcia-Mas, J., Bendahmane, A., and Aranda, M.A. (2004) Advances in understanding recessive resistance to plant viruses. Mol. Plant Pathol. 5, 223–233.

Domier, L.L., Latorre, I.J., Steinlage, T.A., McCoppin, N., and Hartman, G.L. (2003) Variability and transmission by *Aphis glycines* of North American and Asian *Soybean mosaic virus* isolates Arch. Virol. 148.

Eggenberger, A., Hajimorad, M.R., and Hill, J. (2007) Gain of virulence by an avirulent strain of Soybean mosaic virus on Rsv1-genotype soybean requires concurrent mutations in both P3 and HC-Pro. Phytopathology. (In press), Abstract at the Annual Meeting of the American Phytopathology Society, San Diego.

Eggenberger, A.L., and Hill, J.H. (1997) Analysis of resistance-breaking determinants of soybean mosaic virus. Phytopathology. 87, S27.

Espinoza, C., Vega, A., Medina, C., Schlauch, K., Cramer, G., and Arce-Johnson, P. (2007) Gene expression associated with compatible viral disease in grapevine cultivars. Funct. Integr. Genomics. 7, 95–110.

Fasoula, V.A., Harris, D.K., Bailey, M.A., Phillips, D.V., and Boerma, H.R. (2003) Identification, mapping, and confirmation of a soybean gene for bud blight resistance. Crop Sci. 43, 1754–1759.

García-Arenal, F., and McDonald, B. (2003) An analysis of the durability of resistance to plant viruses. Phytopathology. 93, 941–952.

Giesler, L.J., Ghabrial, S.A., Hunt, T.E., and Hill, J.H. (2002) *Bean pod mottle*: A threat to U.S. soybean production. Plant Dis. 86, 1280–1289.

Gillaspie, A.G., Jr., and Hopkins, M.S. (1991) Spread of peanut stripe virus from peanut to soybean and yield effects on soybean. Plant Dis. 75, 1157–1159.

Godwin, M., Tucker, D.M., Jeong, S.C., Hayes, A.J., Schmidt, D., Chaky, J.M., Han, F., and Saghai Maroof, M.A. (2008) Molecular mapping of disease-resistance related sequence tags and resistance gene analogues in soybean. (unpublished).

Goldbach, R., Martelli, G.P., and Milne, R.G. (1995) Family Comvirdae. In: F. A. Murphy, C. M. Fauquet, D. H. L. Bishop, S. A. Ghabiral, A. W. Jarvis, G. P. Martelli, M. A. Mayo and M. D. Summers (Eds.) *Virus Taxonomy*. Springer-Verlag, New York, pp. 341–347

Goodrick, B.J., Kuhn, C.W., and Boerma, H.R. (1991) Inheritance of nonnecrotic resistance to cowpea chlorotic mottle virus in soybean. J. Hered. 82, 512–514.

Gordon, S.G., St Martin, S.K., and Dorrance, A.E. (2006) *Rps8* maps to a resistance gene rich region on soybean molecular linkage group F. Crop Sci. 46, 168–173.

Gore, M.A., Hayes, A.J., Jeong, S.C., Yue, Y.G., Buss, G.R., and Saghai Maroof, M.A. (2002) Mapping tightly linked genes controlling potyvirus infection at the *Rsv1* and *Rpv1* region in soybean. Genome. 45, 592–599.

Gu, H., and Ghabrial, S.A. (2005) The *Bean pod mottle virus* proteinase cofactor and putative helicase are symptom severity determinants. Virology. 333, 271–283.

Gu, H., Zhang, C., and Ghabrial, S.A. (2006) Novel naturally occurring *Bean pod mottle virus* reassortants with mixed heterologous RNA1 genomes. Phytopathology. 97, 79–86.

Gu, H., Clark, A.J., de Sá, P.B., Pfieffer, T.W., Tolin, S., and Ghabrial, S.A. (2002) Diveristy among isolates of *Bean pod mottle virus*. Phytopathology. 92, 446–452.

Gunduz, I., Buss, G.R., Chen, P., and Tolin, S.A. (2002) Characterization of SMV resistance genes in Tousan 140 and Hourei soybean. Crop Sci. 42, 90–95.

Gunduz, I., Buss, G.R., Chen, P., and Tolin, S.A. (2004) Genetic and phenotypic analysis of soybean mosaic virus resistance in PI 88788 soybean. Phytopathology. 94, 687–692.

Gunduz, I., Buss, G.R., Ma, G., Chen, P., and Tolin, S.A. (2001) Genetic analysis of resistance to *Soybean mosaic virus* in OX670 and Harosoy soybean. Crop Sci. 41, 1785–1791.

Hajimorad, M.R., and Hill, J.H. (2001) *Rsv1*-mediated resistance against *Soybean mosaic virus*-N is hypersensitive response-independent at inoculation site, but has the potential to initiate a hypersensitive response-like mechanism. Mol. Plant Microbe Interact. 14, 587–598.

Hajimorad, M.R., Eggenberger, A.L., and Hill, J.H. (2003) Evolution of Soybean mosaic virus-G7 molecularly cloned genome in *Rsv1*-genotype soybean results in emergence of a mutant capable of evading *Rsv1*-mediated recognition. Virology. 2, 497–509.

Hajimorad, M.R., Eggenberger, A.L., and Hill, J.H. (2005) Loss and gain of elicitor function of Soybean mosaic virus G7 provoking *Rsv1*-mediated lethal systemic hypersensitive response maps to P3. J. Virology. 79, 1215–1222.

Hajimorad, M.R., Eggenberger, A.L., and Hill, J.H. (2006) Strain-specific P3 of *Soybean mosaic virus* elicits Rsv1-mediated extreme resistance, but absence of P3 elicitor function alone is insufficient for virulence on Rsv1-genotype soybean. Virology. 345, 156–166.

Hajimorad, M.R., Eggenberger, A.L., and Hill, J.H. (2007) Adaptation of chimeras derived from an avirulent strain of Soybean mosaic virus to *Rsv1*-genotype soybeans is associated with mutations in HC-Pro. American Society for Virology Meeting Abstract, W6–11.

Han, Z., Guo, W., Song, X., and Zhang, T. (2004) Genetic mapping of EST-derived microsatellites from the diploid *Gossypium arboretum* in allotetraploid cotton. Mol. Gen. Genomics. 272, 308–327.

Hartman, G.L., Domier, L.L., Wax, L.M., Helm, C.G., Onstad, D.W., Shaw, J.T., Solter, L.F., Voegtlin, D.J., D'Arcy, C.J., Gray, M.E., Steffey, K.L., Isard, S.A., and Orwick, P.L. (2001) Occurrence and distribution of *Aphis glycines* on soybeans in Illinois in 2000 and its potential control. Plant Health Progess, doi:10.1094/PHP-2001-0205-01-HN.

Hayes, A.J., and Saghai Maroof, M.A. (2000) Targeted resistance gene mapping in soybean using modified AFLPs Theor. Appl. Genet. 100, 1279–1283.

Hayes, A.J., Y.G., Y., and Saghai Maroof, M.A. (2000a) Expression of two soybean resistance gene candidates shows divergence of paralogous single copy genes. Theor. Appl. Genet. 101, 789–795.

Hayes, A.J., Ma, G., Buss, G.R., and Saghai Maroof, M.A. (2000b) Molecular marker mapping of *Rsv4*, a gene conferring resistance to all known strains of soybean mosaic virus. Crop Sci. 40, 1434–1437.

Hayes, A.J., Jeong, S.C., Gore, M.A., Yu, Y.G., Buss, G.R., Tolin, S.A., and Sagahi Maroof, M.A. (2004) Recombination within a nucleotide-binding-site/leucine-rich-repeat gene cluster produces new variants conditioning resistance to soybean mosaic virus in soybeans. Genetics. 166, 493–503.

Hill, J.H., Alleman, R., Hogg, D.B., and Grau, C.R. (2001) First report of transmission of Soybean mosaic virus and Alfalfa mosaic virus by *Aphis glycines* in the New World. Plant Dis. 85, 561.

Hill, J.H., Koval, N.C., Gaska, J.M., and Grau, C.R. (2007) Identification of field tolerance to Bean pod mottle and Soybean mosaic viruses in soybean. Crop Sci. 47, 212–218.

Hwang, T.-Y., Moon, J.-K., Yu, S., Yang, K., Mohankumar, S., Yu, Y.H., Lee, Y.H., Kim, H.S., Kim, H.M., Maroof, S.M., and Jeong, S.C. (2006) Application of comparative genomics in developing molecular markers tightly linked to the virus resistance gene *Rsv4* in soybean. Genome. 49, 380–388.

Ithal, N., Recknor, J., Nettleton, D., Hearne, L., Maier, T., Baum, T.J., and Mitchum, M.G. (2007) Parallel genome-wide expression profiling of host and pathogen during soybean cyst nematode infection of soybean. Mol. Plant Microbe Interact. 20, 293–305.

Iwaki, M., Roechan, M., Hibino, H., Tochihara, H., and Tantera, D.M. (1980) A persistent aphid-borne virus of soybean, Indonesian *Soybean dwarf virus* transmitted by *Aphis glycines*. Plant Dis. 64, 1027–1030.

Jain, R.K., McKern, N.M., Tolin, S.A., Hill, J.H., Barnett, O.W., Tosic, M., Ford, R.E., Beachy, R.N., Yu, M.H., and Ward, C.W. (1992) Confirmation that fourteen potyvirus isolates from soybean are strains of one virus by comparing coat protein peptide profiles. Phytopatholoy. 82, 294–299.

Jayaram, C., Hill, J.H., and Miller, W.A. (1992) Complete nucleotide sequences of two soybean mosaic virus strains differentiated by response of soybean containing the *Rsv* resistance gene. J. Gen. Virology. 73, 2067–2077.

Jenner, C.E., Tomimura, K., Ohshima, K., Hughes, S.L., and Walsh, J.A. (2002) Mutations in *Turnip mosaic virus* P3 and cylindrical inclusion proteins are separately required to overcome two *Brassica napus* resistance genes. Virology. 300, 50–59.

Jenner, C.E., Wang, X., Tomimura, K., Ohshima, K., Ponz, F., and Walsh, J.A. (2003) The dual role of the potyvirus P3 protein of *Turnip mosaic virus* as a symptom and avirulence determinant in brassicas. Mol. Plant-Microbe Interact. 16, 777–784.

Jeong, S.C., and Saghai Maroof, M.A. (2004) Detection and genotyping of SNPs tightly linked to two disease resistance loci, *Rsv1* and *Rsv3*, of soybean. Plant Breeding. 123, 305–310.

Jeong, S.C., Hayes, A.J., Biyashev, R.M., and Maroof, M.A.S. (2001) Diversity and evolution of a non-TIR-NBS sequence family that clusters to a chromosomal "hotspot" for disease resistance genes in soybean. Theor. Appl. Genet. 103, 406–414.

Jeong, S.C., Kristipati, S., Hayes, A.J., Maughan, P.J., Noffsinger, S.L., Gunduz, I., Buss, G.R., and Saghai Maroof, M.A. (2002) Genetic and sequence analysis of markers tightly linked to the soybean mosaic virus resistance gene, *Rsv3*. Crop Sci. 42, 265–270.

Kachroo, A., and Ghabrial, S. (2007) Functional genomics in soybean: elucidating the molecular components of soybean defense signaling pathways. Phytopathology. 97, S129.

Kang, B.C., Yeam, I., and Jahn, M.M. (2005) Genetics of plant virus resistance. Phytopathology. 43, 581–621.

Kantety, R.V., La Rota, M., Matthews, D.E., and Sorrells, M.E. (2002) Data mining for simple sequence repeats in expressed sequence tags from barley, maize, rice, sorghum and wheat. Plant Mol. Bio. 48, 501–510.

Kartaatmadja, S., and Sehgal, O.P. (1990) Decline of *Bean pod mottle virus* specific infectivity in vivo correlates with degradation of encapsidated RNA-1. Phytopathology. 80, 1182–1189.

Kelly, J.D., Afanador, L., and Haley, S.D. (1995) Pyramiding genes for resistance to *Bean common mosaic virus* Euphytica. 82, 207–212.

Kiihl, R.A.S., and Hartwig, E.E. (1979) Inheritance of reaction to soybean mosaic virus in soybeans. Crop Sci. 19, 372–375.

Kim, Y.H., Kim, O.S., Lee, B.C., Moon, J.K., Lee, S.C., and Lee, J.Y. (2003) G7H, a new Soybean mosaic virus strain its virulence and nucleotide sequence of CI gene. Plant Dis. 87, 1372–1375.

Koo, J.M., Choi, B.K., Ahn, H.J., Yum, H.J., and Choi, C.W. (2005) First report of an *Rsv* resistance-breaking isolate of *Soybean mosaic virus* in Korea. Plant Pathol. 54, 573–573.

Kwon, S.H., and Oh, J.H. (1980) Resistance to a necrotic strain of soybean mosaic virus in soybeans. Crop Sci. 20, 403–404.

Lee, J.M., Hartman, G.L., Domier, L.L., and Bent, A.F. (1996) Identification and map location of TTR1, a single locus in *Arabidopsis thaliana* that confers tolerance to tobacco ringspot nepovirus. Mol. Plant Microbe Interact. 8, 729–735.

Liao, L., Chen, P., Buss, G.R., Yang, Q., and Tolin, S.A. (2002) Inheritance and allelism of resistance to soybean mosaic virus in Zao18 soybean from China. J. Hered. 93, 447–452.

Lim, W.S., Kim, Y.H., and Kim, K.H. (2003) Complete genome sequences of the genomic RNA of soybean mosaic virus strains G7H and G5. Plant Pathol. J. 19, 171–176.

Ma, G., Chen, P., Buss, G.R., and Tolin, S.A. (1995) Genetic characteristics of two genes for resistance to soybean mosaic virus in PI486355 soybean. Theor. Appl. Genet. 91, 907–914.

Ma, G., Chen, P., Buss, G.R., and Tolin, S.A. (2002) Complementary action of two independent dominant genes in Columbia soybean for resistance to soybean mosaic virus. J. Hered. 93, 179–184.

Ma, G., Chen, P., Buss, G.R., and Tolin, S.A. (2003) Genetic study of a lethal necrosis to soybean mosaic virus in PI 507389 soybean. J. Hered. 94, 205–211.

Maule, A.J., Caranta, C., and Boulton, M.I. (2007) Sources of natural resistance to plant viruses: status and prospects. Mol. Plant Path. 8, 223–231.

Morgante, M., and Olivieri, A.M. (1993) PCR-amplified microsatellites as markers in plant genetics. Plant J. 3, 175–182.

Omunyin, M.E., Hill, J.H., and Miller, W.A. (1996) Use of unique RNA sequence-specific oligonucleotide primers for RT-PCR to detect and differentiate *Soybean mosaic virus* strain. Plant Dis. 80, 1170–1174.

Paguio, O.R., and Kuhn, C.W. (1974) Survey for peanut mottle virus in peanut in Georgia. Plant Dis. Rep. 58, 107–110.

Paguio, O.R., Kuhn, C.W., and Boerma, H.R. (1988) Resistance-breaking variants of cowpea chlorotic mottle virus in soybean. Plant Dis. 72, 768–770.

Peters, S.A., Verver, J., Nollen, E.A., van Lent, J.W., Wellink, J., and van Kammen, A. (1994) The NTP-binding motif in cowpea mosaic virus B polyprotein is essential for virus replication. J. Gen. Virol. 75, 3167–3176.

Pouwels, J., Carette, J.E., van Lent, J., and Wellink, J. (2002) *Cowpea mosaic virus*: effect on host cell processes. Mol. Plant Pathol. 3, 411–418.

Rector, B.G., All, J.N., Parrott, W.A., and Boerma, H.R. (1999) Quantitative trait loci for antixenosis resistance to corn earworm in soybean. Crop Sci. 39, 531–538.

Rector, B.G., All, J.N., Parrott, W.A., and Boerma, H.R. (2000) Quantitative trait loci for antibiosis resistance to corn earworm in soybean. Crop Sci. 40, 233–238.

Reddy, M.S.S., Ghabrial, S.A., Redmond, C.T., Dinkins, R.D., and Collins, G.B. (2001) Resistance to *Bean pod mottle virus* in transgenic soybean lines expressing the capsid polyprotein. Phytopathology. 91, 831–838.

Reichmann, J.L., Lain, S., and Garcia, J.A. (1992) Highlights and prospects of potyvirus molecular biology. J. Gen. Virology. 73, 1–16.

Roane, C.W., Tolin, S.A., and Buss, G.R. (1983) Inheritance of reaction to two viruses in the soybean cross 'York' X 'Lee 28'. J. Heredity. 74, 289–291.

Rodriguiz-Cerezo, E., Ammar, E.D., Pirone, T.P., and Shaw, J.G. (1993) Association of the non-structural P3 viral protein with cylindrical inclusions in potyvirus-infected cells. J. Gen. Virology. 74, 1945–1949.

Ross, J.P. (1969) Pathogenic variation among isolates of soybean mosaic virus. Phytopathology. 59, 829–832.

Roth, B.M., Pruss, G.J., and Vance, V.B. (2004) Plant viral suppressors of RNA silencing. Virus Res. 102, 97–108.

Sandhu, D., Schallock, K.G., Rivera Velez, N., Lundeen, P., Cianzio, S., and Bhattacharyya, M.K. (2005) Soybean phytophthora resistance gene *Rps8* maps closely to the *Rps3* region J. Heredity 96, 536–541.

Saruta, M., Kikuchi, A., Okabe, A., and Sasaya, T. (2005) Molecular characterization of A_2 and D strains of *Soybean mosaic virus*, which caused a recent virus outbreak in soybean cultivar Sachiyutaka in Chugoku and Shikoku regions of Japan J. Gen. Plant Path. 71, 431–435.

Sekine, K.-T., Nandi, A., Ishihara, T., Hase, S., Ikegami, M., Shah, J., and Takahashi, H. (2004) Enhanced Resistance to *Cucumber mosaic virus* in the *Arabidopsis thaliana* ssi2 mutant is mediated via an SA-independent mechanism. Molecular Plant-Microbe Inter (MPMI). 17, 623–632

Shakiba, E., Chen, P., and Gergerich, R.C. (2006) Reaction of current soybean cultivars in Arkansas to soybean mosaic virus and tobacco ringspot virus. ASA-CSSA-SSSA, Indianapolis, IN.

Shipe, E.R. 1979. Genetics of reactions to peanut mottle virus in soybean. Diss., Virginia Polytechnic Institutes and State University Blacksburg, VA.

Shipe, E.R., Buss, G.R., and Roane, C.W. (1979) Resistance to peanut mottle virus (PMV) in soybean (*Glycine max*) plant introductions. Plant Dis. Rep. 63, 757–760.

Shoemaker, R.C., Polzin, K., Labate, J., Specht, J., Brummer, E.C., Olson, T., Young, N., Concibido, V., Wilcox, J., Tamulonis, J.P., Kochert, G., and Boerma, H.R. (1996) Genome duplication in soybean (*Glycine* subgenus *soja*). Genetics. 144, 329–338.

Shukla, D.D., Frenke, M.J., and Ward, C.W. (1991) Structure and function of the potyvirus genome with special reference to the coat protein coding region. Can. J. Plant Pathol. 13, 178–191.

Steinlage, T.A., Hill, J.H., and Nutter, F.W., Jr. (2001) Temporal and spatial spread of *Soybean mosaic virus* (SMV) in soybeans transformed with the coat protein gene of SMV. Phytopathology. 92, 478–486.

Strömvik, M.K., Latour, F., Archambault, A., and Vodkin, L.O. (2006) Identification and phylogenetic analysis of sequences of *Bean pod mottle virus*, *Soybean mosaic virus*, and *Cowpea chlorotic mottle virus* in expressed sequence tag data from soybean. Can. J. Plant Pathol. 28, 289–301.

Sundararaman, V.P., Stromvik, M.V., and Vodkin, L.O. (2000) A putative defective interfering RNA from *Bean pod mottle virus*. Plant Dis. 84, 1309–1313.

Tamulonis, J.P., Luzzi, B.M., Hussey, R.S., Parrott, W.A., and Boerma, H.R. (1997a) DNA marker analysis of loci conferring resistance to peanut root-knot nematode in soybean. Theor. Appl. Genet. 95, 664–670.

Tamulonis, J.P., Luzzi, B.M., Hussey, R.S., Parrott, W.A., and Boerma, H.R. (1997b) DNA markers associated with resistance to Javanese root-knot nematode in soybean. Crop Sci. 37, 783–788.

Terry, L.I., Chase, K., Jarvik, T., Orf, J., Mansur, L., and Lark, K.G. (2000) Soybean quantitative trait loci for resistance to insects. Crop Sci. 40, 375–382.

Tolin, S.A., and Lacy, G.H. (2004) Viral, Bacterial, and Phytoplasmal Diseases of Soybean *Soybean: Improvement, Production, and Uses*, 3rd ed. ASA-CSSA-SSSA, Madison, WI, pp. 765–819

Tougou, M., Furutani, N., Yamagishi, N., Shizukawa, Y., Takahata, Y., and Hidaka, S. (2002) Development of resistant transgenic soybeans with inverted repeat-coat protein genes of soybean dwarf virus. Plant Cell Rep. 25, 1213–1218.

Vodkin, L.O., Kahnna, A., Shealy, R.T., Clough, S.J., Gonzales, D.O., Philip, R., Zabala, G.C., Thibaud-Nissen, F., Sidarous, M., Stromvik, M.V., Shoop, E.G., Schmidt, C., Retzel, E., Erpelding, J., Shoemaker, R.C., Rodrizues-Huete, A.M., Polacco, J.C., Coryell, V., Keim, P., Gong, G., Lui, L., Pardinas, J.L., and Schweitzer, P. (2004) Microarrays for global expression constructed with a low redundancy set of 27,500 sequenced cDNAs representing an array of developmental stages and physiological conditions of the soybean plant. Biomed Central Genomics. 5, 73.

Walker, D., Boerma, H.R., All, J., and Parrott, W. (2002) Combining *cry1AC* with QTL alleles from PI 229358 to improve soybean resistance to lepidopteran pests. Mol. Breed. 9, 43–51.

Wang, X., Eggenberger, A.L., Nutter Jr, F.W., and Hill, J.H. (2001) Pathogen-derived transgenic resistance to soybean mosaic virus in soybean. Mol. Breed. 8, 119–127.

Wang, Y., Hobbs, H.A., Bowen, C.R., Bernard, R.L., Hill, C.B., Haudenshield, J.S., Domier, L.L., and Hartman, G.L. (2006) Evaluation of soybean cultivars, 'Williams' isogenic lines, and other selected soybean lines for resistance to two *Soybean Mosaic Virus* strains. Crop Sci. 46, 2649–2653.

Warwick, D., and Demski, J.W. (1988) Susceptibility and resistance of soybeans to peanut stripe virus. Plant Dis. 72, 19–21.

Werner, K., Friedt, W., and Ordon, F. (2005) Strategies for pyramiding resistance genes against the barley yellow mosaic virus complex (BaMMV, BaYMV, BaYMV-2) Mol. Breed. 16, 45–55.

Whitham, S.A., Quan, S., Chang, H.-S., Cooper, B., Estes, B., Zhu, T., Wang, X., and Hou, Y.-M. (2003) Diverse RNA viruses elicit the expression of common sets of genes in susceptible *Arabidopsis thaliana* plants. The Plant J. 33, 271–283.

Yang, C., Guo, R., Jie, F., Nettleton, D., Peng, J., Carr, T., Yeakley, J.M., Fan, J.B., and Whitman, S.A. (2007) Spatial analysis of *Arabidopsis thaliana* gene expression in response to *Turnip mosaic virus* infection Molecular Plant-Microbe Inter. 20, 358–370.

Yu, Y.G., Buss, G.R., and Saghai Maroof, M.A. (1996) Isolation of a superfamily of candidate disease-resistance genes in soybean based on a conserved nucleotide-binding site. Proc. Natl. Acad. Sci. (USA). 93, 11751–11756.

Yu, Y.G., Saghai Maroof, M.A., Buss, G.R., Maughan, P.J., and Tolin, S.A. (1994) RFLP and microsatellite mapping of gene for soybean mosaic virus resistance. Phytopathology. 84, 60–64.

Yue, P., Arelli, P.R., and Sleper, D.A. (2001) Molecular characterization of resistance to *Heterodera glycines* in soybean PI 438489B. Theor. Appl. Genet. 102, 921–928.

Zhang, C., and Ghabrial, S.A. (2006) Development of *Bean pod mottle virus*-based vectors for stable protein expression and sequence-specific virus-induced gene silencing in soybean. Virology. 344, 401–411.

Zhang, C., Eggenberger, A., Hajimorad, M.R., Tsang, S., and Hill, J. (2007) The N terminal Soybean mosaic virus CI is required for SMV virulence and is a symptom determinant on *Rsv3* genotype soybean. Phytopathology 97, S129.

Zheng, C., Chen, P., and Gergerich, R. (2005a) Effect of temperature on the expression of necrosis in soybean infected with *Soybean mosaic virus*. Crop Sci. 45, 916–922.

Zheng, C., Chen, P., Hymowitz, T., Wickizer, S., and Gergerich, R. (2005b) Evaluation of Glycine species for resistance to *Bean pod mottle virus*. Crop Protect. 24, 49–56.

Zou, J., Rodriguez-Zas, S., Aldea, M., Li, M., Zhu, J., Gonzalez, D.O., Vodkin, L.O., DeLucia, E., and Clough, S.J. (2005) Expression profiling soybean response to *Pseudomonas syringae* reveals new defense-related genes and rapid HR-specific downregulation of photosynthesis. Mol. Plant Microbe Interact. 11, 1161–1174

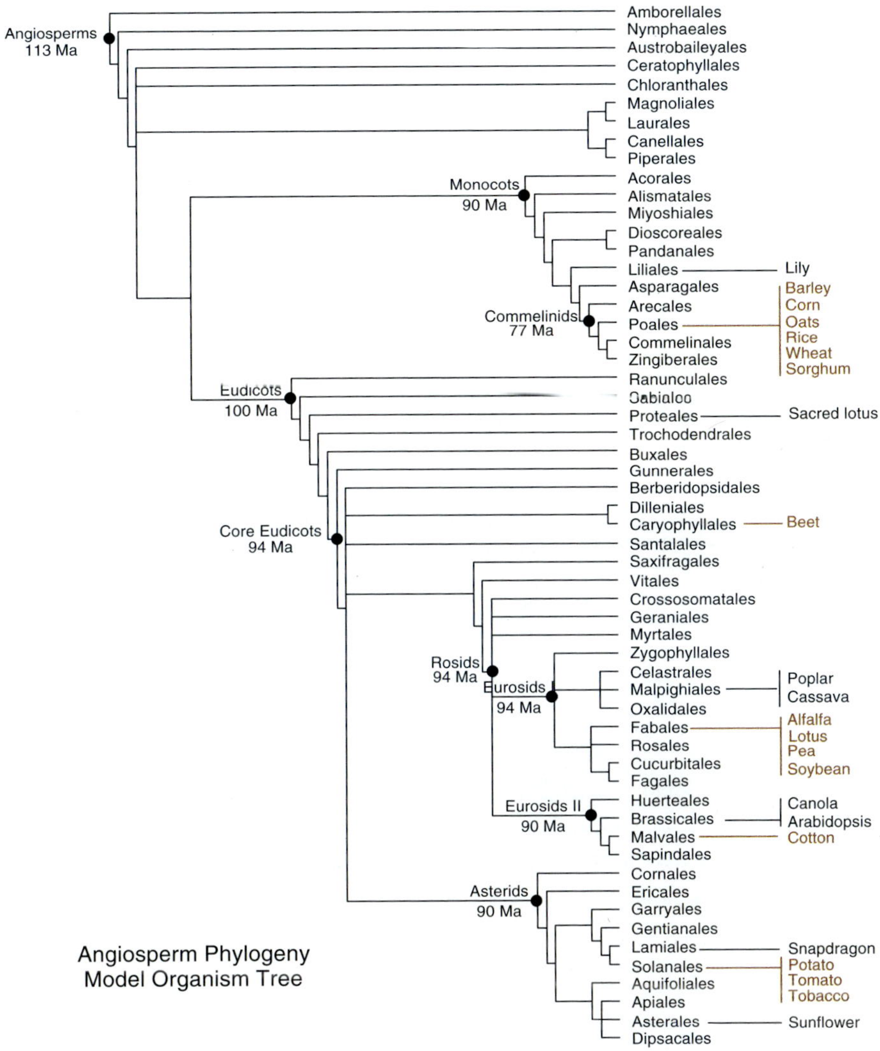

Fig. 4.1 Synopsis of the model organism tree adapted from the Angiosperm Phylogeny web site (Stevens 2001; www.mobot.org/MOBOT/research/APweb). The minimum ages of marked clades in millions of years (Ma) are taken from Crepet et al. (2004), except for the minimum age of the Commelinid clade, which is taken from Magallón and Sanderson (2001). Crepet et al. (2004) take a conservative approach to assigning fossils to extant clades only with unequivocal morphological evidence. Some clades deeper in the phylogeny have equivalent minimum ages to those nested higher in the phylogeny. This is because the fossils used to make these age assignments come from strata with similar minimum ages

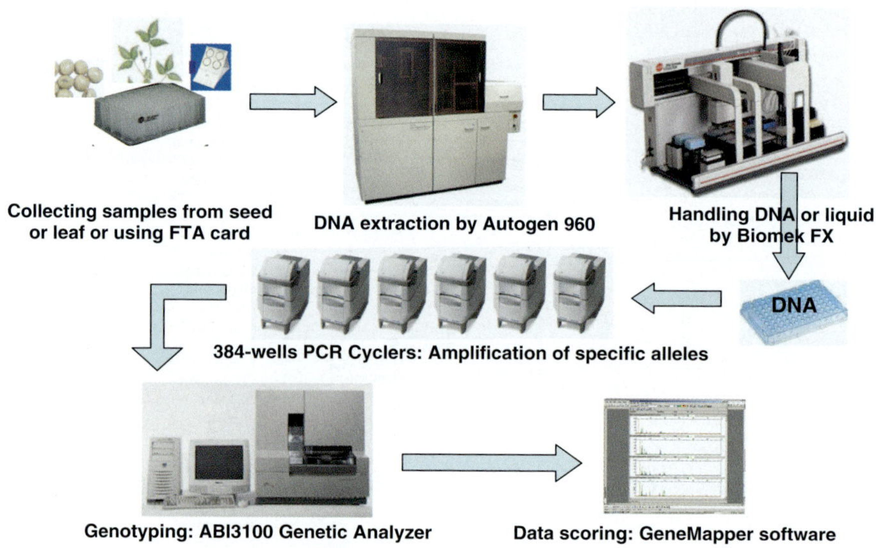

Fig. 8.2 High throughput MAS operation at the Soybean Genome Mapping Lab of the University of Missouri-Columbia

(A)

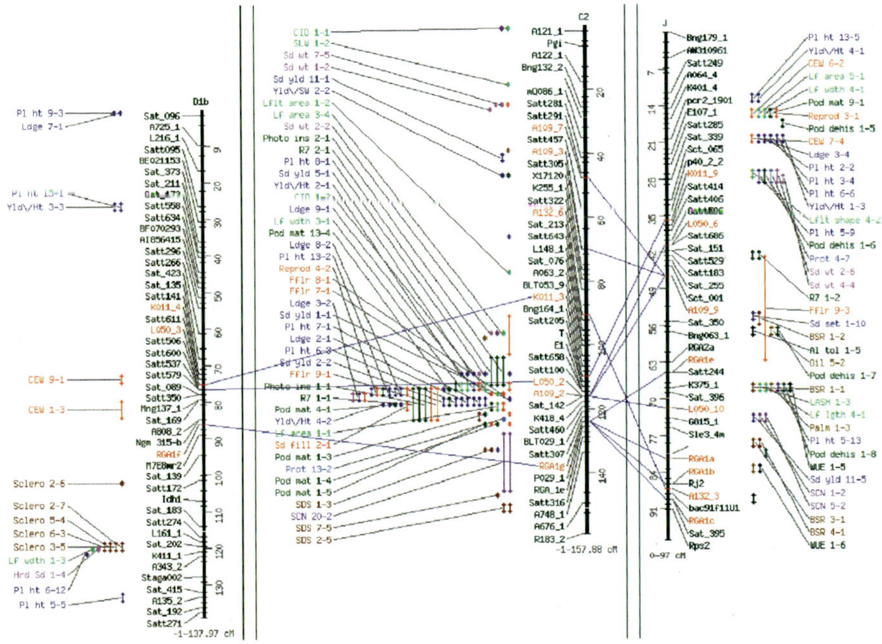

Fig. 10.1 Comparative Map Display Capabilities of the Soybean Breeder's Toolbox. **Panel A**. Comparison of molecular linkage groups D1b, C2 and J. Thick black vertical lines represent the linkage group genetic maps. The positions of molecular markers are designated by black horizontal hashes with their identifiers linked by black lines. Molecular markers that are in common between any two linkage groups are highlighted in red with blue lines connecting them. The positions of various QTL are identified by colored vertical dumbbell shapes. The colors used for QTL group QTL of similar phenotype thus all QTL for fungal resistance are in brick red, resistance to bacterial pathogens are in orange, oil related QTL are in gold etc. A listing of all color codes is included in each CMap legend. The comparison of linkage groups with many mapped QTL can be quite difficult to interpret. CMap allows the user to display individual types of markers in order to make the interpretation of maps with dense markers easier. **Panel B** depicts the same comparison with only QTL for insect, bacterial and fungal visible for ease of interpretation

(B)

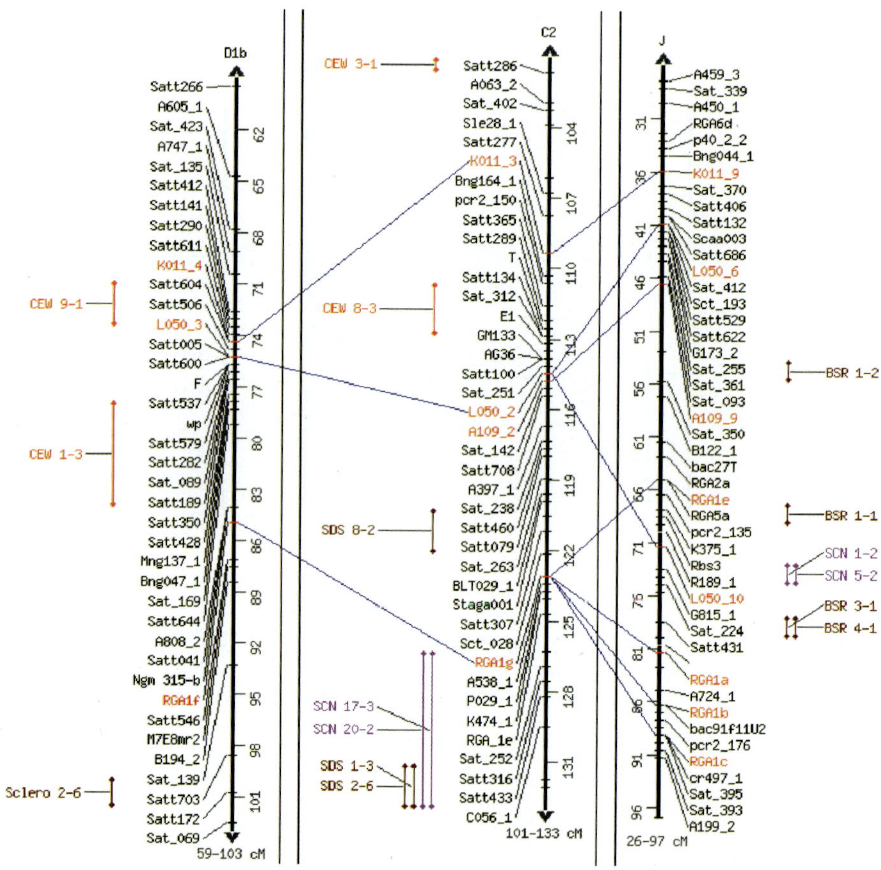

Fig. 10.1 (Continued)

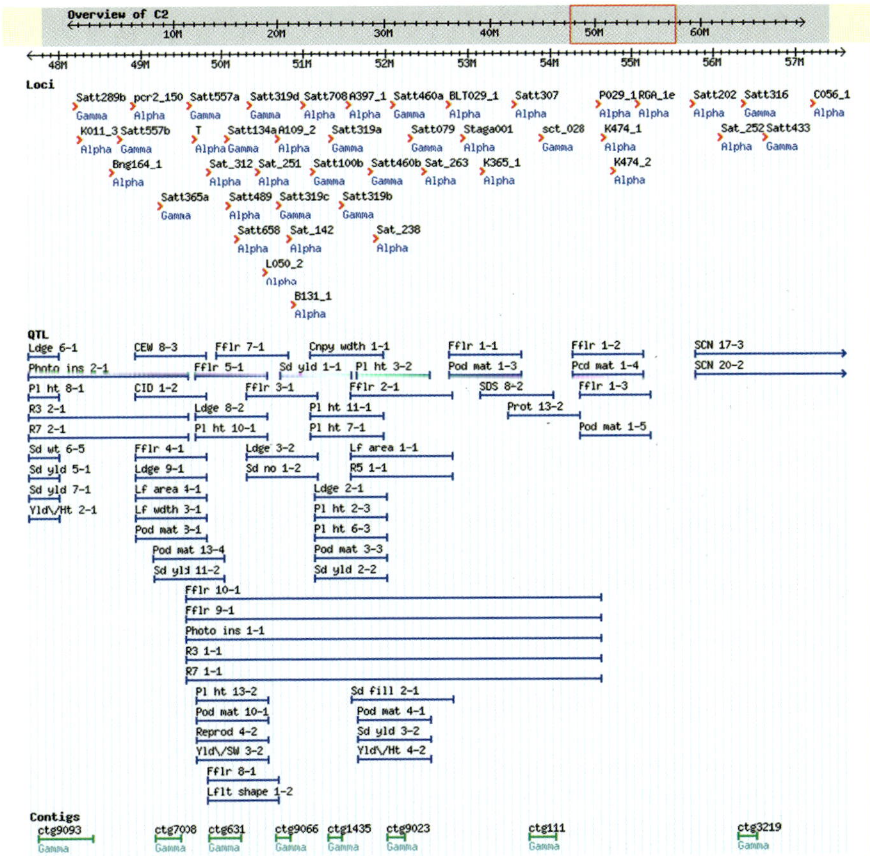

Fig. 10.2 Example of the Forrest Physical Map Display and Gbrowse. An example of the Forrest Physical Map displayed using the Gbrowse display engine. Pictured is a 9.9 Mbp section of molecularlinkage group C2 around the *T* locus. The black horizontal line at the top of the display in the gray box represents the whole linkage group. The region in the red box represents the segment displayed in the light box below it. In the light box with light blue rules, we can see the representation of the linkage group as a black horizontal line ruled in 100 kbp segments. Under the horizontal lines are different "tracks" containing different types of features. Displayed are loci, QTL and BAC contigs (Contigs). More types were available but not displayed for ease of interpretation. As indicated, the Gbrowse engine was designed as a map display and annotation tool thus it allows the presentation of a single map at a time. Features in common with other maps or species could be displayed as a new set of "tracks". Interpretation of synteny or genomic context between the comparisons would be difficult because the different maps could not be directly visualized together as they could be in a CMap display

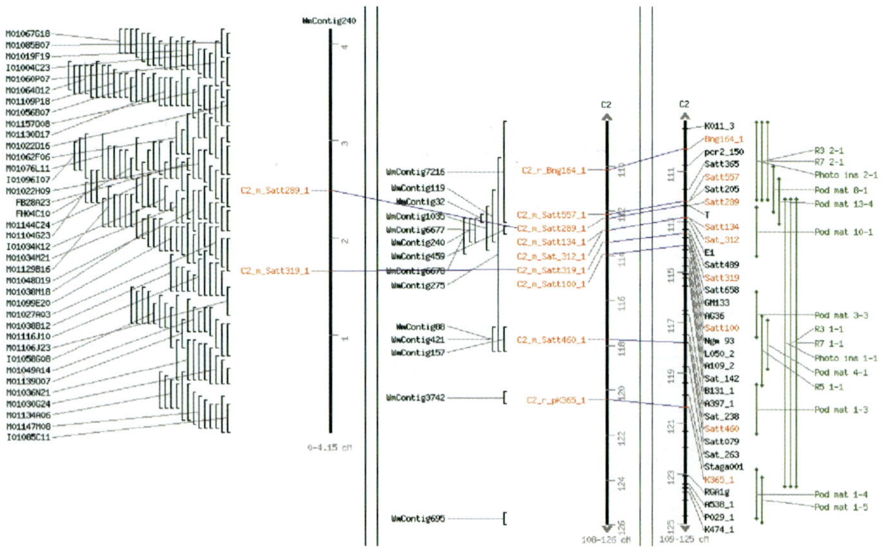

Fig. 10.3 Example of the Williams Physical Map Display using CMap. Pictured is a view of molecular linkage group C2 similar to that depicted in Fig. 10.2 in the Forrest Physical Map. The genetic map is drawn on the right with only QTL for pod maturity visible. The physical map positions and approximate sizes of the Williams physical map contigs are displayed in the middle. On the left is an explosion of a physical map contig that may contain the *T* locus (WmContig240) based on the presence of markers Satt289 and Satt319. Individual BAC clones mapped with those markers are indicated in green (Satt289) and light orange (Satt319). Thus these BACs could serve as a starting point for the fine scale mapping of the *T* locus

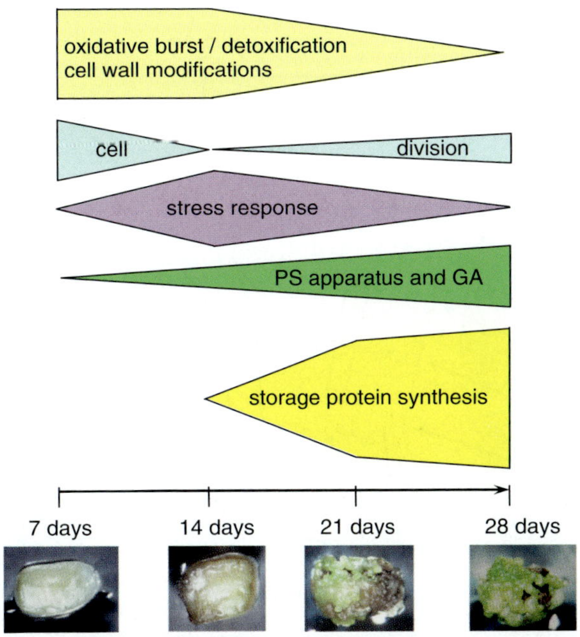

Fig. 11.2 Transcript patterns associated with the early induction phase of somatic embryogenesis in the adaxial sides. The timescale represents the number of days the cotyledons were on high auxin media. The figures underneath illustrate the appearance of the sampled tissue at each timepoint. The heights of the blocks represent the relative levels of the transcript expression (From Thibaud-Nissen et al. 2003 ©American Society Plant Biologists)

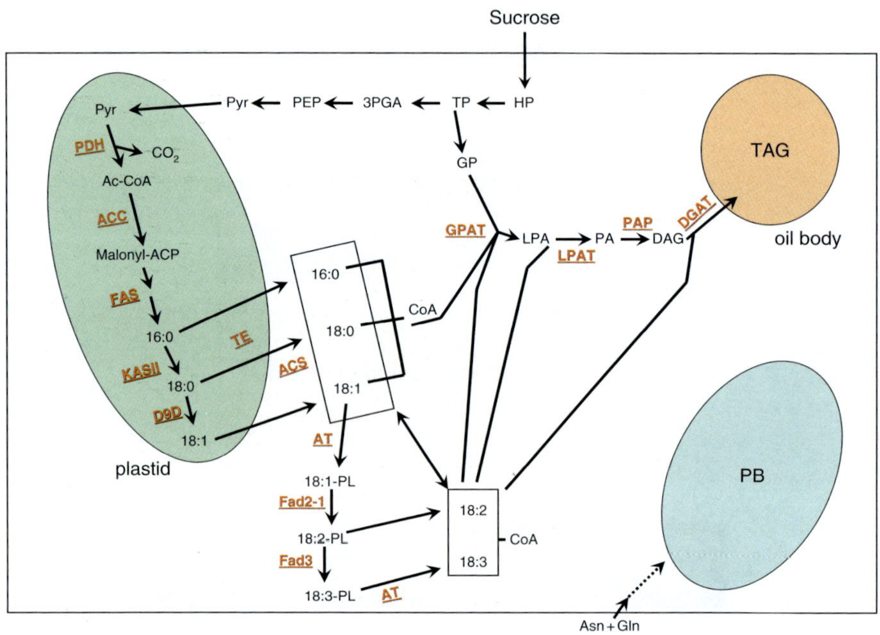

Fig. 12.3 Summary of some of the major steps of oil or triacylglycerol (TAG) biosynthesis in plant cells including developing soybean seeds. PB = protein body. The enzymes discussed are underlined and described in Table 12.2

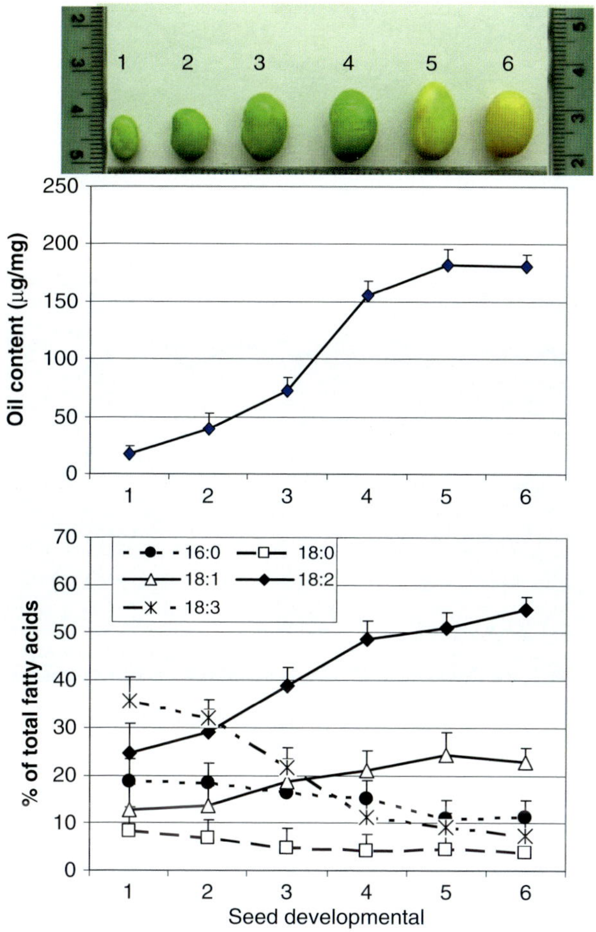

Fig. 12.4 Oil and fatty acid accumulation in soybean seed development

Fig. 15.1 Soybean (*right*) showing antixenotic resistance towards SBL

Fig. 15.2 Corn earworms (*right*) showing an antibiotic effect on their growth

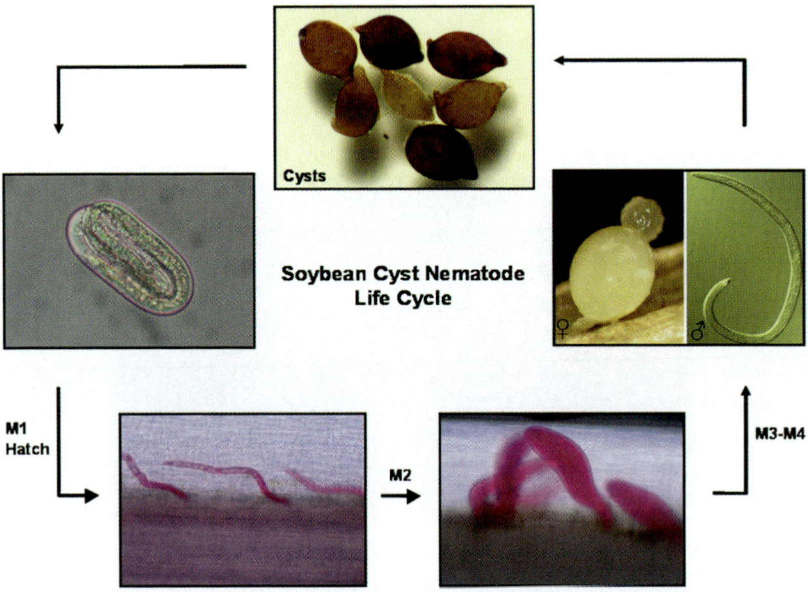

Fig. 17.1 Life cycle of *Heterodera glycines*, the soybean cyst nematode. Second-stage juveniles (J2) hatch from eggs in the soil and are attracted to soybean roots. The J2 penetrates into the roots and migrates towards the vascular cylinder where it selects a cell to initiate the formation of a syncytium. The nematode proceeds through three more molts to the adult male and female life stages. Following fertilization, the adult female secretes a small number of eggs in a gelatinous matrix outside her body while the majority of the eggs are retained within the female uterus. The dead female body forms the cyst to protect the eggs in the soil until favorable conditions arise. The average life cycle takes 25–30 days for completion under optimal conditions. M = molt

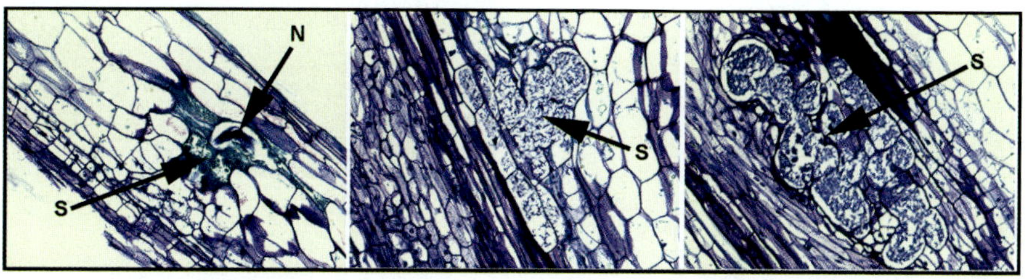

Fig. 17.2 Thin sections showing development of a syncytium at 2, 5, and 10 days post-inoculation of soybean roots with soybean cyst nematode second-stage juveniles. Reproduced with permission from Ithal et al., 2007a, Parallel genome-wide expression profiling of host and pathogen during soybean cyst nematode infection of soybean, *Molecular Plant-Microbe Interactions* 20: 293–305

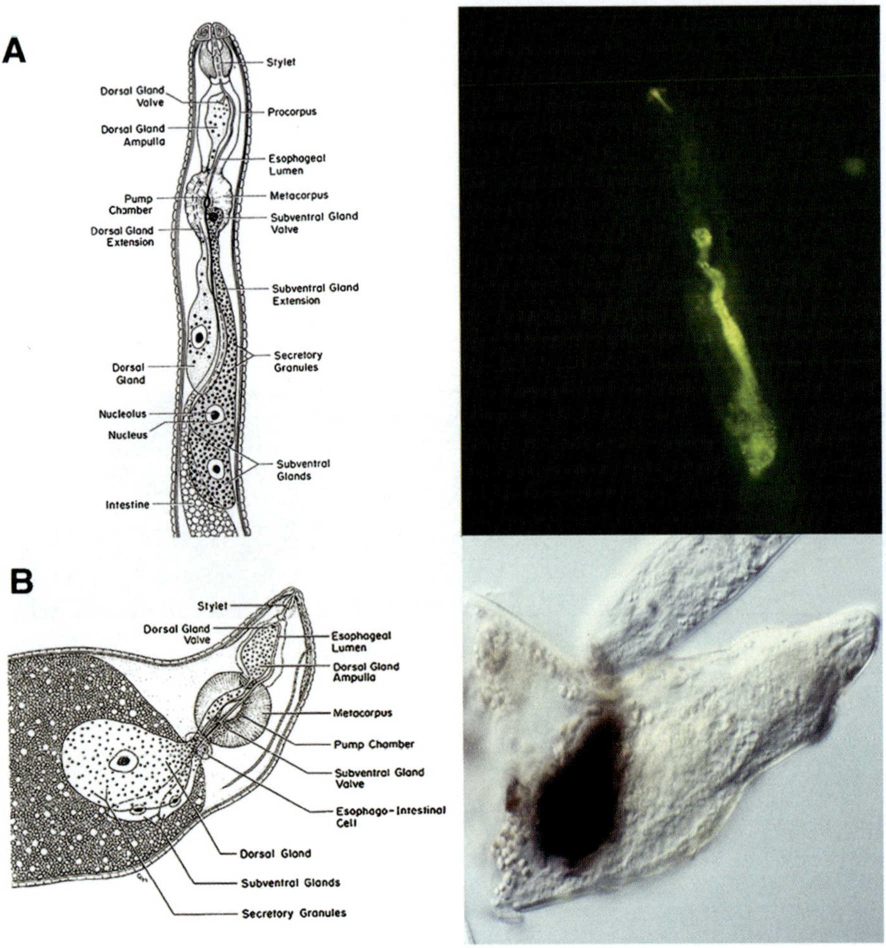

Fig. 17.3 Illustrations of the three soybean cyst nematode esophageal gland cells during second-stage juvenile (**A**, *left panel*) and adult female (**B**, *left panel*) life stages and corresponding parasitism gene expression confirmed by either immunolocalization of the protein to the two subventral glands (**A**, *right panel*) or an in situ hybridization to detect mRNA accumulation in the dorsal gland (**B**, *right panel*). Immunolocalization and in situ pictures courtesy of Eric Davis and Richard Hussey, respectively. Drawings reprinted, with permission, from the *Annual Review of Phytopathology,* Volume 27 ©1989 by Annual Reviews, www.annualreviews.org

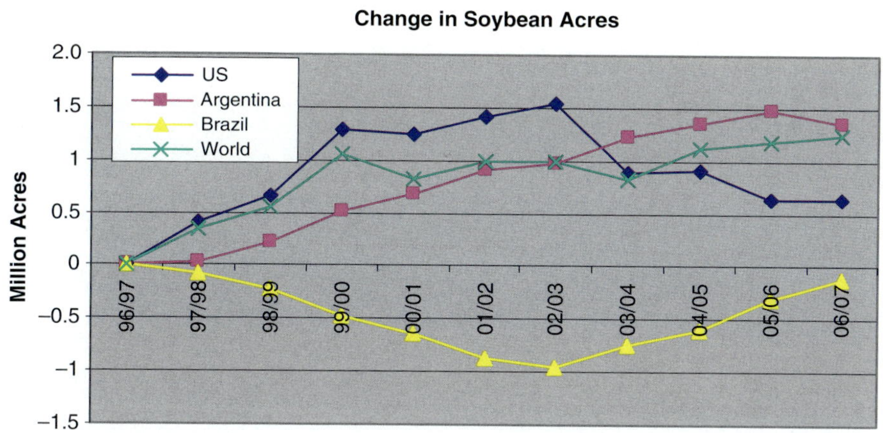

Fig. 19.2 Change in soybean acres worldwide due to the introduction of RR soybeans, 1996–2006

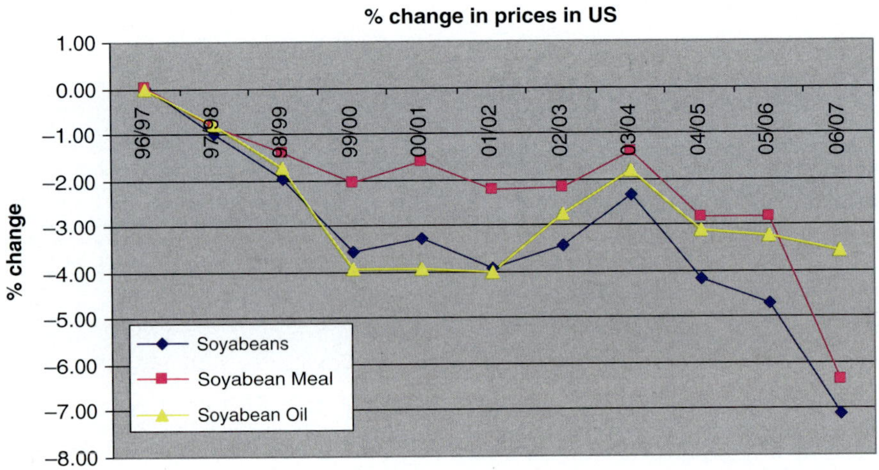

Fig. 19.3 Percent change in US soybean, soyoil and soymeal prices due to the introduction of RR soybeans, 1996–2006

Chapter 17
Genomics of the Soybean Cyst Nematode-Soybean Interaction

Melissa G. Mitchum and Thomas J. Baum

Introduction

The soybean cyst nematode (SCN), *Heterodera glycines*, is a problem worldwide wherever soybean is grown and consistently is the most economically important pathogen of soybean in the United States. It is estimated that this microscopic roundworm causes nearly $1 billion annually in yield losses to US soybean producers. SCN-resistant cultivars serve as the primary means of management; however, genetic heterogeneity of SCN populations allows this pathogen to readily overcome resistance. In addition, a lack of understanding of nematode virulence has hampered the ability of researchers to devise novel management tactics. In recent years, however, an increase in genomic analyses of both soybean and *H. glycines* has begun to provide the necessary tools for developing a better understanding of this complex plant-nematode interaction. Expressed sequence tags (ESTs), generated from cDNA libraries representing different *H. glycines* and soybean developmental stages are a powerful tool for identifying genes important in soybean-SCN interactions. More than 22,000 *H. glycines* and 350,000 soybean ESTs have been deposited to publicly available databases. Moreover, the soybean and *H. glycines* EST datasets were used to develop microarray platforms representing 7,530 SCN and 35,611 soybean probe sets for comprehensive profiling of nematode and plant gene expression changes during parasitism. Soybean and SCN genetic and physical maps are under development and genome sequencing projects are underway for both organisms. These community efforts promise to provide both the tools and genome-wide catalogues of soybean and nematode genes for both functional and comparative analysis that should reveal additional, novel insight into mechanisms of *H. glycines* parasitism of soybean. In this chapter, recent genomics advances in our understanding of the soybean-SCN interaction are highlighted.

M.G. Mitchum
Christopher S. Bond Life Sciences Center, Division of Plant Sciences, University of Missouri-Columbia, MO 65211, USA
e-mail: goellnerm@missouri.edu

G. Stacey (ed.), *Genetics and Genomics of Soybean*,

SCN Biology

The soybean cyst nematode is an obligate sedentary endoparasite. Sedentary endoparasitic nematodes penetrate host roots and migrate to the vascular tissue where they become sedentary and begin feeding (Hussey and Grundler, 1998). As with all plant-parasitic nematodes, SCN uses a hollow, protrusible stylet to pierce the plant cell wall, secrete proteins produced in the esophageal gland cells directly into host roots, and withdraw cellular contents. The details of the soybean cyst nematode life cycle (reviewed in Niblack et al., 2006) are depicted in Fig. 17.1 and are briefly described here. Infective second-stage juveniles (J2) hatch from eggs in the soil and locate host plant roots through attraction to diffusates. J2 mechanically penetrate the cell wall using their stylets while secreting cell wall-hydrolyzing enzymes to facilitate migration towards the root vasculature. Once the juvenile reaches the vasculature of the root, it selects an individual cell to begin the formation of a unique feeding structure called a syncytium (Fig. 17.2), which consists of hundreds of fused and metabolically highly active root cells. As the juvenile feeds, its body swells and at this point in its life cycle the parasitic juvenile is completely dependent

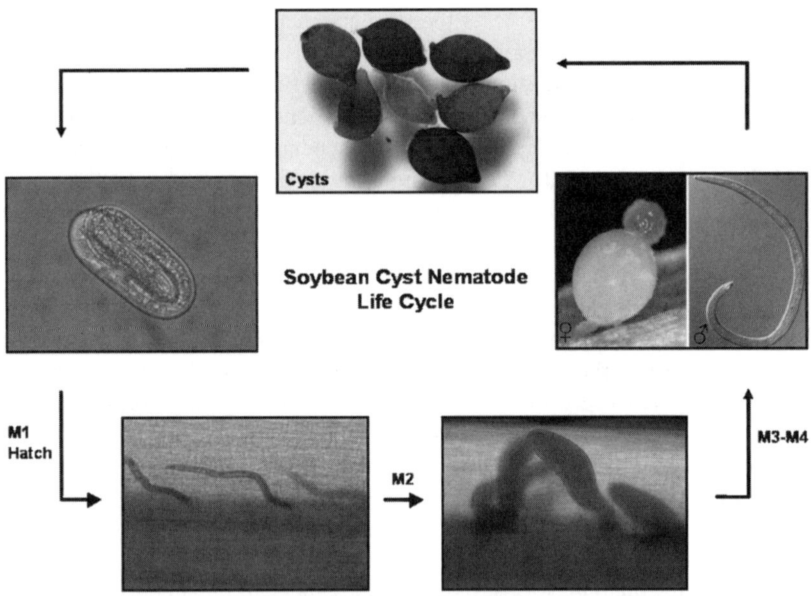

Fig. 17.1 Life cycle of *Heterodera glycines*, the soybean cyst nematode. Second-stage juveniles (J2) hatch from eggs in the soil and are attracted to soybean roots. The J2 penetrates into the roots and migrates towards the vascular cylinder where it selects a cell to initiate the formation of a syncytium. The nematode proceeds through three more molts to the adult male and female life stages. Following fertilization, the adult female secretes a small number of eggs in a gelatinous matrix outside her body while the majority of the eggs are retained within the female uterus. The dead female body forms the cyst to protect the eggs in the soil until favorable conditions arise. The average life cycle takes 25–30 days for completion under optimal conditions. M = molt (*See* also Color Insert)

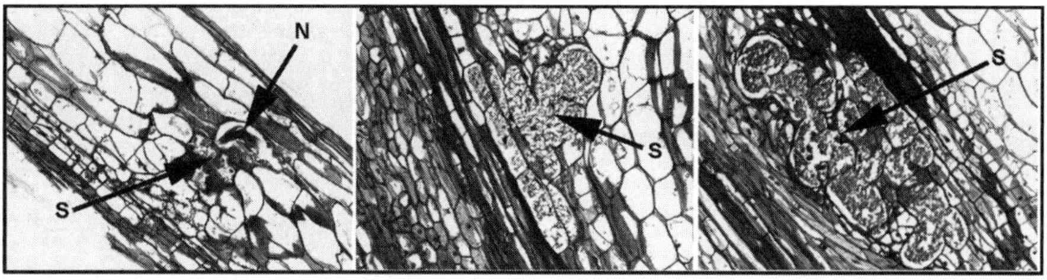

Fig. 17.2 Thin sections showing development of a syncytium at 2, 5, and 10 days post-inoculation of soybean roots with soybean cyst nematode second-stage juveniles. Reproduced with permission from Ithal et al., 2007a, Parallel genome-wide expression profiling of host and pathogen during soybean cyst nematode infection of soybean, *Molecular Plant-Microbe Interactions* 20: 293–305 (*See* also Color Insert)

on the successful formation of feeding cells to obtain the nutrients necessary to meet the increased energy demands required to proceed through three more molts to the adult male or female life stages. The soybean cyst nematode reproduces by obligate amphimixis, which requires that adult males regain their vermiform shape and motility and migrate out of the root to fertilize females protruding from the root surface. Fertilized females produce up to a few hundred eggs, most of which are retained within the uterus. When the female dies, her body serves as a cyst to protect the eggs in the soil from adverse environmental conditions in the absence of a host for many years until conditions are favorable.

SCN has two subventral (SvG) and a single dorsal (DG) esophageal gland cell (Fig. 17.3). Proteins originating in the SCN esophageal gland cells are secreted through the nematode stylet into host roots to metabolically and developmentally reprogram normal soybean root cells for the formation of syncytia (Fig. 17.2; Davis et al., 2000, 2004; Baum et al., 2007). Syncytia serve as major sinks for metabolites that are withdrawn by the feeding nematodes using their stylets. The syncytium forms through coordinated dissolution of plant cell walls, a process that is thought to be mediated by proteins of plant origin (Goellner et al., 2001). Protoplast fusion of adjacent cells results in the formation of a multinucleate syncytium made up of hundreds of cells. The nuclei within the syncytium are polyploid due to repeated rounds of endoreduplication and take on an amoeboid shape with a prominent nucleolus. The large central vacuole is dispersed into several smaller vacuoles, organelles proliferate, cell walls thicken, and there is an observed increase in cytoplasmic density (Hussey and Grundler, 1998). Cell wall ingrowths form along walls adjacent to the vasculature, typical of transfer cells, which increases the plasma membrane surface area for nutrient uptake (Jones and Northcote, 1972).

SCN Parasitism Gene Identification

The observed physiological and molecular changes of host root cells that occur during nematode feeding cell formation have long been considered to be a direct result of proteins secreted from the nematode stylet into plant tissues (Williamson and

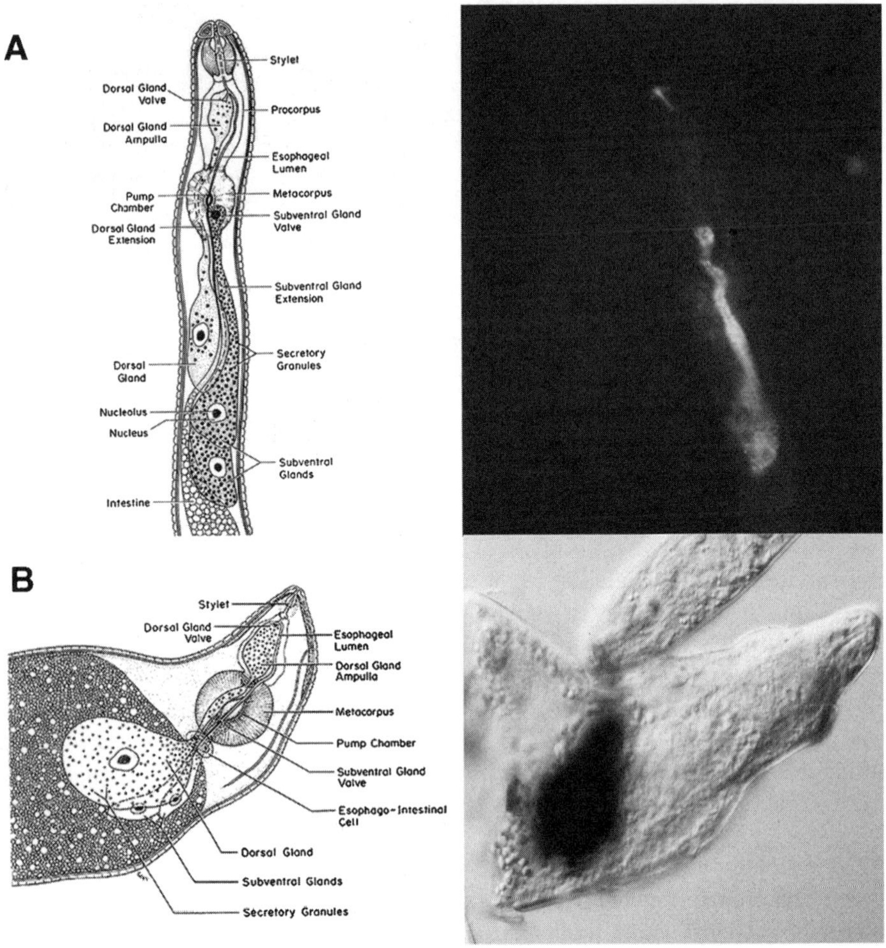

Fig. 17.3 Illustrations of the three soybean cyst nematode esophageal gland cells during second-stage juvenile (**A**, *left panel*) and adult female (**B**, *left panel*) life stages and corresponding parasitism gene expression confirmed by either immunolocalization of the protein to the two subventral glands (**A**, *right panel*) or an in situ hybridization to detect mRNA accumulation in the dorsal gland (**B**, *right panel*). Immunolocalization and in situ pictures courtesy of Eric Davis and Richard Hussey, respectively. Drawings reprinted, with permission, from the *Annual Review of Phytopathology,* Volume 27 ©1989 by Annual Reviews, www.annualreviews.org (*See* also Color Insert)

Hussey, 1996). Stylet-secreted proteins are encoded by nematode parasitism genes that are expressed in the three esophageal gland cells. In general, parasitism genes are defined as genes that encode secretions from a nematode with a direct role in parasitism (Davis et al., 2004). Over the course of the last decade, a combination of genomic and proteomic approaches led to tremendous advancements in identifying nematode parasitism gene candidates (PGC) (Davis et al., 1994; Goverse et al., 1994; Hussey et al., 1990; Ding et al., 1998; Lambert et al., 1999; Ding et al., 2000;

Qin et al., 2000; de Meutter et al., 2001; Gao et al., 2001a,b; Wang et al., 2001; Jaubert et al., 2002; Grenier et al., 2002; Gao et al., 2003; Huang et al., 2003, 2004; Tytgat et al., 2004; Blanchard et al., 2005; Vanholme et al., 2005), resulting in the identification of key molecules involved in parasitism.

Direct targeting of the esophageal gland cells has proven to be the most successful strategy for the identification of SCN PGCs (Wang et al., 2001; Gao et al., 2001a, 2003). Using a microaspiration approach, the gland cell contents from SCN parasitic life stages were extracted and used to generate gland-enriched cDNA libraries. Sequencing and bioinformatic analyses of the gland-enriched cDNA libraries led to the identification of proteins with predicted secretion signal peptides. Following confirmation of gland-specific expression by in situ mRNA hybridization or immunolocalization (Fig. 17.3), more than 60 SCN parasitism gene candidates were identified (Wang et al., 2001; Gao et al., 2001a, 2003). Significantly, this provided the first snapshot of the SCN parasitome, i.e., the transcriptome of the esophageal gland cells coding for putative stylet-secreted proteins.

Bioinformatic Analysis of SCN Parasitism Gene Sequences

Sequence annotations of the more than 60 SCN parasitism gene candidate sequences identified to date, predict functions for a small subset of the protein products including those with roles in cell wall degradation, metabolism, and molecular mimicry of host proteins involved in cellular signaling and protein degradation pathways. Readers are referred to several comprehensive reviews of this subject (Davis et al., 2004; Davis and Mitchum, 2005; Baum et al., 2007). Interestingly, more than 70 % of the identified PGCs lack homology with any sequences in existing databases and were termed "pioneers" (Wang et al., 2001; Gao et al., 2001a, 2003). The following sections provide a brief review of some of the predicted roles of SCN parasitism proteins, which have provided new insight into our understanding of SCN parasitism with regard to cell wall modification during infection, metabolic and developmental reprogramming of host cells, and manipulation of host defense mechanisms. Readers are referred to Baum et al. (2007) for a detailed description of the molecular approaches currently being used to advance our understanding of SCN parasitism protein function.

SCN Parasitism Protein Function

Cell Wall Modification

The structural barrier of the plant cell wall presents a formidable challenge to an SCN infective J2. However, the nematode is well-equipped with a stylet for mechanical penetration and the ability to secrete an array of cell wall degrading enzymes (CWDEs) to facilitate penetration and migration through host root tissues. Genes encoding beta-1,4-endoglucanases (ENGs; EGases) and pectate lyases were cloned from SCN (Smant et al., 1998; Yan et al., 1998, 2001; De Boer et al., 2002; Gao et al., 2002) and secretion of EGases was detected along the migratory path of juveniles invading host roots (Wang et al., 1999). Expression of the SCN *ENG* genes was shown to be developmentally regulated. That is, *ENG* expression is high in juveniles within eggs prior to hatching and in hatched juveniles prior to root penetration. However, expression declines substantially in parasitic juvenile stages as the nematode establishes a feeding site suggesting a minimal role, if any, in syncytium formation (de Boer et al., 1999; Gao et al., 2004a). The *ENG* genes are also expressed in adult vermiform males that regain mobility for migration out of the root to fertilize the females (de Boer et al., 1999). Gene sequences coding for secreted cellulose binding domain (CBD) proteins were also identified in SCN (Gao et al., 2003; Gao et al., 2004b). However, the function of the SCN CBD remains unknown. Recombinant plant and bacterial CBDs, however, were shown to modulate cell elongation and growth in vitro (Shpigel et al., 1998; Safra-Dassa et al., 2006).

Despite the mounting evidence that plant-parasitic nematodes secrete a complex mixture of CWDEs to facilitate penetration and migration through host root tissues, there is very little evidence to support a role for these proteins in the induction and formation of feeding cells. On the contrary, cell wall modifications within feeding cells appear to be the result of direct or indirect activation of plant cell wall modifying proteins (CWMPs) including EGases, expansins, pectin acetylesterases, and xyloglucan endotransglycosylases (Goellner et al., 2001; Vercauteren et al., 2002; Wieczorek et al., 2006; Ithal et al., 2007a,b; Wang et al., 2007) that likely act in a controlled, coordinated fashion to regulate feeding cell formation.

Metabolic Reprogramming

Once a SCN juvenile has selected a cell near the vasculature to feed from, there is evidence to suggest that it secretes molecules to redirect host cell metabolism to establish a syncytium. Chorismate mutase, an enzyme of the shikimate pathway, was shown to be expressed within the esophageal gland cells of SCN (Bekal et al., 2003; Gao et al., 2003). The shikimate pathway in plants produces essential aromatic amino acids required by the nematode that can only be obtained from their diet. Chorismate mutase converts plastid-derived chorismate to prephenate to provide precursors for the synthesis of the aromatic amino acids, phenylalanine, tyrosine and tryptophan. The aromatic amino acids can then serve as precursors for the production of molecules with defined roles in plant-microbe interactions including auxin, flavonoids, salicyclic acid, and phytoalexins. A cytosolic branch of the shikimate pathway for the production of aromatic amino acids from chorismate also exists. Therefore, nematode secretion of chorismate mutase into the host cytoplasm may be used to developmentally reprogram or suppress the accumulation of plant defense compounds produced in the plastidal branch of the shikimate pathway by increasing flux through the cytosolic branch. Consistent with this hypothesis, overexpression of a root-knot nematode (RKN) chorismate mutase gene in soybean hairy roots resulted in phenotypes characteristic of auxin deficiency, including lateral root termination and impaired vasculature development (Doyle and Lambert, 2003). These phenotypes could be rescued by application of exogenous IAA suggesting that the transgenic roots overexpressing the RKN chorismate mutase were indeed auxin deficient. Additional studies will be needed to determine exactly how SCN may be altering the regulation of the shikimate pathway and potentially other metabolic pathways to establish a compatible interaction.

Ligand Mimicry

The dedifferentiation of soybean cells into syncytia likely requires an exchange of signals between SCN and the initial cell selected to establish a feeding site. Recent evidence suggests that SCN has evolved secreted mimics of plant signals to reprogram host cell development for syncytium formation providing the first insight into the "putative" signal or signals that are exchanged between soybean and SCN. The first nematode gene encoding a putative mimic of a plant protein was identified from SCN (Wang et al., 2001). The gene coded for a small, secreted protein that contained a C-terminal domain with similarity to plant CLAVATA3/ESR-related (CLE) peptides (Olsen and Skriver, 2003). The SCN CLE was shown to be expressed in the dorsal gland cell (Wang et al., 2001) when the nematode is actively initiating or feeding from syncytia and is either absent or expressed at extremely low levels in non-feeding life stages (Wang et al., 2006) supporting a role for CLEs in syncytium induction, function, or maintenance. Constitutive overexpression of the SCN CLE gene in Arabidopsis produced a *wuschel* (*wus*)-like phenotype (Brand et al., 2000; Hobe et al., 2003) characteristic of CLV3 overexpression (Wang et al., 2005). Consistent with this result, the SCN CLE could complement a *clv3* mutant (Wang et al., 2005). These studies suggest that nematode CLE peptides may have functional similarity to plant CLE peptides. More recent data indicate that there are two predominant forms of SCN CLEs that differ only in a variable domain immediately upstream of the CLE domain (Gao et al., 2003; Wang et al., 2006). Interestingly, these two SCN CLEs give different overexpression phenotypes in Arabidopsis (Wang et al., 2006) suggesting that they may have evolved different functions. Significantly, this work documents the first identification of a functional CLE domain outside of the plant kingdom and suggests that nematodes may be co-opting host developmental programs for syncytium formation through ligand mimicry (Wang et al., 2005).

Molecular Mimicry of Host Ubiquitination Pathway Components

Candidate SCN parasitism genes coding for secreted proteins with homology to ubiquitin extension, S-phase kinase associated protein 1 (SKP-1)-like, and RING-H2 proteins were also identified (Gao et al., 2003; reviewed in Davis et al., 2004).

Interestingly, these proteins are components of the ubiquitination pathway in eukaryotes; a pathway that targets proteins for degradation and is used as a mechanism for post-translational gene regulation. Recently, it was shown that a number of different pathogens can deliver effectors into host plant cells that can mimic the function of components of host E3 ubiquitin ligase complexes. Ubiquitination-mediated protein degradation was shown to play a role in regulating defense responses in plants and the secretion of host E3 ubiquitin ligase complex mimics may be one mechanism used by pathogens to manipulate plant defense responses to their own advantage (Zeng et al., 2006). The identification of SCN-secreted proteins with homology to proteins of the E3 ubiquitin ligase complex suggests that plant-parasitic cyst nematodes may also be equipped with the ability to mimic the host ubiquitination pathway to establish successful parasitic associations.

(A)virulence Proteins

Bona fide avirulence proteins have not been identified from plant-parasitic nematodes. However, effector molecules coded for by nematode parasitism genes are strong avirulence protein candidates. Recent hypotheses suggest that variants of plant-pathogen effector molecules with a central role in virulence function as avirulence factors depending on the genetic context of the host plant (Birch et al., 2006). Recent data indicate that the *H. glycines* chorismate mutase gene (*Hg-cm-1*) may play a role in virulence. Polymorphisms in *Hg-cm-1* were identified that correlate with virulence on a set of soybean cyst nematode-resistant soybean lines (Bekal et al., 2003) and the *Hg-cm-1A* allele was preferentially selected on the germplasm PI88788, the most common source of soybean cyst nematode resistance (Lambert et al., 2005). Similarly, polymorphisms were identified in the SCN *CLE* genes among soybean cyst nematode inbred populations that differ in their ability to parasitize resistant soybean implicating these proteins in virulence (Wang et al., 2006). In addition, small cysteine-rich proteins (<150aas), well known for their role as avirulence factors and/or elicitors of plant defense responses (Birch et al., 2006), are among the gland-expressed parasitism gene candidates (Gao et al., 2003).

SCN Genomics

In addition to targeting the identification and function of a select set of nematode genes, such as the parasitism genes described above, large-scale genomic approaches to elucidate various aspects of SCN biology and host plant interactions are well underway. The recent construction of SCN life stage-specific cDNA libraries provided the template that generated a collection of 20,100 expressed sequence tags (ESTs) to add to the existing SCN sequences previously deposited in GenBank (Parkinson et al., 2003, 2004; Elling et al., 2007). These publicly known EST sequences form 6,860 contigs, which likely reflect close to a third of all SCN genes assuming the number of genes in SCN to be conserved with that of the free-living nematode *Caenorhabditis elegans* (~19,000 genes). Publicly available EST sequences were used by Affymetrix Inc. to develop the Soybean Genome Array GeneChip, which contains 7,530 SCN probe sets in addition to 37,593 soybean and 15,800 *Phytopthora sojae* probe sets. Use of this genomic resource for the study of SCN biology was first reported by Ithal et al. (2007a) who established expression profiles for known SCN parasitism genes during early events of the SCN life cycle. Subsequently, the Soybean Genome Array Gene Chip was used to determine the first profile of global gene expression changes throughout all major SCN life stages (eggs, infective J2, parasitic J2, J3, J4, and virgin females), excluding adult males (Elling et al., 2007). The authors demonstrated the utility of the dataset by comparing the mechanism of arrested development in SCN infective juveniles with that of the dauer stage of *C. elegans* (Elling et al., 2007). Undoubtedly, this dataset provides a valuable community resource to dissect a variety of other aspects of SCN biology and comparative genomics of nematodes (Mitreva et al., 2005). Several groups also initiated the development of genetically homogeneous SCN populations, molecular markers, and genetic maps to facilitate map-based cloning efforts (Dong and Opperman, 1997; Atibalentja et al., 2005). SCN BAC libraries were constructed and are being used to develop a SCN physical map anchored to the genetic map (Lambert et al., 2006) providing a solid framework for the public SCN genome sequencing projects that are in progress (Lambert et al., 2006). These efforts will no doubt distinguish SCN as an invaluable model plant-parasitic cyst nematode.

Genetic Analysis of the Soybean-SCN Interaction

Our understanding of SCN virulence (i.e., the ability of SCN to evolve to either evade or overcome host plant resistance) is limited and continues to present a challenge for SCN management which relies heavily on natural host plant resistance. Extensive genetic variability exists within field populations of SCN. This variability is currently described as HG types (for <u>H</u>*eterodera* <u>g</u>*lycines* type) which define SCN populations' virulence on a set of seven resistant soybean indicator lines (Niblack et al., 2002, 2006). Genetic analysis of both soybean and SCN suggest that there are multiple genes in both the plant and nematode for resistance and virulence, respectively. However, on a genetic level there are several examples of plant-nematode interactions that were shown to follow the gene-for-gene hypothesis. This was demonstrated for the potato cyst nematode, *G. rostochiensis*, interaction with potato carrying the *H1* resistance gene (Janssen et al., 1991) and the root-knot nematode, *Meloidogyne incognita*, interaction with tomato carrying the *Mi* resistance gene (Milligan et al., 1998). Several plant resistance genes that confer resistance to plant-parasitic nematodes were cloned and characterized (Cai et al., 1997; Milligan et al., 1998; van der Vossen et al., 2000; Ernst et al., 2002; Paal et al., 2004). Despite these successes, soybean genes that confer resistance to SCN have not been cloned. Nevertheless, multiple quantitative trait loci (QTLs) conferring resistance to individual SCN populations in numerous soybean germplasm sources were mapped (reviewed in Concibido et al., 2004). Two major QTLs controlling resistance to SCN that were mapped in multiple germplasm sources are located at the *rhg1* locus on linkage group G and the *Rhg4* locus on linkage group A2. Resistance to SCN in the soybean cultivar Forrest was shown to be bigenic, requiring both *Rhg1* and *Rhg4* (Meksem et al., 2001). *Rhg1* is codominant and *Rhg4* is dominant. To achieve full resistance, at least one copy of the Forrest *Rhg4* allele and two copies of the Forrest *Rhg1* allele are required. If one copy of the Forrest *Rhg1* allele is present, only partial resistance is conferred. Fine mapping of both loci is currently underway and a number of candidate genes for resistance have been identified (Meksem and Lightfoot, personal communication). Complementation and reverse genetic approaches in soybean will be required to confirm their roles in resistance to SCN. Similarly, positional cloning of other soybean QTLs for SCN resistance is underway (see Chapter 8 in this volume). The cloning and characterization of soybean genes for resistance to SCN will provide a much needed understanding of the molecular basis of soybean resistance to SCN.

The identification of the corresponding avirulence genes from nematodes, including SCN, using map-based cloning approaches has not been trivial. The inability to transform plant-parasitic nematodes has slowed progress in this area compared with other plant pathogens. In addition, the obligate nature of plant-parasitic nematodes and a lack of tools for genetic mapping have presented two major setbacks for using forward genetic strategies for the identification of SCN avirulence genes. Consequently, this approach has only been used in a few studies to date. A classical genetic approach identified several dominant and recessive *H. glycines ror* (for *r*eproduction *o*n a *r*esistant host) genes in SCN inbred lines carrying fixed genes for parasitic ability on soybean (Dong and Opperman, 1997; Dong et al., 2005). However, the SCN genes controlling parasitism on resistant soybean at these loci were not cloned. As mentioned earlier, the recent development of a collection of SCN genomic tools will certainly facilitate future map-based cloning efforts (Atibalentja et al.,2005). The utility of the SCN genetic map was already demonstrated by mapping the *Hg-Cm-1* gene (Atibalentja et al., 2005). Thus, researchers will soon be equipped with both the genomic and genetic tools necessary to identify the genes controlling a variety of different traits such as nematode virulence and host range providing essential knowledge for both the development and deployment of nematode resistant soybean.

Genomics of the Soybean Response to SCN Parasitism

The complex plant response to SCN infection spans the early migratory stages during penetration and migration through roots into the later sedentary stages of syncytium induction and feeding, which ultimately facilitates nematode development and reproduction. During the early migratory phase of the nematode life cycle in either a susceptible or resistant host, SCN penetrates through the epidermis of the root and migrates intracellularly towards the root vasculature, leaving in its wake, a path of damaged tissue. Transition from the migratory phase to the sedentary phase of the nematode life cycle occurs during syncytium induction. Syncytium formation is the result of and accompanied by drastic changes in plant gene expression giving rise to a cell type that is so far unique to the plant-nematode interaction. These localized cellular changes near the head of the nematode make up only a small fraction of the soybean root tissue. In nematode-resistant plants, syncytium development is compromised such that syncytia degenerate within just days after nematodes initiate feeding. As a result, nematode development is slowed and eventually terminated (Endo, 1965). Just the opposite occurs in susceptible host roots where well-developed syncytia provide the nutritional requirements of developing nematodes. With each molt, the nematode body expands putting stress on the surrounding cells that likely incites additional plant responses. Ultimately, adult females break through the epidermis of the root while adult males migrate back out of the root for copulation.

The complexity of the host plant response to cyst nematodes and the uniqueness of the feeding cells that are initiated have inspired numerous studies to identify plant gene expression changes in response to nematode infection. Early pioneering studies to identify genes that change their expression levels during the early compatible and incompatible interactions of cyst nematodes were conducted using differential display technology (Hermsmeier et al., 1998, 2000; Mahalingam et al., 1999; Vaghchhipawala et al., 2001). In one study, the approach successfully identified fifteen cDNA clones corresponding to mRNAs with different abundances in susceptible SCN-infected versus uninfected soybean roots (Hermsmeier et al., 1998). One of the soybean genes was identified as a phosphoglycerate mutase/biphosphoglycerate mutase enzyme, a key catalyst of glycolysis that was not previously characterized in plants (Mazarei et al., 2003). Identification of a homologous gene in Arabidopsis (*AtPGM*) and characterization of promoter-reporter gene fusions showed that this gene was induced in both root-knot and cyst nematode induced feeding cells (Mazarei et al., 2003). In conjunction with findings in other nematode pathosystems (Favery et al., 1998; Juergensen et al., 2003; Ithal et al., 2007b), these data suggest that increased rates of metabolism through glycolysis and the pentose phosphate pathway may elevate sugar levels within feeding cells to meet increased energy demands of the nematode. Another possible scenario presented by the authors, is that the induction of key enzymes of glycolysis and pentose phosphate pathways provides chorismate, the key substrate for the shikimate pathway, which is actively involved in producing compounds known to play important roles in plant-pathogen interactions. Chorismate would also serve as a substrate for nematode-secreted chorismate mutases described earlier (Popeijus et al., 2000; Bekal et al., 2003; Gao et al., 2003).

Other SCN-responsive soybean genes identified using differential display included two genes coding for polygalacturonases (PGs) whose transcript abundance increased during the early compatible interaction (Mahalingam et al., 1999), and a gene coding for an ethylene-responsive element-binding protein (GmEREBP), whose mRNA abundance decreased during a compatible interaction (Mazarei et al.,

2002). PGs are cell wall modifying enzymes that catalyze the hydrolysis of α-1,4 linked D-galacturons and were shown to play important roles during plant development. Although the spatial expression pattern of PGs in infected roots remains to be shown, the up-regulation of PGs in compatible interactions and down-regulation in incompatible interactions suggests a possible role for PGs in syncytium formation, possibly in conjunction with other plant CWMPs to regulate the controlled cell wall architectural changes that occur during syncytium formation (Goellner et al., 2001; Wieczorek et al., 2006). GmEREBP belongs to a large class of plant-specific transcription factors that were shown to play important roles in regulating plant defense responses (Gutterson and Reuber, 2004). Further study of *GmEREBP* showed that its transcript abundance not only decreased during compatible plant-nematode interactions, but increased during SCN infection of resistant cultivars (Hermsmeier et al., 2000; Mazarei et al., 2002) suggesting that it might play a key regulatory role in SCN resistance. This hypothesis was tested by constitutive over-expression of *GmEREBP* in either Arabidopsis or a susceptible soybean cultivar (Mazarei et al., 2007). Transgenic plants displayed activated pathogenesis-related (PR) gene expression. However, plants inoculated with nematodes did not exhibit an altered susceptibility phenotype (Mazarei et al., 2007).

Attempts also were made to position SCN-responsive soybean genes on an available public linkage map of soybean to test for associations of any of these genes with known soybean SCN resistance QTLs (Vaghchhipawala et al., 2001). One soybean gene, phosphoribosylformyl-glycinamidine (FGAM) synthase, was shown to be tightly linked to a major QTL for resistance, *rhg1*, on linkage group G. FGAM synthase catalyzes an important step in *de novo* purine biosynthesis and is active in rapidly dividing cells. Further studies confirmed that the promoters of two soybean FGAM synthases directed expression of GFP within syncytia induced by *Heterodera schachtii* in Arabidopsis (Vaghchhipawala et al., 2004). Nevertheless, the potential role of FGAM synthase in syncytium formation remains to be shown. Despite these findings, the early approaches identified only a handful of soybean genes responding to the nematode and little is known regarding their role, if any, in syncytium formation.

With the advent of functional genomic methodologies, plant molecular biologists and nematologists quickly set out to apply these new techniques to understand responses to cyst nematode infection. While first efforts focused on the sugar beet cyst nematode (Puthoff et al., 2003), attention quickly shifted to SCN. Initial efforts focused on generating ESTs from soybean roots infected with SCN (Alkharouf et al., 2004) and subsequently custom-made soybean cDNA microarrays were developed to examine host responses to infection (Khan et al., 2004; Alkarouf et al., 2006). In a study to examine the susceptible soybean response of whole roots at 2 days post-inoculation (2 dpi) with SCN using a soybean cDNA microarray representing 1,300 transcripts, 99 genes (8 %) were found to be induced in response to nematode infection (Alkarouf et al., 2004). The majority of these genes coded for stress-responsive proteins including proline-rich glycoproteins, SAM22, β-1,3-endoglucanase, peroxidase, and others associated with plant defense. It is not surprising that SCN juveniles induce plant defense and wound-related responses at the early timepoint of 2 dpi. During this phase of the nematode life cycle, infective J2 are still migrating intracellularly or have just completed the migratory phase of their life cycle and begun to initiate a feeding cell. Similar responses were observed at early timepoints in a time course analysis of whole soybean roots infected with SCN using a cDNA microarray representing 6,000 transcripts, (Alkharouf et al., 2006). Differentially expressed soybean genes were classified into several groups based on their expression profiles over the course of infection. A subset of wound, stress, and defense-related genes were elevated at all time points. At later time points, however, there was an observed increase in genes involved in transcription, protein synthesis, and metabolism likely reflecting alterations in plant gene expression associated with the development of highly metabolically active feeding cells in roots. With the advent of the Affymetrix Soybean Genome Array GeneChip representing 35,611 soybean transcripts and 7,431 SCN transcripts, a much more comprehensive expression profiling experiment of both soybean and SCN during a time course of the infection process was undertaken (Ithal et al., 2007a). This study provided the first global profile of gene expression changes in both soybean and SCN during different stages of nematode parasitism of roots. The simultaneous analysis of changes in both soybean and SCN mRNA levels identified 429 soybean genes with statistically significant differential expression between mock-infected and SCN-infected root tissues at three time points and 1,850 SCN genes with significant changes in expression during parasitism (Ithal et al., 2007a). Consistent with Alkarouf et al. (2006), this study observed a general activation of plant stress and defense mechanisms in response to SCN infection. Additional soybean genes differentially regulated by SCN infection included those involved in metabolism, cell wall modification, phytohormone response, and cellular signaling. The 1,850 differentially expressed SCN genes were grouped into 9 clusters based on their expression during parasitism. Of significant interest was the fact that the majority of previously identified SCN parasitism gene candidates was differentially expressed and displayed coordinated regulation during parasitism (Ithal et al., 2007a).

Despite the many efforts to profile global soybean responses to SCN infection using differential display and microarray approaches, these studies have contributed little to our understanding of gene expression changes occurring within developing feeding cells. This is due to the fact that the feeding cells make up only a small fraction of whole root tissues; thus, it is difficult to differentiate global gene expression changes from those occurring specifically within developing feeding cells. As a result, each gene identified as differentially expressed in a whole root study must then be confirmed for feeding cell expression. Alterations in plant gene expression within developing feeding cells were studied extensively in several different plant-nematode pathosystems to gain insight into the molecular mechanisms underlying their formation (reviewed in Gheysen and Fenoll, 2002). To date, however, only a handful of plant genes were shown to be expressed within developing feeding cells using promoter-reporter gene fusions, in situ hybridization, and immunolocalization techniques. Of these, even fewer were shown to play a role in feeding cell formation using reverse genetic approaches. Soybean is no exception. To date, only a few soybean genes were shown to be expressed in developing syncytia induced by

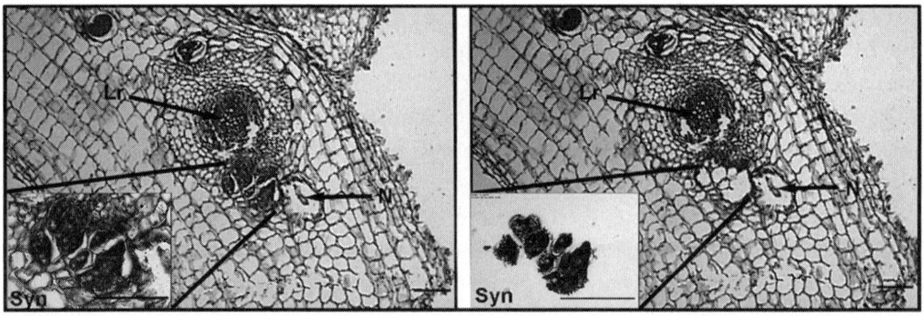

Fig. 17.4 Laser capture microdissection of a soybean cyst nematode-induced syncytium in soybean roots at 10 days post-inoculation. Reproduced with permission from Ithal et al., 2007b, Developmental transcript profiling of cyst nematode feeding cells in soybean roots, *Molecular Plant-Microbe Interactions* 20: 510–525

SCN. Nevertheless, technological advancements such as laser capture microdissection (LCM) have made cell-specific analyses feasible in plants (Day et al., 2005) and its application for isolating nematode feeding cells was recently demonstrated (Ramsay et al., 2004; Klink et al., 2005; Ithal et al., 2007b). This approach has opened the door for large-scale nematode feeding cell-specific analyses using functional genomic approaches. In a recent study, laser capture microdissection was coupled with microarray profiling to determine the first developmental transcript profile of cyst nematode feeding cells in susceptible soybean roots (Ithal et al., 2007b; Fig. 17.4). The authors identified 1,765 soybean genes expressed within syncytia that change expression within the first 2 days after syncytium induction and identified metabolic and regulatory pathways that likely play important roles in syncytium development.

Reverse Genetic Strategies to Dissect the SCN-Soybean Interaction

Until recently, a major bottleneck in interpreting large-scale SCN and soybean functional genomics data was a lack of reverse genetic approaches in both organisms. Functional analysis of gene products identified in genome-wide studies of the SCN-soybean interaction, such as those described above, is essential to advance our understanding of this plant-nematode interaction. A major breakthrough for functional biology of soybean and SCN was the application of RNA interference (RNAi) technology (Fire et al., 1998; Wesley et al., 2001) for post-transcriptional gene silencing of target genes. RNAi "soaking" methodologies to knock down genes in *C. elegans* and other nematodes were recently adapted for the plant-parasitic nematodes (Bakhetia et al., 2005; Urwin et al., 2002). This approach was used to demonstrate the roles of several SCN gland-expressed genes in nematode pathogenicity (Chen et al., 2005; Bakhetia et al., 2007). Moreover, a transgenic *in planta*-based RNAi approach (Wesley et al., 2001) to assess gene function by delivering dsRNA or siRNA to nematodes during feeding was demonstrated for root-knot nematodes (Huang et al., 2006; Yadav et al., 2006). Similarly, *in planta*-based RNAi approaches to assess gene function in SCN are yielding promising results (Baum, T.J. and Davis, E.L., personal communication). Nematode targets have included genes required for fundamental aspects of nematode biology and essential parasitism genes, thus demonstrating the potential of RNAi not only for understanding gene function but for developing transgenic crop plant resistance to nematodes (Huang et al., 2006; Yadav et al., 2006).

In addition to *in planta*-based RNAi to assess gene function in soybean, approaches such as TILLING (Targeting Induced Local Lesions in Genomes), fast neutron mutagenesis, Virus-Induced Gene Silencing (VIGS), and transposon-tagging have been developed or are under development (see Chapter 9 in this volume). Thus, it will not be long before characterizing the role of soybean genes in various aspects of soybean-SCN interactions such as resistance and nematode feeding cell formation will be a much more straightforward process aiding in the rapid identification of plant targets for engineered resistance.

Conclusions

Advances in genomics during the last decade have contributed substantial new knowledge to our understanding of the complex soybean-SCN pathosystem. As a result, SCN is at the forefront of being developed into a model genetic system for plant-parasitic nematodes. Significant progress was made to identify the genes coding for nematode stylet secretions and progress towards elucidating the functions of secreted proteins has revealed fascinating new insight into the mechanisms of nematode parasitism of plants. Nematode transcriptome and genome sequencing projects provide unprecedented opportunities for comparative nematode biology that should shed light on the nature of the evolution of plant parasitism. Refined approaches for cell-specific analyses are enabling the discovery of the genes and pathways contributing to the formation of unique, highly specialized feeding cells induced within plant roots by parasitic nematodes, and emerging tools, such as RNA interference, hold tremendous potential not only for answering interesting biological questions related to nematode parasitism and feeding cell formation, but for identifying novel targets and mechanisms for engineering nematode resistant crop plants.

Acknowledgments The authors gratefully acknowledge current and past members of their laboratories for their contributions to this work. We would also like to thank our colleagues at University of Georgia and North Carolina State University for sharing unpublished data and providing images in Fig. 17.3. Research support for the authors has been awarded by the National Research Initiative of the United States Department of Agriculture Cooperative State Research, Education, and Extension Service, United Soybean Board, Missouri Soybean Merchandising Council, Iowa Soybean Promotion Board, Illinois-Missouri Biotechnology Alliance, and the Experiment Stations of the University of Missouri and Iowa State University.

References

Alkharouf, N.W., Khan, R., and Matthews, B. (2004) Analysis of expressed sequence tags from roots of resistant soybean infected by the soybean cyst nematode. *Genome* 47, 380–388.

Alkharouf, N.W., Klink, V.P., Chouikha, I.B., Beard, H.S., MacDonald, M.H., Meyer, A., Knap, H.T., Khan, R., and Matthews, B.F. (2006) Time course microarray analyses reveal global changes in gene expression of susceptible *Glycine max* (soybean) roots during infection by *Heterodera glycines* (soybean cyst nematode). *Planta* 224, 838–852.

Atibalentja, N., Bekal, S., Domier, L.L., Niblack, T.L., Noel, G.R., and Lambert, K.N. (2005) A genetic linkage map of the soybean cyst nematode *Heterodera glycines*. *Molecular Genetics and Genomics* 273, 273–281.

Bakhetia, M., Charlton, W.L., Urwin, P.E., McPherson, M.J., and Atkinson, H.J. (2005) RNA interference and plant-parasitic nematodes. *Trends in Plant Science* 10, 362–367.

Bakhetia, M., Urwin, P.E., and Atkinson, H.J. (2007) qPCR analysis and RNAi define pharyngeal gland cell-expressed genes of *Heterodera glycines* required for initial interactions with the host. *Molecular Plant-Microbe Interactions* 20, 306–312.

Baum, T.J., Hussey, R.S., and Davis, E.L. 2007. Root-knot and cyst nematode parasitism genes: the molecular basis of plant parasitism. Pp. 17–43 In: Setlow, J.K., ed. *Genetic Engineering, Volume 28*, Springer, Heidelberg.

Bekal, S., Niblack, T.L., and Lambert, K.N. (2003) A chorismate mutase from the soybean cyst nematode *Heterodera glycines* shows polymorphisms that correlate with virulence. *Molecular Plant-Microbe Interactions* 16, 439–446.

Birch, P.R.J., Rehmany, A.P., Pritchard, L., Kamoun, S., and Beynon, J.L. (2006) Trafficking arms: oomycete effectors enter host plant cells. *Trends in Microbiology* 14, 8–11.

Blanchard, A., Esquibet, M., Fouville, D., and Grenier, E. (2005) Ranbpm homologue genes characterised in the cyst nematodes *Globodera pallida* and *Globodera 'mexicana'*. *Physiological and Molecular Plant Pathology* 67, 15–22.

Brand, U., Fletcher, J.C., Hobe, M., Meyerowitz, E.M., and Simon, R. (2000) Dependence of stem cell fate in *Arabidopsis* on a feedback loop regulated by *CLV3* activity. *Science* 289, 617–619.

Cai, D., Kleine, M., Kifle, S., Harloff, H-J., Sandal, N.N., Marcker, K.A., Klein-Lankhorst, R.M., Salentijn, E.M.J., Lange, W., Stiekema, W.J., Wyss, U., Grundler, F.M.W., and Jung, C. (1997) Positional cloning of a gene for nematode resistance in sugar beet. *Science* 275, 832–834.

Chen, Q., Rehman, S., Smant, G., and Jones, J.T. (2005) Functional analysis of pathogenicity proteins of the potato cyst nematode *Globodera rostochiensis* using RNAi. *Molecular Plant-Microbe Interactions* 18, 621–625.

Concibido, V.C., Diers, B.W., and Arelli, P.R. (2004) A decade of QTL mapping for cyst nematode resistance in soybean. *Crop Science* 44, 1121–1131.

Davis, E.L., Allen, R., and Hussey, R.S. (1994) Developmental expression of esophageal gland antigens and their detection in stylet secretions of *Meloidogyne incognita*. *Fundamental and Applied Nematology* 17, 255–262.

Davis, E.L., Hussey, R.S., Baum, T.J., Bakker, J., and Schots, A. (2000) Nematode parasitism genes. *Annual Review of Phytopathology* 38, 365–396.

Davis, E.L., Hussey, R.S., and Baum, T.J. (2004) Getting to the roots of parasitism by nematodes. *Trends in Parasitology* 20, 134–141.

Davis, E.L. and Mitchum, M.G. (2005) Nematodes: Sophisticated parasites of legumes. *Plant Physiology* 137, 1182–1188.

Day, R.C., Grossnicklaus, U., and MacKnight, R.C. (2005) Be more specific! Laser-assisted microdissection of plant cells. Trends in Plant Science 10, 397–406.

de Boer, J.M., Yan, Y.T., Wang, X.H., Smant, G., Hussey, R.S., Davis, E.L., and Baum, T.J. (1999) Developmental expression of secretory beta-1,4-endoglucanases in the subventral esophageal glands of *Heterodera glycines*. *Molecular Plant-Microbe Interactions* 12, 663–669.

de Boer, J.M., McDermott, J.P., Davis, E.L., Hussey, R.S., Popeijus, H., Smant, G., and Baum, T.J. (2002) Cloning of a putative pectate lysase gene expressed in he subventral esophageal glands of *Heterodera glycines*. *Journal of Nematology* 34, 9–11.

de Meutter, J., Vanholme, B., Bauw, G., Tygat, T., Gheysen, G., and Gheysen, G. (2001) Preparation and sequencing of secreted proteins from the pharyngeal glands of the plant-parasitic nematode *Heterodera schachtii*. *Molecular Plant Pathology* 2, 297–301

Ding, X., Shields, J., Allen, R., and Hussey, R.S. (1998) A secretory cellulose-binding protein cDNA cloned from the root-knot nematode (*Meloidogyne incognita*). *Molecular Plant-Microbe Interactions* 11, 952–959.

Ding, X., Shields, J., Allen, R., and Hussey, R.S. (2000) Molecular cloning and characterisation of a venom allergen AG5-like cDNA from *Meloidogyne incognita*. *International Journal for Parasitology* 30, 77–81.

Dong, K., and Opperman, C.H. (1997) Genetic analysis of parasitism in the soybean cyst nematode *Heterodera glycines*. *Genetics* 146, 1311–1318.

Dong, K., Barker, K.R., and Opperman, C.H. (2005) Virulence genes in *Heterodera glycines*: Allele frequencies and ror gene groups among field isolates and inbred lines. *Phytopathology* 95, 186–191.

Doyle, E.A., and Lambert, K.N. (2003) *Meloidogyne javanica* chorismate mutase 1 alters plant cell development. *Molecular Plant-Microbe Interactions* 16, 123–131.

Elling, A.A, Mitreva, M., Recknor, J., Gai, X., Martin, J., Maier, T.R., McDermott, J.P., Hewezi, T., Bird, D., Davis, E.L., Hussey, R.S., Nettleton, D.S., McCarter, J.P., and Baum, T.J. (2007) Divergent evolution of arrested development in the dauer stage of *Caenorhabditis elegans* and the infective stage of *Heterodera glycines*. *Genome Biology* 8, R211

Endo, B. (1965) Histological responses of resistant and susceptible soybeans varieties, and back-cross progeny to entry and development of *Heterodera glycines* (unpublished) *Phytopathology* 55, 375–381.

Ernst, K., Kumar, A., Kriseleit, D., Kloos, D.U., Phillips, M.S., and Ganal, M.W. (2002) The broad-spectrum potato cyst nematode resistance gene (*Hero*) from tomato is the only member of a large gene family of NBS-LRR genes with an unusual amino acid repeat in the LRR region. *Plant Journal* 31, 127–136.

Favery, B., Lecomte, P., Gil, N., Bechtold, N., Bouchez, D., Dalmasso, A., and Abad, P. (1998) *RPE*, a plant gene involved in early developmental steps of nematode feeding cells. *EMBO Journal* 17, 6799–6811.

Fire, A., Xu, S.Q., Montgomery, M.K., Kostas, S.A., Driver, S.E., and Mello, C.C. (1998) Potent and specific genetic interference by double-stranded RNA in *Caenorhabditis elegans*. *Nature* 391, 806–811.

Gao, B.L., Allen, R., Maier, T., Davis, E.L., Baum, T.J., and Hussey, R.S. (2001a) Identification of putative parasitism genes expressed in the esophageal gland cells of the soybean cyst nematode *Heterodera glycines*. *Molecular Plant-Microbe Interactions* 14, 1247–1254.

Gao, B., Allen, R., Maier, T., Davis, E.L., Baum, T.J., and Hussey, R.S. (2001b) Molecular characterisation and expression of two venom allergen-like protein genes in *Heterodera glycines*. *International Journal of Parasitology* 31, 1617–1625.

Gao, B., Allen, R., Maier, T., Davis, E.L., Baum, T.J., and Hussey, R.S. (2002) Identification of a new beta-1,4-endoglucanase gene expressed in the esophageal subventral gland cells of *Heterodera glycines*. *Journal of Nematology* 34, 12–15.

Gao, B.L., Allen, R., Maier, T., Davis, E.L., Baum, T.J., and Hussey, R.S. (2003) The parasitome of the phytonematode *Heterodera glycines*. *Molecular Plant-Microbe Interactions* 16, 720–726.

Gao, B.L., Allen, R., Davis, E.L., Baum, T.J., and Hussey, R.S. (2004a) Developmental expression and biochemical properties of a beta-1,4-endoglucanase family in the soybean cyst nematode, *Heterodera glycines*. *Molecular Plant Pathology* 5, 93–104.

Gao, B., Allen, R., Davis, E.L., Baum, T.J., and Hussey, R.S. (2004b) Molecular characterisation and developmental expression of a cellulose-binding protein gene in the soybean cyst nematode *Heterodera glycines*. *International Journal of Parasitology* 34, 1377–1383.

Gheysen, G. and Fenoll, C. (2002) Gene expression in nematode feeding sites. *Annual Review of Phytopathology*. 40, 124–168.

Goellner, M., Wang, X.H., and Davis, E.L. (2001) Endo-beta-1,4-glucanase expression in compatible plant-nematode interactions. *Plant Cell* 13, 2241–2255.

Goverse, A., Davis E.L., and Hussey, R.S. (1994) Monoclonal antibodies to the esophageal glands and stylet secretions of *Heterodera glycines*. *Journal of Nematology* 26, 251–259.

Grenier, E., Blok, V.C., Jones, J.T., Fouville, D., and Mugniery, D. (2002) Identification of gene expression differences between *Globodera pallida* and *G-'mexicana'* by suppression subtractive hybridization. *Molecular Plant Pathology* 3, 217–226.

Gutterson, N. and Reuber, T.L. (2004) Regulation of disease resistance pathways by AP2/ERF transcription factors. *Current Opinion in Plant Biology* 7, 465–471.

Hermsmeier, D., Mazarei, M., and Baum, T.J. (1998) Differential display analysis of the early compatible interaction between soybean and the soybean cyst nematode. *Molecular Plant-Microbe Interactions* 11, 1258–1263.

Hermsmeier, D., Hart, J.K., Byzova, M., Rodermel, S.R., and Baum, T.J. (2000) Changes in mRNA abundance within *Heterodera schachtii*-infected roots of *Arabidopsis thaliana. Molecular Plant-Microbe Interactions* 13, 309–315.

Hobe, M., Muller, R., Grunewald, M., Brand, U., and Simon, R. (2003) Loss of CLE40, a protein functionally equivalent to the stem cell restricting signal CLV3, enhances root waving in *Arabidopsis. Development Genes and Evolution* 213, 371–381.

Huang, G.Z., Gao, B.L., Maier, T., Allen, R., Davis, E.L., Baum, T.J., and Hussey, R.S. (2003) A profile of putative parasitism genes expressed in the esophageal gland cells of the root-knot nematode *Meloidogyne incognita. Molecular Plant-Microbe Interactions* 16, 376–381.

Huang, G., Dong, R., Maier, T., Allen, R., Davis, E.L., Baum, T.J., and Hussey, R.S. (2004) Use of solid-phase subtractive hybridization for the identification of parasitism gene candidates from the root-knot nematode *Meloidogyne incognita. Molecular Plant Pathology* 5, 217–222.

Huang, G., Dong, R., Allen, R., Davis, E.L., Baum, T.J., and Hussey, R.S. (2006). Engineering broad root-knot resistance in transgenic plants by RNAi silencing of a conserved and essential root-knot nematode parasitism gene. *Proceedings of the National Academy of Sciences* 103, 14302–14306.

Hussey, R.S., Pagio, O.R., and Seabury, F. (1990) Localization and purification of a secretory protein from the esophageal glands of *Meloidogyne incognita* with a monoclonal antibody. *Phytopathology* 80, 709–714.

Hussey, R.S., and Grundler, F.M. (1998) Nematode parasitism of plants. In *Physiology and biochemistry of free-living and plant-parasitic nematodes*, R.N. Perry and J. Wright, eds (England: CAB International Press), pp. 213–243.

Ithal, N., Recknor, J., Nettleton, D., Hearne, L., Maier, T., Baum, T.J., and Mitchum, M.G. (2007a) Parallel genome-wide expression profiling of host and pathogen during soybean cyst nematode infection of soybean. *Molecular Plant-Microbe Interactions* 20, 293–305.

Ithal, N., Recknor, J., Nettleton, D., Maier, T., Baum, T.J., and Mitchum, M.G. (2007b) Developmental transcript profiling of cyst nematode feeding cells in soybean roots. *Molecular Plant-Microbe Interactions* 20, 510–525.

Janssen, J., Bakker, J., and Gommers, F.J.G. (1991) Mendelian proof for a gene-for-gene relationship between virulence of *Globodera rostochiensis* and the *H*1 resistance gene in *Solanum tuberosum* ssp. *andigena* CPC 1673. *Rev. Nematology* 14, 207–211.

Jaubert, S., Ledger, T.N., Laffaire, J.B., Piotte, C., Abad, P., and Rosso, M.N. (2002) Direct identification of stylet secreted proteins from root-knot nematodes by a proteomic approach. *Molecular and Biochemical Parasitology* 121, 205–211.

Jones, M.G.K. and Northcote, D.H. (1972) Nematode induced syncytium – multinucleate transfer cell. *Journal of Cell Science* 10, 789–809.

Juergensen, K., Scholz-Starke, J., Sauer, N., Hess, P., van Bel, A.J.E., and Grundler, F.M.W. (2003) The companion cell-specific Arabidopsis disaccharide carrier AtSUC2 is expressed in nematode-induced syncytia. *Plant Physiology* 131, 61–69.

Khan, R., Alkharouf, N., Beard, H., MacDonald, M., Chouikha, I., Meyer, S., Grefenstette, J., Knap, H., and Matthews, B. (2004) Microarray analysis of gene expression in soybean roots susceptible to the soybean cyst nematode two days post-invasion. *Journal of Nematology* 36, 241–248.

Klink, V.P., Alkharouf, N., MacDonald, M., and Matthews, B. (2005) Laser capture microdissection (LCM) and expression analyses of *Glycine max* (soybean) syncytium containing root regions formed by the plant pathogen *Heterodera glycines* (soybean cyst nematode). *Plant Molecular Biology* 59, 965–979.

Lambert, K.N., Allen, K.D., and Sussex, I.M. (1999) Cloning and characterization of an esophageal-gland-specific chorismate mutase from the phytoparasitic nematode *Meloidogyne javanica. Molecular Plant-Microbe Interactions* 12, 328–336.

Lambert, K.N., Bekal, S., Domier, L.L., Niblack, T.L., Noel, G.R., and Smyth, C.A. (2005) Selection of *Heterodera glycines* chorismate mutase-1 alleles on nematode-resistant soybean. *Molecular Plant-Microbe Interactions* 18, 593–601.

Lambert, K.N., Bekal, S., Niblack, T.L., Hudson, M.E., and Domier, L.L. (2006) Nanotechnology and the genome of *Heterodera glycines. Phytopathology* 96, S149.

Mahalingam, R., Wang, G., and Knap, H.T. (1999) Polygalacturonase and polygalacturonase inhibitor protein: gene isolation and transcription in *Glycine max-Heterodera glycines* interactions. *Molecular Plant-Microbe Interactions* 12, 490–498.

Mazarei, M., Puthoff, D.P., Hart, J.K., Rodermel, S.R., and Baum, T.J. (2002) Identification and characterization of a soybean ethylene-responsive element-binding protein gene whose mRNA expression changes during soybean cyst nematode infection. *Molecular Plant-Microbe Interactions* 15, 577–586.

Mazarei, M., Lennon, K.A., Puthoff, D.P., Rodermel, S.R., and Baum, T.J. (2003) Expression of an Arabidopsis phosphoglycerate mutase homologue is localized to apical meristems, regulated by hormones, and induced by sedentary plant-parasitic nematodes. *Plant Molecular Biology* 53, 513–530.

Mazarei, M., Elling, A., Maier, T.R., Puthoff, D.P., and Baum, T.J. (2007) GmEREBP1 is a transcription factor activating defense genes in soybean and Arabidopsis. *Molecular Plant-Microbe Interactions* 20, 107–119.

Meksem, K., Pantazopoulos, P., Njiti, V.N., Hyten, L.D., Arelli, P.R., and Lightfoot, D.A. (2001) 'Forrest' resistance to the soybean cyst nematode is bigenic: saturation mapping of the *Rhg1* and *Rhg4* loci. *Theoretical and Applied Genetics* 103, 710–717.

Milligan, S.B., Bodeau, J., Yaghoobi, J., Kaloshian, I., Zabel, P., and Williamson, V.M. (1998) The root-knot nematode resistance gene *Mi* from tomato is a member of the leucine zipper, nucleotide binding, leucine-rich repeat family of plant genes *Plant Cell* 10, 1307–1319.

Mitreva, M., Blaxter, M.L., Bird, D.M., and McCarter, J.P. (2005) Comparative genomics of nematodes. *Trends in Genetics* 21, 573–581.

Niblack, T.L., Arelli, P.R., Noel, G.R., Opperman, C.H., Orf, J.H., Schmitt, D.P., Shannon, J.G., and Tylka, G.L. (2002) A revised classification scheme for genetically diverse populations of *Heterodera glycines. Journal of Nematology* 34, 279–288.

Niblack, T.L., Lambert, K.N., and Tylka, G.L. (2006) A model plant pathogen from the kingdom animalia: Heterodera glycines, the soybean cyst nematode. *Annual Review of Phytopathology* 44, 283–303.

Olsen, A.N., and Skriver, K. (2003) Ligand mimicry? Plant-parasitic nematode polypeptide with similarity to CLAVATA3. *Trends in Plant Science* 8, 55–57.

Paal, J., Henselewski, H., Muth, J., Meksem, K., Menendez, C.M., Salamini, F., Ballvora, A., and Gebhardt, C. (2004) Molecular cloning of the potato *Gro1-4* gene conferring resistance to pathotype Ro1 of the root cyst nematode *Globodera rostochiensis*, based on a candidate gene approach. *Plant Journal* 38, 285–297.

Parkinson, J., Mitreva, M., Hall, N., Blaxter, M., and McCarter, J.P. (2003). 400,000 nematode ESTs on the Net. *Trends in Parasitology* 19, 283–286.

Parkinson, J., Mitreva, M., Whitton, C., Thomson, M., Daub, J., Martin, J., Schmid, R., Hall, N., Barrell, B., Waterston, R.H., McCarter, J.P., and Blaxter, M.L. (2004) A transcriptomic analysis of the phylum Nematoda. *Nature Genetics* 36, 1259–1267.

Popeijus, M., Blok, V.C., Cardle, L., Bakker, E., Phillips, M.S., Helder, J., Smant, G., and Jones, J.T. (2000) Analysis of genes expressed in second stage juveniles of the potato cyst nematodes

Globodera rostochiensis and *Globodera pallida* using the expressed sequence tag approach. *Nematology* 2, 567–574.

Puthoff, D. P., Nettleton, D., Rodermel, S.R., and Baum, T.J. (2003) *Arabidopsis* gene expression changes during cyst nematode parasitism revealed by statistical analyses of microarray expression profiles. *The Plant Journal* 33, 911–921.

Qin, L., Overmars, B., Helder, J., Popeijus, H., van der Voort, J.R., Groenink, W., van Koert, P., Schots, A., Bakker, J., and Smant, G. (2000) An efficient cDNA-AFLP-based strategy for the identification of putative pathogenicity factors from the potato cyst nematode *Globodera rostochiensis*. *Molecular Plant-Microbe Interactions* 13, 830–836.

Ramsay, K., Wang, Z., and Jones, G.K. (2004) Using laser capture microdissection to study gene expression in early stages of giant cells induced by root-knot nematodes. *Molecular Plant Pathology* 5, 587–592.

Safra-Dassa, L., Shani, Z., Danin, A., Roiz, L., Shoseyov, O., and Wolf, S. (2006) Growth modulation of transgenic potato plants by heterologous expression of bacterial carbohydrate-binding module. *Molecular Breeding* 17, 355–364.

Shpigel, E. Roiz, L., Goren, R., and Shoseyov, O. (1998) Bacterial cellulose-binding domain modulates *in vitro* elongation of different plant cells. *Plant Physiology* 117, 1185–1194.

Smant, G., Stokkermans, J.P.W.G., Yan, Y.T., de Boer, J.M., Baum, T.J., Wang, X.H., Hussey, R.S., Gommers, F.J., Henrissat, B., Davis, E.L., Helder, J., Schots, A., and Bakker, J. (1998) Endogenous cellulases in animals: Isolation of beta-1,4-endoglucanase genes from two species of plant-parasitic cyst nematodes. *Proceedings of the National Academy of Sciences of the United States of America* 95, 4906–4911.

Tytgat, T., Vanholme, B., De Meutter, J., Claeys, M., Couvreur, M., Vanhoutte, I., Gheysen, G., Van Criekinge, W., Borgonie, G., Coomans, A., and Gheysen, G. (2004) A new class of ubiquitin extension proteins secreted by the dorsal pharyngeal gland in plant-parasitic cyst nematodes. *Molecular Plant-Microbe Interactions* 17, 846–852.

Urwin, P.E., Lilley, C.J., and Atkinson, H.J. (2002) Ingestion of double-stranded RNA by preparasitic juvenile cyst nematodes leads to RNA interference. *Molecular Plant-Microbe Interactions* 15, 747–752.

Vaghchhipawala, Z., Bassuner, R., Clayton, K., Lewers, K., Shoemaker, R., and Mackenzie, S.A. (2001) Modulations in gene expression and mapping of genes associated with cyst nematode infection of soybean. *Molecular Plant-Microbe Interactions* 14, 42–54.

Vaghchhipawala, Z.E., Schlueter, J.A., Shoemaker, R.C., and Mackenzie, S.A. (2004) Soybean FGAM synthase promoters direct ectopic nematode feeding site activity. *Genome* 47, 404–413.

Vanholme, B., Mitreva, M.D., Van Criekinge, W., Logghe, M., Bird, D., McCarter, J.P., and Gheysen, G. (2005) Detection of putative secreted proteins in the plant-parasitic nematode *Heterodera schachtii*. *Parasitology Research* 98, 414–424.

van der Vossen, E.A.G., van der Voort, J.N.A.M.R., Kanyuka, K., Bendahmane, A., Sandbrink, H., Baulcombe, D.C., Bakker, J., Stiekema, W.J., and Klein-Lankhorst, R.M. (2000) Homologues of a single resistance-gene cluster in potato confer resistance to distinct pathogens: a virus and a nematode. *Plant Journal* 23, 567–576.

Vercauteren, I., de Almeida Engler, J., De Groot, R., and Gheysen, G. (2002) An *Arabidopsis thaliana* pectin acetylesterase gene is upregulated in nematode feeding sites induced by root-knot and cyst nematodes. *Molecular Plant-Microbe Interactions* 15, 404–407.

Wang, J., Replogle, A., Wang, X., Davis, E.L., and Mitchum, M.G. (2006) Functional analysis of nematode secreted CLAVATA3/ESR (CLE)-like peptides of the genus *Heterodera*. *Phytopathology* 96, S120.

Wang, X.H., Meyers, D., Yan, Y.T., Baum, T., Smant, G., Hussey, R., and Davis, E. (1999) *In planta* localization of a beta-1,4-endoglucanase secreted by *Heterodera glycines*. *Molecular Plant-Microbe Interactions* 12, 64–67.

Wang, X.H., Allen, R., Ding, X.F., Goellner, M., Maier, T., de Boer, J.M., Baum, T.J., Hussey, R.S., and Davis, E.L. (2001) Signal peptide-selection of cDNA cloned directly from the esophageal gland cells of the soybean cyst nematode *Heterodera glycines*. *Molecular Plant-Microbe Interactions* 14, 536–544.

Wang, X.H., Mitchum, M.G., Gao, B.L., Li, C.Y., Diab, H., Baum, T.J., Hussey, R.S., and Davis, E.L. (2005) A parasitism gene from a plant-parasitic nematode with function similar to CLAVATA3/ESR (CLE) of Arabidopsis thaliana. *Molecular Plant Pathology* 6, 187–191.

Wang, X.H., Replogle, A.R., Davis, E.L., and Mitchum, M.G. (2007) The tobacco *Cel7* gene promoter is auxin-responsive and locally induced in nematode feeding sites of heterologous plants. *Molecular Plant Pathology* 8, 423–436.

Wesley, S.V., Helliwell, C.A., Smith, N.A., Wang, M., Rouse, D.T., Liu, Q., Gooding, P.S., Singh, S.P., Abbott, D., Stoutjesdijk, P.A., Robinson, S.P., Gleave, A.P., Green, A.G., and Waterhouse, P.M. (2001) Construct design for efficient, effective and high-throughput gene silencing in plants. *Plant Journal* 27, 581–590.

Wieczorek, K., Golecki, B., Gerdes, L., Heinen, P., Szakasits, D., Durachko, D.M., Cosgrove, D.J., Kreil, D.P., Puzio, P.S., Bohlmann, H., and Grundler, F.M.W. (2006) Expansins are involved in the formation of nematode-induced syncytia in roots of *Arabidopsis thaliana*. *Plant Journal* 48, 98–112.

Williamson, V.M. and Hussey, R.S. (1996) Nematode pathogenesis and resistance in plants. *Plant Cell* 8, 1735–1745.

Yadav, B.C., Veluthambi, K., and Subramaniam, K. (2006). Host-generated double-stranded RNA induces RNAi in plant-parasitic nematodes and protects the host from infection. *Molecular and Biochemical Parasitology* 148, 219–222.

Yan, Y.T., Smant, G., Stokkermans, J., Qin, L., Helder, J., Baum, T., Schots, A., and Davis, E. (1998) Genomic organization of four beta-1,4-endoglucanase genes in plant-parasitic cyst nematodes and its evolutionary implications. *Gene* 220, 61–70.

Yan, Y.T., Smant, G., and Davis, E. (2001) Functional screening yields a new beta-1,4-endoglucanase gene from Heterodera glycines that may be the product of recent gene duplication. *Molecular Plant-Microbe Interactions* 14, 63–71.

Zeng, L-R., Vega-Sanchez, M.E., Zhu, T., and Wang, G-L. (2006) Ubiquitination-mediated protein degradation and modification: an emerging theme in plant-microbe interactions. *Cell Research* 16, 413–426.

Chapter 18
Genomics of Abiotic Stress in Soybean

Babu Valliyodan and Henry T. Nguyen

Introduction

Legumes represent the most utilized plant family with 20,000 species and their usage ranges from forage and pasture crops to animal feed and human food. Human consumption of seeds of grain legumes is important world wide, including common beans (*Phaseolus* sp.), soybean (*Glycine max*), peas (*Pisum sativum*), peanut (*Arachis hypogaea*), cowpea (*Vigna unguiculata*), chickpea (*Cicer arientinum*), and pigeon pea (*Cajanus cajan*). Legumes account for 27 % of the world's primary crop production, with grain legumes alone contributing 33 % of the dietary protein nitrogen (N) needs of humans (Graham and Vance 2003). Their capacity for symbiotic nitrogen fixation is unique and because of this biological process legumes are valuable asset to cropping systems. Among these legumes, soybean [*Glycine max* (L.) Merr] is an important crop in terms of its wide usage as edible vegetable oil and as a high-protein feed supplement for livestock. Soybean seed products are widely used in industrial and pharmaceutical applications. Above all, soybean biodiesel was recognized as one of the alternatives to fossil fuels for the future (Hill et al. 2006).

Soybean was domesticated in East Asia before the first century A.D. and later several landraces were established based on different environmental adaptations (Abe et al. 2003). The soybean was introduced in North America in 1765 and by late 1800 this crop was grown widely throughout the areas of its adaptation (Wilcox 2001). By selection and hybridization breeders in the USA increased soybean yield considerably and efforts for yield improvement are still continuing to meet the world's need (Fehr 1984; Specht et al. 1999). The world soybean production in the year 2006 was ~218 Million metric tones with the USA the largest producer (83.37 Million metric tones). The contribution from other major countries such as Brazil, Argentina, and China were 55, 40 and 16 Million metric tones, respectively (http://www.fas.usda.gov).

H.T. Nguyen
National Center for Soybean Biotechnology, Division of Plant Sciences, University of Missouri-Columbia, MO 65211, USA
e-mail: nguyenhenry@missouri.edu

G. Stacey (ed.), *Genetics and Genomics of Soybean*,

Major abiotic stresses such as drought, heat, water logging and salinity have major implications to the adaptability and productivity of crop plants worldwide. It was reported that the average yield losses due to abiotic stresses are more than 50 % for major crops (Boyer 1982). Crop improvement with traits that confer tolerance to abiotic stresses is commonly practiced using traditional and modern breeding, as well as through biotechnology. Global soybean production and crop quality are severely affected by various environmental stresses and drought is the most devastating. To develop soybean plants with enhanced tolerance to stress, a basic understanding of the physiological, biochemical and gene regulatory networks is essential. Various genomic and functional genomics tools have helped to advance our understanding of stress signal perception and transduction and the associated molecular regulatory network. These tools revealed several stress-inducible genes and various transcription factors that regulate the stress inducible systems. The objectives of this book chapter are to briefly review recent efforts to better understand soybean responses to several common abiotic stresses and to highlight efforts to integrate results of advances in physiology, genetics and molecular biology into soybean improvement programs. This chapter gives a snapshot of the magnitude of various abiotic stress impacts on soybean and updates the genomic approaches and recent progress in understanding the mechanisms of gene regulation and the progress in genetic or metabolic engineering for enhanced stress tolerance in soybean.

Synopsis of Genomic Resources and Tools

The reduction in yield attributable to the various physiochemical factors (abiotic stresses) has been the subject of much study by plant biologists (Nilsen and Orcutt 1996; Ludlow and Muchow 1990; Raper and Kramer 1987; Araus et al. 2002). However, progress in improving plant tolerance to environmental stresses has been slow because of the complexity of the trait. Advances in genomic technologies provide unprecedented opportunities to understand global patterns of gene expression and their association with the development of specific phenotypes. However the highly complex, non-linear interactions of the different environmental factors affecting gene expression for a particular trait are difficult to understand. The identification of genomic regions responsible for tolerance to abiotic stresses in soybean will be useful for marker-assisted selection (MAS) to expedite the development of high-yielding stress tolerant soybean cultivars. With MAS, it is possible to combine multiple favorable alleles into a single cultivar through marker assisted selection.

Soybean has a genome size of 1115 Mb/1C (Arumuganathan and Earle 1991) and 40–60 % of its genome is repetitive sequence and heterochromatic (Goldberg 1978; Singh and Hymowitz 1988). Genes and quantitative trait loci (QTLs) in soybean segregate in simple Mendelian fashion, facilitating genetic analysis. The soybean composite genetic map is well developed. The classical map contains only 67 loci on 19 linkage groups (Hedges and Palmer 1993), while the molecular map encompasses all 20 linkage groups and ~2500 cM based on SSRs and RFLPs mapped in five populations. The composite map contains 1849 markers including 1015 SSRs, 709 RFLPs, 73 RAPDs, 24 classical traits, 10 isozymes and 12 others (Song et al. 2004).

A number of public-sector BAC libraries are available providing >35-fold genome coverage (Danesh et al. 1998; Marek and Shoemaker 1997; Tomkins et al. 1999; Wu et al. 2004a,b); also available is the 12×Williams 82 Bsty1 library developed by Tomkins, Nguyen and Stacey (unpublished). Also, the author's research group developed a 6-dimensional pool array of the *BstyI* soybean cv. Williams 82 BAC library comprising of 208 pools and has mapped over 1,000 genes to the soybean physical map. This map is being integrated to the genetic map.

The public EST (expressed sequence tag) project for soybean led by different Universities, public and private sectors, developed several cDNA libraries and gene sequences of economic importance from many developmental stages and organs of soybean (Shoemaker et al. 2002). These efforts generated 323,439 EST sequences (http://www.ncbi.nlm.nih.gov/UniGene/UGOrg.cgi?TAXID=3847) from roots, shoots, leaves, stems, pods, cotyledons, germinating shoot tips, flower meristems, nodules, tissue culture-derived embryos, and pathogen challenged tissues. The cluster analysis revealed that the entire public EST collection yields 61,127 unigenes of which 36,357 are contigs and 24,770 are singletons (Vodkin et al. 2004). Similarly, The Institute for Genomic Research (TIGR) has a total of 63,676 unigenes consisting of 31,918 TC sequences and 31,748 singletons (http://www.tigr.org/tdb/tgi). As a community resource, the Vodkin group (University of Ilinois, Urbana-Champaign) printed two microarray slide sets each consisting of 18,432 single-spotted PCR products derived from the low redundancy cDNA sets. The GmcDNA18kA set (representing sequence-driven unigene clone libraries Gm-r1021, Gm-r1083, and Gm-r1070) is highly representative of genes expressed in the developing flowers and buds, young pods, developing seed coats, and immature cotyledons, as well as from roots of seedlings and adult plants, including roots infected with the nodulating bacterium, *Bradyrhizobium japonicum*. The GmcDNA18kB microarray slide (unigene clone libraries Gm-r1088 and Gm-r1089) is highly representative of clones selected from libraries derived from tissue-culture embryos, germinating cotyledons, and seedlings subjected to various stresses including some challenged by pathogens. Completion of both sets brings the total number of cDNAs represented to 36,864 (Vodkin et al. 2004). Also, the same group also developed long oligo arrays (70mer) preferentially using the 3′ regions of the cDNA to design oligonucleotides that distinguish among gene family members. Recent GeneBank submitted EST sequences from the subtracted library of drought stressed soybean root tips contribute to the stress specific unigenes for the further functional genomics approach. A mixed Soybean GeneChip (http://www.affymetrix.com) is commercially available for studying gene expression of soybean comprising ~60,000 transcripts. Size of the soybean probeset on the GeneChip is an 11-probe pair and 11-micron. These GeneChips contains ~37,500 *G. max* transcripts, ~15,800 *Phytophthora* root and stem rot transcripts, and over 7,500 soybean cyst nematode transcripts.

The collaborative efforts on the sequencing of soybean genome by the Department of Energy-Joint Genome Institute (DOE-JGI) and the soybean research community will open up new dimensions for soybean gene discovery and functional genomics. The sequence based gene discovery will be highly helpful for further understanding of the molecular networks involved in various biological process involved in plant development and stress biology.

Physiology and Genetics of Stress Responses

Several abiotic stresses such as drought, flooding, temperature, salinity and mineral stresses severely affect soybean yield through various physiological and genetic mechanisms. In general, advances in plant physiology, genetics and functional genomics have enhanced our understanding of plant responses to these abiotic stresses and the basis of varietal differences in tolerance. But in the case of soybean, the efforts to understand the physiological mechanisms and the genetic dissection of the abiotic stress responses are still in the early stages. This section highlights the physiological factors associated with stress tolerance and related genetic information.

Drought

Drought is the major abiotic stress factor limiting crop productivity worldwide. Water is an increasingly limited resource and its availability limits crop productivity in many parts of the world. In the USA, parts of the country have experienced severe drought conditions over the past few years. Irrigation is not an economically viable option for most soybean farming states in the USA (Boyer 1983). Thus, alternative ways for alleviating plant moisture stress are needed. Drought stress during flowering and early pod development significantly increases the rate of flower and pod abortion, thus decreasing final yield (Boyer 1983; Westgate and Peterson 1993). Soybean yield reductions of 40 % from drought are common (Muchow and Sinclair 1986; Spect al. 1999). The water status of the crop plant is usually discussed in terms of its water potential or the components of water potential (Turner 1986). Major components of water potential are osmotic potential, pressure potential, matric potential and gravitational potential (Begg and Turner 1976). Mechanisms of drought resistance fall into three general categories such as drought escape, drought avoidance or dehydration postponement and dehydration tolerance (Turner et al. 2001). Drought escape is defined as the ability of a plant to complete its life cycle before serious soil and plant water deficits develop where the selection is aimed at those developmental and maturation traits that match seasonal water availability with crop needs. Drought avoidance is a mechanism for avoiding lower water status in tissues during drought by maintaining relatively high tissue water potential, cell turgor and cell volume and, in this case, the selection is focused on traits that reduce evaporatory water loss from plant surfaces or maintain water uptake and turgor during drought via a deeper and more extensive root system. The third category, drought tolerance is the ability of plants to withstand water-deficit and maintain metabolism at low tissue water potential. In this category the selection is directed at maintaining cell turgor or enhancing cellular constituents that protect cytoplasmic proteins and membranes from drying. The genetic basis of drought avoidance and drought (dehydration) tolerance is not well understood, and understanding how plant growth responses to drought are regulated is vital for efforts to modify the impact of water supply on soybean plants.

Improved selection techniques to identify germplasm with improved performance (e.g. grain yield in crop plants like soybean) under water deficit conditions are required. So, the other frame work for the improved drought resistance is the yield component framework which considers the yield variation in terms of characteristics affecting water use efficiency and the harvest index (Ludlow and Muchow 1990; Turner 2000). Both a drought resistance frame work and a yield component frame work were extensively used for the drought resistance improvement in cereal crops. However, the physiological and biochemical traits important for breeding improved yield in grain legumes for water deficit environment are less characterized (Turner et al. 2001).

Over 2000 Plant Introductions (PIs) from the USDA-ARS national soybean germplasm collection were evaluated over the past 20 years in North Carolina to search for drought resistance utilizing special fields where drought occurs each year. PI's and breeding lines were identified or developed which wilted more slowly than existing varieties. Two PIs, PI 416937 and PI 471938 were slow wilting and had drought tolerance (Shannon and Carter 2003). Major physiological traits that may affect performance of soybean when soil water availability is limiting are water use efficiency (WUE), regulation of whole plant water use in response to soil water content, and leaf epidermal conductance (g_e) when stomata are closed (Hufstetler et al. 2007). The authors stated that their results open the possibility that slow wilting is associated primarily with deep rooting or deep rooting in combination with the other traits such as WUE and stomatal conductance.

Environmental factors affecting plant water status have a major effect on plant transpiration efficiency and photosynthesis. Leaf transpiration efficiency describes the ratio of photosynthesis to transpiration rates. Genetic variation in transpiration efficiency is affected by either of these components (Masle et al. 2005). Screening for lines for high water use efficiency (WUE) is a technique that can lead to soybean lines with higher drought tolerance. Genetic variability for WUE was reported in various crop species including legumes, such as peanut (Hubick et al. 1986, 1988) and soybean (Mian et al. 1996, 1998). A restriction fragment length polymorphism (RFLP) map was constructed from a soybean population of 120 F4-derived lines from a cross of Young × P1416937 to identify quantitative trait loci (QTL) associated with WUE and LASH (leaf ash) in 36-d-old, greenhouse-grown plants. Four QTLs associated with both WUE and LASH at two genomic regions was identified and together explained 16 and 21 % of the variation in WUE and LASH respectively (Mian et al. 1996). Another study by the same group (Mian et al. 1998) identified an additional QTL for WUE on Linkage Group (LG) L using a F2-derived soybean population from a cross of "S100"× "Tokyo" and determined the consistency of this WUE QTL across the two soybean population. Marker A489H, on LG L, was unique to the S100 × Tokyo population and explained 14 % of the variation in WUE in this soybean population.

Specht et al. (2001) studied the genetic basis of beta and carbon isotope discrimination (CID), a theorized indicator of transpiration efficiency (TE) in soybean. A Minsoy × Noir 1 population of 236 recombinant inbred lines (RILs), genotyped at 665 loci, was evaluated in six water treatments (100, 80, 60, 40, 20, and 0 % ET) for two years in this study. This was the first study using a large RIL population to examine the relationship, at both the phenotypic and genomic levels, between a yield *beta* estimate of WUE for a genotype and a leaf δ^{13}C estimate of its TE. The data suggested that the major quantitative trait loci (QTL) for yield *beta*, yield, and CID were coincident with maturity and/or determinacy QTLs, except for a CID QTL in linkage group U09-C2, but it had no effect on *beta*. The finding indicates that yield *beta* and leaf δ^{13}C can be added to the list of traits affected by the major QTLs described in prior studies (Orf et al. 1999). Hufstetler et al. (2007) compared soybean genotypes for their regulation of whole plant water use under drought stress. Substantial variation was found among genotypes for WUE, fraction of transpirable soil water ($FTSW_C$), g_e, and also the extent to which normalized transpiration rate recovered on re-watering. Generally, adapted cultivars had greater WUE and lower g_e than did PIs. However, PI 471938 and its progeny N98-7264 were clear exceptions to this trend. An unexpected finding in the above study was that WUE was significantly negatively correlated with g_e across genotypes.

Root Growth Responses Under Drought

Plant tissue may maintain turgor during drought by either dehydration avoidance, dehydration tolerance or both (Kramer 1980). These forms of stress resistance are controlled by root traits such as root thickness, root penetration ability through compacted soil layers, and root depth and mass (Nguyen et al. 1997; Price et al. 2002). Those are constitutive phenotypic traits and do not require stress conditions to occur. In contrast, adaptive traits such as osmotic adjustment and dehydration tolerance arise in response to water deficit (Ingram and Bartels 1996). There is evidence for genetic variability in soybean for dehydration avoidance, primarily associated with increased rooting depth and volume (Boyer et al. 1980; Kaspar et al. 1984). Sloane et al. (1990) compared the relative drought tolerance of PI 416937, a slow wilting accession with Forrest and reported that the slow wilting accession maintained lower level of solute potential and higher levels of pressure potential and RWC. The key features of molecular mechanism and the signaling systems that are required for the deep root growth and development are not completely illustrated in previous studies.

There are two features of root system responses to water deficit conditions. The roots can grow and elongate at a substantial rates when the water potential level is low enough to inhibit the shoot growth (Sharp and Davies 1989). The early establishment of seedling under water deficit conditions occurs in primary roots of several species (Spollen et al. 1993) is a good example. The second response is the phenotypic plasticity in roots in response to water deficits which helps to increase the water absorbing surface area of the root system. Enhanced lateral root growth (Read and Bartlett 1972; Jupp and Newman 1987) and more root hairs (Vasellati et al. 2001) are the major plasticity responses in root system under soil drying conditions. The field study of common beans conducted by Sponchiado et al. (1989) shows an association between genetic variability of the root system response to soil drying and the plant performance under water deficit conditions. In this study the authors tested four lines under irrigated conditions and the root system development and the final yield were similar in these lines under well watered conditions. But under soil drying conditions two lines exhibited significant increase in root proliferation and three times more yield when compared to other two lines. Another follow up grafting study conducted using the same lines showed the capacity of root proliferation and yield increase under soil drying conditions (White and Castillo 1989). These results indicate the role of some form of internal regulation in the root system under water deficit conditions.

Other traits were also explored as possible mechanisms to increase the amount of water available for transpiration during drought (Purcell and Specht 2004). Genes that increase rooting depth in a particular environment, however, are most likely to increase the amount of water available for transpiration and have a favorable agronomic and economic return. Taylor et al. (1978) reported differences in root growth rates in 29 soybean cultivars after the genotypic evaluation under controlled conditions. The evaluation of the taproot penetration rate of soybean cultivars suggested that a more rapid root extension rate would result in a more extensive root system, which constitutes an efficient mechanism for water extraction during drought conditions, perhaps leading to drought avoidance. In another approach using more cultivars from different maturity groups, Kaspar et al. (1984) indicated that the tap root elongation rate within a maturity group differed among cultivars by 1.3 cm/day. Maturity group II was selected for field studies and roots of faster elongation cultivars were 10 cm deeper than slower elongation cultivars within the maturity group. Studies were conducted to compare the ability of taproots and basal roots of different maize and soybean cultivars in growth chambers to penetrate a compact soil layer (Bushhamuka and Zobel 1998). There were large genotypic differences among the cultivars in both tap root length and basal-root length. QTL for root penetration capability were identified in rice (Ray et al. 1996). In soybean, with the existence of genetic variation, as shown in previous studies, it is possible that molecular marker analysis may allow soybean breeding programs to accomplish marker assisted selection for this trait.

N_2 *Fixation and Drought Effect*

It was widely reported that biological N_2 fixation in legume nodules declines under drought and other environmental stresses (Zahran 1999; Gálvez et al. 2005; Marino et al. 2007). The effect of drought on biological nitrogen fixation was extensively studied and is considered to be the most important environmental factor resulting in crop yield loss (Boyer 1982; Marino et al. 2007). It was proposed that increased tolerance of N_2 fixation to water deficits would increase overall water-deficit tolerance in soybean (King and Purcell 2005). Several studies were conducted to understand the molecular and physiological mechanism of N_2 fixation under water deficit conditions. Several factors such as reduced oxygen availability (Durand et al. 1987), reduction of carbon flux in nodules (Arrese-Igor et al. 1999), increases in ureides and free amino acids in soybean tissues (King and Purcell 2005) were related to the inhibition of nitrogen fixation under drought.

Durand et al. (1987) reported that during a week of water deprivation there was a close relationship between decreases in leaf and nodule water potential. Nitrogenase activity showed a 70 % decrease during the first 4 days, while photosynthesis only declined by 5 %. The data suggested that water stress exerts an influence on nitrogenase activity that is independent of the rate of photosynthesis. It also acts directly on nodule activity through increases in the resistance to oxygen diffusion to the bacteroids. These results suggest that a linear relationship between oxygen diffusion resistance and water potential is more important than any reduction in photosynthate supply. The increase in O_2 diffusion resistance, the decrease in nitrogenase-linked respiration and nitrogenase proteins, the accumulation of respiratory substrates and oxidized lipids and proteins, and the up-regulation of antioxidant genes reveal that respiratory activity of bacteroids is impaired and that oxidative stress occurs in nodules under drought conditions prior to any detectable effect on sucrose synthase or leghemoglobin (Naya et al. 2007).

The ureides, allantoin and allantoate, are the final products of N_2 fixation that are exported from soybean nodules to the shoot (McClure and Israel 1979), where they are catabolized. Under water deficit conditions, the ureide catabolism rate in leaves decreases and the shoot uriede concentration increases (Vadez and Sinclair 2001). Several soybean genotypes exhibiting substantial tolerance of N_2 fixation to water deficit were identified. Sall and Sinclair (1991) identified the cultivar Jackson as having N_2 fixation sensitivity to water deficit that was no worse than that of mass accumulation. These results were followed by a screen of a large collection of soybean plant introduction lines (>3,000 lines) to identify lines that exhibited N_2 fixation drought tolerance (Sinclair et al. 2000). This study resulted in the identification of eight plant introduction lines that had N_2 fixation that was more tolerant of soil drying than was leaf gas exchange. Studies conducted by Sinclair et al. (2003) indicated that ureide catabolism independent of manganese was active in six of the eight soybean plant introduction lines identified to express N_2 fixation tolerance to soil drying. Genes converting uric acid to allantoin in soybean nodules and allantoin to glyoxylate and ammonia in soybean leaves were cloned (Yang and Han 2004) but nothing is known of the genes involved in the final steps of uriede degradation, signaling and transport (Todd et al. 2006). A more precise understanding at expression and molecular genetics of the factors limiting and regulating the response of nitrogen fixation to drought is lacking.

Effect of Drought on Seed Development

Water availability plays a major role in the regulation of seed filling and development. Drought stress imposed during the pod lengthening and seed filling stages had the greatest effect on the number of pods produced per unit of dry matter, and decreased the seed weight and led to some seed abortion in the pod (Momen et al. 1985; Desclaux et al. 2000). According to Desclaux and Raumet (1996), each reproductive stage was shorter under drought stress, mainly because of the appearance of new organs that prevented the emergence of organs belonging to the earlier ontogenetic stages. The seed-filling stage and the final stage in seed abortion began earlier in stressed plants and the duration of the maturation period was significantly reduced by drought stress during seed filling, leading to accelerated senescence. Drought stress significantly altered the content of some oil components in plants such as *Tagetes minuta* (Mohamed et al. 2002). There is considerable interest in the possible environmental effects on seed composition and, hence, the nutritional value of soybean seeds. Although it is well-established that the amounts of protein and oil as well as fatty acid distribution in soybean seeds are affected by temperature, the effects of drought on seed components has been little studied.

The anatomical and physiological changes during soybean reproduction and seed development have been described in detail (Carlson and Lersten 1987). In soybean, reproductive potential may be reduced considerably due to abscission of developing flowers and pods soon after anthesis during pro-embryo development (Peterson et al. 1990) even under optimal environmental conditions (Carlson 1987). The pro-embryo developmental stage is one of active cell division in the young ovules, coinciding with a rapid pod expansion (Peterson et al. 1992), which is specifically sensitive to water deficit conditions (Westgate and Peterson 1993). The period from anthesis to maturity lasts from 18 to 60 days depending upon maturity group of the variety and growing conditions. Seed development consists of an early period of rapid cell division of the globular and heart stage embryos. The next phase is characterized by rapid cell elongation in the cotyledons, a rapid metabolic rate, and an increase in dry weight as storage oils and proteins accumulate. The late maturation period is characterized by dehydration and a cessation of storage protein synthesis. A typical mature soybean seed will consist of 20 % oil, 40 % protein and 12 % non-structural carbohydrates, although some varieties have as high as 50 % protein (Nielsen 1996). Generally, there is an inverse relationship between protein and oil composition.

The period of soybean seed fill beginning at the cotyledon stage and continuing to maturity has been well studied by physiologists, biochemists, and molecular biologists. Although the metabolic rate can be changed at the biophysical level without changing the cellular biological pathways (Gillooly et al. 2001), the rapid increase in storage protein synthesis is largely attributable to increases in transcription of storage protein gene families that code for subunits of the major storage proteins, glycinin and conglycinin (Goldberg et al. 1989; Harada et al. 1989; Nielsen 1996). The expression of storage protein genes is considered a marker for seed maturation and a gradient of storage protein gene expression is generally observed in developing seeds. In soybean cotyledons, the expression of several storage protein genes is developmentally regulated, and the expression pattern is spread in a wave from the outer to the inner surface (Perez-Grau and Goldberg 1989).

During seed development, the major assimilates are transported to the developing embryos from the sites of photosynthesis in the form of sucrose. This is facilitated by the phloem transport system which supplies about 98 % of carbon, 89 % of nitrogen and 40 % water from the parent plant to the embryos (Pate 1980). Phloem transport may be directly involved in the remote control of reserve mobilization to the seeds, because sugars themselves regulate gene expression in sinks (Rolland et al. 2002). It was shown that the root originated xylem sap abscisic acid (ABA) can move to crop reproductive structures and accumulate under drought conditions. The accumulation of ABA in reproductive structures involved in the regulation of regulates kernel/pod abortion, presumably via inhibition of cell division in young ovaries (Liu et al. 2003). Also, low pod water potential, which might lead to disruptions in metabolic activitie, is important in determining pod abortion in soybean. The import of amino acids into the seeds occurs mainly by this transport system, which requires efficient exchange mechanisms between the xylem and the phloem (Pate et al. 1977). Okumoto et al. (2002) recently reported an *Arabidopsis* amino acid transporter (AAP6) with an expression pattern and biochemical properties for xylem-to-phloem transfer of amino acids. High affinity transport of all available amino acids is an essential prerequisite for the accumulation of seed storage proteins. In peas, it was demonstrated that PsAAP1 is expressed in the transfer cell layer of cotyledons and might be responsible for transport of amino acids released from the seed coat (Tegeder et al. 2000). In addition to amino acid transport, peptide transport seems to play an important role in protein deposition during seed development and storage protein hydrolysis during germination (Stacey et al. 2002). In an extensive study, Miranda et al. (2003) reported that in developing seeds, the highest levels of peptide transporter VfPTR1 transcripts were reached during mid-cotyledon development, whereas the VfAAP1 transcripts were most abundant during early cotyledon development, before the appearance of storage protein gene transcripts. The amount of protein accumulation during seed development is regulated at different levels, including the availability of assimilates and genetic background (Golombek et al. 2001). It is reported that the drought-induced decreases in photosynthetic rate are significant in inducing pod abortion, probably as a consequence of carbohydrate deprivation. The effects of ABA and brassinosteroids (BA) on pod set may be partially due to their effects on photosynthate supply (Liu et al. 2004). The correlation between the reserve deposition and changes in transcript profile and protein expression in the different developmental stages under drought stress conditions are not known.

Flooding

Flooding due to excess soil water is the second most damaging constraint on crop growth, after drought, and affects about 16 % of the production areas worldwide (Boyer 1982). Soil can become flooded when it is poorly drained or when rainfall or irrigation is excessive. Lack of oxygen was proposed as the main problem associated with flooding (Kozlowski 1984). During the last two decades, more information has accumulated from research on the molecular, biochemical and physiological responses of plants to the lack of oxygen than to flooding (Ricard et al. 1994; Ismond et al. 2003). Klok et al. (2002) used gene expression profiling to characterize the anaerobic response in experiments with root cultures. Analysis of the 5′ regulatory regions of the differentially expressed genes after microarray profiling revealed common sequence motifs, suggesting the expression of gene groups may be regulated by common regulatory factors (Klok et al. 2002). Recently, a comprehensive analysis of the hypoxia-responsive transcriptional networks in the model plant Arabidopsis was conducted by Liu et al. (2005) using whole-genome DNA amplicon microarrays.

Flooding causes premature senescence that results in leaf chlorosis, necrosis, defoliation, cessation of growth and reduced yield (VanToai et al. 1994). It is reported that the cytokinin, the anti-senescence hormone, content in the xylem sap falls lower within one day of flooding (Burrows and Carr 1969). Cytokinins are synthesized at the root apical meristem (Short and Torrey 1972) where depressed metabolic activity and cell death due to flooding occur much earlier than in other tissues (VanToai et al. 2001). Flooding can be divided into either waterlogging, where only the roots are flooded or complete submergence where the entire plants are under water. Waterlogging is more common than complete submergence stress and is also less damaging. While soybeans die within one or two days of complete submergence (Sullivan et al. 2001), plants develop adaptive mechanisms that allow them to survive long term waterlogging (Bacanamwo and Purcell 1999). Soybean responses to waterlogging can be divided into two phases such as the "reactive" phase, which occurs immediately after the stress and is followed by the second "acclimatized" phase. The duration of the reactive phase was 2–4 days shorter in the flood-tolerant genotypes than flood-susceptible genotypes, indicating that flood-tolerant genotypes acclimatized more quickly to flooding stress than the flood-susceptible genotypes (VanToai et al. 2003). Bacanamwo and Purcell (1999) reported that the rates of biomass accumulation during the first week of flooding were negligible. One to two weeks after flooding, biomass accumulated at approximately one half of the control rates and these observations coincided with increased root porosity due to aerenchyma and adventitious root development. Waterlogging can reduce soybean yield 17–43 % at the vegetative growth stage and 50–56 % at the reproductive stage (Oosterhuis et al. 1990). Yield losses are the result of reduced root and shoot growth, nodulation, nitrogen fixation, photosynthesis, biomass accumulation, stomatal conductance, and plant death due to diseases and physiological stress (Scott et al. 1989; Oosterhuis et al. 1990).

Various reports on watgerlogging tolerance in crop plants concluded that this trait is quantitatively inherited (Sripongpangkul et al. 2000; Setter and Waters 2003; Boru et al. 2003). Breeding for stress tolerance controlled by quantitative traits is difficult because of low heritability, variability among stress treatments, and the difficulty of screening large numbers of progeny in field or greenhouse assays of tolerance (Reyna et al. 2003). QTL for waterlogging tolerance associated with improved soybean growth (18 %) and grain yields (180 %) in waterlogged environments were identified (VanToai et al. 2001). In their studies, soybean genotype "Archer" was used as the source of water logging tolerance and marker Sat_064. The Sat_064 QTL, which was mapped to LG G of the USDA soybean linkage map (Cregan et al. 1999), was uniquely associated with waterlogging tolerance and was not associated with maturity, normal plant height or grain yields. Also, **?** mentioned that the water logging QTL associated with Sat_064 was not related to the effects of the *Rps4* gene. Reyna et al. (2003) evaluated the effect of this QTL on waterlogging tolerance in the Southern USA and assessed the variability for waterlogging tolerance in Archer-derived populations. They used Sat_064 genotype data to create seven sets of NILs from the populations A5403 × Archer and 9641 × Archer and no significant variations was observed between Sat_064 and water logging tolerance in their study. Other major QTL for water-logging tolerance from Archer on LG A1 and LG F were discovered in the US Mid-south environment (Cornelious et al. 2005). Recombinant inbred lines (RILs) from A5403 × Archer (Population 1) and P9641 × Archer (Population 2), respectively, were used for mapping these QTLs.

Temperature Stress

Soybean plant growth, development, seed composition and quality of soybean are dependent on many factors including genetic background, growing season, geographical location and environmental stresses (Liu et al. 1995; Piper and Boote 1999). It is known that temperature variation has a significant effect on reserve mobilization and partitioning during seed development, but the molecular and regulatory mechanisms are not well understood. Several studies showed a difference in seed protein and oil composition due to varied temperature regimes from 15/12 to 40/30°C (Carver et al. 1986; Dornbos and Mullen 1992; Rebetzke et al. 1996). In addition to changes in the seed oil concentration, the ratio of fatty acids in soybean oil changes when seeds develop under elevated temperature (Thomas et al. 2003). In soybeans, oleic acid content is most affected by night temperatures rather than day temperatures. Higher oleic acid content is obtained in the southeast (e.g. North Carolina) than in the northern midwest (e.g. Iowa). A recent phytotron study showed a variation in oleic acid content from 17 to 30 % in soybean cultivar "Brim" under the day/night temperature regime of 22/18°C vs 38/27°C. Under the same conditions, oleic acid content varied from 36 to 66 % in the germplasm line N98-4445A (Burton et al. 2006). This germplasm will be a useful genetic resource for breeding mid-oleic soybean varieties; that is, those with concentrations of oleic acid between 400 and 700 g/kg. Increased oleic acid in this line causes a correlated decrease in polyunsaturated fatty acids giving the added advantage of linolenic acid concentrations of less than 30 g/kg (Burton et al. 2006).

The protein concentration of soybean produced in the southern USA is generally greater than that of soybean produced in northern regions (Breened et al. 1988; Hurburgh et al. 1990). As one of the major environmental factors, temperature has a direct role on the protein and oil accumulation during seed development. Wolf et al. (1982) observed the drastic reduction of linolenic acid and stachyose content and increased total oil, methionine and protein content when soybean plants were grown under warm temperature 33/28°C (day/night). In contrast, a negative correlation between oil and protein was reported (Hymowitz et al. 1972; Burton 1987; Watanabe and Nagasawa 1990) where oil concentration tends to decrease as the protein concentration increases. This response was attributed to both environmental and genotypic variations (Watanabe and Nagasawa 1990). Keirstead (1952) and Kane et al. (1997) found a positive correlation between oil content and temperature, i.e. considerable increase in total oil content with increase in temperature in many cultivars of soybean. Recently, in a controlled growth chamber and greenhouse experiments, Piper and Boote (1999) reported an increase in oil and protein content of different soybean cultivars under temperature stress. Their data clearly showed genotypic differences in response to temperature stress.

Tsukamoto et al. (1995) reported that elevated temperature dramatically decreased isoflavones in soybean seeds, but had no effect on saponins. Vlahakis and Hazebroek (2000) reported that high temperature increased sterols in soybean seeds, with relative increases in campesterol and decreases in stigmasterol and β-sitosterol. Almonor et al. (1998) found that elevated temperatures caused small (~10 %) increases in total tocopherols (both free and esterified), partly as a result of increased γ-tocopherol, the tocopherol species found most abundantly in soybean seed. The other main soybean tocopherols, α-tocopherol and δ-tocopherol, did not change consistently. Recently it was shown that that weather or climate can significantly affect seed tocopherols (Britz and Kremer 2002).

Temperatures of 33/28°C (day/night) (Keigley and Mullen 1986), 35°C (Dornbos and Mullen 1991), 35/30°C (Gibson and Mullen 1996), 38/33°C (Spears et al. 1997), and 38/27°C (TeKrony et al. 2000; Egli et al. 2005) during seed filling reduced germination of seed from several cultivars. Seed vigor was often more sensitive to high temperature than standard germination and the seed vigor was reduced at 33/28°C, while 38/33°C was required to reduce standard germination (Spears et al. 1997). Gibson and Mullen (1996) reported no difference in the temperature required to reduce standard germination and accelerated-aging germination. Pollen abnormalities were reported in soybean grown in elevated temperature and it was reported that the morphology of pollen was affected more severely and significantly in the heat-sensitive and heat-intermediate genotypes than in the heat-tolerant genotype at the high temperature (Koti et al. 2005; Salem et al. 2007).

Lobell and Asner (2003) studied the relationship between climate variation and soybean and corn crop production by synthesizing data on temperature, precipitation, solar radiation, and crop yields in the United States between 1982 and 1998. In their studies, two regions with a distinct correlation between yield and climate anomalies were observed: a large Midwest region where yields were favored by cooler, wetter years and a smaller region including the Northern Great Plains favored by hotter, drier years. They concluded that gradual temperature changes had a measurable impact on crop yield trends i.e. the slope of regression (r_y) indicated a roughly 17 % relative decrease in both corn and soybean yield for each degree increase in growing season temperature. Recent simulation studies showed that in a future warmer climate with increased mean temperatures, heat waves would become more intense, longer lasting, and/or more frequent (Karl and Trenberth 2003; Meehl and Tebaldi 2004) and this certainly will reduce soybean production. Soybean yield reductions of 27 % were reported when the plants were exposed to 35°C for 10 hr during the day (Gibson and Mullen 1996).

Diers et al. (1992) mapped two major QTL controlling protein and oil concentration with RFLP markers in a population of F2-derived lines developed by crossing a *G. max* experimental line and a *G. soja* plant introduction. The QTL mapped were then labeled LG A and K of the soybean map. Chung et al. (2003) reported the mapping of a protein QTL allele to LG I from PI 437088A, a *G. max* accession with a high protein concentration. The results from the Set 1 and Set 2 evaluations and mapping studies conducted by Nichols et al. (2006) confirmed the findings of Sebolt et al. (2000), Diers et al. (1992), and Chung et al. (2003), who demonstrated that a major QTL for seed protein and oil concentration is located on LG I. This protein QTL is designated cqPRO-003 and the oil QTL cqOIL-004 under the category of confirmed QTL at the Soybase website. All these results show that seed component traits can be successfully modified through genetic mapping coupled with marker-assisted selection.

Salinity

Salinity is one of the major threats to crop yields and about one third of the world's irrigated lands are affected by salt (Moore 1984). The development of salt tolerant crops is always a challenge since this trait is complex genetically and physiologically. Soybean is considered to be a salt sensitive crop (Lauchli 1984) and soybean cultivars exhibit differential tolerance to salinity during seed germination and plant growth (Abel and MacKenzie 1964). It is reported that the differential effects of salinity on growth rates and photosynthesis might be due to specific ion toxicities in crops (Kingsbury and Epstein 1986). Chloride toxicity was recognized as a problem in soybean fields (Myron et al. 1983). The major symptoms of chloride induced toxicity in soybean plants grown in saline conditions are leaf chlorosis, stunted growth and biomass reduction (Abel and MacKenzie 1964; Kurniadie and Redmann 1999).

The salt tolerance of a plant is closely associated with the efficiency of regulation of Na^+/Cl^- transport from root system to shoot. A chloride tolerant soybean line showed elevated levels of chloride in roots when compared to the leaf tissue under salt stressed conditions. The sensitive cultivar had higher absorption rates even under lower levels of salt concentrations and the release of Cl^- from the symplast into the xylem was less controlled (Wieneke and Läuchli 1979). To develop salt tolerant crops, the evaluation for salinity tolerance threshhold within the crop and their wild-type relatives should be conducted. Chloride tolerant germplasm identified with their accession numbers will be a valuable resource for the soybean research community (Hymowitz and Bernard 1991). Initial screening of half dozen soybean varieties at various salinity levels showed no varietal difference in chloride accumulation in roots (Abel and MacKenzie 1964). Another evaluation with 65 cultivars showed that more than 50 % of the lines developed leaf scorch, higher levels of chloride in leaf and seed tissues, and significantly reduced seed yield (Parker et al. 1986). Yang and Blanchar (1993) also showed differential grain yield and chloride accumulation among 60 soybean cultivars. Also, considerably greater variation in sodium chloride tolerance was reported among the perennial *Glycine* accessions than among the *G. max* cultivars. The variability for chloride tolerance among these accessions has potential utility for developing enhanced salt tolerance in soybean (Pantalone et al. 1997). Perennial *Glycine* to *G. max* gene transfer is considered to be difficult (Hymowitz and Bernard 1991), but the gene introduction from related species is an achievable approach to modify the soybean genome (Luo et al. 1994).

The identification of major QTL in two contrasting environments indicates a consistent expression of a salt tolerance QTL in the soybean population (Lee et al. 2004). Cation transport is thought to be an important process for ion homeostasis in plant cells. Luo et al. (2005) reported that a soybean putative cation/proton antiporter GmCAX1 may be a mediator of this process. GmCAX1 is expressed in all tissues of the soybean plants but at a lower level in roots.

Soil Mineral Stress

Sub-optimal availability of mineral nutrients and ion toxicities are major constraints for several natural and agricultural ecosystems. Plants take up essential mineral nutrients from soil along with water uptake. N, P, K, Ca, Mg, and S are considered as macro-nutrients, because they are required in large quantities that range between 1 and 150 g/kg of plant dry matter. Fe, Zn, Mn, Cu, B, Mo and Cl are minor or micro-nutrients that are required at rates of 0.1–100 mg/kg of plant dry matter (Marschner 1997). In this section, we summarize the major physiological factors and QTL related to some of the important mineral stresses and ion toxicity.

Phosphorus (P) deficiency is one of the soil mineral stresses that limit plant growth and crop productivity (Sanchez and Salinas 1981). The ability to fix nitrogen makes P the most limiting elements for growth and yield in soybean. One of the severe problems of P deficiency in soybean is limited nodule growth relative to shoot growth (Ribet and Drevon 1995). Decrease in shoot growth is the earliest and most pronounced responses to P-deficiency, specifically in leaf number and leaf size (Lynch et al. 1991; Chiera et al. 2002)). Suboptimal P supply to plants in vivo generally results in enhanced levels of foliar starch. Fredeen et al. (1989) observed increased levels of starch in young growing leaves, mature leaves, and fibrous roots of low-P plants. Li et al. (2005) reported 7 major QTLs for three traits (shoot fresh weight, phosphorus content in root, and phosphorus content in leaf) associated with P deficiency tolerance in soybean and these QTLs were mapped on F linkage group, two on F1, and five on F2.

Aluminum (Al) toxicity and P deficiency often coexist in acid soils and are most important due to their ubiquitous existence and overwhelming impact on plant growth (Kochian et al. 2004). Dong et al. (2004) showed interactions between root Al and P on soybean growth and the Al toxicity and P deficiency influenced root organic acid exudation patterns. This study reported that the oxalate and malate release was induced by P deficiency, while Al activated root citrate exudation. Liao et al. (2006), through a comprehensive study using different soybean germplasm, concluded that the phosphorus-efficient genotypes may be able to enhance Al tolerance not only through direct Al-P interactions but also through indirect interactions associated with stimulated exudation of different Al-chelating organic acids in specific roots and root regions, which in turn enhances plant tolerance to Al toxicity. At the gene level, Al activated the threonine-oriented phosphorylation of a plasma membrane H^+-ATPase in a dose- and time-dependent manner and it was demonstrated that up-regulation of plasma membrane H^+-ATPase activity was associated with the secretion of citrate from soybean roots (Shen et al. 2005). Recent genetic analysis using 120 F_4-derived progeny from Young × PI 416937 revealed five QTL from independent linkage groups that conditioned root extension under high aluminum stress (Bianchi-Hall et al. 2000).

Genes and Genetic Engineering for Abiotic Stress Tolerance

Plants respond and adapt to abiotic stresses through various biochemical and physiological processes. Studies using functional genomics tools, such as transcriptomics and proteomics, revealed several stress-inducible genes and various transcription factors that regulate drought-stress-inducible systems. Several classes of genes induced by drought stress include genes for osmolyte accumulation, amino acid transporters, membrane protection, early signaling and transcription factors (Valliyodan and Nguyen 2006). Most of these genes or regulatory networks were identified in plants such as *Arabidopsis*, rice and maize. Efforts to identify drought stress related genes/proteins and transcription factors in soybean are underway. Translational genomics of these candidate genes using model plants provided encouraging results, but the field testing of transgenic crop plants for better performance and yield is still minimal.

One of the major physiological events during water stress in plants is the production of the phytohormone abscisic acid (ABA), which in turn causes stomatal closure and induces expression of stress-related genes (Shinozaki and Yamaguchi-Shinozaki 2007). There are at least four independent regulatory systems for gene expression in response to water stress. Two are ABA-dependent and two are ABA-independent (Yamaguchi-Shinozaki and Shinozaki 2005, 2006). A *cis*-acting element is involved in the ABA-independent regulatory systems, Dehydration-responsive element/C-repeat (DRE/CRT) which also functions in cold- and high-salt-responsive gene expression. When the DRE/CRT-binding protein DREB1/CBF was over-expressed in transgenic *Arabidopsis* plants, changes in the expression of more than 40 stress-inducible genes were identified, and these changes led to increased freezing, salt, and drought tolerance (Seki et al. 2001; Fowler and Thomashow 2002; Maruyama et al. 2004). The trans-acting factor, DRE-binding (DREB), protein can bind to DRE to activate gene expression in the stress-signaling pathway of plants. Many DREB proteins were found in *Arabidopsis* (Liu et al. 1998; Medina et al. 1999). The *DREB/CBF* genes can be regulated by various transcription factors (Chinnusamy et al. 2003; Novillo et al. 2004) and Ca^{2+} related transporters and proteins (Catala et al. 2003).

Li et al. (2005) isolated three *DREB* genes, *GmDREBa*, *GmDREBb*, and *GmDREBc* from soybean and analyzed their DRE-binding activities and also examined their transcriptional activation activities, genomic organizations, and expression patterns under different stress conditions. All of the DREB genes encode proteins with the conserved AP2 domain of 64 amino acids. Their results indicate that expression of both *GmDREBa* and *GmDREBb* was induced by salt, drought, and low temperature in leaves of soybean seedlings. Expression of *GmDREBc* was not induced by salt, drought, low temperature, or ABA in leaves but induced in the roots of soybean seedlings. The authors' research group has expressed the arabidopsis DREB gene in soybean in a stress inducible manner and found a higher degree of drought stress resistance in soybean plants (unpublished data). Recently, Chen et al. (2007) isolated a DREB homologous gene, GmDREB2, from soybean. Based on its similarity with AP2 domains, GmDREB2 was classified into the A-5 subgroup of the DREB subfamily in the AP2/EREBP family. Similar to the earlier reports from Li et al. (2005), the expression of GmDREB2 gene was induced by drought, high salt, and low temperature stresses and ABA treatment. In this study, a stress inducible (Rd29A) and constitutive (CaMV35S) promoters were used to control expression of *GmDREB2* gene in transgenic Arabidopsis and tobacco plants. Both 35S::GmDREB2 and Rd29A::GmDREB2 transgenic plants had no phenotypic changes during the seedling stage or mature stage and the results suggested that the survival rate of 35S::GmDREB2 *Arabidopsis* plants was higher than that of Rd29A::GmDREB2 plants under water deficit and salinity stress conditions.

The NAC family is another major group of transcription factors (TF) that have role in root development and stress tolerance in plants (Tran et al. 2004). The name NAC comes from proteins of Petunia no apical meristem (NAM), Arabidopsis ATAF1, ATAF2 and *CUC2* (cup-shaped cotyledon) (Aida et al. 1997; Souer et al. 1996). Considering the important roles of these transcriptional factors, NAC proteins are proposed to play vital roles during soybean development. Meng (2006) isolated and cloned six NAC-like genes in soybean to illustrate their functions in soybean. They found that all six NAC-like genes share similar genomic organization, and showed high sequence similarity, especially within the NAC domain. These transcription factors (TFs) were categorized into five subgroups based on phylogenetic analysis. Tissue-specific expression analysis showed that these genes exhibited different expression patterns, especially during seed filling and seed development. However, still lacking is a functional characterization of the soybean NAC TFs to understand their specific role in root development and abiotic stress responses.

It was reported that plants can adapt to a stress condition by inducing a specific group of genes encoding late embryogenesis abundant proteins (LEA) (Moons 1997). During cellular dehydration, LEA proteins play an important role in maintenance of the structure of other proteins, vesicles, or endomembrane structures, in the sequestration of ions such as calcium, in binding or replacement of water, and functioning as molecular chaperones (Heyen et al. 2002; Koag et al. 2003). Two dehydrin-encoding genes were cloned from *Glycine max* (*gmlea* 8 and *gmlea* 10) and were analyzed for their contribution to the response against drought in mycorrhizal soybean plants (Porcel et al. 2005). In this experiment, the soybean plants showed that most of the treatments did not result in LEA gene expression under well-watered conditions but higher gene expression was found in non-inoculated plants subjected to drought. It was also found that plants singly inoculated with *Bradyrhizobium japonicum* showed a higher level of LEA gene expression under well-watered conditions and a reduced level under drought-stress conditions. These results demonstrate that the LEA genes respond to drought stress and their mRNA accumulated under drought conditions in roots and nodules of soybean contributing to their protection against drought. The same research group cloned two soybean aquaporin genes, GmPIP1 and GmPIP2 and studied their role in water deficit conditions (Porcel et al. 2006). The results suggested that arbuscular mycorrhizal plants respond to drought stress by down-regulating the expression of the two PIP genes studied. This down-regulation of PIP genes is likely to be a mechanism to decrease membrane water permeability and to allow cellular water conservation.

Rodrigues et al. (2006) cloned a soybean antiquitin homologue gene, designated GmTP55, which encodes a dehydrogenase motif-containing 55 kDa protein induced by dehydration and salt stress. Antiquitin genes are aldehyde dehydrogenase superfamily enzymes catalyzing the conversion of various endogenous and exogenous aldehydes to the corresponding carboxylic acids using the coenzyme NAD^+ or $NADP^+$ (Yoshida et al. 1998). The physiological function of antiquitin is believed to be related to the regulation of turgor pressure or to a general stress response (Rodrigues et al. 2006). It was reported that the plant antiquitin homologue genes are induced by water deprivation and exposure to high salinity in pea (*Pisum sativum*), and in Arabidopsis (Guerrero et al. 1990; Kirch et al. 2005). When they expressed the soybean ALDH7 in tobacco and *Arabidopsis* plants, the transgenic lines producing the soybean enzyme displayed a lower concentration of reactive aldehydes and enhanced tolerance to drought, salinity, and ROS-producing chemical treatments. It was suggested that the accumulation of amino acids such as proline plays a protective role in plants under drought, temperature and salinity stress conditions (Good and Zaplachinski 1994; Valliyodan and Nguyen 2006). It was reported that the suppression of proline synthesis gene pyrroline-5-carboxylate reductase (P5CS) in transgenic soybean plants by antisense silencing resulted in an increased sensitivity to water stress (de Ronde et al. 2000). The transgenic soybean plants over-expressing P5CS showed faster proline accumulation and experienced the least water loss when compared to the antisense transformants, which possessed slower proline accumulation (Simon-Sarkadi et al. 2005).

Reprogramming of the transcriptional machinery is the central mechanism of heat shock responses and that is characterized by the synthesis of heat shock proteins (HSP) and the acquisition of thermotolerance (Vierling 1991; Klueva 2001; Volkov et al. 2003). The transcription activation of heat shock genes is generally associated with the decline of transcriptional activity of other genes in plants. In soybean seedlings, the complexity and abundance of mRNAs was shown to be significantly reduced after heat stress at 40°C compared to tissue incubated under normal conditions at 28°C (Schöffl and Key 1982). In another study, it was shown that the transcription of constitutively expressed non-heat shock genes was down-regulated during heat stress (Schöffl et al. 1987), but at the same conditions the transcription of heat shock genes was strongly activated. The specific role of HSPs, the heat stress transcription factors (HSFs) and the transcriptional cascade regulating expression of heat stress proteins during seed development in model plants were well documented (Kotak et al. 2004, 2007a,b). However, there is no report showing the role of specific HSPs or HSFs in soybean adaptation to heat stress in the field conditions. Another implication of higher temperature in oil seeds including soybean is the alteration of fatty acid composition. In soybean, growth at higher temperatures results in decreased linoleic and linolenic acid concentrations of seed triacylglycerols and a corresponding increase in oleic acid concentration (Rennie and Tanner 1989; Rebetzke et al. 1996). The studies on the mechanisms of temperature-dependent alterations of fatty acid composition of plant membrane lipids provided evidence of control at both transcriptional and post-transcriptional levels for genes encoding the *omega*-3 desaturases responsible for converting linoleic acid to linolenic acid (Tang et al. 2005). It was reported that the soybean genome possesses two seed-specific isoforms of FAD2, FAD2-1A and FAD2-1B and their expression studies revealed that the FAD2-1A isoform was more unstable than FAD2-1B at elevated growth temperatures (Tang et al. 2005). Temperature regulated changes in the poly unsaturated fatty acid content help to maintain the right membrane fluidity which play crucial role in genotype environmental interactions.

Future Perspectives

A better understanding of the molecular mechanisms of abiotic stress responses will allow an efficient application of genomics technology in crop improvement and sustainable agriculture. Several studies focusing on the genetic variability of abiotic stress responses were conducted in soybean, but the key molecular regulation of the abiotic stress responsive genes/proteins and their interactive networks associated with plant performance (i.e. yield) is lacking. The authors' research group has conducted extensive gene expression and protein expression profiling in root and leaf tissues to identify the candidates genes/regulatory switches involved in the drought responsive mechanisms in soybean (Valliyodan et al. unpublished data). Because of the nutritional and commercial importance of soybean protein and oil, it is also desirable to understand the molecular basis of phenotypic variation in seed composition associated with drought and temperature stress.

Recently, genomic technologies have emerged as promising tools to overcome abiotic stress effects in plants, but progress has been limited in food legumes, especially in soybean. Major achievements and advances in soybean transformation and marker-assisted selection, together with the application of functional genomics, offer great potential to soybean improvement. Availability of mutant populations in soybean is another resource to study and characterize the abiotic stress gene functions leading to better soybean productivity. The collaborative effort between the soybean research community and DOE-JGI on soybean genome sequencing will improve the process of gene discovery and the elucidation of regulatory pathways involved in stress responses and soybean development. Also, the functional characterization of abiotic stress genes and transcription factors should be translated toward the production of better soybean plants. Another major need for the soybean improvement program is further characterization of traits associated with abiotic stress avoidance/tolerance and plant performance such as root architecture and plasticity, water-use efficiency, and seed filling. More efforts must be put on the genetic screening of soybean germplasm resources for stress resistance traits. Development of high throughput phenotyping assays and suitable screening facilities is key to the genetic characterization and dissection of genes/QTL controlling stress resistance and adaptation. Moreover, secondary germplasm pools composed of perennial soybean species such as *G. tomentella* and *G. canescens*, which provide an untapped resource for abiotic stress resistance genes, should also be exploited. The integration of genetics and knowledge generated by various genomic and functional genomic approaches should lead to more accurate and efficient breeding of soybean crops. In the context of soybean improvement it is important to link basic "stress tolerance" findings in the laboratory and controlled conditions to field performance.

References

Abe, J., Xu, D., Suzuki, Y., Kanazawa, A. and Shimamoto, Y. (2003) Soybean germplasm pools in Asia revealed by nuclear SSRs. Theor. Appl. Gen. 106, 445–453.

Abel, G.H. and MacKenzie, A.J. (1964) Salt tolerance of soybean varieties (*Glycine max* L. Merill) during germination and later growth. Crop Sci 4:, 157–161.

Aida, M., Ishida, T., Fukaki, H., Fujisawa, H. and Tasaka, M. (1997) Genes involved in organ separation in *Arabidopsis*, An analysis of the cup-shaped cotyledon mutant. Plant Cell 9, 841–857.

Almonor, G.O., Fenner, G.P. and Wilson, R. F. (1998) Temperature effects on tocopherol composition in soybeans with genetically improved oil quality. J. Am. Oil Chem. Soc. 75, 91–596

Araus, J.L., Slafer, G.A., Reynolds, M.P. and Royo, C. (2002) Plant breeding and drought in C_3 cereals: what should we breed for? Ann.Bot. 89, 925–940.

Arrese-Igor, C., González, E.M., Gordon, A.J., Minchin, F.R., Gálvez, L., Royuela, M., Cabrerizo, P.M. and Aparicio-Tejo, P.M. (1999) Sucrose synthase and nodule nitrogen fixation under drought and other environmental stresses. Symbiosis 27, 189–212

Arumuganathan, K. and Earle, E.D. (1991) Nuclear DNA content of some important plant species. Plant Mol. Biol. Rep. 9, 208–219.

Bacanamwo, M. and Purcell, L. (1999) Soybean dry matter and N accumulation responses to flooding stress, N sources and hypoxia. J. Exp. Bot. 50, 689–696.

Begg, J. E. and Turner, N. C. (1976) Crop water deficits. Adv. Agron. 28, 161–217.

Bianchi-Hall, C.M., Carter, Jr. T.C., Bailey, M.A., Mian, M.A.R., Rufty, T.W., Ashley, D.A., Boerma, H.R., Arellano, C., Hussey, R.S. and Parrott, W.A. (2000) Aluminum tolerance associated with quantitative trait loci derived from soybean PI 416937 in hydroponics. Crop Sci. 40, 538–545.

Boru, G., Vantoai, T., Alves, J., Hua, D. and Knee, M. (2003) Responses of soybean to oxygen deficiency and elevated root-zone carbon dioxide concentration. Ann. Bot. 91, pp. 447–453.

Boyer, J.S. (1982) Plant productivity and environment. Science 218, 443–448.

Boyer, J.S. (1983) Environmental stress and crop yields. In: C.D. Raper and P.J. Kramer (Eds.), *Crop reactions to water and temperature stresses in humid, temperate climates*. Westview press, Boulder, Colorado, pp. 3–7.

Boyer, J.S., Johnson, R.R. and Saupe, S.G. (1980) Afternoon water deficits and grain yields in old and new soybean cultivars. Agron J. 72, 981–986.

Breened, W.M., Lin, S., Hardman, L., and Orf, J. (1988) Protein and oil content of soybeans from different geographic locations. J. Am. Oil. Chem. Soc. 65, 1927–1931

Britz, S.J. and Kremer, D.F. (2002) Warm Temperatures or Drought during Seed Maturation Increase Free -Tocopherol in Seeds of Soybean (*Glycine max* [L.] Merr.). J. Agric. Food Chem. 50, 6058–6063.

Burrows, W.J. and Carr, D.J. (1969) Effects of flooding the root system of sunflower plants on the cytokinin content of the xylem sap. Plant Physiol. 22, 1105–1112.

Burton, J.W. (1987) Quantitative genetics; Results relavent to soybean breeding. In: J.R. Wilcox (ed.), Soybeans: Improvement, production and uses. Agronomy Monograph 16. 2nd ed. ASA, CSSA, and SSSA, Madison, WI

Burton, J.W., Wilson, R.F., Rebetzke, G.J. and Pantalone,V.R. (2006) Registration of N98-4445A Mid-Oleic Soybean Germplasm Line. Crop Sci. 46, 1010–1012.

Bushhamuka, V.N., and Zobel, R.W. (1998) Different genotypic and root type penetration of compacted soil layers. Crop Sci. 38, 776–781.

Carlson, D.R., Dyer, D.J., Cotterman, C.D. and Durley, R. (1987) The physiological basis for cytokinin induced increases in pod set in IX93-100 soybeans. Plant Physiol. 84, 233–239.

Carlson, J.B. and Lersten, N.R. (1987) Reproductive morphology. In: J.R. Wilcox (Ed.) *Soybeans: Improvement, Production and Uses*. American Society of Agronomy, Madison, WI, pp. 95–134.

Carver, B.F, Burton, J.W, Carter, T.E Jr., and Wilson, R.F. (1986) Response to environmental variation of soybean lines selected for altered unsaturated fattyacid composition. Crop Sci. 26, 1176–1180.

Catala, R., Santos, E., Alonso, J.M., Ecker, J.R., Martinez-Zapater, J.M. and Salinas, J. (2003) Mutations in the Ca^{2+}/H^{+} transporter CAX1 increase CBF/DREB1 expression and the cold-acclimation response in *Arabidopsis*. Plant Cell 15, 2940–2951.

Chen, M., Wang, Q.Y., Cheng, X.G., Xu, Z.S., Li, L.C., Ye, X.G., Xia, L.Q. and Ma, Y.Z. (2007) GmDREB2, a soybean DRE-binding transcription factor, conferred drought and high-salt tolerance in transgenic plants. Biochem. Biophys. Res. Commun. 353, 299–305.

Chiera, J., Thomas, J. and Rufty, T. (2002) Leaf initiation and development in soybean under phosphorus stress. J Exp Bot 53, 473–481.

Chinnusamy, V., Ohta, M., Kanrar, S., Lee, B.H., Hong, X.H., Agarwal, M. and Zhu, J.K. (2003) ICE1: a regulator of cold-induced transcriptome and freezing tolerance in *Arabidopsis*. Genes Dev 17, 1043–1054.

Chung, J., Babka, H.L., Greaf, G.L., Staswick, P.E., Lee, D.J., Cregan, P.B., Shoemaker, R.C. and Specht, J.E. (2003) The seed protein, oil, and yield QTL on soybean linkage group I. Crop Sci. 43, 1053–1067.

Cornelious, B., Chen, P., Chen, Y., de Leon, N., Shannon, J.G. and Wan, D. (2005) Identification of QTLs Underlying Water-Logging Tolerance in Soybean. Mol Breeding 16, 103–112.

Cregan, P.B., Jarvik, T., Bush, A.L., Shoemaker, R.C., Lark, K.G., Kahler, VanToai, T.T., Lohnes, D.G., Chung, J. and Specht J.E. (1999) An integrated genetic linkage map of the soybean. Crop Sci. 39, 1464–1490.

Danesh, D., Penuela, S., Mudge, J., Denny, R.L., Nordstrom, H., Martinez, J.P. and Young, N.D. (1998) A bacterial artificial chromosome library for soybean and identification of clones near a major cyst nematode resistance gene. Theor. Appl. Genet. 96, 196–202.

de Ronde, J.A., Spreeth, M.H. and Cres, W.A. (2000) Effect of antisense l-Δ1-pyrroline-5-carboxylate reductase transgenic soybean plants subjected to osmotic and drought stress. Plant Growth Regul. 32, 13–26.

Desclaux, D., Huynh, T. and Roumet P. (2000) Identification of soybean plant characteristics that indicate the timing of drought stress. Crop Sci. 40, 716–722.

Diers, B.W., Keim, P., Fehr, W.R. and Shoemaker,R.C. (1992) RFLP analysis of soybean seed protein and oil content. Theor. Appl. Genet. 83, 608–612.

Dong,D., Peng, X. and Yan, X. (2004) Organic acid exudation induced by phosphorus deficiency and/or aluminum toxicity in two contrasting soybean genotypes. Physiol. Plant. 122, 190–199.

Dornbos, D.L., and Mullen, R.E. (1992) Soybean seed protein and oil contents and fattyacid composition adjustments by drought and temperature. J Am. Oil Chem. Soc. 69, 228–231.

Dornbos, D.L.J. and Mullen, R.E. (1991) Influence of stress during soybean seed fill on seed weight, germination and seedling growth rate. Can. J. Plant Sci. 35, 373–383.

Durand, J.L., Sheehy, J.E. and Minchin, F.R. (1987) Nitrogenase activity, photosynthesis and nodule water potential in soybean plants experiencing water deprivation. J Exp. Bot. 38, 311–321.

Egli, D.B., TeKrony, D. M., Heitholt, J. J. and Rupe, J. (2005) Air temperature during seed Filling and soybean seed germination and vigor. Crop Sci. 45, 1329–1335.

Fehr, W.R. (1984) *Genetic contribution to yield gains of five major crop plants*. CSSA Special Publ. 7. CSSA and ASA, Madison, USA.

Fowler, S. and Thomashow, M.F. (2002) Arabidopsis transcriptome profiling indicates that multiple regulatory pathways are activated during cold acclimation in addition to the CBF cold response pathway, Plant Cell 14, 1675–1690.

Fredeen, A.L., Rao, I.M. and Terry, N. (1989) Influence of phosphorus nutrition on growth and carbon partitioning in Glycine max. Plant Physiol. 89, 225–230.

Gálvez, L., González, E.M. and Arrese-Igor, C. (2005) Evidence for carbon flux shortage and strong carbon/nitrogen interactions in pea nodules at early stages of water stress. J Exp. Bot. 419, 2551–61.

Gibson, L.R., and Mullen, R.E. (1996) Soybean seed quality reductions by high day and night temperature. Crop Sci. 36:1615–1619.

Gillooly, J.F., Brown, J.H., West, G.B., Savage, V.M. and Charnov, E.L. (2001) Effects of size and temperature on metabolic rate. Science 293, 2248–2251.

Goldberg, R.B. (1978) DNA sequence organization in the soybean plant. Biochem. Genet. 16, 45–68.

Golombek, S., Rolletschek, H., Wobus, U. and Weber, H. (2001) Control of storage protein accumulation during legume seed development. J Plant Physiol. 158, 457–464.

Good, A.G. and Zaplachinski, S.T. (1994) The effects of drought stress on free amino acid accumulation and protein synthesis in *Brassica napus*. Physiol. Plant. 90, 9–14.

Graham, P.H. and Vance, C.P. (2003) Legumes: importance and constraints to greater use. Plant Physiol. 131, 872–827.

Guerrero, F.D., Jones, J.T. and Mullet, J.E. (1990) Turgor-responsive gene transcription and RNA levels increase rapidly when pea shoots are wilted. Sequence and expression of three inducible genes. Plant Mol. Biol. 15, 11–26.

Harada, J.J., Barker, S.J. and Goldberg, R.B. (1989) Soybean b-conglycinin genes are clustered in several DNA regions and are regulated by transcriptional and post-transcriptional processes. Plant Cell 1, 415–425.

Hedges, B.R. and Palmer, R.G. (1993) Mapping the w4 locus in soybean. Soybean Genetics Newsletter 20, 20–26.

Heyen, B.J., Aslsheikh, M.K., Smith, E.A., Torvik, C.F., Seals, D.F. and Randall, S.K. (2002) The calcium-binding activity of a vacuole-associated dehydrin-like protein is regulated by phosphorylation. Plant Physiol. 130, 675–687.

Hill, J., Nelson, E., Tilman, D., Polasky, S. and Tiffany, D. (2006) Environmental, economic, and energetic costs and benefits of biodiesel and ethanol biofuels. Proc. Natl. Acad. Sci. USA 103, 11206–11210.

Hubick, K.T., Shorter, R. and Farquhar, G.D. (1988) Heritability and genotype x environment interactions of carbon isotope discrimination and transpiration efficiency in peanut (Arachis hypogaea L.). Aust. J. Plant Physiol. 15, 799–813.

Hufstetler, E.V., Boerma, H.R., Carter, Jr. T.E. and Earl, H.J. (2007) Genotypic variation for three physiological traits affecting drought tolerance in soybean. Crop Sci. 47, 25–35.

Hurburgh, C.R., Brumm, T.J. Jr., Guinn, J.M. and Hartwig, R.A. (1990) Protein and oil patterns in US and world soybean markets. J Am. Oil. Chem. Soc. 67, 966–973.

Hymowitz, T. and Bernard, R.L. (1991) Origin of the soybean and germplasm introduction and development in North America. In: H.L. Shands and L.E. Wiesner (Eds.), Use of Plant introductions in cultivar development, pp. 147–164. Part 1. CSSA Spec. Publ. No. 17. Madison, WI.

Hymowitz, T., Collins, F.I., Panczar, J. and Walker, W.M. (1972) Relationship between the content of oil, protein and sugar in soybean seed. Agron. J. 64, 613–616.

Ingram, J. and Bartels, D. (1996) The molecular basis of dehydration tolerance in plabnts. Annu Rev Plant Physiol Plant Mol Biol. 47, 377–403.

Ismond, K.P., Dolferus, R., De Pauw, M., Dennis, E.S. and Good A.G. (2003) Enhanced low oxygen survival in arabidopsis through increased metabolic flux in the fermentative pathway. Plant Physiol. 132, 1292–1302.

Jupp, A.P and Newman, E.I. (1987) Morphological and anatomical effects of severe drought on the roots of *Lolium perenne* L. Ann. Bot. 105, 393–402.

Kane, M.V., Steele, C.C., Grabau, L.J., MacKown, C.T. and Hildebrand, F.D. (1997) Early- maturing soybean cropping system. III. Protein and oil contents and oil composition. Agron. J. 89, 464–469.

Karl, T. R. and Trenberth, K. E. (2003) Science 302, 1719.

Kaspar, T.C., Taylor, H.M. and Shibles, R.C. (1984) Taproot elongation rates of soybean cultivars in the glasshouse and their relation to field rooting depth. Crop Sci. 24, 916–920.

Keigley, P.J., and Mullen, R.E. (1986). Changes in soybean seed quality from high temperature during seed fill and maturation. Crop Sci. 26, 1212–1216.

Keirstead, C.H. (1952) Marketing study of factors affecting the quantity and value of products obtained from soybeans. USDA, Production and Marketing Administration, Washington DC, USA, 35pp

King, C.A. and Purcell, L.C. (2005). Inhibition of N_2 fixation in soybean is associated with elevated ureides and amino acids. Plant Physiol. 137, 1389–1396.

Kingsbury, Ralph, W. and Emanuel Epstein. (1986) "Salt Sensitivity in Wheat : A Case for Specific Ion Toxicity. Plant Physiol. 80.3, 651–654.

Kirch, H.H., Schlingensiepen, S., Kotchoni, S., Sunkar, R. and Bartels, D. (2005) Detailed expression analysis of selected genes of the aldehyde dehydrogenase (*ALDH*) gene superfamily in *Arabidopsis thaliana*. Plant Mol. Biol. 57, 315–332.

Klok, E.J., Wilson, I.W., Wilson, D., Chapman, S.C., Ewing, R.M., Somerville, S.C., Peacock,W.J., Dolferus, R. and Dennis, E.S. (2002) Expression Profile Analysis of the Low-Oxygen Response in Arabidopsis Root Cultures. Plant Cell 14, 2481–2494.

Klueva, N.Y., Maestri, E., Marmiroli, N., Nguyen, H.T. (2001) Mechanisms of thermotolerance in crops. In AS Basra (Ed) *Crop Responses and Adaptations to Temperature Stress*. Food Products Press, Binghamton, NY, pp 177–217.

Koag, M.C., Fenton. R.D., Wilkens, S. and Close, T. (2003) The binding of maize DHN1 to lipid vesicles. Gain of structure and lipid specificity. Plant Physiol. 131, 309–316.

Kochian, L.V., Hoekenga, O.A. and Piñeros, M.A. (2004) How do crop plants tolerate acid soils? Mechanisms of aluminum tolerance and phosphorous efficiency. Annu. Rev. Plant Biol. 55, 459–493.

Kotak, S., Larkindale, J., Lee, U., von Koskull-Döring, P., Vierling, E. and Scharf, K.D. (2007b) Complexity of the heat stress response in plants. Curr. Opin. Plant Biol. 10, 310–316.

Kotak, S., Port, M., Ganguli, A., Bicker, F., and von Koskull-Döring, P. (2004). Characterization of C-terminal domains of Arabidopsis heat stress transcription factors (Hsfs) and identification of a new signature combination of plant class A Hsfs with AHA and NES motifs essential for activator function and intracellular localization. Plant J. 39, 98–112.

Kotak, S., Vierling, E., Bäumlein, H. and von Koskull-Döring, P. (2007a) A Novel Transcriptional Cascade Regulating Expression of Heat Stress Proteins during Seed Development of Arabidopsis. Plant Cell 19, 182–195.

Koti, S., Reddy, K.R., Reddy, V.R., Kakani, V.G. and Zhao, D. (2005) Interactive effects of carbon dioxide, temperature, and ultraviolet-B radiation on soybean (*Glycine max* L.) flower and pollen morphology, pollen production, germination, and tube lengths. J. Exp. Bot. 56, 725–736.

Kozlowski, T.T. (1984) Extent, Causes, and Impacts of Flooding. In T. T. Kozlowski (Ed.) *Flooding and Plant Growth*. Academic Press, Inc. Florida, pp. 1–8.

Kramer, P. J. (1980). Drought stress, and the origin of adaptations. In: *Adaptation of plants to water and high temperature stress*. N. C. Turner and P. J. Kramer. Wiley: New York, pp. 7–20.

Kurniadie, D. and Redmann, R.E. (1999) Growth and chloride accumulation in *Glycine max* treated with excess KCl in solution culture. Commun. Soil Sci. Plant Ana. 30, 699–709.

Lauchli, A. (1984) Salt exclusion: an adaptation of legumes for crops and pastures under saline conditions. . In: Salinity Tolerance in Plants, Strategies for Crop Improvement. R.C. Staples and G.H. Toenniessen (eds.), John Wieley and Sons, NY. pp171–187.

Lee, G.J., Boerma, H.R., Villagarcia, M.R., Zhou, X., Carter, Jr.T.E., Li, Z. and Gibbs, M.O. (2004) A major QTL conditioning salt tolerance in S-100 soybean and descendent cultivars Theor. Appl. Gene. 109, 1610–1619.

Li, X.P., Tian, A.G., Luo, G.Z., Gong, Z.Z., Zhang, J.S. and Chen, S.Y. (2005) Soybean DRE-binding transcription factors that are responsive to abiotic stresses. Theor. Appl. Genet. 110, 1355–1362.

Li, Y D., Wang, Y. J., Tong, Y. P., Gao, J. G., Zhang, J. S. and Chen, S. Y. (2005) QTL Mapping of phosphorus deficiency tolerance in soybean (*Glycine max* L. Merr.). Euphytica 142, 137–142.

Liao, H., Wan, H.Y., Shaff, J., Wang, X.R., Yan, X.L. and Kochian, L.V. (2006) Phosphorus and aluminum interactions in soybean in relation to aluminum tolerance. Exudation of specific organic acids from different regions of the intact root system. Plant Physiol. 141, 674–684.

Liu, F., Andersen, M.N. and Jensen, C.R. (2003) Loss of pod set caused by drought stress is associated with water status and ABA content of reproductive structures in soybean. Funct. Plant. Biol. 30, 271–280.

Liu, F., Jensen, C.R. and Andersen, M.N. (2004) Pod set related to photosynthetic rate and endogenous ABA in soybeans subjected to different water regimes and exogenous ABA and BA at early reproductive stages. Ann. Bot. 94, 405–411.

Liu, K.S., Ortheofer, F., and Brown, E.A. (1995) Association of seed size with genotypic variation in the chemical constituents of soybeans. J Am. Oil. Chem. Soc. 72, 191–193.

Liu, Q., Kasuga, M., Sakuma, Y., Abe, H., Miura, S., Yamaguchi-Shinozaki. Y. and Shinozaki, K. (1998) Two transcription factors, DREB1 and DREB2, with an EREBP/AP2 binding domain separate two cellular signal transduction pathways in drought- and low-temperature-responsive gene expression, respectively, in *Arabidopsis*. Plant Cell 10, 1391–1406.

Lobell, D.B., and Asner, G.P. (2003) Climate and management contributions to recent trends in U.S. agricultural yields. Science 299, 1032.

Ludlow, M.M. and Muchow, R.C. (1990) A critical evaluation of traits for improving crop yields in water limited environments. Adv. Agron. 43, 107–153.

Luo, G., Hepburn, A. and Widholm, J. (1994) A simple procedure for the expression of genes in transgenic soybean callus tissue. Plant Cell Reports 13, 632–636.

Luo, G.Z., Wang, H.W., Huang, J., Tian, A.G., Wang, Y.J., Zhang, J.S. and Chen, S.Y. (2005) A putative plasma membrane cation/proton antiporter from soybean confers salt tolerance in Arabidopsis. Plant Mol. Biol. 59, 809–820.

Lynch, J., Läuchli, A. and Epstein, E. (1991) Vegetative growth of the common bean in response to phosphorus nutrition. Crop Sci. 31, 380–387.

Marek, L.F. and Shoemaker, R.C. (1997) BAC contig development by fingerprint analysis in soybean. Genome 40, 420–427.

Marino, D., Frendo, P., Ladrera, R., Zabalza, A., Puppo, A., Arrese-Igor, C. and González, E.M. (2007) Nitrogen fixation control under drought stress. Localized or systemic? Plant Physiol. 143, 1968–1974.

Marschner, H. (1997) Introduction, Definition, and Classification of Mineral Nutrients. In: *Mineral Nutrition of Higher Plants* 2nd Ed. Academic Press, London. pp. 3–5.

Maruyama, K., Sakuma, Y., Kasuga, M., Ito, Y., Seki, M., Goda, H., Shimada, Y., Yoshida, S., Shinozaki, K. and Yamaguchi-Shinozaki, K. (2004) Identification of cold-inducible downstream genes of the Arabidopsis DREB1A/CBF3 transcriptional factor using two microarray systems, Plant J. 38, 982–993.

Masle, J., Gilmore, S.R. and Farquhar, G.D. (2005) The *ERECTA* gene regulates plant transpiration efficiency in arabidopsis. Nature 436, 866–870.

McClure, P.R. and Israel, D.W. (1979) Transport of nitrogen in the xylem of soybean plants. Plant Physiol. 64, 411–416.

Medina, J., Bargues, M., Terol, J., Perez-Alonso, M., and Salinas, J. (1999). The Arabidopsis CBF gene family is composed of three genes encoding AP2 domain-containing proteins whose expression is regulated by low temperature but not by abscisic acid or dehydration. Plant Physiol. 119, 463–469.

Meehl, T.G.A. and Tebaldi, C. (2004) More intense, more frequent, and longer lasting heat waves in the 21st century. Science 305:994–997.

Meng, Q., Zhang, C., Gai, J., Yu, D. (2006) Molecular cloning, sequence characterization and tissue-specific expression of six NAC-like genes in soybean (Glycine max (L.) Merr.). J Plant Physiol. doi:10.1016/j.jplph.2006.05.019.

Mian, M.A.R., Ashley, D.A. and Boerma, H.R. (1998) An additional QTL for water use efficiency in soybean. Crop Sci. 38, 390–393.

Mian, M.A.R., Mailey, M.A., Ashley, D.A., Wells, R., Carter, T.E., Parrot, W.A. and Boerma, H.R. (1996) Molecular markers associated with water use efficiency and leaf ash in soybean. Crop Sci. 36, 1252–1257.

Mohamed, M.A,, Harris, P.J., Henderson, J. and Senatore, F. (2002) Effect of Drought Stress on the Yield and Composition of Volatile Oils of Drought-Tolerant and Non-Drought-Tolerant Clones of *Tagetes minuta*. Planta Med. 68, 472–474.

Moons, A., De Keyser, A., Van Montagu, M. (1997) A group 3 LEA cDNA of rice, responsive to abscisic acid, but not to jasmonic acid, shows variety-specific differences in salt stress response. Gene 191, 197–204.

Moore, C.V. (1984) An economic analysis of plant improvement strategies for saline conditions. In: Salinity Tolerance in Plants, Strategies for Crop Improvement. R.C. Staples and G.H. Toenniessen (eds.), John Wieley and Sons, NY. pp381–397.

Muchow, R.C. and Sinclair, T.R. (1986) Water and nitrogen limitations in soybean grain production. II. Field and model analyses. Field Crop Res. 15, 143–156.

Myron, B.P., Gascho, G.J. and Gaines (1983) TP Chloride toxicity of soybean grown on Atlantic coast flatwoods soil. Agron J. 75, 439–443.

Naya, L., Ladrera, R., Ramos, J., Gonzalez, E.M., Arrese-Igor, C., Minchin, F.R. and Becana, M. (2007) The Response of Carbon Metabolism and Antioxidant Defenses of Alfalfa Nodules to Drought Stress and to the Subsequent Recovery of Plants. Plant Physiol. 144, 1104–1114.

Nguyen, H. T., Babu, R. C. and Blum, A. (1997) Breeding for drought resistance in rice: Physiology and molecular genetics considerations. Crop Sci. 37, 1426–1434.

Nichols, D.M., Glover, K.D., Carlson, S.R., Specht, J.E. and Diers, D.W. (2006) Fine mapping of a seed protein QTL on soybean linkage group and its correlated effects on agronomic data. Crop Sci. 46, 834–839.

Nilsen, E.T. and Orcutt, D.M. (1996) Stable isotopes and plant stress physiology. In: E.T. Nilsen and D.M. Orcutt (Eds.), *Physiology of plants under stress. Abiotic factors*. John Wiley & Sons, Inc. New York, pp. 199–230.

Novillo, F., Alonso, J.M., Ecker, J.R. and Salinas, J. (2004) CBF2/DREB1C is a negative regulator of CBF1/DREB1B and CBF3/DREB1A expression and plays a central role in stress tolerance in *Arabidopsis*. Proc. Natl. Acad. Sci. USA 101, 3985–3990.

Okumoto, S., Schmidt, R., Tegeder, M., Fische,r N.W., Rentsch, D., Frommer, B.W. and KochHigh, W. (2002) Affinity amino acid transporters specifically expressed in xylem parenchyma and developing seeds of *Arabidopsis*. J Biol. Chem. 277, 45338–45346.

Oosterhuis, D.M., Scott, H.D., Hampton, R.E. and S.D. and Wullschleger (1990) Physiological response of two soybean [*Glycine max* (L.) Merr] cultivars to short-term flooding. Environ. Exp. Bot. 30, 85–92.

Orf, J.H., Chase, K., Jarvik, T., Mansur, L.M., Cregan, P.B., Adler, F.R., and Lark, K.G. (1999) Genetics of soybean agronomic traits: I. Comparison of three related recombinant inbred populations. Crop Sci. 39, 1642–1651.

Pantalone, V.R., Kenworthy, W.J., Slaughter, L.H. and James, B.R. (1997) Chloride tolerance in soybean and perennial *Glycine* accessions. Euphytica 97, 235–239.

Parker, M.B., Gaines, T.P. and Gascho, G.J. (1986) Sensitivity of soybean cultivars to soil chloride. Georgia Agric. Exp. Sta. Res. Bull. 347.

Pate, J.S. (1980) Transport and partitioning of nitrogenous solutes. Annu. Rev. Plant. Physiol. 31, 313–340.

Pate, J.S., Sharkey, P.J. and Atkin, C.A. (1977) Nutrition of a developing legume fruit. Plant Physiol. 59, 506–510.

Perez-Grau, L. and Goldberg, R.B. (1989) Soybean Seed Protein Genes Are Regulated Spatially during Embryogenesis. Plant Cell 1, 1095–1109.

Peterson, C.M., Mosjidis, C.O.H., Dute, R.R. and Westgate, M.E. (1992) A flower and pod staging system for soybean. Ann. Bot. 69, 59–67.

Peterson, C.M., Williams, J.C. and Kuang, A. (1990) Increased pod set of determinate cultivars of soybean, *Glycine max*, by 6-benzylaminapurine. Botanical Gazette 151, 322–330.

Piper, E.L., and Boote, J. (1999) Temperature and cultivar effects on soybean seed oil and protein concentrations. J Am. Oil. Chem. Soc. 76, 1233–1241.

Porcel, R., Aroca, R., Azcon, R. and Ruiz-Lozano, J.M. (2006) PIP aquaporin gene expression in arbuscular mycorrhizal *Glycine max* and *Lactuca sativa* plants in relation to drought stress tolerance. Plant. Mol. Biol. 60, 389–404.

Porcel, R., Azcón, R. and Ruiz-Lozano, J.M. (2005) Evaluation of the role of genes encoding for dehydrin proteins (LEA D-11) during drought stress in arbuscular mycorrhizal Glycine max and Lactuca sativa plants. J Exp. Bot. 56, 1933–1942.

Price, A.H., Steele, K. A., Gorham, J., Bridges, J. M., Moore, B.J., Evans, J., Richardson, P. and Wyn-Jones, G. (2002) Upland rice grown in soil-filled chamfers exposed to contrasting water-deficit regimes. I. Root distribution, water use and plant water status. Field Crop. Res. 76, 11–24.

Purcell, L.C. and Specht, J.E. (2004) Physiological traits for ameliorating drought stress. In: H.R. Boerma and J.E. Specht (Ed.) *Soybeans: Improvement, production, and uses*. Third edition. Agron. Monograph 16. Amer. Soc. Agron., Madison, WI, pp. 569–620.

Raper, C. D. Jr. and Kramer, P. J. (1987) Stress physiology. In: J. R. Wilcox (Ed.), *Soybeans: Improvement, Production and Uses*. ASA-CSSA-SSSA, Madison, pp. 589–641.

Ray, J.D., Yu, L.X., McCouch, S.R., Champoux, M.C., Wang, G., and Nguyen, H.T. (1996) Mapping quantitative trait loci associated with root penetration ability in rice (*Oryza sativa* L.). Theor. Appl. Genet. 42, 627–636.

Read, D.J. and Bartlett, E.M. (1972) The physiology of drought resistance in the soy-bean plant (*Glycine max*). I. The relationship between drought resistance and growth. J Appl. Ecol. 9, 487–499.

Rebetzke, G.J., Pantalone, V.R., Burton, J.W., Craver, B.F., and Wilson, R.F. (1996) Phenotypic variation for saturated fatty acid content in soybean. Euphytica 91, 281–295.

Reyna,N., Cornelious,B., Shannon,J.G. and Sneller,C.H. (2003) Evaluation of a QTL for Waterlogging Tolerance in Southern Soybean Germplasm. Crop Sci. 43, 2077–2082.

Ribet, J. and Drevon, J.J. (1995) Phosphorus deficiency increases the acetylene-induced decline in nitrogenase activity in soybean (*Glycine max* (L.) Merr.). J Exp.l Bot. 46, 1479–1486.

Ricard, B., Couee, I., Raymond, P., Saglio, P.H., Saint-Ges, V. and Pradet, A. (1994) Plant Metabolism under Hypoxia and Anoxia. Plant Biochem. 32,1–10.

Rodrigues, S.M., Andrade, M.O., Gomes, A.P., Damatta, F.M., Baracat-Pereira, M.C. and Fontes, E.P. (2006) Arabidopsis and tobacco plants ectopically expressing the soybean antiquitin-like ALDH7 gene display enhanced tolerance to drought, salinity, and oxidative stress. J Exp. Bot. 57:1909–1918.

Rolland, F., Moore, B. and Sheen, J. (2002) Sugar sensing and signaling in plants. Plant Cell 14, S185–S205.

Salem, M. A., Kakani, V. G., Koti, S. and K. R. Reddy. (2007) Pollen-Based Screening of Soybean Genotypes for High Temperatures. Crop Sci. 47, 219–231.

Sall, K. and Sinclair, T.R. (1991) Soybean genotypic differences in sensitivity of symbiotic nitrogen fixation to soil dehydration. Plant Soil 133, 31–37.

Sanchez, P.A. and Salinas, J.G. (1981) Low-input technology for managing oxisols and ultisols in tropical America. Adv. Agron. 34, 279–406.

Schöffl, F., Key, J.L. (1982) An analysis of mRNAs for a group of heat shock proteins of soybean using cloned cDNAs. J Mol. Appl.Gene. 1, 301–314.

Schöffl, F., Rossol, I. and Angermüller, S. (1987) Regulation of the transcription of heat shock genes in nuclei of soybean (*Glycine max*) seedlings. Plant, Cell and Environ. 10, 113–119.

Scott, H.D., DeAngulo, J., Daniels, M.B. and Wood, L.S. (1989) Flood duration effects on soybean growth and yield. Agron. J. 81, 631–636.

Sebolt, A.M., Shoemaker, R.C. and Diers, B.W. (2000) Analysis of a quantitative trait locus allele from wild soybean that increases seed protein concentration in soybean. Crop Sci. 40, 1438–1444.

Seki, M., Narusaka, M., Abe, H., Kasuga, M., Yamaguchi-Shinozaki, k., Carninci, P., Hayashizaki, Y. and Shinozaki, K. (2001) Monitoring the expression pattern of 1300 Arabidopsis genes under drought and cold stresses by using a full-length cDNA microarray, Plant Cell 13, 61–72.

Setter, T.L and Waters, I. (2003) Review of prospects for germplasm improvement for waterlogging tolerance in wheat, barley and oats. Plant and Soil 253, 1–34.

Shannon, J.G. and Carter, T. (2003) Breeding soybeans for tolerance to abiotic stress. Proc. of the eleventh national tillage conference-symposium on international soybean research. pp139–150. Aug. 26–29, 2003 Rosario, Santa Fe- Argentina.

Sharp, R.E. and Davies, W.J. (1989) Regulation of growth and development of plants growing with a restricted supply of water. In: *Plant Under Stress*. H.G. Jones, T.J. Flowers, M.B. Jones (Eds), Cambridge Univ. Press, pp71–93.

Shen, H., He, L.F., Sasaki, T., Yamamoto, Y., Zheng, S.J., Ligaba, A., Yan, X.L., Ahn, S.J., Yamaguchi, M., Hideo, S., and Matsumoto, H. (2005). Citrate secretion coupled with the modulation of soybean root tip under aluminum stress. Up-regulation of transcription, translation, and threonine-oriented phosphorylation of plasma membrane H+-ATPase. Plant Physiol. 138, 287–296.

Shinozaki, K. and Yamaguchi-Shinozaki, K. (2007) Gene networks involved in drought stress response and tolerance. J Exp. Bot. 58, 221–227.

Shoemaker, R., Keim, P., Vodkin, L., Retzel, E., Clifton, S.W., Waterston, R., Smoller, D., Coryell, V., Khanna, A., Erpelding, J., Gai, X., Brendel, V., Raph-Schmidt, C., Shoop, EG., Vielweber, C.J., Schmatz, M., Pape, D., Bowers, Y, Theising, B., Martin, J., Dante, M., Wylie, T. and Granger, C. (2002) A compilation of soybean ESTs: generation and analysis. Genome 45, 329–338.

Short, K.C., Torrey, J.G. (1972) Cytokinins in seedling roots of pea. Plant Physiol. 49, 155–160.

Simon-Sarkadi, L., Kocsy, G., Várhegyi, A., Galiba, G. and de Ronde, J.A. (2005) Genetic Manipulation of Proline Accumulation Influences the Concentrations of Other Amino Acids in Soybean Subjected to Simultaneous Drought and Heat Stress. J. Agric. Food Chem. 53, 7512–7517.

Sinclair, T.R., Purcell, L.C., Vadez, V., Serraj, R., King, C.A. and Nelson, R. (2000) Identification of soybean genotypes with N_2 fixation tolerance to water deficits. Crop Sci. 40, 1803–1809.

Sinclair, T.R., Vadez, V. and Chenu, K. (2003) Ureide accumulation inresponse to Mn nutrition by eight soybean genotypes with N2 fixation tolerance to soil drying. Crop Sci. 43, 592–597.

Singh, R.J. and Hymowitz, T. (1988) The genomic relationship between *Glycine max* (L.) Merr. and *G. soja* Sieb. and Zucc. as revealed by pachytene chromosomal analysis. Theor. Appl. Genet. 76, 705–711.

Sloane, R.J., Patterson, R.P. and Carter, T.E. Jr. (1990) Field drought tolerance of a soybean plant introduction. Crop Sci. 30, 118–123.

Song, Q.J., Marek, L.F., Shoemaker, R.C., Lark, K.G., Concibido, V.C., Delannay, X., Specht, J.E. and Cregan, P.B. (2004) A new integrated genetic linkage map of the soybean. Theor. Appl. Genet. 109, 122–128.

Souer, E., van Houwelingen, A., Kloos, D., Mol, J. and Koes, R. (1996) The *No Apical Meristem* gene of Petunia is required for pattern formation in embryos and flowers and is expressed at meristem and primordia boundaries. Cell 85, 159–170.

Spears, J.F., TeKrony, D.M. and Egli, D.B. (1997) Temperature during seed filling and soybean seed germination and vigour. Seed Sci. Technol. 25, 233–244.

Specht, J.E., Chase, K., Macrander, M., Graef, G.L., Chung, J., Markwell, J.P., Germann, M., Orf, J.H. and Lark, K.G. (2001) Soybean response to water: A QTL analysis of drought tolerance. Crop Sci. 41, 493–509.

Specht, J.E., Hume, D.J. and Kumudini, S.V. (1999) Soybean yield potential—a genetic and physiological perspective. Crop Sci. 39, 1560–1570.

Specht,J.E., Hume, D.J. and Kumudini, S.V. (1999) Soybean Yield Potential–A Genetic and Physiological Perspective. Crop Sci. 39, 1560–1570.

Spollen, W.G., Sharp, R.E., Saab, I.N. and Wu, Y. (1993) Regulation of cell expansion in roots and shoots at low water potentials. In: *Water Deficits: Plant Responses from Cell to Community*. J.A.C. Smith, H. Griffiths (Eds), BIOS Sci Publ, Oxford, pp37–52

Sponchiado, B.N., White, J.W., Castillo, J.A. and Jones, P.G. (1989) Root growth of four common bean cultivars in relation to drought tolerance in environments with contrasting soil types. Exp. Agri. 25, 249–257.

Sripongpangkul, K., Posa, G.B.T., Senadhira, D.W., Brar, D., Huang, N., Khush, G.S. and Li, Z.K. (2000) Genes/QTLs affecting flood tolerance in rice. Theor. Appl. Genet. 101, 1074–1081.

Stacey, G., Koh, S., Granger, C. and Becker, J.M. (2002) Peptide transport in plants. Trends Plant Sci. 7, 257–263.

Sullivan. M., VanToai, T.T., Fausey, N., Beuerlein, J., Parkinson, R. and Soboyejo, A. (2001) Evaluating on-farm flooding impacts on soybean. Crop Sci. 41, 93–100.

Tang, G-Q., Novitzky, W.P., Griffin, H.C., Huber S.C. and Dewey, R.E. (2005) Oleate desaturase enzymes of soybean: evidence of regulation through differential stability and phosphorylation. Plant J. 44, 433–446.

Taylor, H.M., Burnett, E., and Booth, G.D. (1978) Taproot elongation rates of soybeans. Cont. North Central and Southern Regions, United staes Department of Agricutlure, Agron. Department, Iowa state University and Texas Agricutlure Exepriment Station.

Tegeder, M., Offler, C.E., Frommer, W.B. and Patrick, J.W. (2000) Amino acid transporters are localized to transfer cells of developing pea seeds. Plant Physiol. 122, 319–325.

TeKrony, D.M., Egli, D.B. and Spears, J.L. (2000) Seed quality and the early soybean production system. In: *Proc. 30th Soybean Seed Res.* Conf., Chicago, IL. 6–8 Dec. 2000. American Seed Trade Assoc., Alexandra, VA, pp. 45–57

Thomas, J.M.G., Boote, K.J., Allen, L.H. Jr., Gallo-Meagher, M., and Davis, J.M. (2003) Elevated temperature and carbondioxide effects on soybean seed composition and transcript abundance. Crop Sci. 43:1548–1557

Todd, C.D., Tipton, P.A., Blevins, D.G., Piedras, P., Pineda, M. and Polacco, J.C. (2006) Update on ureide degradation in legumes. J. Exp. Bot. 57, 5–12.

Tomkins, J.P., Mahalingham, R., Miller-Smith, H., Goicoechea, J.L., Knapp, H.T. and Wing, R.A. (1999) A soybean bacterial artificial chromosome library for PI 437654 and the identification of clones associated with cyst nematode resistance. Plant Mol. Biol. 41, 25–32.

Tran, Nakashima, K., Sakuma, Y., Simpson, S.D., Fujita, Y. and Maruyama, K. *et al.* (2004) Isolation and functional analysis of *Arabidopsis* stress-inducible NAC transcription factors that bind to a drought-responsive *cis*-element in the *early responsive to dehydration stress 1* promoter. Plant Cell 16, 2481–2497.

Tsukamoto, C., Shimada, S., Igita, K., Kudou, S., Kokubun, M., Okubo, K. and Kitamura, K. (1995) Factors affecting isoflavone content in soybean seeds: changes in isoflavones, saponins, and composition of fatty acids at different temperatures during seed development. J. Agric. Food Chem. 43, 1184–1192.

Turner, N.C., Wright, G.C. and Siddique, K.H.M. (2001) Adaptation of grain legumes (pulses) to water limited environment. Adv. Agro. 71, 193–231.

Turner, N.C. (1986) Adaptation to water deficit: a changing perspective. Aust. J. Plant Physiol. 13, 175–190.

Vadez, V., Sinclair, T.R. (2001) Leaf ureide degradation and N_2 fixation tolerance to water deficit in soybean. J Exp. Bot. 52, 153–159.

Valliyodan, B. and Nguyen, H.T. (2006) Understanding regulatory networks and engineering for enhanced drought tolerance in plants. Curr. Opin. Plant Biol. 9, 189–95.

VanToai, T.T., St. Martin, S.K., Chase, K., Boru, G., Schnipke, V., Schmitthenner, A.F., Lark, K.G. (2001) Identification of a QTL associated with tolerance of soybean to soil waterlogging. Crop Sci. 41, 1247–1252.

VanToai, T. T., Beuerlein, J. E., Schmitthenner A. F. and St. Martin, S. K. (1994) Genetic variability for flooding tolerance in soybeans. Crop Sci. 34, 1112–1115.

VanToai, T., Yang, Y., Ling, P., Boru, G., Karica, M., Roberts, V., Hua, D. and Bishop. B. (2003) Monitoring soybean tolerance to flooding stress by image processing technique. In T.T. VanToai, et al. (Ed.) *Digital Imaging and Spectral Techniques: Applications to Precision Agriculture and Crop Physiology*. ASA Special Publication No 66. The American Society of Agronomy. Madison, WI. pp 43–51.

Vasellati, V., Oesterheld, M., Medan, D. and Loreti, J. (2001) Effects of flooding and drought on the anatomy of *Paspalum dilatum*. Ann. Bot. 88, 355–360.

Vierling, E. (1991) The Roles of Heat Shock Proteins in Plants. Ann. Rev. Plant Biol. 42, 579–620.

Vlahakis, C., Hazebroek, J. (2000) Phytosterol accumulation in canola, sunflower, and soybean oils: effects of genetics, planting location, and temperature. J. Am. Oil Chem. Soc. 77, 49–53.

Vodkin, L.O., Khanna, A., Shealy, R., Clough, S.J., Gonzalez, D.O., Philip, R., Zabala, G., Thibaud-Nissen, F., Sidarous, M., Stromvik, M.V., Shoop, E., Schmidt, C., Retzel, E., Erpelding, J., Shoemaker, R.C., Rodriguez-Huete, A.M., Polacco, J.C., Coryell, V., Keim, P., Gong, G., Liu, L., Pardinas, J. and Schweitzer, P. (2004) Microarrays for global expression constructed with a low redundancy set of 27,500 sequenced cDNAs representing an array of developmental stages and physiological conditions of the soybean plant. BMC Genomics 5, 73.

Volkov, R.A., Panchuk, I.I. and Schöffl, F. (2003) Heat-stress-dependency and developmental modulation of gene expression: the potential of house-keeping genes as internal standards in mRNA expression profiling using real-time RT-PCR. J Exp. Bot. 54, 2343–2349.

Wang, D. and Shannon, M.C. (1999) Emergence and seedling growth of soybean cultivars and maturity groups under salinity. Plant Soil 214, 117–124.

Watanabe, I., and Nagasawa, T. (1990) Appearance and chemical composition of soybean seeds in germplasm collection of Japan. II. Correlation among protein, lipid and carbohydrate percentage. Japan J Crop Sci. 59, 661–666.

Westgate, M.E. and Peterson, C.M. (1993) Flower and pod development in water-deficient soybean. J. Exp. Bot. 258, 109–117.

White, J.W. and Castillo, J.A. (1989) Relative effect of root and shoot genotypes on yield of common bean under drought stress. Crop Sci. 29, 360–362.

Wieneke, J. and Läuchli, A. (1979) Short-term studies on the uptake and transport of Cl^- by Glycine max differing in salt tolerance. Z Pflanzenernähr Bodenk 142, 799–814.

Wilcox, J.R. (2001) Sixty years of improvement in publicly developed elite soybean lines. Crop Sci. 49, 1711–1716.

Wolf, R.B., Cavins, J.F., Kleiman, R., and Black, L.T. (1982) Effect of temperature on soybean seed constituents: Oil, protein, moisture, fattyacids, amino acids and sugars. J Am. Oil Chem. Soc. 59, 230–232.

Wu, C., Nimmakayala, P., Santos, F.A., Springman, R., Scheuring, C., Meksem, K., Lightfoot, D.A. and Zhang, H.B. (2004a) Construction and characterization of a soybean bacterial artificial chromosome library and use of multiple complementary libraries for genome physical mapping. Theor. Appl. Genet. 109, 1041–1050.

Wu, C., Sun, S., Nimmakayala, P., Santos, F.A., Meksem, K., Springman, R., Ding, K., Lightfoot, D.A. and Zhang, H.B. (2004b) A BAC- and BIBAC-based physical map of the soybean genome. Genome Res. 14, 319–326.

Yamaguchi-Shinozaki, K. and Shinozaki, K. (2005) Organization of *cis*-acting regulatory elements in osmotic- and cold-stress-responsive promoters. Trends in Plant Sci. 10, 88–94.

Yamaguchi-Shinozaki, K. and Shinozaki, K. (2006) Transcriptional regulatory networks in cellular responses and tolerance to dehydration and cold stress. Annu. Rev. Plant Biol. 57, 781–803.

Yang, J. and Blanchar, R.W. (1993) Differentiating chloride susceptibility in soybean cultivars. Agron. J. 85, 880–885.

Yang, J. and Han, K.H. (2004) Functional characterization of allantoinase genes from *Arabidopsis* and a non-ureide-type legume black locust. Plant Physiol. 134, 1039–1049.

Yoshida, A., Rzhetsky, A., Hsu, L.C. and Chang, C. (1998) Human aldehyde dehydrogenase gene family. Euro. J. Biochem, 251, 549–557.

Zahran, H.H. (1999) *Rhizobium*–legume symbiosis and nitrogen fixation under severe conditions and in an arid climate. Micro. Mole. Biol. Rev.63, 968–989.

Part IV
Early Messages

Chapter 19
The Global Economic Impacts of Roundup Ready Soybeans

Srinivasa Konduru, John Kruse, and Nicholas Kalaitzandonakes[1]

Introduction

In 1996, Monsanto released Roundup Ready (RR) soybeans and kicked off drastic and swift changes in soybean production around the world. RR soybeans have been genetically engineered to include a gene from the soil bacterium *Agrobacterium tumefaciens*, which makes them tolerant to the broad-spectrum herbicide Roundup^{TM} (glyphosate). By using RR soybeans, weeds can be controlled through over the top application of glyphosate. Use of the RR soybean technology typically results in the replacement of multiple applications with selective post emergence herbicides by one or two over the top applications with glyphosate.

Adoption of RR soybeans in key producing countries has been fast. Within ten years from their introduction, RR soybeans were used on almost 60 % of the world soybean acreage (see Table 19.1). In this chapter, we examine the economic impacts that have followed this broad adoption of RR soybeans around the world. We first measure the changes in land use, production volumes and prices of soybeans and other oilseeds that resulted from the use of RR soybeans over the 1996–2006 period. We then measure the resulting changes in welfare and their distribution among adopters and non-adopters, producers of other oilseeds, as well as consumers.

The Potential Impacts of RR Technology on Soybean Production

Introduction of RR soybeans has, first and foremost, augmented input substitution possibilities in soybean production. Producers can continue to use conventional

S. Konduru
The Economics and Management of Agrobiotechnology Center, Department of Agricultural Economics, University of Missouri-Columbia, MO 65211-6200, USA
e-mail: Spkdkc@mizzou.edu

[1] S. Konduru is assistant professor of agricultural economics, California State University at Fresno; J. Kruse is managing director of Global Insight; N. Kalaitzandonakes is MSMC endowed professor of agribusiness, University of Missouri-Columbia.

G. Stacey (ed.), *Genetics and Genomics of Soybean*,

Table 19.1 Global soybean acreage, RR adoption, conservation and reduced tillage acreage

	1996	1997	1998	1999	2000	2001	2002	2003	2004	2005	2006
US											
Total Soybean (million acres)[1]	64.1	69.9	71.3	73.3	73.3	73.8	73.3	73.3	74.8	72.1	75.5
RR adoption (%)[2]	7.4	17.0	44.2	55.8	54.0	68.0	75.0	81.0	85.0	87.0	89.0
Conservation Till under RR (million acres)[3]	2.6	7.8	18.1	25.4	24.3	24.5	34.7	34.7	39.2	37.8	39.5
Reduced (million acres) under RR[3]	0.6	1.5	5.6	7.9	8.2	8.3	11.8	11.8	10.6	10.2	10.7
Argentina											
Total Soybean (million acres)[4]	16.7	17.9	21.0	22.0	26.7	29.1	31.5	36.3	36.0	38.3	39.9
RR adoption (%)[5]	5.55	24.47	57.14	75.54	84.39	93.86	98.72	91.07	97.63	99.16	99.16
Conservation Till under RR (million acres)[6]	0	1.6	5.7	12.2	16.4	20.8	24.3	26.5	31.3	33.5	34.9
Brazil											
Total Soybean (million acres)[7]	28.1	32.5	32.1	33.4	34.5	40.3	45.7	52.8	57.6	54.9	50.9
RR adoption (%)[7]	0.00	0.00	0.00	0.00	0.21	2.01	4.67	13.57	20.36	36.00	48.00
Conservation Till under RR (million acres)[8]	0	0	0	0	0.1	0.6	1.5	5.1	8.6	15.8	19.5
Reduced Till under RR (million acres)[8]	0	0	0	0	0.0	0.1	0.4	1.0	1.9	2.7	3.4
Rest of World											
Total Soybean (million acres)[9]	47.5	52.1	54.8	51.6	55.0	56.0	55.3	58.9	65.3	65.8	66.2
RR adoption (%)[10]	0	0	0	0.6	1.1	1.1	1.8	2.7	3.8	8.8	8.5

[1]Source: USDA;
[2]Source: USDA (2000–06), Fernandez-Cornejo, Klotz-Ingram and Jans and McBride (2002), pp 16 (1996–99);
[3]Compiled from CTIC (Figures for 1999, 2001, 2003, 2005 and 2006 extrapolated from previous years) & ARMS (Figures for 2001, 2003–06, extrapolated from previous years);
[4]Source: SAGPyA(Trigo and Cap 2006) (for 2006, extrapolated from previous year);
[5]Source: Argenbio;
[6]Source: Brookes and Barfoot (2006);
[7]Source: USDA;
[8]Compiled from previous source. Figures for 2006 were extrapolated from previous year;
[9]USDA PSD database;
[10]Compiled from website of Monsanto.

weed control methods or adopt RR technology, which facilitates substitution of one class of herbicides (e.g., broad-spectrum) for another (e.g., selective post-emergence) and other inputs, such as management, labor, and capital.

Furthermore, due to its broad spectrum, over the top use of glyphosate on RR soybeans controls weeds more effectively while it extends the application window, thereby increasing convenience and reducing production risk. Through more effective weed control and reduced production risk, adoption of RR soybeans could over time, although not necessarily in every season, lead to higher yields.

More subtle, but also more intriguing, are the impacts of RR soybeans on agronomic practices and cropping systems. RR soybeans have been associated with increased adoption of conservation and reduced tillage practices (Fawcett and Towery (2002), Marra et al. (2004), Trigo and Cap (2006), though also see Fernandez-Cornejo et al. (2002, 2003)). If tillage before planting is eliminated or reduced, weeds must be controlled with herbicides before, at, or after planting. Usually chemical weed control is best and RR soybeans were found to be well-suited in such circumstances.

When adoption of RR soybeans encourages parallel adoption of conservation and reduced tillage practices, soil and water conservation can also be impacted. Conservation and reduced tillage can improve the availability of organic matter and minerals in the soil leading to enhanced soil structure and fertility. Soil erosion and water runoff are also reduced, sustaining the productive capacity of land and minimizing surface and groundwater contamination. Due to improved soil moisture content and retention capacity, efficiencies in water use may also be achieved.

Adoption of RR soybeans has also been associated with shifts towards early planting, increased use of double cropping, changes in crop rotations, and increasing adoption of narrow row planting (Carpenter and Gianessi (2003), Fernandez-Cornejo et al. (2003)). Narrow row spacing and double cropping practices can lead to efficiencies in land use. Overall, the shifts in agronomic practices can result in complex reallocation of essential farm resources – land, labor and capital – and significantly broaden the scope of the production impacts of RR technology in soybean production.

A distinguishing feature of RR soybean technology is that its use and adoption has also been associated with many non-pecuniary (not priced or traded in the market) benefits. Those include ease of use and convenience, as well as an improved safety profile for the handler and for the environment. A number of studies found that RR soybean technology leads to a lower amount of active ingredient applied (e.g. Gianessi and Carpenter, 2000). Also, the toxicity of glyphosate is lower than that of other herbicides and its persistence in the environment is short (Heimlich et al. (2000), Ervin et al. (2000), EPA). When comprehensive measures of human and environmental impact are calculated, such as those that take into consideration toxicity and environmental exposure data (e.g. the Environmental Impact Quotient), RR soybeans compare well to conventional soybean systems (Brookes and Barfoot (2006), Nelson and Bullock (2003)). Marra et al. (2004) concluded from their national survey that some soybean growers placed significant economic value on the improved operator and worker safety, and environmental impact profile of the technology.

The (pecuniary and non-pecuniary) production and environmental impacts of RR soybeans frame the incentives for the producers who consider adoption. In this context, economic theory is helpful for understanding why producers might choose to adopt or not adopt RR soybeans and at what level. When RR soybeans augment input substitution possibilities in weed control, producers observe the relative prices of conventional and RR seeds, relevant herbicides, labor, capital and other inputs and choose their mix so that they minimize production costs. They do similarly when they compare conventional tillage to conservation and reduced tillage systems that are synergistic with RR soybeans. Potential savings in labor, fuel and other variable costs due to the adoption of conservation and reduced tillage systems is taken into account. When adoption of RR soybeans is expected to lead to lower input costs per unit of output, producers will be inclined to adopt the technology. Soybean producers also value the possibility of reduced production risk and higher revenue from potential yield increases. When adoption of RR soybeans is expected to lead to higher output per unit of input and higher expected profits, producers will be more likely to adopt. Producers that value the convenience as well as the improved human and environmental safety of the technology will, again, be inclined to adopt it.

Measuring Economic Impact and Its Distribution

Economic theory can be useful not only in explaining why some producers might be more inclined to adopt RR soybeans but also in quantifying the economic impact that might follow from such adoption. As a first step, the farm-level impacts of the technology must be carefully quantified. Any differences in input use, costs, yields, revenue and profitability between RR and conventional soybeans must be assessed and measured. This requires a proper benchmark which, ideally, is defined by the following counterfactual: what would have been the costs (yields, revenues, profits, environmental impacts, etc.) of the adopters if RR soybean technology was not available? That is, since it is adopters that generate impacts, as Frisvold and Tronstad (2003) put it, the object of impact analysis is not to explain why everyone did not adopt. Rather, the question of interest is: "what if everyone who did adopt these innovations couldn't?" The counterfactual clarifies that the performance of RR soybeans is measured against a benchmark that is not directly observable – the optimal solution that would have been chosen by the adopters in the absence of the technology. Economic theory therefore provides guidance on how to approximate the unobserved benchmark by measurable indicators.

In the early stages of RR soybean commercialization, experimental field trials were used to assess the farm-level impacts of RR soybeans. Comparing the performance of RR and conventional soybeans in yield or weed control trials, provides useful but partial measures. The first trial type is designed to maximize yield performance and does not account for the potential advantage of RR soybeans in weed control (Carpenter and Gianessi, 2003). Similarly, weed trials compare various weed control programs but overlook yield potential (ibid). Importantly, neither seeks to maximize profits – the sort of behavior that drives producer adoption decisions. Hence, yields, herbicide and other input use, as well as implied costs and revenues from field trial data, might not be representative of actual farm practices.

Some studies compared the economic performance of adopters against that of non-adopters or against the population average, which includes both adopters and non-adopters. Again, those indicators can provide useful, if partial insights, on the relative farm performance of RR soybeans as they do not account for (unobserved) differences in land productivity, weed incidence and pressure, differential managerial skills, and other factors that influence producer adoption/non adoption decisions in the sample and can lead to systematic biases.

It might be possible to control for the influence of management and other unobserved factors by assessing the performance of RR soybeans against that of conventional ones on the farms of partial adopters alone (Marra, 2001). This method captures how partial adopters optimally allocate their lands resources to RR soybeans and conventional soybeans on the basis of their relative advantages. Nevertheless, such measures can also be subject to biases from other unobserved factors that influence partial adoption, such as risk behavior, learning, differential quality of land and others. The key point here is that a proper benchmark in the assessment of the farm level impact of RR soybeans is conceptually but not empirically simple.

While the farm level production and economic impacts of RR soybeans are of interest, of greater interest is how they, in turn, influence aggregate production, consumption, trade, and social welfare in different national and regional economies. Accordingly, an appropriate method for aggregating farm level impacts across producers to a national and regional level is necessary.

Farm level impacts can be extrapolated to national or regional levels by multiplying the microlevel impacts with observed acreage. But such extrapolated values are only incomplete indicators of aggregate impacts, as they do not account for price variability. While individual producers cannot affect input and output prices with their actions, groups of producers can. Similarly, innovators with market power and governments that can influence prices, shape both the size of the aggregate impacts and their distribution. As farm level adoption of RR soybeans leads to long term productivity and output growth, individual producer responses collectively translate into increased aggregate production from existing land resources and/or through diversion of land from less profitable crops. Such shifts in aggregate output supply will typically result in reduced real output prices and changes in consumer patterns. From these price and quantity changes, welfare changes for consumers and producers can be computed. It is therefore important to account for such price endogeneity while estimating the aggregate or market level impact of RR soybeans.

In this context, Alston et al. (1995) economic surplus method provides an aggregate supply and demand framework for estimating the welfare changes arising from the introduction of RR soybeans. This method is commonly used for estimating economic impacts and associated welfare changes from new technologies. Application of this model assumes that the introduction of RR soybeans occurs in a large open economy and allows for technology spillovers to encompass cases of usage of RR soybean seeds in other countries with or without payments of technology fees. It also allows for representation of trade among countries so that equilibrium can be attained by equating their excess supply and demand.

To illustrate the basic workings of this framework consider a possible scenario of innovation within a simplified case where only two regions are assumed to exist. Figure 19.1 illustrates this simplified case. Panels A, B and C represent respectively the home country where the innovation is initiated (Country A), the interaction of excess supply and excess demand and the Rest of World (ROW). All the supply and demand curves are assumed linear. The international equilibrium price P_0 is obtained from the intersection of the excess supply ($ES_{A,0}$) and excess demand curves ($ED_{B,0}$). Due to the introduction of the innovation in country A, there is a parallel shift in domestic supply from $S_{A,0}$ to $S_{A,1}$, and in consequence the excess supply shifts from $ES_{A,0}$ to $ES_{A,1}$. But, due to a technology spillover to the ROW, there are additional supply shifts from $S_{B,0}$ to $S_{B,1}$ and correspondingly there is a reduction in excess demand from $ED_{B,0}$ to $ED_{B,1}$. So, the world price falls from P_0 to P_1. Due to the innovation and its spillover, producers in country A benefit as long as the overall price reduction (from P_0 to P_1) is smaller than the initial vertical supply shift in country A. The benefits to producers in country A are given by area P_1bcd in panel A. ROW producers are net losers, even after the adoption of the innovation since the area P_1ij is less than the area P_0hk, but in general they could gain or lose. The benefits to consumers in country A are shown in the figure as P_0aeP_1 in panel A and the benefits to consumers in ROW is shown as P_0fgP_1 in panel C. This model can be generalized to include multiple countries and commodities adding realism and detail in the analysis of economic surpluses from innovation and their distribution.

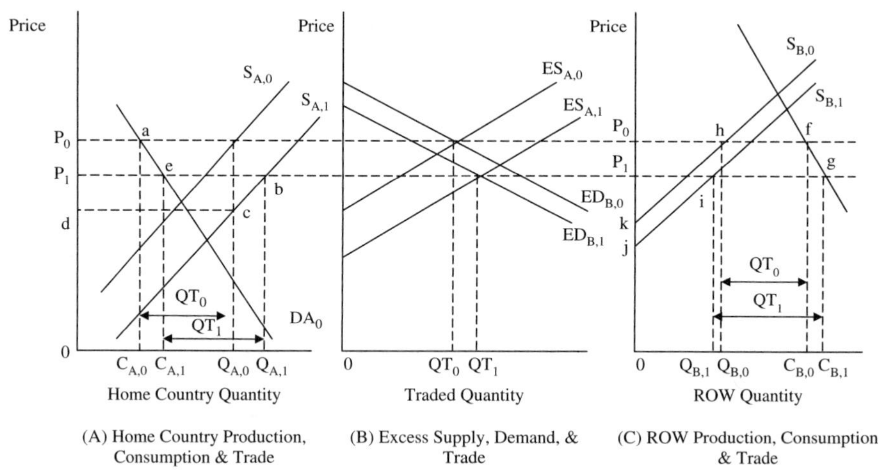

Fig. 19.1 Welfare changes from introduction of RR Soybeans

Existing Evidence on Economic Impact of RR Soybeans

So what do we know so far about the farm-level and aggregate economic impacts of RR soybeans? RR soybeans have been grown for over ten years and empirical evidence on their economic impacts at the farm and at an aggregate level has continued to accumulate. A number of studies measured such impacts at different points in time and locations using a variety of methods and datasets. Their results have begun to paint a picture of the overall economic contribution of RR soybeans and its underlying sources.

Sankula and Blumenthal (2004) and Sankula (2006) studied the farm level performance differences between RR and conventional soybeans by surveying experts (state and university weed specialists) in the USA. They gathered data on weed control and associated expenditures from all producing states in USA, thereby representing various production systems. They concluded that RR soybeans allowed 53 and 46 % cost savings for weed control in 2003 and 2005, respectively.

Fernandez-Cornejo and McBride (2000) used data from the 1997 Agricultural Resource Management Survey (ARMS) survey of USDA and concluded that adoption of RR soybeans in that year did not have a significant impact on net farm returns. However, they did find that use of RR soybeans was quite profitable in some regions of the US (e.g. 17 % more in Heartland).

Lin et al. (2001) studied the farm level effects of RR soybeans in the United States using again the 1997 ARMS survey. Their elasticity based estimates showed that across all regions adopters of RR soybeans spent 1–34 % less than non-adopters on soybean weed control and 11 less in the heartland of US (where 70 % of soybeans are grown).

Marra et al. (2004) analyzed, through a national producer survey, the reasons USA soybean growers chose RR soybeans over conventional varieties and their associated impacts. Their survey evaluated various potential incentives, both pecuniary and non-pecuniary. They estimated that the net benefit of RR soybeans along with benefits from a parallel adoption of reduced tillage can be substantial, up to $37 per acre for some farmers.

A few studies also assessed the economic impact of RR soybeans at an aggregate level. Some used extrapolation from farm-level estimates to reach conclusions for the market effects of the technology. Brookes and Barfoot (2005) studied the global economic and environmental impact of genetically modified crops since their commercial introduction in 1996. Their economic impact analysis focused on farm income effects and their environmental impact analysis focused on changes in the use of insecticides and herbicides with GM crops and the resulting impact on the environmental load from crop production. They reported that the global economic benefits from RR soybeans over the 1996–2004 period were approximately $9.3 billion.

Other studies used equilibrium models to assess the aggregate economic impacts of RR soybeans. Falck-Zepeda et al. (2000a) modeled the change in welfare effects from the adoption of Bt cotton and RR soybeans using a two-region framework (USA and rest of world [ROW]), based on the approach in Alston et al. (1995).

They estimated the total world surplus from the use of RR soybeans for the year of their calculations at $1.06 billion.

Similarly, Moschini et al. (2000) modeled the global welfare effects of RR soybeans through a three-region world model that included a monopolist technology seller, as well as consumers and producers. They assumed that the technology resulted in a US $20 per hectare savings in costs based on 1997–1998 farm-level estimates in Iowa. They estimated that the total efficiency gains world wide were approximately $804 million in the crop year 1999–2000. In another study based on a three region model, Qaim and Traxler (2002) analyzed the impact of RR soybeans in Argentina, the USA and the ROW. They found that the RR technology increased total factor productivity by 10 % on average from a producer survey. They then estimated global welfare effects of RR soybeans in 2001 at $1.2 billion.

Trigo and Cap (2006) estimated that the total accumulated benefit of herbicide tolerant soybeans in Argentina to be approximately $19.7 billion from 1996–2005. They attributed about 36 % of the total increase in employment gains over this period. Finally, they analyzed the impact of the increase in soybean production in Argentina on consumers world-wide, through an increase in price that the commodity would have reached in absence of that additional output. The result was an accumulation of savings in global consumer spending in the tune of $26 billion.

In many of the aggregate economic impact studies (Falck-Zepeda et al. (2000a, b), Moschini et al. (2000), Qaim and Traxler (2002)), it was observed that the shifts in input and output prices affect the size of the impacts, as well as their distribution. For instance, an increase in the price of RR soybean seed transfers a portion of the innovation rents from the adopters to the innovator. Similarly, a reduction in the output price transfers a portion of the economic gains to consumers. Furthermore, given that input and output prices determine the incentives for producer adoption, shifts in input and output prices cause variations in the adoption and the aggregate diffusion levels and, ultimately, in the actual impacts of RR soybeans. Hence, at an aggregate level, prices become endogenous making both the size and the distribution of the impacts also endogenously determined.

Following this line of thought, Falck-Zepeda et al. (2000a) estimated that the total world surplus is shared among USA farmers, Monsanto, seed suppliers, USA consumers and the ROW in the ratio of 76, 7, 3, 4 and 9 % respectively. The figures obtained in Argentina by Trigo and Cap (2006) were similar. The benefits were distributed in the ratio of 77, 4, 5, and 13 among the farmers, seed suppliers, herbicide suppliers and national government, respectively. In Moschini et al. (2000), the share of the innovator-monopolist was reported to be as high as 45 % of the total surplus. Lin et al (2001) reported that the share to USA farmers was approximately 20 % of the total innovation benefits in 1997.

Overall, there is agreement across most studies that the farm-level and aggregate economic impacts of RR soybeans have been significant, though there is still significant variation in their size estimates. This is, in part, attributable to the differences in the methodologies used; the data sets employed; the different dimensions of production and environmental impacts taken into account; and the different times and places where such assessments occurred. Year-to-year and place-to-place variations in the economic impact of RR soybeans are expected since key underlying factors, which determine such impacts, vary with time and space (e.g. weed infestation).

We add to this literature by evaluating here the global economic impacts of RR soybeans over the 1996–2006 period. We also make a number of methodological improvements that can further clarify the size and distribution of such impacts. For instance, prior studies overlooked the fact that innovation in soybean production must have influenced the supply and demand conditions of other oilseeds, like canola, sunflower, and palm oil; thereby, affecting the overall surplus in the market and its distribution. Similarly, the structure of the soybean complex and the underlying separate demand for oils and meals was not accounted for in previous studies. We address these shortcomings in this study by estimating the aggregate worldwide economic impact of RR soybeans within the context of a multi-country, multi-crop, multi-sector model.

Empirical Measurement of RR Soybean Economic Impact in this Study

In order to estimate the aggregate economic impacts of RR soybeans over the 1996–2006 period, we developed an econometric model that captures the global interrelationships among oilseeds and competing crops, competing oilseed products, as well as between oilseeds and the rest of the agricultural sector. For instance, the relationship between feed demand and the livestock sector is captured by theoretically derived specifications. The model allows estimation of the price effects, land reallocation patterns and substitution among oilseeds that result from the adoption of RR soybeans. The simulation of the scenario that RR soybeans are not available is created by suppressing any yield changes and cost savings caused by the introduction of RR soybeans. The resulting counterfactual prices, quantities and acreages are then compared with the actual historical values (baseline) to account for the effects of RR soybean adoption. The difference between them permits estimation of the benefits of RR soybeans in the form of consumer and producer surpluses.

Model Structure

The structural equation model developed in this study includes a set of supply and demand equations for each commodity of interest. Separate supply and demand functions are specified for the various oilseeds and their derivative oils and meals for each of the twelve countries represented in the model (US, Canada, Mexico, Brazil, Argentina, China, India, Indonesia, Malaysia, Japan, EU 25, and ROW). The general structure of the model for a given commodity and country is described by the following equations:

Beginning Stocks = Ending stocks$_{(t-1)}$ (Oilseeds, Meals and Oils)
Production = Harvested Area* Yield (Oilseeds)
Production = Crush* Crushing Yield (Meals and Oils)
Total Supply = Beginning Stocks + Production + Imports (Oilseeds, Meals and Oils)
Total Demand = Crush + Food Use + Other Use + Exports + Ending Stocks (Oilseeds)
Total Demand = Food Use + Feed Use + Industrial Use + Exports + Ending Stocks (Meals and Oils)
Domestic Use = Crush + Food Use + Other Use + Ending Stocks (Oilseeds)
Domestic Use = Food Use + Feed Use + Industrial Use+ Ending Stocks (Meals and Oils)

Within our model the prices of all countries and commodities are linked to those of every other trading country using price linkage equations that include import tariffs, taxes and other relevant shifters.

The percent vertical shift (K) in the supply function of soybeans due to introduction of RR technology is obtained by multiplying the net change in production costs with the rate of adoption of RR soybeans in each region. The net change in production costs reflects savings in weed control and shifts in tillage practices due to adoption of RR technology. The savings are net of any technology fees and seed premiums paid for the usage of RR technology. When there are changes in yields due to the use of RR technology, they are also accounted for in the calculation of the vertical shift of soybean supply.

For the purpose of our surplus calculations we aggregate our results to four countries/regions: USA, Argentina, Brazil and ROW. The formulas for calculating the change in producer and consumer surpluses follow Alston et al. (1995) and are specified as follows:

$$\begin{aligned}\Delta PS_{R,S} &= P_0 Q_{A,0}(K - Z)(1 + 0.5Z\varepsilon_S)\\ \Delta PS_{R,O} &= -P_0 Q_{A,0} Z(1 + 0.5Z\varepsilon_O)\\ \Delta CS_{R,O} &= (P_o - P_1)C_{A,0} + 0.5(C_{A,1} - C_{A,0})(P_0 - P_1)\end{aligned}$$

where ΔPS is the change in producer surplus; ΔCS is the change in consumer surplus; R denotes USA, Argentina, Brazil or ROW; S denotes soybeans; O denotes sunflower, rapeseed or palm oil; P_0 is a counterfactual price; P_1 is an actual price; $C_{A,0}$ is a counterfactual quantity; $C_{A,1}$ denotes an actual quantity; ε_S is the supply elasticity of soybeans; ε_O denotes supply elasticities of other oilseeds and Z is the relative price change given by $-(P_1 - P_0)/P_0$.

Model Calibration and Assumptions

We do not estimate the farm-level impacts of RR soybeans in this study. Instead, we depend on estimates from previous studies to calculate the net change in the costs of production and yield changes caused by the introduction of RR soybeans. The specific assumptions that have been built into our aggregate model are given below.

Influence of RR Technology on Soybean Yields

A number of studies measured the impacts of RR technology on soybean yields (see Table 19.2). Some studies reported yield suppression from the adoption of RR soybeans (Elmore et al., 2001, Oplinger et al., 1998, Qaim and Traxler, 2002, Duffy and Ernst, 1999). Others observed that yield reductions in some areas in the early stages of adoption disappeared over time as the transgene was inserted into more elite varieties across various soybean maturities (Marra et al. 2002). There are also some studies which reported a positive influence of RR technology on soybean yields (Lin et al. (2001), Gianessi and Carpenter, 2000, Falck-Zepeda et al. (2000a)). Hence, drawing firm conclusions from previous studies on the impact of RR soybeans on yields is somewhat difficult.

If significant yield changes were, in fact, caused by the adoption of RR soybeans a departure from historical yield trends observed prior to the introduction of the new technology should be visible in aggregate data. For the purpose of our study, we considered such a possibility and tested for a possible structural change

Table 19.2 Yield differences between RR and conventional soybeans reported in literature

	Yield effect (%)
US	
Falck-Zepeda et al. (2000a), McBride and Brooks (2000)	Based on 1997 survey; Corn belt: 13; South-east: 18.2; Delta: −14.7; N. Plains: 15.4
Sankula and Blumenthal (2004), Sankula (2006), Marra et al. (2004), Moschini et al. (2000)	No yield advantage or disadvantage
Gianessi and Carpenter (2000)	(1997, 1998) Heartland: 13.6, 4.4; MS portal: −6.2, −3.6; N. Crescent: −4.6, 4.8; Prairie Gateway: 21, 24.2; S. Seaboard: 13.3, 21.4; E. Uplands: –, −8
Qaim and Traxler (2002)	−2.9
Fernandez-Cornejo and McBride (2002)	Elasticity of yields: 0.03 (1997), 0 (1998)
Lin et al. (2001)	From ARMS 1997 survey; Heartland: 14.23; MS portal: −0.09; N. Crescent: −0.01; Prairie Gateway: 20; S. Seaboard: 16.13; E. Uplands: 5
Duffy and Ernst (1999)	−3.81 (1998), −3.56 (2000)
Benbrook (1999)	Based on Oplinger's summary (1998); IL: 3.4; IA: −6.6; MI: −3; MN: −7.6; NB: −12.1; OH: −3.3; SD: −10.2; WI: −2.8
Argentina: Qaim and Traxler (2002)	−0.2
Canada: Council for Biotechnology Information, Canada (2002)	2.4 (2000), 4.3 (2001)
Romania: Brookes (2003)	29.2 (<5000 ha), 32.6 (>5000 ha)

in yield trends in the three countries that RR soybeans were adapted (USA, Brazil and Argentina) using PSD-USDA data over the 1983–2006 period. The adoption curve for each country and a 0–1 dummy variable indicating positive adoption were used in alternative specifications of regression models used to test the possibility of a structural brake in historical yield trends. The results of one such regression (shift = 0-1 dummy indicating adoption) are reported below. Irrespectively of specification in the regression models, we could not find any evidence of a shift in the historical yield trends caused by the introduction of the RR technology in any of the countries.[2]

$$\begin{aligned} Yield_{US} &= a - 0.004Trend^{*}Shift + 0.019Trend^{\wedge} \\ Yield_{Arg} &= a + 0.004Trend^{*}Shift + 0.01Trend \\ Yield_{Brz} &= a - 0.004Trend^{*}Shift + 0.267Trend^{\wedge\wedge} \end{aligned}$$

^ = Sig. at 0.05, ^^ = Sig. at 0.00

Based on these results and our review of previous studies, we assumed in our structural model that RR and conventional soybeans have comparable yields. This assumption is similar to that made in Moschini et al. (1999), Sankula and Blumenthal, 2004, Sankula (2006), and Marra et al., 2004.

Influence of RR Technology on Costs of Weed Control

The evidence that RR soybeans reduce the number of applications for weed control and cause savings in the costs of weed control is strong (see Table 19.3). Lin et al. (2001) concluded that weed control costs, which include cost of herbicides, herbicide application, scouting, and cultivation, for RR soybean adopters were lower than for non-adopters in 1997. They estimated the savings at 11 % over conventional varieties in thc Midwest region of the US. Sankula and Blumenthal (2004) and Sankula (2006) estimated that, after accounting for technology fees, savings from the use of RR soybeans were 53 and 46 % of the weed control costs (herbicide costs) relative to those of conventional soybeans in 2004 and 2006. Moschini et al. (2000) assumed that the RR technology results in 8 % advantage in total production costs (both fixed and variable) over conventional soybean systems, based on farm-level estimates in Iowa in 1997–1998. Qaim and Traxler (2002) made a similar assumption (i.e. 8 %

---

[2] We note here that we also analyzed yield differences between the RR and conventional soybean varieties using data from on-farm yield trials performed over the 1998–2006 period in the states of Illinois, Iowa, Missouri, Minnesota and Ohio and across different maturity groups. These included hundreds of observations that, when averaged out, resulted in yield differences of RR and conventional soybeans varying from a high of 6.9 % in 2006 to a low of −1.53 % in 2003. However, there was no evidence that such differences were statistically significantly at any conventional level; thereby, further strengthening the conclusion that no significant differences in yields from the introduction of RR technology could be deduced.

Table 19.3 Cost of cultivation differences between RR soybeans and conventional soybeans

	Cost differences (% over conventional varieties)
USA	
Falck-Zepeda et al. (2000a) – [Percent Pesticide, Tillage, Cultivation and Other Cost Change (RR – conventional)]	Based on 1997 survey; Corn belt: 3.7; South–east: 4.9; Delta: 6.2; N. Plains: 2.5
Sankula and Blumenthal (2004) – (Production costs including herbicide costs and technology fees)	−53.16
Sankula (2006) – (Production costs including herbicide costs and technology fees)	−45.95
Qaim and Traxler (2002) – (Production costs includes both fixed and variable costs)	−7.9
Couvillion et al. (2000) – (Total specified cost of production less chemical and seed control costs)	16.64 (1997), −0.34 (1998)
Fernandez-Cornejo and McBride (2000) – (Total seed and weed control costs)	4.18
Marra et al. (2004)	
Lin et al. (2001) – (Weed control costs includes herbicides, herbicide application, scouting, and cultivation)	From ARMS 1997 survey (Elasticity based estimates); Heartland: −10.65; MS portal: −4.45; N. Crescent: −12.21; Prairie Gateway: −0.89; S. Seaboard: −3.91; E. Uplands: −33.92; From ARMS 1997 survey (Means based estimates); Heartland: −30.89; MS portal: −27.18; N. Crescent: 8.77; Prairie Gateway: −0.89; S. Seaboard: −44.78; E. Uplands: −33.92
Moschini et al. (2000) – (Production costs includes both fixed and variable costs)	−7.84
Duffy and Ernst (1999) (Total costs without land and labor)	−7.25
Argentina: Qaim and Traxler (2002) – (Variable costs including seed, herbicides, other chemicals, own machinery, hired labor and commercialization)	−9.72
Canada: Council for Biotechnology Information, Canada (2002) – (Total variable costs)	−8.65

advantage) in their USA study and found similar impacts in Argentina through their producer survey (10 % savings in variable costs over conventional soybean systems). They recognized that cost savings might still be higher because RR soybeans were found to reduce the time for weed scouting, but such potential costs savings were not included in their estimates.

Based on previous studies we assumed for this study that use of RR soybeans in the USA has led to 8 % savings in total production costs over conventional soybeans. For Argentina and Brazil, we adopt the same values for such costs savings used in Brookes and Barfoot (2005) and Qaim and Traxler (2002) (see Table 19.4).

Table 19.4 Assumptions of cost advantages and tillage benefits for RR soybeans (US $)

	1996	1997	1998	1999	2000	2001	2002	2003	2004	2005	2006
US											
RR cost advantages[1]	6.58	6.51	6.50	6.27	6.34	6.71	6.04	6.36	6.70	7.38	7.93
Tillage benefits[2]	12.80	12.62	11.86	12.73	13.19	10.94	11.94	11.45	12.05	12.58	12.99
	19.38	19.13	18.36	19	19.53	17.65	17.98	17.81	18.75	19.96	20.92
Argentina											
RR cost advantages[3]	9.00	8.68	8.44	8.32	8.28	8.28	10.40	10.40	10.80	10.80	10.80
Tillage benefits[4]	0.00	7.05	8.98	14.89	13.09	14.03	11.84	12.48	15.50	16.84	17.55
	9.00	15.73	17.42	23.21	21.37	22.31	22.24	22.88	26.3	27.64	28.35
Brazil											
RR cost advantages[5]	0	0	0	0	8.28	10.40	10.40	10.80	10.80	10.80	10.80
Tillage benefits[4]	0	0	0	0	7.72	6.89	6.35	5.55	6.12	6.86	7.32
	0	0	0	0	16.00	17.29	16.75	16.35	16.92	17.66	18.12
Rest of World											
RR cost advantage[6] (%)		8	8	8	8	8	8	8	8	8	8

[1]Compiled from assumptions made in section on cost advantages;
[2]Compiled based on costs of cultivation of soybean under different tillage systems from 2002 ARMS survey, applied in the same proportion to all other years;
[3]Source: Brookes and Barfoot (2005), extrapolated from previous years for 2005–06. Calculated from Qaim and Traxler (2002);
[4]Differences in costs of cultivation of soybean under different tillage systems from 2002 ARMS survey, cost of cultivation of soybean in Argentina from FAPRI database;
[5]Same as applied for Argentina;
[6]Same as for applied for USA.

The Role of Conservation and Reduced Tillage Systems

While the synergistic relationship between RR soybeans and the parallel adoption of conservation and reduced tillage systems in the US and elsewhere has been documented in the literature, specific estimates on the proportion of minimum tillage systems use due to the introduction of the technology are not readily available. To remain conservative with our assumptions, we do not assume additional acreage in reduced and conservation tillage due to the adoption of RR soybeans in our model. However, we do account for the cost advantages of RR over conventional soybeans that are cultivated under those minimum tillage systems. Such cost differentials in the USA are based on data from USDA's ARMS 2002 survey. In the absence of other data, we apply the same ratio of cost savings observed in the US to all countries we consider in our model. Our assumptions on the cost savings due to the use of reduced and conservation tillage systems with RR soybeans for different countries over the 1996–2006 period are reported in Table 19.4.

Results

The economic impacts of the RR technology estimated in our model are driven by the adoption rate. The United States and Argentina were the quickest to adopt RR technology and subsequently expanded their soybean acreage at the expense of

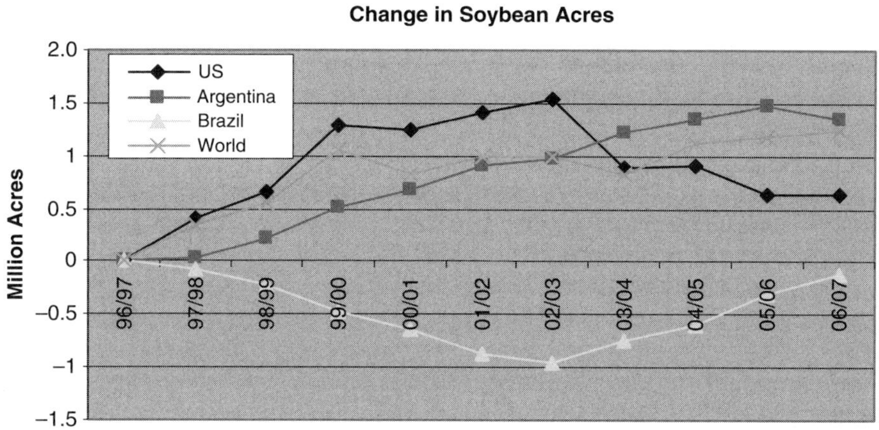

Fig. 19.2 Change in soybean acres worldwide due to the introduction of RR soybeans, 1996–2006 (*See* also Color Insert)

Brazil and to a lesser extent China (Fig. 19.2). In the United States, soybean acreage in 1997 is estimated to be 0.4 million acres higher and grows to 1.53 million acres higher in 2002 before Brazil begins to adopt the technology and slow down the growth in USA acres (Table 19.5). In Argentina, low initial adoption rates result in only 0.04 million additional hectares in 1997 but, by 2005, Argentina soybean hectares were estimated at 1.46 million acres higher with adoption reaching 99 %. Brazil adopted the technology at a much slower pace initially losing 0.08 million acres in 1997, which grew to 0.95 million acres by 2002 when they begin to adopt the technology. By 2006, the Brazilian adoption rate climbed to 48 %, and their soybean area lost to other countries was down to 0.11 million acres. Chinese farmers did not adopt the technology and also experienced limited acreage displacement. They begin with a displacement of 0.01 million acres in 1997 which grew to 0.43 million acres by 2006.

Table 19.5 Average percent change in prices and acreage due to RR soybeans 1996–2006

Prices*	US	Argentina	Brazil	EU 25
Soybeans	−3.2	−3.4	−3.3	−3.2
Soymeal	−2.6	−2.8	−2.6	−2.5
Soyoil	−2.2	−2.3	−2.2	−1.9
ACREAGE	US	Argentina	Brazil	World
Soybean	1.2	2.5	−1	0.4
Sunflower	0	2.0	0	−0.7
Rapeseed	0	0	0	−0.1

*The prices for soybean in US, Argentina, Brazil and EU 25 are Farm Price, Farm Price-Buenos Aires, Port-Rio Grande and CIF Rotterdam respectively. The prices for Soymeal in that order are Decatur, FOB, FOB and Hamburg respectively. The prices for soyoil in that order are Decatur, FOB, FOB Rio Grande and FOB Dutch respectively.

Table 19.6 Average percent change in prices of other oilseeds and their products due to RR soybeans 1996–2006

Product	Price change (%)
Sunflower (Farm Price, Buenos Aires)	−3.4
Sunflower Oil (Argentina FOB)	−1.7
Sunflower Meal (Argentina FOB)	−1.5
Rapeseed (Farm Price, Canada)	−1
Rapeseed Oil (Canada FOB)	−1.3
Rapeseed Meal (Canada FOB)	−0.7
Palm Oil (Malaysia FOB)	−0.9

Cross commodity impacts are most pronounced for sunflowers due to their greater degree of competition with soybean acreage (Table 19.6). Since rapeseed/canola and palm oil are produced in areas without strong competition from soybean acreage (Canada for rapeseed, Indonesia and Malaysia for palm oil), the impact on acreage is smaller because they are more affected by other commodity prices. Sunflower area is down an average of 0.7 % over the 1996–2006 period, or 0.35 million acres. Rapeseed area is down an average of 0.1 % over the 1996–2006 period, or 0.05 million acres. Palm area is virtually unaffected (a decline of 0.03 %) due to the long investment cycle associated with the perennial nature of the crop.

USA soybean variable cost reductions averaged about $18.95 per acre over the 1996–2006 period, while Argentina and Brazil average $21.50 and $17.01, respectively. Utilizing adoption rate assumptions and historical acreage estimates, USA soybean producers reduced their cost of production by $1.4 billion in 2006 using RR technology on 89 % of their acreage. Of the $1.4 billion in savings, $0.9 billion was associated with reduction in tillage and $0.5 billion was associated with reduction in chemicals and other costs. In Argentina and Brazil, soybean producers reduced their 2006 cost of production by $1.1 billion, of which $0.7 billion was associated with reduced tillage and $0.4 billion was associated with reduction in chemicals and other costs. In Brazil, soybean producers reduced their 2006 cost of production by $0.4 billion, of which $0.2 billion was associated with reduction in tillage and $0.2 billion was associated with reduction in chemicals and other costs.

As expected, the largest price impact of the RR technology was on soybean prices with a decline of 1 % in 1997 when the USA adoption rate was only 17 % compared with 7.6 % in 2006 when USA adoption reached 89 % and Argentina approached 100 %. While soybean gross returns averaged $7.50 per acre lower over the forecast period, average variable cost was reduced by $18.95 per acre making technology adopters better off. The average price of soybeans decreased by 3.2, 3.3 and 3.4 % in USA (Farm Price), Brazil (Rio Grande port price) and Argentina (Buenos Aires Farm price), respectively. The average price decrease in soybean meal was about 2.6 % in USA (Decatur) and Brazil (Rio Grande FOB), and 2.8 % in Argentina (FOB price) for the period 1996–2006. In 2006, the tight supplies of meal in the world market made the RR technology even more valuable to livestock feeders reducing soybean meal prices an average of 6.2 %. The soybean oil prices also declined by

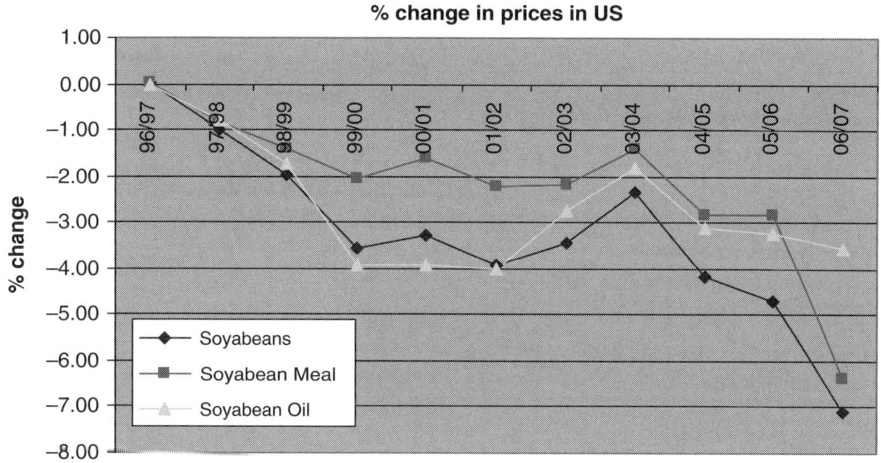

Fig. 19.3 Percent change in US soybean, soyoil and soymeal prices due to the introduction of RR soybeans, 1996–2006 (*See* also Color Insert)

2.2 % in USA (Decatur) and Brazil (FOB price), 2.3 % in Argentina (FOB price) and 1.9 % in EU 25 (CIF Rotterdam) during the same period. The shifts in price of soybeans, soymeal and soyoil in US from 1996–2006 can be seen in Fig. 19.3. The average price changes of other oilseeds and their products are also negative. The average price of sunflower, sunflower oil and sunflower meal in the Argentinean market (residual supplier for sunflower in our model) decreased by 3.4, 1.7 and 1.5 % during the period 1996–2006. Similar observations for rapeseed and its products and palm oil can be seen in Table 19.6.

For consumers, RR technology reduced the price of soybean oil an average of 2.7 % over the 1996–2006 period. Sunflower oil, canola oil, and palm oil were reduced by an average of 1.8 %, 1.3 %, and 1 %, respectively. Consumer and producer surplus estimates are given in Table 19.7, (in 2000 dollars.) The global consumer surplus from soybean complex is $11.2 billion, and the consumer surplus from other oilseed crops due to the decrease in their prices caused by RR soybean adoption is $3.6 billion. Of the total consumer surplus of about $14.8 billion, US consumers received 18 % of this surplus, where as consumers in Argentina, Brazil and ROW enjoyed about 2, 6, and 74 %, respectively. The analysis also showed that the calculated surplus for soybeans directed for crush is larger than the estimated consumer surplus soymeal and soyoil together by about $2.8 billion worldwide. This implies that the soybean supply chain (e.g. traders, crushers) retained this portion of surplus and they too benefited from the innovation.

The global producer surplus from soybeans due to the adoption of RR soybeans is estimated to be $16.5 billion over the 1996–2006 period. USA and Argentina soybean producers gained $14.4 billion and $8.8 billion in producer surplus, while Brazil and the rest of the world lost $1.1 billion and $2.2 billion, respectively. Clearly the US and Argentina benefited from the quick adoption of the technology

Table 19.7 Estimates of economic surplus from the adoption of RR soybeans from 1996–2006

($ Millions (2000=100)	US	Argentina	Brazil	ROW	Total
Consumer surplus					
(a) Soybeans for food & other uses	316	85	181	1370	1952
(b) Soybeans for crush	3199	1672	2059	5150	12080
(c) Soymeal	1200	20	357	3864	5440
(d) Soyoil	998	56	378	2385	3817
Soybean Complex (a+c+d)	2514	157	916	7620	11210
Sunflower Complex, Rapeseed Complex, Palm oil	212	46	10	3380	3647
Producer surplus					
Soybean	14379	8786	−1104	−2191	19870
Sunflower, Rapeseed, palm oil	−113	−268	−9	−2973	−3364
Total world surplus	16992	8721	−187	5836	31363

while Brazil and the rest of the world lost producer surplus due to the price effect. (However, it is important to note that to the extent that Brazil was able to extract any premium from international markets for non-GMO soybeans, the impact on their producer surplus was reduced.)

Producer surplus for the other oils was, however, reduced due to lower price without any offsetting reduction in variable costs. Combining sunflowers, rapeseed and palm oil, the total loss for these crops is $3.3 billion.

Concluding Comments

In this chapter, we examined the global economic impacts of RR soybeans in their first 10 years of adoption. Our analysis showed that there were substantial economic benefits, amounting to a total of about $31 billion. These include benefits to consumers and producers all over the world as well as benefits to the supply chain of the soybean complex. We were not able to separate the share of the innovator (Monsanto) as we have only incomplete data on the premiums they were able to extract in the three different countries were most adoption took place. Our estimates account for the surplus gained/lost by consumer and producer due to supply and demand shifts in other oilseed markets caused by the introduction of RR soybeans.

Our analysis reinforces results and lessons learned in previous studies. First, early adopters benefit most from innovation. Out of the total world surplus, producers and consumers in the USA and Argentina captured a bit over 80 % of it. Brazil as a late adopter just started to make up for the losses it experienced in the early years of the innovation cycle. Similarly, producers of competitive oilseeds that did not benefited from parallel innovation also experienced economic losses. Second, consumers benefited from the technology almost as much as producers both through the use of soybean oils and meals, as well as through oils and meals of competing oilseeds that were affected. Third, the aggregate economic impacts of RR soybeans are large, dynamic and sustained.

We should note in closing that as large as our measured economic impacts of the technology have been, they might understate the actual economic impacts of RR soybeans over the last ten years. Our assumptions about the farm-level impacts of the technology on yields and cost savings are conservative and our measures do not account for any potential increases in the adoption of minimum tillage systems or for the economic value of environmental and other non-pecuniary impacts that have been documented in the literature. Clearly, a more accurate assessment of the global economic impacts of RR soybeans must rectify such limitations.

References

Alston, Julian M., G. W. Norton and P.G. Pardey (1995). Science under scarcity: Principles and practice for agricultural research evaluation and priority petting, Cornell University Press, Ithaca, NY.

Benbrook, Charles (1999). "Evidence of the magnitude and consequences of the Roundup Ready soybean yield drag from university-based variety tests in 1998, Ag BioTech Info Net Technical Paper 1, www.biotech-info.net/RR_yield_drag_98.pdf

Brookes G. (2003). *The farm level impact of using Roundup Ready soybeans in Romania*. Frampton, UK: PG Economics Limited. Available on the World Wide Web: http://www.pgeconomics.co.uk/pdf/GM_soybeans_Romania.pdf.

Brookes, Graham., & Barfoot, P. (2005). GM crops: The global economic and environmental impact-the first nine years 1996–2004. *AgBioForum*, *8*(2&3), 187–196. Available on the World Wide Web: http://www.agbioforum.org.

Brookes, Graham., & Barfoot, P. (2006). :GM Crops: The first ten years-Global socio-economic and environmental impacts,". *ISAAA Brief* No. 36. ISAAA: Ithaca, NY.

Carpenter, Janet, E., and L. P. Gianessi (2003), "Trends in pesticide use since the introduction of genetically engineered crops," in "The economic and environmental impacts of agbiotech: a global perspective," Kalaitzandonakes, N., editor, Kluwer Academic Publishers, New York.

Council for Biotechnology Information (2002). Agronomic, economic and environmental impacts of the commercial cultivation of Glyphosate tolerant soybeans in Ontario, (unpublished report). Guelph, Ontario.

Couvillion, Warren C., F. Kari, D. Hudson, and A. Allen (2000), "A preliminary economic assessment of Roundup Ready soybeans in Mississippi," Research Report 2000–005, Mississippi State University

Duffy, M., and M. Ernst, (1999). "Does Planting GMO Seed Boost Farmers' Profits?", Leopold Letter, Vol. 11, No. 3, Leopold Center for Sustainable Agriculture, Iowa State University, Ames, IA.

Elmore, R. W., Roeth, F. W., Nelson, L. A., Shapiro, C. A., Klein, R. N., Knezevic, S. Z., Martin, A. (2001). "Glyphosate-Resistant Soybean Cultivar Yields Compared with Sister Lines," Agronomy Journal, 93: 408–412.

Ervin, D.E., S.S. Batie, R. Welsh, C.L. Carpentier, J.I. Fern, N.J. Richman, and M.A. Schulz (2000) *Transgenic crops: An environmental assessment*. Henry Wallace Center for Agricultural & Environmental Policy at Winrock International. *http://www.winrock.org/transgenic.pdf.*

Falck-Zepeda, Jose B., G. Traxler, R.G. Nelson, (2000a). "Rent creation and distribution from biotechnology innovations: The case of Bt cotton and herbicide-tolerant soybeans in 1997," Agribusiness, Vol. 16, 1, 21–32.

Falck-Zepeda, Jose B., G. Traxler, R.G. Nelson, (2000b). "Surplus distribution from the introduction of a biotechnology innovation," American Journal of Agricultural Economics, 82: 360–369.

Fawcett, R., and D. Towery (2002). "Conservation tillage and plant biotechnology: How new technologies can improve the environment by reducing the need to plow," Conservation Technology Information Center (CTIC), West Lafayette, IA.

Fernandez-Cornejo, J. and W. McBride (2000), with contributions from Cassandra Klotz-Ingram, Sharon Jans, and Nora Brooks, Genetically engineered crops for pest management in U.S. agriculture, Agricultural Economics Report No. (AER786) 28 pp, May 2000, http://www.ers.usda.gov/publications/aer786/

Fernandez-Cornejo J & McBride W (2002) Adoption of bio-engineered crops, USDA, ERS Agricultural Economics Report No 810

Fernandez-Cornejo J., C. Klotz-Ingram, and S. Jans, (2002) "Farm-level effects of adopting herbicide-tolerant soybeans in the USA," Journal of Agricultural and Applied Economics.

Fernandez-Cornejo, Jorge., C. K. KLotz-Ingram, R. Heimlich, M. Soule, W. McBride and S. Jans (2003), "Economic and environmental impacts of herbicide tolerant and insect resistant crops in the United States," in "The economic and environmental impacts of agbiotech: a global perspective," Kalaitzandonakes, N., editor, Kluwer Academic Publishers, New York.

Frisvold, G., and R. Tronstad, (2002). "Economic effects of Bt cotton adoption and the impact of Government Programs," in N. Kalaitzandonakes (Ed.), The economic and environmental impacts of agbiotech: A global perspective, Kluwer-Plenum, New York.

Frisvold, G., & Tronstad, R. (2003). "Economic impacts of Bt cotton adoption in the US" in "The economic and environmental impacts of agbiotech: a global perspective," Kalaitzandonakes, N., editor, Kluwer Academic Publishers, New York.

Gianessi, L.P., and J. E. Carpenter (2000). "Agricultural Biotechnology: Benefits of Transgenic Soybeans," A Report by National Center for Food and Agricultural Policy, Washington DC.

Heimlich, R. E., J. Fernandez-Cornejo, W. McBride, C. Klotz-Ingram, S. Jans, and N. Brooks. "Genetically engineered crops: Has adoption reduced pesticide use?" *Agricultural Outlook* AGO-273, August 2000.

Lin, W., G. K. Price, and J. Fernandez-Cornejo (2001). "Estimating farm-level effects of adopting herbicide-tolerant soybeans," Oil crops situation and outlook yearbook, OCS-2001, USDA, ERS, Washington, DC.

Marra, M. (2001, January). Economic impacts of transgenic crops: A critical review of the evidence to date. Paper presented at *Agricultural Biotechnology: Markets, and Policies in an International Setting Workshop*, International Food Policy Research Institute, Adelaide, South Australia.

Marra, M., Pardey, P., & Alston, J. (2002). The payoffs to transgenic field crops: an assessment of the evidence. *AgBioForum, 5*(2), 43–50.

Marra, Michele C., N. E. Piggott, and G. A. Carlson (2004), "The net benefits, including convenience, of Roundup Ready soybeans: Results from a national Survey," NSF Center for IPM Technical Bulletin 2004–3, 39pp, Raleigh, NC.

McBride, William, D., N. Brooks, "Survey evidence on producer use and costs of genetically modified seed," Agribusiness, Vol. 16, No. 1, 6–20 (2000)

Moschini, Giancarlo, H. Lapan, and A. Sobolevsky (2000). "Roundup Ready soybeans and welfare effects in the soybean complex," *Agribusiness*, 16 (1): 33–55.

Nelson, G., and D. Bullock, (2003) "Environmental effects of glyphosate resistant soybeans in the United States" in "The economic and environmental impacts of agbiotech: a global perspective," Kalaitzandonakes, N., editor, Kluwer Academic Publishers, New York.

Oplinger, E.S., M. J. Martinka, J. M. Gaska, K. A. Schmitz and C. R. Grau (1998). Wisconsin soybean variety test results, A3654, Department of Agronomy and Plant Pathology, University of Wisconsin-Madison, http://www.uwex.edu/ces/soybean/research/98soyvar/98soyvar.pdf.

Qaim, M., and G. Traxler (2002). "Roundup Ready Soybeans in Argentina: Farm Level, Environmental, and Welfare Effects," 6th International ICABR Conference, Ravello, Italy, July 11–14.

Sankula, Sujatha., and Blumenthal, E. (2004). Impacts on US agriculture of biotechnology-derived crops planted in 2003: An update of eleven case studies. Washington, DC: National Center for Food and Agriculture Policy. Available on the World Wide Web: http://www.ncfap.org/whatwedo/pdf/2004finalreport.pdf.

Sankula, Sujatha (2006). "Quantification of the impacts on US agriculture of biotechnology-derived crops planted in 2005," Report by National Center for Food and Agricultural Policy, Washington DC.

Trigo, E.J., and Cap, E.J. (2003). The impact of the introduction of transgenic crops in Argentinean agriculture. *AgBioForum*, 6(3), 87–94. Available on the World Wide Web: http://www.agbioforum.org.

Trigo, E.J., and Cap, E.J. (2006). "Ten years of genetically modified crops in Argentine agriculture," Research Report, Argentine Council for Information and Development of Biotechnology – ArgenBio.

Index

Printed in The United States of America